Product Innovation Toolbox

Product Innovation Toolbox

A Field Guide to Consumer Understanding and Research

Second Edition

Edited by

Kannapon Lopetcharat

Nouveau Centric Co. Ltd.
Bangkok, Thailand

Dulce Paredes

Takasago International Corporation (USA)
Rockleigh, New Jersey, USA

Jacqueline H. Beckley

The Understanding & Insight Group LLC
Denville, New Jersey, USA

WILEY

This second edition first published 2023

Edition History
John Wiley & Sons, Inc (1e, 2012)

Registered Office(s)
John Wiley & Sons, Inc., 111 River Street, Hoboken, NJ 07030, USA
John Wiley & Sons Ltd, The Atrium, Southern Gate, Chichester, West Sussex, PO19 8SQ, UK

Editorial Office
The Atrium, Southern Gate, Chichester, West Sussex, PO19 8SQ, UK

For details of our global editorial offices, customer services, and more information about Wiley products visit us at www.wiley.com.

Wiley also publishes its books in a variety of electronic formats and by print-on-demand. Some content that appears in standard print versions of this book may not be available in other formats.

Library of Congress Cataloging-in-Publication Data applied for
Hardback ISBN 9781119712848

Cover Design: Wiley
Cover Image: Wiley

Set in 10/12pt Interstate by Straive, Pondicherry, India

Contents

List of Contributors

Kathryn Ambroze
Manager of Behavioral and Marketing Sciences, HCD Research
Flemington, New Jersey USA

Jacqueline H. Beckley
Owner/Founder, The Understanding & Insight Group LLC
Denville, New Jersey USA

Stacey Cox
Director, Consumer Insights NA Core Capabilities and Accelerators, The Kraft Heinz Company
Chicago, Illinois USA

Daniel M. Ennis, Ph.D.
President
The Institute for Perception
Richmond, Virginia USA

Michele M. Foley
Retired Director, Sensory & Consumer Insights and Knowledge Management
Nestle
Fremont, Michigan USA

Jeff Garza
Principal
Garza Consulting, LLC
Grand Rapids, Michigan USA

Carter Green, Ph.D.
Consultant, Takasago International Corporation (USA)
Stony Point, New York USA

Anthony Jackel
Director, Consumer Analytics, Ferrara,
Chicago, Illinois USA

Melissa Jeltema, Ph.D.
U&I Collaboration LLC
Jetersville, Virginia USA

Jennifer Lewis
Principal Research Analyst, Altria Client Services
Richmond, Virginia USA

Kannapon Lopetcharat, Ph.D.
Executive Director, Nouveau Centric Co. Ltd.
Bangkok, Thailand

Paulina Morquecho-Campos, Ph.D.
Senior Analyst, Consumer Insights and Market Research
Takasago Europe GmbH
Zulpich, Germany

David Moskowitz
Vice President Data Analytics
Internal Audit,
Bank of China USA
New York, New York USA

Howard Moskowitz, Ph.D.
Co-Founder and Scientific Director at World Institute of Competitive Excellence, LLC (WICEINT)
White Plains, New York USA

James Mwai
Managing Director Consumer & Marketplace Insights
Altria Client Services
Richmond, Virginia USA

Michelle Niedziela, Ph.D.
Vice President of Research & Innovation, HCD Research
Flemington, New Jersey USA

Dulce Paredes, Ph.D.
Emeritus Global Leader, Consumer Insights and Market Research, Flavor Division, Takasago International Corporation (USA)
Rockleigh, New Jersey USA

Stephanie Plunkett, Ph.D.
Director of Behavioral Science, Consumer Strategy
Altria Client Services
Richmond, Virginia USA

Katherine C.S. Rhodes
Formerly with Takasago International Corporation (USA)
Anchorage, Alaska USA

Mary Sonnen
Global Senior Product Development Manager, Happy Family Brands (Nurture Inc), Boise, Idaho USA

Alina Stelick
Manager, Sensory Evaluation Center
Cornell University
Ithaca, New York USA

Veronica Symon
Director, Sensory and Consumer Guidance, Americas
Diageo
Stamford, Connecticut USA

Ratapol Teratanavat, Ph.D.
Global Leader, Consumer Insights & Market Research, Takasago International Corporation (USA)
Rockleigh, New Jersey USA

Jennifer Vahalik
General Manager & Chief Research Officer, U&I Collaboration LLC
Randolph, New Jersey USA

Vanessa Zuccoli
Director, Marketing & Consumer Insights and Market Research EMEA
Takasago Europe GmbH
Zulpich, Germany

Acknowledgements

First, I would like to thank Dulce and Jackie for many opportunities and their dedications to coach and mentor me throughout the years. I also want to thank my parents (Mrs. Preeya Suwankul and Mr. Somkirt Lopetcharat), Professor Mina MacDaniel, Professor Jae Park, and Dr. Siriporn Pipatsattayanuwong Chua for their support and for opening opportunities for me to study and work in the United States, where I met Dulce and Jackie. I also need to thank all my clients who gave me opportunities to hone my skills and accumulate experiences.

Kannapon Lopetcharat

Kannapon and Jackie, you are my forever innovation buddies, thank you! To my loving family, Rollie, Nathalie, and Robert Tadena, you are my daily source of joy! And deepest gratitude to Brian Buck, who empowered me to put all that was written in the first edition into practice at Takasago.

Dulce Paredes

Kannapon and Dulce, what a trio we have made. I hope that our readers will enjoy the updated book and will carry our passion for understanding people further. It has been a fun ride. I want to thank each of you for wanting to do such an awesome job to help others. I would also like to thank Leslie, my husband, for his patience with me and my "projects." When we started this transition to a different way of thinking about how to innovate more than 20 years ago, we knew that we had to help people on the path. Technology has made it somewhat easier. In a complex world - we all have a lot to learn and stay focused on. My best wishes to all our authors and to those who will read this book and be inspired.

Jacqueline H. Beckley

Finally, we want to thank our book contributors who are all excellent practitioners and are willing to share their knowledge for future Consumer Explorers and Product Researchers. We thank our family and friends who helped provide materials and finishing touches to the book: Divina Paredes (business technology editor), Nadine Paredes, Joanna Clifton, Joanna Bantilo, Ruta Lesniauskas and Dan Diehn.

The Editors

Introduction

The View from Pixel to Picture

Kannapon Lopetcharat, Dulce Paredes, and Jacqueline H. Beckley

The editors want to thank you for deciding to read this book. Whether you read a chapter or mark up the whole book, we are thankful and delighted. And if you are buying this for a second time because you really enjoyed the first edition, thank you even more. While some of the chapters have stayed the same, we have refreshed this edition with a lot of new materials, and we hope you find the wisdom good.

Let us explain our goals:

(1) **Think of this book as a mentor:** Someone who wants to give you some help, so you don't have to struggle with some of the issues they went through before you.
(2) **Think of this book as a cookbook:** If you like to dabble or work on your own Innovation pathway. We have asked a range of successful "Consumer Explorers" to help us craft chapters that give you tools you can put to work immediately or feel strong enough to modify as if you were adding your own herbs to an interesting recipe.
(3) **Think of this book as a coach, one you have paid a lot of money to:** So that you may have your thinking challenged because some of the authors present ideas or approaches you have not considered.
(4) We hope through every chapter that we have done one thing well: If we embraced ideas from another person, scientist, or other professional, we have acknowledged their contribution and made sure they are receiving the credit that belongs to them. Appreciating someone else's work is great - yet recognizing how you have built upon their thinking is extremely important.

Why point #4? We have found in this field of innovation, consumer design, and product development, some people feel it is okay to not acknowledge another person's work. We strongly disagree. Just because you did not take the time to realize that your "new & cool" idea built on something someone else did, spoke about, presented, in our world of writing and journalism, there is By - **before you** (credited to Bill Maher, a social commentator and sometime comedian, on his HBO show in May 2021). Pay attention and give recognition - whether you improve upon the researcher's ideas, steal, and use completely, or somewhere in-between. It is the right and important thing to do.

Since we are dealing with a field guide for understanding consumers (and people), processes to enable that understanding and the research you might want to consider, we are presenting material in a way that allows you to go to your outcomes quickly.

The three editors and 22 authors are all passionate about grounding innovation from the viewpoint of humans. To do that, we are bringing together a group of authors who have a range of thinking, some more radical or boundary testing to those who are much more classic or traditionally based. People, humans, "consumers" are having a moment. Researchers who believed that classic survey and product research would help solve most questions have come to embrace a more complicated and complex world. To get great outcomes, more and more researchers are realizing they need a much deeper understanding of how people are looking at products, why they are considering the products that they select, and also that there may be a growing or a large group, of individuals in this world, who appear to have ideas about products that behave much more differently than the norms and assumptions we have long held. We hope that the chapters we have provided for your consideration give you strategies that will go beyond the norm and enable realistic outcomes for you.

The book is laid out as a field guide that equips the reader, the consumer researcher, to become a "Consumer Explorer" (CE), an insight leader and strategic innovator who can infuse and apply deep consumer understanding throughout the product innovation process.

A CE differs from ordinary consumer researchers because, in addition to understanding consumers deeply, the CE can guide project teams to connect the dots and comprehend the big picture faster in a systematic manner. We advocate following an "Iterative Experimentation Qual-Quant process" (IEQQR) that will enable companies to understand their product environment comprehensively through knowledge mapping exercises, hypothesis testing, consumer deep dives, clear action standards and key performance criteria, and action-oriented results. The approaches described in the field guide will enable the CE to select and deploy appropriate and efficient consumer understanding and research tools for their current situation. They could also help guide the project team to successful innovation and new product introductions that have the opportunity of taking insight to income.

We use two visual principles to ground you on how to navigate the journey of the Consumer Explorer:

(1) The View (What to Why)

Above and below the surface

(2) Pixel to Picture (How you are looking at it and Who sees it first)

From pixel to picture, winners are the ones who connect the dots faster

We used the image of an agricultural field with an aquifer to show that what can be seen above the surface is often nourished by what lies hidden underneath. Capturing both gives you **the What** and **the Why**. An aquifer is any underground layer of water-bearing rock or porous geological formation from which water can move freely providing groundwater for springs and wells. When you relate the aquifer concept to research, what you see when you first scope out the innovation landscape may not always be what it seems. And to get a deeper perspective on what it all means from what is readily visible to what lies below the surface requires different tools.

Think of the Consumer Explorer taking a digital picture of the consumer and/or category landscape to identify breakthrough insights. A digital image is composed of many pixels (pixel is short for picture element; a pixel is a single point in a graphic image) that are close together so that the pixels appear connected. The number of pixels can define the quality of the image resolution, where the more pixels, the more exact the image. But also, a whole image is the sum of its parts, so the individual pixels in a picture make up what we see as the whole image. To gain a competitive edge, the CE must scope the landscape, capture the Aha! moment quicker and connect the dots faster to reveal the breakthrough insight faster than everyone else. Being well-versed in both the pixel and the picture can make all the difference in the strategic direction of a project. Our imagery throughout is to move from pixels, or pieces of the picture, to the whole cross section of the picture to show what is underneath supporting what is readily seen and why it matters. Many legacy and conventional tools can be used to capture the overt space while you need specialized tools to probe for deeper understanding for what lies below the surface, the unknown unknowns and unspoken truths. Another skill for the CE to master is staying vigilant and flexible throughout the learning process, constantly sifting, and becoming aware of deep fakes and misinformation.

This field guide contains 17 major chapters divided in three parts that will transform a consumer researcher into a Consumer Explorer by providing a step-by-step guide that shows how to design an innovative consumer research program from start to finish (pixel to picture).

Part one
Prepare for your journey

Chapter 1: Setting the Direction: First, Know Where You are

Chapter 2: The Consumer Explorer: Key to Delivering the Innovation Strategy

Chapter 3: Invention and Innovation

Chapter 4: Designing the Research Model

Chapter 5: What You Must Look for – Signs of High Potential Insights

Part two
Gear up for your journey

Chapter 6: Tools for Upfront Research on Consumer Triggers And Barriers Qualitative Tools

Chapter 7: Tools for Upfront Research on Consumer Triggers and Barriers Qualitative-Quantitative Tools

Chapter 8: Tools for Up-Front Research on Understanding Consumer Values

Chapter 9: New Tools Beyond Conventional Qualitative and Quantitative Meanings

Chapter 10: Tools to Refine and Screen Product Ideas in New Product Development

Part three
Word of the wise:
Wisdom from experienced explorers

Chapter 11: Putting it All Together: Driving Consumer-Centric Innovation in an Organization

Chapter 12: Above Averages: Use of Statistics and Design of Experiments in Product Innovation Applications

Chapter 13: How to Work With Industry Experts and Influencers for Innovation

Chapter 14: Virtual Staff

Chapter 15: Found in Translation: The Adventure of Conducting Multi-Cultural Consumer Research

Chapter 16: Sturdy Bridges to Future Trends

Chapter 17: Future Trends and Directions

Part I

Preparing for your journey (Chapters 1-5) defines the foundations and definitions to anyone who is responsible to set up, practice, and direct a sustainable consumer research program.

Part II

Gearing up for your consumer journey (Chapters 6-10) contains time-proven and cutting-edge tools from upfront consumer research where providing direction is key when you get closer in to product launch where finetuning and precision are valued.

Mastering the topics in parts one and two arms Consumer Explorers to set the direction, prepare, and be ready for any product innovation project.

Part III

Words of the wise (Chapters 11-17) covers practical tips and steps to bring the learnings from the earlier sections to life. Seasoned Consumer Explorers share their experiences in working with cross-functional teams and where they see future directions in building consumer-centric business decisions.

In each chapter, we begin with a summary of what the chapter is about and wrap up with suggested discussion questions for you to think through what you just read and what it means to you, your team, and your business.

From the first edition, we predicted that emerging research trends such as blurring the lines between qualitative-quantitative research tools, increasing use of digital technology and the rise of hybrid data will dominate the consumer research world. Ten years later, all of these are now mainstream practices. The future to understanding humans, people, and consumers is happening now as we continue to capture their shifting behaviors, mindsets, and attitudes toward products, done in context and enabled by technology. There are numerous ways to capture consumer data. The challenges for the Consumer Explorer are threefold: understand the why and not just the what; differentiate quality from volume; and convert what they capture to actionable insights that improves the status quo. We hope that this field guide helps you overcome the challenges you might face as an insight leader.

This book was originally designed to be a "how to" for individuals who want or need to get behind some of the leading approaches in new methods of understanding the consumer in today's market environment. We expect what has been written to evolve over time with contributions from readers and practitioners such as yourself. It is up to you to follow the detailed steps or improve and create your own imprint. Please enjoy!

Part I
Prepare for Your Journey

1

We're not lucky.
We win because we work hard.

Roger Penske
Head of one of the most successful racing team groups for the last 40 years

What this is about

To think outside the box, one must know where the box is and where you are relative to the box. Chapter 1 honestly discusses how you can transform yourself to an "insight leader" through self-education and on-the-job training and "playing" in the company sandbox. The authors share their experiences in delivering and communicating impactful research.

Product Innovation Toolbox: A Field Guide to Consumer Understanding and Research, Second Edition. Edited by Kannapon Lopetcharat, Dulce Paredes and Jacqueline H. Beckley.

Chapter 1

Setting the Direction: First, Know Where You Are

Howard Moskowitz and Jacqueline H. Beckley

Key learnings

- ✓ Knowledge worker: what is your job? Does it matter what profession you affiliate with?
- ✓ Prepare yourself to be an insight leader
- ✓ Dealing with management and your clients
- ✓ Two types of heroes in a corporation
- ✓ Knowing your limits
- ✓ How to report your data

1.1 Roles in the corporation – the dance of the knowledge worker

The great management theorist and consultant, Peter Drucker (1999), coined a term for a class of professionals in industry whose job was to understand and manipulate symbols. He called them "knowledge workers," to distinguish them from other workers who produced things. Knowledge workers had the job to manipulate symbols and to create knowledge and manipulation of symbols. In Drucker's mind, the knowledge worker was the worker of the future, as machines augmented, and even replaced people. Today's world of computers, business intelligence, artificial intelligence, algorithms, and the utter connectivity between people and corporations makes the knowledge worker even more important than Drucker imagined.

In corporations, a special group of knowledge workers practices the job of understanding the products a company makes and at the same time the mind of the customer who uses the product. There is not one group, but rather at least three groups who share this grand responsibility of knowing the product/the user/and the product-user interface.

Product Innovation Toolbox: A Field Guide to Consumer Understanding and Research, Second Edition. Edited by Kannapon Lopetcharat, Dulce Paredes and Jacqueline H. Beckley.

These groups fall into at least three classes: *market, product*, and *sensory professionals*. Of the three classes, two (market professionals and sensory professionals) have well-defined responsibilities. The third class, the "product professional" has eluded a clear definition of who they are and what they do. Over the years, the titles have changed for some of these jobs, However, whether you are called an insight professional or a market researcher or a data analytics specialist, the tasks are the same - the source material may have changed, yet the end result is essentially the same. So we will use the traditional language of the researchers for our discussion.

The *market researcher* studies the market to understand market trends and to discover how the consumer (citizen, person if we want to disentangle commerce requirement from the individual) fits into that market. When it comes to products, market researchers focus on how the product and consumer come together for market benefit.

The job of the *sensory professional* is to understand the product, and specifically the product as perceived by and used by the consumer, generally from an analytical perspective. The consumer is merely one of several instruments used to measure the properties of the product. The sensory field has been fairly clear in its definition of "sensory" professional. It defines the job by certain analytical methods that many sensory professionals' value: descriptive analysis, discrimination testing, and product understanding through a vast array of survey research tools. All these tools tend to be heavily anchored in analytics. The consumer represents a tool much like an instrument.

The evolving job of the *product researcher* (sometimes called a consumer technical leader) focuses on the product and the consumer interaction with the product. The product researcher has evolved out of the confluence of the sensory researcher and the market researcher, respectively. The product researcher studies the product but within the context of the consumers' world. With the product researcher, the heavy reliance on explicit analytics is not always useful nor always required, but in many cases, a background in analytics helps one do the job. Today, this hybrid between market and sensory is becoming more prevalent as demands for understanding both explicit and implicit drivers of choice are becoming more well understood.

It is important to understand the nature of these three groups in a company that sells products or services. We focus on these three professional groups in this chapter, for one major reason, relevance. In years gone by, it was enough to label oneself a sensory professional and to focus on the evaluation of the product from the point of view of description, describing each of the facets of the product, in agonizing detail. It was also enough for the market researcher, to label themselves as just that, a market researcher. One didn't need to be technical. There were consultants and research companies that did this job, as well as the plethora of computer services, so-called "tab-houses," ready to run the data and provide pretty tables. One simply needed to evaluate products in the unactionable, evaluative language of the marketer (the product tastes sweet; the product is sophisticated, and so forth).

But what about this new profession, this product researcher? Who is this person? Is the product researcher, a marketer or a scientist (sensory professional)? Neither or both?

It is all about what one knows and the world in which one feels comfortable. It is the focus of specific knowledge for these professionals that end up deciding what job the person does in a company. And it is the focus of knowledge, *your* knowledge, that ultimately decides what role *you* will play, who *you* really are, what is expected of *you*.

Let us look for a minute at the two foci of knowledge, that for the marketing researcher and that for the sensory professional. The marketing researcher traces the intellectual history back to sociology, with its focus on the behavior of people in groups. To the sociologist, or more properly to the market researcher, it is not the product per se, or perhaps only the product in passing. It is rather the behavior of the person. The product is just one facet of behavior. Other behaviors need not involve the product at all, but might involve what media the individual "consumes."

Now turn to the sensory professional, where science of the product, not sociology of the consumer, takes front stage. The sensory professional traces his or her history back to the scientist or product developer who created the product, to the person who knows the innards of the product, and knows what makes the product "tick." Company asks the sensory professional to link the presumed behavior of the consumer to the physical characteristics of the product. The research specialties include understanding how the person interacts with the product, and how the different characteristics of the product "drive" perceptions.

The focus of the product researcher is to blend the science of product design with the marketing science elements (emotional, kinesthetic, and behavioral). The product researcher must allow the product to tell its story as a person uses it and link that story with the marketers' job to connect to the product. Of all companies that have all three professions, Procter & Gamble has had the longest running products research group. How the model came to be is part of myth and folklore, but this early creation of disciplines foreshadowed what has been evolving for most product/service companies.

1.2 Insights leader – learning on the job vs. learning in school

In recent years, there has been a curious shift in the role of many individuals in the worlds of both sensory analysis and market research.

The new names have something to do with the word "insight." Although one might cynically say this is just a name change of a department, the truth of the matter is that something else is going on. The focus is shifting, away from what the professional does (studies the product, studies the market), and toward what the product delivers. The product now is named "insights." It's not completely clear yet whether these insights are simply another way of talking about the same thing (old wine in new bottles), or whether we are witnessing a change in these knowledge workers, from what one did to what one contributed. Interestingly, product research has not had experience or, more cynically, "endured" this change.

In the spirit of the new change, let us investigate the new requirements of this job called insights manager or insights provider or analytics researcher. We will call this person the "insight leader" because the role involves beyond "managing" insights or "providing" insights but leading the quest to get both insights and understanding, and more specifically, what constitutes an insight.

In our new Internet-connected world, definitions of "insight" reflect many different approaches:

The *Random House Dictionary* (1971) defines "insight" as an instance of apprehending the true nature of a thing... penetrating mental vision or discernment of the underlying truth.

For business: Nigel Bradley (2007) points out a recent development in the world of business. There has been the emergence of new departments in corporations which carry the word "insight" in their titles. We have customer insight departments, insight management unit, consumer insight, and so on. This extends to the job titles of executives working in those areas. One reason for this development was a realization that the emphasis of results from individual research projects needed to be shifted to a wider understanding of the dynamics operating in the full marketplace. Another reason was the impact of information technology. Progress in technology gave way to the availability of enormous masses of information found in databases from a variety of sources. The advantages of insight management are numerous. Making use of all existing information is a great way to get customer insights faster and at more minimal costs.

For marketing: Lee et al. (2009) point out that an insight is a statement based on a deep understanding of your target consumers' attitudes and beliefs, which connect at an emotional level with your consumer, provoking a clear response (This brand understands me! That is exactly how I feel! - even if they've never thought about it quite like that) which, when leveraged, has the power to change consumer behavior. Insights must affect a change in consumer behavior that benefits your *brand, leading to the achievement of the marketing objective*.

Insights can be based on real or perceived weakness to be exploited in competitive product performance or value:

(1) Attitudinal or perceived barrier in the minds of consumers, regarding your brand
(2) Untapped or compelling belief or practice
(3) Insights are most effective when they
    (a) are unexpected
    (b) create a disequilibrium
    (c) change momentum
    (d) are exploited via a benefit or point of difference that your brand can deliver.

It is, therefore, critical to think carefully about who you are and what path of professionalism you are trying to pursue. As you do, try to answer the question that we just posed: Can the job of insights leader be learned or must it emerge from education? Or can it be something that is just "picked up," on the fly and requires little training or knowledge about behavior, decision-making, and perception?

1.3 Being the authentic you

We don't believe we can teach you curiosity or make you passionate about an idea or skeptical about what you are taught. Some of these behaviors are mindsets that we believe you start with from birth. They may have been nudged along by parents or teachers when you are young. But if you aspire to be an insight leader, you must be on a path of continuous learning. Even if the organization you work in *is not* a learning organization, you have to get into the fray, into the center, where the "action is." You cannot lead insights from the sidelines. To that end, you must. . . *educate yourself.*

Now, just how do you go about educating yourself?

(1) **Look around you:** Look, observe, and take note. What is happening and how is it happening and do you make sense of the happenings?

(2) **Be curious about the trappings of the "set":** For example, a good idea is to go into a person's office, and simply sidle up to the bookcase. (In our new age of nonbooks, this does take a cleverer approach! By book we do not mean only Google searches and Wikipedia references!) Is there a bookcase? What's in the bookcase? How are the books arranged? Does it look like the books have been read, or are they just there for show? Do you get a sense of connection between the books you see on the shelf and the person to whom you speak in the office? During the pandemic year of 2020, when we had a chance to peer into a lot of people's homes and offices, we found that books, bookshelves, and personal libraries were still popular with many and attracted our attention. So whether it is an e-book, an auditory book, or a physical one, be curious and "read"!

(3) **Ask people what they are reading, right now:** The reason for asking about one's books (or reading) rather than one's education is simple. It helps you understand what they are thinking about.

(4) **Books are different from a résumé:** Résumés give you a sense of a magnificent achievement of a life, perhaps a unique life. There may not be any single accomplishment, but the résumé itself is a work of magnificence. Every achievement is highlighted, written, described in glowing terms, perhaps an action that signifies the worthiness of the person. There are companies and services which, for the right price, will polish one's résumé, making one's achievements look far more impressive than they actually are. And, since résumés are the coin of the job hunter, the key to the new job, it's important that the résumé looks, lives, breathes, and speaks like the junior partner the résumé owner is destined to be at this moment in time. But those books and the bookshelf? Well, that's an entirely different world. The books that a person buys are friendlier. There are no services, no organization, whose express desire is to polish an executive's bookshelf (in the past this was also an area of prestige, but in business today, not so much).

1.4 What should you read?

(1) Reading is an investment in time. But it's more than time. Reading, at least reading books, gives you a private space, a private conversation with the masters who wrote the books. When you read, you get a sense of how people think. When you read good literature, good history, you may be sufficiently fortunate to come across someone who has a felicitous style, who writes well. Relish that writing. Stop for a moment and look to see how the writer constructs the sentences. Verbalize to yourself what you see. It's a worthwhile exercise. It helps make you a better thinker. We hope you are seeing how this process is creating personal *insight*.

(2) When reading, it's tempting to pick up novels and easy reads, stuff that passes the time, but doesn't force you into an active contemplation. If it's novels, then read classic ones, novels by the great authors, which probe topics. Or plays, or poetry.

(3) A more productive read, for insight training, is history, something with a structure, something that happened. Pick up a good history, something well written, such as Gibbon's *Decline and Fall of the Roman Empire*. It's long, yes, three well- written volumes. Look at the Modern Library Edition in three somewhat condensed volumes. Pick a section, and read it. Get a sense of how history unfolds. It's worth the effort; you'll end up fascinated, and your mind will be exercised.

(4) It is important to also read broadly. Read newspapers. Read your local newspaper along with those of the world. For Americans, reading The *New York Times*, The *Boston Globe*, The *Washington Post* exposes you to different ways of thinking about more contemporary topics. Every now and then, read a real physical newspaper. The way it is laid out matters and may rev up your sense of wonder about people and the world. Juxtaposing contemporary reads like these to scholarly, academic papers is helpful. There is a very different structure to a paper from Food Quality and Preference or the *Journal of Consumer Psychology*. Reading broadly enables your mind to stretch and takes you from tactics to wider, relevant thinking.

1.5 What else do you need to do to prepare to be an insight leader?

(1) **Know who you are and where you need to go to be you:** Take every opportunity to use those personality tests and surveys (Myers-Briggs, DISC, Personalysis, etc.). Why? your biases and your mindset shape how you see insights and make sense of them. The less time you spend "faking it," the faster you get to the authentic person you are and then you can really dive into insights.

(2) **Look at who you are and be honest with yourself:** When all is said and done, you are stuck with you. If there are issues you struggle with (you are dyslexic or you have attention deficit hyperactivity disorder, you suffered trauma as a child, you have an addiction to something), you need to own these factors and realize they will shape your view of people, places and

things, aka *insights*, and the meaningful impact you can have being an insight leader will be shaped by *you*. To avoid being just another sensory or market research professional, but rather a unique insight leader, grappling with your demons and angels will make you a strong leader who will find those defining insights we hope you will seek out.

(3) **Heraclitus tells us:** "Lovers of wisdom must open their minds to very many things. I searched into myself. Knowing many things doesn't teach insight" (Von Oech, 2001).

1.6 Dealing with management and your clients/customers

(1) As a knowledge professional, inevitably you deal with clients and/or customers. Clients are not the individuals to whom you report. Reporting in that fashion is a line in an organization chart. Everyone in a company has "clients" of one sort or another. Clients are the individuals to whom you owe a "professional" something. To clients, you owe the work efforts of your professionalism. You may owe clients your thoughts (also called inputs) on how to approach a problem, or the design of a study, execution, analysis, and recommendations. Consider this idea a requirement of the job.

(2) Often, we believe that when we report results, we have to present the results in the same form that we might present results from a graduate or professional study. Read the scientific literature to get an idea what this means. All too often, these papers in journals are difficult to read, filled with table after table of results, with significance tests, with the actual results reported in a convoluted way, defying everyone but graduate students and perhaps the reviewer. The journals are not particularly reader-friendly. They don't need to be. They are archival, for the advancement of the profession. You, however, cannot afford that when your job is to report results that people will want to engage with actively.

(3) We have seen the evolution of reporting, moving from text to video. From text to images. The watch-out is, you can get "too cute." Don't go there unless you truly understand the imagery, the language you are signaling. It is very possible to shorten information streams and lose all content. Or to be culturally insensitive, at a given point in time. So while there is a much broader range of technology to help convey information, you need to be aware of what you are really saying and what that can mean to individuals or groups.

(a) A meaningful case study is the one that occurred for Prada, the global luxury fashion design company during the Christmas holiday season of 2018. As reported by Vox and *The Washington Post*, figurines on display at Prada's Soho New York City store were reviewed as having insensitive racial stereotyping. Those viewing the display window suggested cultural insensitivity. The figurines were supposed to be part of a design collective called Pradamalia - fantasy charms composed of elements of the Prada oeuvre. And in this case, Prada claimed the imagery was supposed to be animals, not people. After the public outcry, the figurines were removed. In 2020, Prada agreed to a settlement with the New York City attorney to provide racial equity training to its employees in New York City and its executives in Milan, Italy. As the *Washington Post* article from 2018 asked, "Seriously, Prada, what were you thinking? (Givhan 2018)".

(b) Along the same visual area, in February of 2019, Gucci was also accused of racial insensitivity due to the sale of $890 "blackface" sweaters (black sweater with a high balaclava turtleneck which when pulled up over the nose had a cut-out for the mouth, with the area around the cut-out being a broad rim of red - looking somewhat like large red lips) (Givhan 2019).

The take-home message is imagery and pictures can create very reactive emotional responses, so you have to pay attention to what you are visualizing and how it may be taken by someone not involved in a project.

1.7 Guidelines to success

When dealing with your management and your clients/customers, it's important to follow a few guidelines. Follow them and the outcome will be more positive than could be imagined:

(1) **Clarity:** When first meeting to discuss the research issues, make every effort to simplify the discussion, to repeat the goals, and where possible, to verbalize potential solutions. Those early times, the start of project discussions, are the best times to ensure clarity. It is inevitable that over time the issues will get muddier, less clear, as details begin to crowd in and practical considerations begin to merge with technical and research issues.

(2) **Verbalization:** Practical experience shows that time after time it helps to verbalize what is going on. Simple, direct sentences are the order of the day here. You might almost consider this to be a running abstract of the topic. You may be tempted to keep quiet, to let things evolve, to summarize at the end. The truth of the matter is that by restating the objectives and strategies in simple, declarative terms throughout the meeting, you make everyone's job easier. You will be clarifying the goals again and again until the structure is crystal clear to everyone. There's a far greater likelihood of success after that clarification than before. Everyone knows the ground rules.

(3) **Feelings:** We often begin work with a combination of insecurity and bravado. In our most private moments, especially as we start out, we realize we don't profoundly understand what we are going to do. And, at the same time, we feel the focus of an external audience, not our parents, as we do our work. The result can lead to withdrawal and flight, or to bravado. Fleeing doesn't do any good. We merely remove ourselves from the fray, either momentarily, or more likely permanently. Bravado doesn't do much good either. Professionals see through us. One can't "fluff" one's way through business issues. Our lack of knowledge catches up. The best thing to do here is simply to admit one is nervous, and move on. We'll talk in Section 1.12 about what to do.

(4) **Imagery:** What's the best way to capture the essence of the meetings, about what's happening, about what you are learning and, of course, share that essence with others? How can you take the verbalization and the emotion and capture the imagery? Is it a sketch? What about an image? Or maybe it is a creative chart. People like William Tufte (1997), Andrew Abela (2008), Tony Buzan (1996), David Byrne (2003) and David Hockney (2001) have all explored meaningful ways to provide a visualization of ideas. The best way to approach this is to expose yourself to a number of approaches and then find the styles that help you communicate more precisely.

(5) **"Wave 0":** Every researcher knows that he or she is engaging in a dialogue with nature. And the smarter ones realize that it's good to know what one is doing. Microbiologist pioneer Louis Pasteur put it best, "Chance favors the prepared mind." But how does one prepare for one's job? One is expected to know what one is doing. Such knowledge comes from years of experience. But what does one do when starting out? The answer to this is really rather simple. Run a small-scale experiment, and do so publicly, not privately. The experiment is not to "crib," to "cram," to learn one's profession quickly. The small-scale experiment (we call it wave 0′) is a public test, to see what the data will bring, before anyone commits. It's not a shameful thing to do. In fact, most clients are grateful. If they don't cost a lot, if it is small scale, people welcome pretests and trials. Wave 0s give everyone a sense of relief that before any commitment, we're going to try it out. The wave 0 is like a biopsy before a major operation; map out the territory, check what's going on. Wave 0s, pretests, pilots, whatever you call them work, and they work easily and well.

(6) **Listening:** Deeply listening is hard. We still haven't found training that many people should get in how to listen without bias. Your clients/customers tell you a lot. By taking your "yes, but. . ." out of a conversation and listening to what is being said to you, it becomes clear what you need to understand about a situation and what needs to be done. Listening requires practice, and many of the previous guidelines are more successful the more you learn how to listen through your bias.

1.8 Reporting results

It may seem a little too much to deal with the notion of reporting results. After all, most companies have either formal or informal guidelines about presenting results to management. For example, Procter & Gamble prescribes a specific form that research uses to present the results. Traditionally, the researchers were told to remain within the confines of the actual results, rather than to speculate.

General Foods (now Kraft Foods), in turn, also had specific formats to use, and the list could go on. The idea is not to create a new form that would probably not be well appreciated in most companies. But the idea is to present results in the most cogent form. This form should combine tables, words, and appropriate imagery. The tables should be simple to read. The language should not be a reiteration of all that the table shows. Rather, the form reporting the results should highlight the key findings and the sound bites that the audience should take away, and the imagery should help clarify what is meant by the entire summary. Infographic approaches are adaptations of this recommendation, as is a whole body of visualizations by those experienced in R or python. What is critical is to remember what you are attempting to summarize. While it has become popular to present very short videos of results, keep in mind that many of your readers can capture results from an infographic or traditional report much more quickly than spending two to five minutes listening/watching your video.

In the archival literature in most fields, there is a certain style. The style varies with the particular journal. For the most part, the journal style is formal, and written to give the sense of gravitas to the results. The truth is most journal articles will go unread. That's not the case with your reports issued in the privacy of the company, dealing with the results of studies. The work that most readers of this book will do will be used by different groups in the corporation, ranging from bench chemists to vice presidents, and even higher, and sometimes, not often, the work may be reviewed a number of years later for an adjacent purpose, so the cleaner and clearer you can make the results, the more successful you are in communicating.

The work that you do is important. Unlike the academe, there is precious little room in a corporation for "exploratory research" for the sake of one's interest. (That's a problem, but it's still reality.) One's work typically revolves around answering questions so that others may read and use the results. It's important to be crisp, clear, succinct, and yet provide the necessary detail. As an insight leader, do have a point of view. Be intellectually honest, yet find a way to work within the company veil that most organizations have in place.

1.9 Do not "winstonize"

The term "winstonize" means taking a table of numbers, converting it to simplified text so that it is "boiled down" to its essence and presenting the meaningful story to the client (for marketing and nonmath people).

Some years ago, in the early 1990s, one of the coauthors of this chapter (HRM) had the pleasure of working with a well-known food company for the development of a new dairy product. The project involved the use of rule developing experimentation (RDE), a variation of conjoint analysis (a more detailed description of conjoint analysis can be found in Chapter 8.2). What's important here is the test stimuli comprised about 90 phrases, each dealing with an aspect of a dairy product. Respondents evaluated combinations of phrases in short sentences. At the end of the evaluation, the computer program deconstructed the combinations to show how each element drove the response. Every one of the 90 phrases generated its own impact or utility value.

We were novices at the time. We had just expanded the use of an approach called IdeaMap.Net (RDE) to many elements. Of course, the RDE study itself worked fine. What was very interesting was the reaction of the client. In the client's mind (he was a consumer researcher, not a marketer), the most appropriate way to present the results was to give the marketer the "big picture." There was a sense that the marketers would not appreciate table after table of data. We were told that the client wanted the results predigested, with the implications strongly presented, along with clear next steps. It took us a week of hard work, but finally we boiled out all of the numbers, and had the best-written prose that we could create to "tell the story."

The presentation was something else altogether. Perhaps we were not particularly dynamic presenters. We kept noticing something which disturbed us. Our research client was attentive, but the marketers and the product developers seemed to wander, and so, we proceeded with the presentation, droning on and on, with what we had first believed to be impeccable powerful prose, but by now we're seeing it to be verbal drivel. We thought the results were there... but they weren't.

All of a sudden, we hit upon a table of data, quite accidentally. We had failed to boil off all of the results, and so, a table of data remained. It was as if we injected our Lazarus audience, bringing them back from the dead. They focused on the data, became animated, discussed the results among themselves and at the end warmly thanked us for what turned out to be a wonderful presentation.

Here are key takeaways from our experience of "winstonizing" data:

(1) **We wanted the "approval" of the client:** As outside suppliers, that is, not part of the client's corporate family, we felt that the client's word was sacrosanct. We were wrong. We didn't realize that the client was as ignorant as we were about the needs of their own clients. In our quest to get approval, because of being outsiders, we sacrificed what really mattered in our role, clarity, and direction.

(2) **Clients are fallible:** Clients are people. Clients are not omniscient. It is we, the outsiders, who give them the power. Clients are nervous (hey they need their jobs; they want their jobs!). But, at the same time, they are misled. They believe what suppliers tell them. Perhaps not at first, but eventually they do. They are fed a diet of positives from the outside, and this diet makes them fallible.

(3) **Avoid "presentertainment":** Yes, it's nice to have pretty pictures or words, to avoid charts. It's so very tempting to believe that the answer is in the presentation. We call it in our earlier works "presentertainment." The term was coined by those audience members who sat through numerous PowerPoint presentations which employed an overwhelming number of images and flash with very minimal verbal content. The emphasis of those presenters seemed to be entertainment over information. It's not the message, but rather the elegance of the presentation that convinces, or so we would like to believe. When we did the "winstonizing," we were just at the beginning era of presentertainment. So tables gave way, mistakenly, to words, to prose, to description. Now that same presentertainment would jettison the tables in favor of pretty pictures, of simple visual entertainments, and the "winstonizing" would be complete; no text, no tables, just images which would convey (it is hoped) what we were trying to do, and we'd be dead wrong.

(4) **Clients like data:** We discovered from the presentation of the dairy product that clients did welcome real data. Pure and simple, honest. But they wanted

data in a form that they could use. They did not want polished data in a form that would disguise. They wanted simple tables.

(5) **And most important, clients want simple, clear presentation:** When it was time to end the presentation, we were to end it. No ands, ifs, buts. No inserted meaningless words such as "... it's quite interesting that..." and the like. Rather, simply hard-hitting points. We noticed that the clients were responsible for making something happen. They just wanted the results. They did not want a presentation that went on forever, to justify the money that they spent. That was irrelevant. They were paying for answers to move the business forward, and that's all.

1.10 Making it public – helpful hints to grow from student to professional

It's a good idea to understand the corporation (or client) in which you work, especially if you are involved in the evaluation of subjective aspects of products or services, whether this is through panel work with experts, community panels, or a range of one or many consumer evaluations. As a student, you may have done your own experiments, all the way from design to analysis, and to some extent reporting. You may have published your results, either yourself or with a professor. And of course, if you've been through the publishing world, you know all about the back and forth of editorial reviews, the occasional nasty rejection, the seemingly infinite number of revisions, and so forth. From these exercises you may have formed an idea of what constitutes a good report of results, and of course what constitutes a skimpy report.

As you begin to work in business/corporations, be prepared to abandon your preconceptions, and face what could be called *Realpolitik* by some, or compromising principles by others. The truth of the matter is that most companies are not in the knowledge business. Companies do studies to find out key information that they want to know. Companies are still not regularly run by scientists (and while engineers are highly competent analytic individuals, engineers are not scientists, either!), and if they happen to be (a rare coincidence), it's not going to be the science of consumer research, descriptive panels or ethnographic discoveries, for example.

But just what does that disturbing information mean, in real terms? Does it mean that no one in the corporation is a real scientist? The answer to that is probably "yes," no one in the corporation is a scientist in the way that an academic is a scientist.

Now that this ugly secret is out, that the company is not a university or an open society, what are you to do? Of course, you don't do slapdash research. You don't sacrifice quality. You may have to reduce the sample size. (Or in many cases, you actually will end up working with far more people than you would have, had you remained an academic. Academics are notoriously impoverished.)

Your big differences will be in the way you report the results. We've already dealt with this above, at the start of this section, but it always helps to reiterate how to communicate. Like real estate, the answer in business is "communicate, communicate, communicate."

(1) **You'll have to make your reports readable:** No more hiding behind the academic jargon, the tightly written, almost unreadable and certainly mostly forgettable results. You can be sure that when you work on relevant topics, your reports will be read.
(2) **You will have to avoid the flight to statistics:** You can't expect your client to wade through a mountain of statistical effects. The company needs to know what you found, and what you did not. Pure and simple. You have to write to be read, not to defend your insecurity about not being worthy of the job. Do the job; it's yours.
(3) **In many instances people are afraid to step up to the plate, to take a chance:** This happens in all professional arenas, and unfortunately, it happens far more frequently than you realize. For market researchers, the conservative fear manifests itself in a 300-page report, with unbearable numbers of tables. That report, the death of uncountable trees, belies the sheer inability to say what's really going on, preventing in turn a junior brand manager from having that all-important "Aha!" experience. This is called bird's eye shot insights or results by the pound. For a sensory analyst, it can be reams of tests, of spider plots for descriptive analysis, cross referenced to massive quantities of analytical data, along with maps in two- and three-dimensional space which look impressive but don't say anything, and for some in consumer research, it can be pages of wonderful images that again, are beautiful and interesting but say nothing.
(4) **Avoid number three above like the plague:** It will turn you into a corporate drone, and perhaps eventually cost you your job when things get tough. It's hard to believe that doing one's job, dutifully reporting results in an objective way, could be counterproductive, but it is. It's nice when you have a staff position and believe that you merely need to grind out the data in a facsimile of a professional, but you're just fooling yourself. The real issue you will face is that the company did not hire you to run studies and report data. That's part of your job, yes, but that's not really the job. The job is to act as an intelligence source, using the data to identify what the company should do to solve a particular problem. The insight leader.

1.11 The two types of professionals in the world of evaluating products (and studying consumers/people)

In a book published by one of the coauthors of this chapter (Moskowitz, 2010), a point was raised about the world of knowledge workers; that there are at least two types. One type prides itself on being able to solve the problem. This is the hero, not in the denigrated sense, but in the sense of Joseph Campbell (the esteemed mythologist). The hero moves out to the luminal space and returns with something transformed. The story is about the hero. It's about what the hero does. All the focus is on changes in the hero, on the trials and tribulations, and in the end, the hero is the one who helps move society along.

There's a duality to the hero. That is the individual who works in business, doing the job. The individual fits in. There is no need to prove oneself, at least at a professional level because the job is not about fulfilling one's scientific education.

There is a need to do the job and to establish oneself as a professional worthy of advancement and reward, but we're not seeing the hero's journey here. We're seeing the individual in business, not the individual fulfilling their own destiny.

The first person, the individual who is in the "hero" role, may correspond to the professional who has been educated in the world of science. The role of this professional is to bring science to the corporation, to bring the benefits of the scientific education (or training) to the job. This is the typical role of the sensory professional and the career consumer researcher. Advancement in the corporation is a by-product of the job. The real nature of the job is to fulfill one's education purpose, and to improve the understanding of the product (or subjects evaluated) through science.

The second person, the individual who is doing the "job," corresponds to many individuals who have the role in the business, but the particular job is one role of many. The person is in the role, but the person is not the role. This individual is the corporate employee who is rotated through different roles to give them both training and exposure. This second person is not so much doing this job for the sake of the profession as doing this job as part of corporate training. Whereas the first hero is anchored in the profession, this second individual is in trouble as an insight leader when he or she has failed to craft an authentic self, prior to engaging in corporate rotation by training. Without authenticating somehow, this type of training can leave this second person an empty, rudderless shell of a person, honestly confused after a few years as to what their purpose in business should be.

1.12 Knowing your limits and inviting others in

In every business of a reasonable size, it sooner or later becomes necessary to learn how to deal with others, not as adversaries but as collaborators. It's one thing to run a small, one or two-person consulting/expertise shop and do everything oneself. That's a perfectly good way to spend one's business career. There are many thousands of one-person shops, of virtuosi who can do all aspects of their business. They really don't have a business, per se, they have a practice. There is nothing wrong with a practice, once people realize that's what it is, an essentially solo operation. The client is buying the expertise of the individual. It's the standard operating mode of individual experts outside the corporation.

The corporate world, larger or smaller, is a different "kettle of fish." The very essence of a corporation is built on the division of labor, the assignment to individual's jobs that they can do better than others. Division of labor moves beyond the practice. It requires that the individual in the corporation be part of a smoothly running business. It's no longer about the individual as the star, and the cynosure, as the lone wolf practitioner. It could be, of course, in one's mind, but that's not what makes the company really work. Rather, it is cooperation.

At first it all seems so simple. After all, who doesn't want their business to succeed? And cooperation is such a wonderful word. There's only one problem. When people cooperate, they have to share the power and the spotlight with others. It's here that the problems arise.

Let's look at two types of people in business, to reiterate in business, and not in solo practice.

(1) **Those who have come up through the scientific ranks:** Who base their self-esteem on their own personal accomplishments, have to share the spotlight of capabilities with others. These individuals who have come up through the ranks on the basis of their own merit soon find that they no longer shine as the younger superstars, the wunderkind, or even the competent. It's a team effort. That's hard to swallow sometimes, no matter how often they get bombarded with motivational messages such as "there's no I in team." Continuing to stay current with science may be a way to stay relevant and valuable. Yet, it is important here to recognize in a professional career it is necessary transition from being a young and rising contributor to being either a recognized member of a successful team or being an esteemed seasoned contributor, who, while not young and flashy, is known for your capacity to do good to great reliable work that is highly valued.

(2) **Those who have come up through the ranks in business carry a different burden:** They do not carry the burden of accomplishment through one's own merits, in a world of scholars and professionals. Rather, they are accustomed to carrying out orders, being cogs in a wheel, doing the job within or under budget, and of course doing the job to which they were assigned. Going outside the job assignment is perceived by many to be risky. They are in some ways clerical-minded. Yet when they rise to a certain level, they are expected to be independent, to think creatively, strategically, and tactically. It's not enough to bring in the project under budget and on time. It's important to move the business ahead. Without working on the authentic self, these people can suffer, often silently, but deeply. They may get all the rewards due them, yet since they have not cultivated a point of view, they have nowhere "inside" to enjoy the fruits of this success. They may be wildly "successful" but empty.

1.13 The bottom line – what's it all about?

You're hired to be a knowledge expert in your corporation. Hooray, you've made it. Yet, you're only at the beginning. A lot of your job will be to unlearn the wonderful habits of your undergraduate, graduate, or university teaching days. Yes, it's important to do things correctly, to be a scientist, to be faithful to the discipline from which you sprang. But there's more:

(1) **Your moral character as a professional:** What is the journey on which the company has placed you? What is the path you need to travel?
(2) **The ethics of the situation:** How do you behave with others?
(3) **Your work:** Just exactly what are your duty bound to return to your employer?
(4) **You:** Changing jobs doesn't change you. So maybe start with being authentic.

Discussion questions

- Why are the authors suggesting that to manage insight for companies, one has to have more than classic science training?
- How is the idea of reading broadly relevant for a good innovation toolbox?
- Are you the reader finding the idea of your authentic self, exciting or terrifying?

References

Abela, A. (2008) *Advanced Presentations by Design*. San Francisco, CA: Pfeiffer, a Wiley Imprint.

Bradley, N. (2007) *Marketing Research: Tools and Techniques*. Oxford: Oxford University Press.

Buzan, A. (1996) *The Mind Map Book: How to Use Radiant Thinking to Maximize your Brain's Untapped Potential*. NYC, NY: Plume.

Byrne, D. (2003) *Envisioning Emotional Epistemological Information*. Göttingen, Germany: Steidl.

Drucker, P.F. (1999) *Management Challenges of the 21st Century*. New York, NY: Harper Business.

Givhan,R. (2019, May 7). 'I was the person who made the mistake': How Gucci is trying to recover from its blackface sweater controversy. The Washingon Post, Style. https://www.washingtonpost.com/lifestyle/style/i-was-the-person-who-made-the-mistake-how-gucci-is-trying-to-recover-from-its-blackface-sweater-controversy/2019/05/06/04eccbb6-6f7d-11e9-8be0-ca575670e91c_story.html (accessed 9 May, 2019).

Hockney, D. (2001) *Secret Knowledge*. NYC, NY: Viking Studio.

Lee, M.S.W., Motion, J. and Conroy, D. (2009) "Anti-consumption and brand avoidance". *Journal of Business Research*, 62(2), 169–180.

Moskowitz, H. (2010) *YOU! What you MUST Know to Start your Career as a Professional*. S. Charleston, SC: CreateSpace.

The Random House Dictionary of the English Language (1971) The Unabridged Edition. NYC, NY: Random House.

Tufte, E. (1997) *Visual Explanations*. Cheshire, CT: Graphics Press.

Von Oech, R. (2001) *Expect the Unexpected (or You Won't Find it)*. New York, NY: Free Press.

2

We've all been understandably racing to master the new technologies in this ever-changing machine. I have a little secret for you: we will never master all these technologies. As long as we try, we will forever be on our heels. I try to simplify by taking the mystery out of the new world and telling our people to look beyond the obsession of technology and turn our attention to what matters – the consumer experience.

Marc Pritchard
P&G's Chief Brand Officer

What this is about

This chapter introduces the concept of the **Consumer Explorer (CE)** as key to delivering innovation strategy. The CE is defined as the ***insight leader*** and ***strategic innovator*** who brings consumer focus into the different goals and activities of the organization. The CE accomplishes multiple goals for both themselves and the organization and has a mindfulness about doing them. Before learning to lead others and accomplishing goals for the organization you work for, you as a Consumer Explorer must first establish a strategic goal for yourself, then identify your core values, motivations, and purpose. It is important not to sacrifice one for the other.

Product Innovation Toolbox: A Field Guide to Consumer Understanding and Research, Second Edition. Edited by Kannapon Lopetcharat, Dulce Paredes and Jacqueline H. Beckley.

Chapter 2

2

The Consumer Explorer: Key to Delivering the Innovation Strategy

Dulce Paredes and Kannapon Lopetcharat

Key learnings

✓ Becoming a Consumer Explorer
✓ Working as a strategic innovator
✓ Having a seat at the innovation table

2.1 The rise of the Consumer Explorer

Whether you are a knowledge worker in product research, sensory sciences, and marketing research and/or managing a consumer research team, you are expected to become valuable contributors to the company's strategic vision and bottom line. Chapter 1 talked about the importance of knowing the mindset of your authentic self and the principles that guide you before you can contribute to the broader company strategy. In this mobile environment, it is important to "build your brand" (Smith, 2011) and not let the company define who you are. You can play multiple roles such as strategist, innovator, tactician, and take those skills with you as you move through the organization or move from one company to the next. You usually get hired or promoted for what you know and what you can bring to the organization. At the same time, there is a fine balance between practicing your internal goals, e.g. your "creative and cognitive brand," with matching the company's fiscal objectives and long-term goals.

Product Innovation Toolbox: A Field Guide to Consumer Understanding and Research, Second Edition. Edited by Kannapon Lopetcharat, Dulce Paredes and Jacqueline H. Beckley.

2.2 The roles of the Consumer Explorer

Elon Musk, Founder and CEO of Tesla, SpaceX and PayPal, once said, "Constantly think about how you could be doing things better and keep questioning yourself." To do so, one must hone their observation skills to the max and that is a primary trait of a Consumer Explorer (CE).

The Consumer Explorer has two important roles in the company's innovation journey:

- Setting the strategic research agenda and team dynamics; and
- Connecting the consumer learnings from one study to the next to facilitate building the business.

The CE's most visible role is putting together the consumer and product research strategy from project initiation to completion in partnership with various functions in the organization like R&D, marketing, marketing research, sales, and supply chain. The CE captures the essence or key learnings from the various qualitative, quantitative, and hybrid studies to enable the research team to generate bigger wins and learn from its failures early in the development.

To be effective, the CE should be a core member of the Innovation Team from the beginning and should not be brought in along the way for tactical reasons or when a consumer research test needs to get done. Howard Moskowitz (2005) referred to this scenario of becoming solely a testing center as "Tests R Us" instead of creating a system of building knowledge for continuous growth. With the rise of open innovation programs, Moskowitz and Saguy (2013), proposed the transformation of the consumer research function from a corporate testing service that measures, consumer reactions to products and concepts using a wide range of analytics and measures into creating new knowledge of the product and understanding the corporation's current and future consumers. Most recently, the market and consumer research function have been challenged to guide the company to be better prepared for the future by learning from past engagements, dealing with current and changing uncertainties and estimating future directions.

To be a valuable member of the Innovation Team, the CE has to be recognized as the technical expert or knowledge worker that can design, execute, and identify actionable consumer learnings to help build the business. In addition, the CE must have the passion and enthusiasm to guide the team to continuously infuse the "voice, eye, touch and brain of the consumer" throughout the project and/or business strategy. Finally, an effective CE has to have a sense of humility to admit what they don't know or can't recognize at different parts of the journey and is open to new approaches from other internal and external sources.

2.2.1 As a research agenda strategist

The role of the CE is to design the consumer-learning plan from upfront innovation to final stage consumer and product research validation. This includes identifying the necessary skills to drive the research agenda and filling in the gaps through new hires and/or building research partnerships and collaborations with internal and external agencies. The CE leads the back and forth discussions between the research team or client and those chartered with designing and executing the research studies. This ensures smooth translation of what is expected and what can be done so there are no big gaps or unwanted surprises in the end.

Although having a strong portfolio of research tools is important, the CE should not be defined solely by the research tools that they know and practice but by the skill of knowing which research tools are appropriate to use and be guided by the principles of what the organization is trying to accomplish to further build and future proof the business. How often have you seen new leaders push their "tool of choice" without regard to whether it is good for the organization long-term?

We have renamed the Consumer Researcher for Innovation to Consumer Explorer to highlight its leadership and strategic role. The role may not always reside in the consumer insights, product research, and/or sensory group. Innovative companies have changed the consumer researcher role from gatekeeper of product testing results to that of a strategic partner in designing and leading the consumer and product research learning plan from upfront idea development to product feasibility and consumer validation.

Most recently, some companies and research organizations have used the word "humans" instead of consumers when referring to the people that they are servicing and have renamed their Market Research and Insights Groups to Human Insights. For this chapter and the rest of the book, we will refer to those performing this strategic innovation role as Consumer Explorers.

Key enablers for the strategist role are the following:

- Critical and creative thinking from hypothesis testing to idea validation (*Chapters 4–6 identify tools and practices to turn inventions to innovations and to recognize insights*).
- Technical skills to understand and digest data balanced with the communication skills to present them clearly and simply (*Chapter 2 discusses data presentation*).
- Access to an extensive tool box of research capabilities and knowing when to use them appropriately (*Chapters 7–10 highlight cutting edge tools that have been proven to provide effective and efficient consumer-based decision-making*).
- Extensive reach of cross-functional internal and external research partners and experts that can be tapped as needed (*Chapters 11–17 feature practical advice from internal and external strategic research partners*).

2.2.2 As a connector

The CE needs to bridge and integrate learnings from one study to the next to enable the innovation team to build bigger opportunities that can come from the learnings. Key enablers for the connector role are the CE's ability to do the following:

- Incorporate internal knowledge mapping and trends analysis for hypothesis testing and model development;
- Extract relevant information from syndicated and external data including unsolicited social media posts;

- Generate relevant consumer understanding and responses to product and communication stimuli;
- And most importantly, integrate the vast consumer and product information and working understanding of the business model and financial inputs to provide actionable business results that could generate future revenues.

The role of the CE is analogous to a ball juggler who is balancing multiple objectives or balls (Figure 2.1):

- Sustainable **Business Strategy**
- Healthy **Innovation** pipeline
- **Process** that works
- Voice of the **People at all touchpoints**
- **Unknown Unknowns.**

Juggling the first four balls: **Business Strategy**, **Innovation**, **Process**, and **People** are not new to CEs. The fifth ball, **Unknown unknowns** are unexpected events that can be a forceful fly by, or it can stick for a while. This information abyss is different from the missing details in your data collection and hindering your ability to provide definitive recommendations. Those can be managed through risk profile assessments. We are referring to dynamic events that start small and quickly magnify and dismantle the status quo from all angles.

Figure 2.1 The Consumer Explorer, just like the experienced juggler, knows where all the balls are going and ensures the organization keeps all of them in continuous motion. Source: Artist: Nadine Paredes.

Unknown unknowns are events that could lead to an unpredictable future, both short and long term. The COVID-19 2020 pandemic, for instance, showed how a localized event can quickly escalate and upend the world. Virus researchers would say that the signs were already lurking but were ignored or we failed to see the potential implications. Others will argue that this could be just the beginning of other pandemics including the potential damages from climate change. The latter points to imagining the food and energy needs in the future and what we must do now to cushion the blow. This worldwide trajectory impacted companies who relied on short-term predictions such as clothing retailers who project trends for their next seasonal show. When physical shops closed, CEs turned to social media for future predictions (Maheshari, 2021).

Many companies talk about innovation but are grounded on controlled processes that could hinder change and adaptation. Business strategies can be articulated through clear objectives and action standards. People's reactions at different stages of product development can be sensed with qualitative, quantitative, and hybrid research that requires connection from one study to the next to frame the real story.

In a recent webinar on sparking innovation and creativity, Thomas Maiorana (2021), University of California Davis professor of design, talked about engaging people with "low resolution" prototypes or versions of the final design idea early in the process. A low-resolution prototype is an incomplete, nonfancy, inexpensive version that is built quickly to allow the researcher to explore and learn for more complex systems later on. Whether we use actual prototypes or not, it is important to get consumer reactions to the new idea concept. Incremental learnings even during times of uncertainty and rapid speed of change can add up to creating a new viable product, but true innovation can only be experienced after the fact.

The definition of a successful innovation can range from success in the marketplace and increase in company revenue, to revolutionizing the market and opening a new competitive framework. High functioning and innovative companies have learned how to keep all the balls in motion. A successful CE has three considerations to keep all the balls moving:

(1) Do I become the lead and lone juggler for all the balls and hire more people to help keep all of them in motion?
(2) Do I get external help so I can concentrate on one to two balls and let others do the rest?
(3) Is our innovation process robust to withstand unknown unknowns, willfully or not?

The first choice requires company investment in resources and people. The second choice relied on open innovation, shared responsibilities with strategic research partners and strong collaboration from all parties involved. Chapters 11–15 show how you can leverage internal and external experts to expedite the research process. The third should be more of an ingrained thought process than a consideration. Lessons from the COVID-19 pandemic have shown us that those who successfully maneuvered or started new businesses and services had common things in their organizational DNA, such as a strong sense of purpose, a working technology, a reliable yet fluid process and a collaboration network that can be triggered, accelerated, or collapsed. We will explore some examples in the subsequent chapters.

2.3 Taking the lead

So, what does the CE have to do to have a prominent seat in the innovation table?

2.3.1 Understand the problem

The objective should always drive the research methodology. As C.K. Prahalad, a foremost strategic business strategist, said in one of his last interviews before his death in 2009, "In developing all of these ideas, I learned not to start with the methodology, but with the problem. A lot of times, research tends to start with the methodology. I prefer to start with a problem that's of interest and apply whatever methodology is appropriate".

Understanding the overall project objective and what it means for the business is key before starting a research. The CE could have a big role in meeting the overall project objective or have one of shared responsibilities with other functions in the Innovation team.

Although the objectives could change as the project progresses, it is important to understand the overall project objective to set you on firm ground. You may need to adjust as information and decision changes, but it is critical to know the overall project objective.

2.3.2 Get the questions right

The CE should be grounded on asking first-order questions or the questions that really matter (Moskowitz, 2004). The questions could either be strategic or tactical. Strategic questions are those that the business wants to get answers to keep the project moving and would be used for key business decisions like investing in additional resources, creating a new line, or cutting short an investment. Tactical questions are related to information needed to generate robust learnings like identifying core consumer targets, global or regional reach, or key metrics/key product indicators (KPIs).

As researchers, we are grounded on Type I and Type II errors to define the research boundaries and manage the risks of generating erroneous conclusions. Technically, Type I Error is the mistake made when a researcher rejects the null hypothesis and accepts the alternative hypothesis, when the null hypothesis is correct. Type II Error is the mistake made when a researcher accepts the null hypothesis and rejects the alternative hypothesis, when the null hypothesis is incorrect.

From the business side, Type I error is sometimes referred to as "producer risk or manufacturer risk." From the point of view of companies or manufacturers, committing Type I error will result in losing market share or consumer confidence. New product launch, line extension, and claim substantiation (especially superiority and differentiating claims) for existing brands are activities that need attention to avoid Type I error. Launching new products (new brand, new product line for an existing brand or upgrading an existing product) requires companies to claim what makes their product better than what is currently in the market; therefore, superiority or differentiating claims must be made by the companies. If consumers do not agree with those new or differentiating claims, they will lose their interest in the brand and ultimately choose other brands.

Type II error is often called "consumer risk." Companies often commit this mistake when they (1) conduct cost reduction activities (replace current formulation with a cheaper one), (2) change suppliers (to reduce cost of raw materials or the old supplier is no longer available) and (3) changing processes (upgrading a machine or change from an old process to a more efficient process). All these activities result in "a new formula" that companies do not want core consumers to notice. Therefore, the risk of wasting money is on the consumer side. Companies must conduct at least a test to ensure that the changes they made are "unnoticeable" by consumers. If companies fail to do so, consumers could lose their trust in the brand and ultimately switch to other products.

To avoid committing either Type I or Type II errors, one must assume that they conduct a perfect study to begin with by (1) using the right tools, (2) asking the right questions, and (3) measuring with right metrics. However, this is not usually the case. At the 2009 8th International Pangborn Sensory conference, Harry Lawless, Cornell University professor, talked about the notion of having Type 0 error or asking the wrong question to begin with (Lawless 2009).

In essence, Type 0 error is the most dangerous mistake that the CE can commit. This means:

- **Running a study without knowing real objective(s) or pretty much running blind (not knowing the right questions):** The CE may have the most precise measurement or method but does not know the real target.
- **Applying inappropriate methods or protocols to wrong situations or questions:** This means knowing the right questions but the CE keeps using the wrong tool vs. leveraging an approach that is most suitable for the situation. This could also include using a metric that has a historical significance to the company but has outweighed its usefulness.

2.3.3 Designing the strategy (learning modules) to answer the questions

Once the CE is grounded on the key questions that need to be answered, he/she is chartered to design a series of learning modules to guide the business and research teams that will provide actionable results from project initiation to completion and postlaunch review.

The CE should be prepared to design efficient testing that can weed out failures early and will enable the team to react quickly and leverage successive consumer and product learnings to build and validate bigger ideas. Chapter 4 discussed the benefits of running "Wave 0" or small-scale pilot studies to test hypotheses, fine-tune protocols, or as a dress rehearsal for larger studies.

The research strategy should cover the work needed to identify consumer insights and feedback from upfront idea generation, to identify viable product ideas for new product development (NPD) and finalize consumer validation for go to market products. Two important aspects that the CE has to consider in this planning stage are timeline and available resources (i.e., people and testing dollars).

2.3.4 Assessing what you know and what you don't know

The first step in the learning module is for the team to assess the current situation and the information available to the team from both internal and external resources. This step has been referred to by various contributors in this book as **Knowledge Mapping**. Moskowitz et al. (2006) defined knowledge mapping as converting company-owned data and existing information into knowledge that can lead to actionable results. The CE has to guide the team in generating the information from both reported data and the information that resides in the research team's mindsets.

At this step, the CE has to think about what they need to learn and know. This could include identifying and challenging the "sacred cows" or pillars of information that are valuable to the company and are held in high esteem and/or treated as "untouchable". Key questions to ask are the following:

- Is there data that exists?
- What hypotheses have been known and proven?
- Is it belief or folklore, i.e. so many people believe the information that it has become the norm?

After grounding the team on current collective knowledge and existing data, the CE has to capture the essence of the available information as baseline or background to form the hypothesis for each of the subsequent learning modules.

Having an engaged team is critical to the project success and the CE's ability to lead the team in identifying key consumer and product insights. The CE has to require that the team members "have their skin in the game." This means requiring the team to be highly involved at all stages starting with delivering project inputs (e.g. knowledge mapping, technology, and prototype development) and digesting key learnings throughout the research cycle. Each team member should have key roles and responsibilities identified upfront. The CE has to set ground rules on how they should participate in consumer learning sessions, e.g. no e-mails and web surfing during consumer research group sessions and project debrief sessions.

2.3.5 Establishing the hypothesis

Creating a hypothesis or the information that needs to be answered or validated will enable the CE to design a focused and purposeful research plan and to avoid a "fishing expedition" that could lead to conflicting results and/or generate loops of research studies.

2.3.6 Creating and executing the learning plan

After the CE has designed the learning modules or stages of research and has identified the key questions and corresponding hypothesis that needs to be validated at each stage, the focus shifts to the tactical aspects of implementing the research.

The following information is needed to successfully implement each learning module and provides the framework for the project dossier:

- **Overall project objective:** The key question that needs to be answered.
- **Research objective:** Specific to the learning module.
- **Introduction/background:** Context of the project.
- **Action standard:** How results will be used; KPIs by which research findings will be judged.
- **Methodology:** Qualitative and quantitative tools, analytics (*see* Chapters 6-10).
- **Design:** Tactical information to research plan, e.g. consumer screening criteria, sampling scheme, questionnaire flow.
- **Results/key deliverables:** How the results will be used by the business.
- **Applications:** Implications for further research and research nuggets for building category learning and deeper understanding of consumers.

During the results analyses, it is important for the CE to understand the reasons behind the numbers so he/she can confidently explain the results and connect the information to the desirable outcome. When presenting results in verbal or written form, the onus is on the CE as an insight leader to "connect the key learnings" digested from the data to make it understandable and actionable to the clients or recipients of the information.

The ultimate challenge in sharing results is when the audience lacks the ability to understand the information or the story stemming from the data. Another scenario is when the audience is not ready to accept the results. Both could pose a big problem. In many cases, the same information requires several versions depending on the audience. Do not hesitate to tweak and edit as you present the same information from one group of recipients to the next. But as we learned in Chapter 1, it is important to keep the relevant information that makes the story come alive and not just "simplify" the story.

The CE has to constantly adapt the key learnings on consumer insights and product feedback from each learning module to the next. At this stage, the CE has to skillfully balance the desired deliverables for each learning module: quality, time, and cost. Quality of the research refers to the robustness of the information that would be used to provide consumer-based business decision-making. It does not always mean that studies with large sample sizes provide robust data. Quality can also be defined as the strength of the insight and its projectionable impact to larger groups of consumers.

It is useful to quote two visionaries that define the research boundaries that CEs operate in to deliver quality research on point and on time. Warren Buffett, one of the most successful investors in the world had said, "I would rather be approximately right than precisely wrong" (Rose, 2008). Jeff Ewald, a leading market research guru, said, "I'd rather have some information that boosts the decision-making process to 80+% but have it FAST, rather than getting it 95% right but too late to be useful" (personal communication with the author [DP], October 2008).

2.3.7 Documentation and project dossier

The work is never done until the documentation is finished. Finally, the effective CE has to write two reports: A topline executive summary report and a full technical research report.

The topline executive summary has to be written simply and concisely with the highest-level executive as the target audience. It has to be delivered on time to impact a business decision and/or before additional research investments need to be made. It could be as short as a single page or a five-pager that captures the key learnings, relevant data, conclusions, implications, and recommendations from the collective research.

The full technical report should include the detailed research plans, results, and recommendations for each learning module. It should include the protocol and questionnaires, which could by itself be used in generating best practices for future research. The final analytics should include deep dive analysis and cross tabulations to support the conclusions and recommendations.

The project dossier is the final compilation of the research. It should include the two versions of the report - the topline/executive summary and the full technical report along with the outcome of the intended research. The outcome should spell out if the research objectives were met and how the company used the results. The outcome should highlight the action and next steps resulting from the learning plan and the actual name of the new product launch vs. the project code name (this would help with future knowledge mapping). The project dossier should be stored in a secure and confidential company site that has appropriate controls and searchable functions. The project dossier is the intellectual property of the company and should remain in its possession for subsequent researchers.

Finally, the CE has to be the "strong closer" and prevent the temptation of the "tired" researcher after a long and tumultuous journey to just wrap it up and file the information so they can move on to the next big project. A carefully documented project dossier serves three main purposes: (1) To capture the knowledge and learnings from the company research investment; (2) To build the category knowledge and best practices for future research; and (3) To provide a solid document framework in the event of a legal challenge for claims and patent infringement cases. Future researchers and the company will thank you for delivering a solid project dossier.

With digital technologies enabling constant flow of massive data, research organizations can easily tap vast amounts of sponsored, and unsponsored research which further magnifies the importance of capturing and cataloguing relevant data for current and future strategic activation. This further shifts the value-added role of the Consumer Explorer from data curatorship to knowledge management. It can start as a simple shared folder and progress to a digital library that can easily be mined, analyzed, and visualized with artificial intelligence/machine learning (AI/ML) tools.

2

2.4 Practical advice from seasoned explorers

Many books have been written about innovation and management models, but this book centers the discussion on the Consumer Explorer, the knowledge worker and insight leader, and their role in ensuring the consumer is at the center of innovation.

The past year of working remotely has enabled the editors to listen to experts across disciplines and identify practices relevant to CEs. Here are some tips from seasoned CEs that we interviewed in 2012 and the ones we listened to in 2021:

- **Think like a scientist**
  Adam Grant, author of *Think Again: The Power of Knowing What You Don't Know* and as keynote speaker at the 2021 SAS Global Forum, said that we can benefit from thinking like a scientist who is more likely to pivot as they learn from testing their hypothesis in a series of experiments (Grant 2021). He states, "Don't let your ideas become your identity. Value humility over pride, curiosity over conviction, and look for reasons where you might be wrong, and not just the reasons you must be right."
- **Be there when it starts**
  All projects have a genesis. New product ideas germinate from innovation teams or core product development teams. Downstream product teams have more defined members. Early innovation teams may not always have a consumer researcher in their midst.
- **Get the team engaged upfront**
  It is critical that you get a multifunctional team with their "skin on the game" who respect each other's point of view, can make decisions during work sessions and champion the collective outcomes with senior sponsors and stakeholders.
- **Don't hesitate to take the lead**
  Taking the lead in designing a product testing strategy means being proactive in providing relevant consumer and product information at key decision points instead of being reactive to a timeline or to a specific task objective.
- **More is not always better. . . edit, edit, edit**
  What you take out is as important as what you put in the test design and questionnaire and data report.
- **Don't get lost in the data**
  Stick to your objective and what you want to find out. Identify the point of the study vs. looking at all different data relationships and looking at data in many ways.
- **Stay curious**
  Tim Harford, author of *The Data Detective: Ten Easy Rules to make Sense of Statistics*, and as a panelist at the 2021 Financial Times Digital Festival states, "I want to use curiosity to help me understand the world and fill the gaps in my knowledge." (Harford 2021) He adds, "Use statistics to understand the world around you versus using statistics as a prop or selectively using it to win an argument or prove a point. Statistics is like a radar. . . use it in a curious way."

- **Stand your ground**
  Know your data and be passionate about it. You need to have the conviction to defend your statement and the data and understanding to back it up. You need to balance the rational and explainable parts of the research with the fear of divulging inadequate or misleading information.
- **You are a piece to the puzzle**
  You provide the voice, eye, touch, and brain of the consumer in the research team. Champion the insight but let product development deliver the goods. Let the business solve the problem when consumers do not like the original idea, but there is an opportunity to create the demand.
- **It is not just about building a story**
  Jeff Bezos, the founder of Amazon.com, summarized why having a story or imagery is not enough: "Put the vast majority of your energy, attention, and dollars into building a great product and a smaller amount into shouting about it."
- **BY???**
  We started the first two chapters in this book to talk about you. In the words of one of America's political commentators, Bill Maher, "There is BCE Common Era, CE Common Era, and now, BY Before You." This is respecting that there is history and knowledge that should be explored before starting something new. Learn what you don't know from those who came before you. Build and add on the knowledge for the next person after you. Reading the rest of the book is a start. You can connect the past to the future.

Discussion questions

- Can you identify a Consumer Explorer in your organization?
- Once you have a seat at the innovation table as a CE, what do you have to do?
- How can you expand your journey as a Consumer Explorer?

References

Grant, A (2021) *Keynote SAS Global Forum*, 18 May 2021. https://www.sas.com/en_us/events/sas-global-forum.html

Harford, T. (2021) *The Data Detective: Ten Easy Rules to Make Sense of Statistics*. New York: Riverhead Books.

Lawless, H. (2009) Keynote Speech. *Presented in 8th Pangborn Sensory Science Symposium*, 30 July, 2007, Florence, Italy.

Maheshari, S. (2021, June 20) *What Are We Going to Wear?* New York Times. https://www.nytimes.com/2021/06/18/business/fashion-forecasting.html (accessed 20 January 2021).

Maiorana, T. (2021, January 13) *UC Davis Virtual Events Series: Sparking Innovation: Human Centered Design* [panel discussion]. https://arts.ucdavis.edu/event/sparking-innovation-and-creativity-human-centered-design (accessed 15 January 2021).

Moskowitz, H. (2004) "From psychophysics the world . . . data acquired, lessons learned". *Food Quality and Preference*, 15(7-8), 633-644.

Moskowitz, H. (2005) "Commentary: whither now the grand sensory "project" in an age of improving methodology". *Journal of Sensory Studies*, 20(1), 93-95.

Moskowitz, H. R., Beckley, J. H. and Resurreccion, A. V. A. (2006) "Sensory and Consumer Research in Food Product Design and Development". Wiley-Blackwell, New York.

Moskowitz, H. and Saguy, I. (2013) "Reinventing the role of consumer research in today's open innovation ecosystem". *Critical Reviews in Food Science and Nutrition*, 53(7): 682-93.

Rose, C. (2008) *An Exclusive Conversation With Warren Buffett*. Business. Wednesday, 1 October 2008. https://charlierose.com/videos/11747 (accessed 24 June 2009).

Smith, E. (2011) Journalism Convocation Speech at Medill, Northwestern University, 18 June 2011, Evanston, IL.

3

The first person to make a novel and prospectively useful product or process is an inventor and the first person or enterprise to exploit that invention in a commercially viable product or service is an innovator.

Schmookler (1966)

Economic leadership in particular must hence be distinguished from "invention." As long as they are not carried into practice, inventions are economically irrelevant. To carry an improvement into effect is a task entirely different from the inventing of it, and a task, moreover, requiring entirely different kinds of aptitudes. Although entrepreneurs of course may be inventors just as they may be capitalists, they are inventors not by nature of their function but by coincidence and vice versa.

Schumpeter (1934)

What this is about

These two quotes contain the essence of most of what will be discovered in this chapter. To begin that exploration, we will start with some established historical cases. The examples will show the transition of inventors to innovators and the transformation of inventions to innovations.

Product Innovation Toolbox: A Field Guide to Consumer Understanding and Research, Second Edition. Edited by Kannapon Lopetcharat, Dulce Paredes and Jacqueline H. Beckley.

Chapter 3

Invention and Innovation

Daniel M. Ennis

3

Key learnings

- ✓ Aptitudes needed for innovation
- ✓ What really matters is "consumer perceived benefits"
- ✓ The Invention-Innovation Paradigm in Science
- ✓ The Time Scale of Innovations

3.1 Dual aptitudes needed for innovation

3.1.1 The steam engine: Watt and Boulton (cf. Scherer, 1984)

According to his own account, in an afternoon in the winter of 1764-1765, James Watt, the mathematical instrument maker at Glasgow University, strolled around the grounds to work out a problem. Earlier he had been given a Newcomen engine to repair, a type of engine invented in 1712. During his walk, he became conscious of a change that would significantly improve the efficiency of the machine by using a separate condensing vessel. Apparently, the time it took for him to come up with this concept was very short - a matter of hours - and it was dwarfed by the time it took before his insight led to a workable machine in 1780. He may not, in that afternoon, have appreciated the future contributions of John Roebuck, who went bankrupt, and then Matthew Boulton, who played the roles of unsuccessful and successful innovators, respectively. In the development of the Watt-Boulton steam engine neither of them could have seen that their machine would be the overture to the Industrial Revolution when large industrial cities were built far from rivers, reducing reliance on hydropower. By way of contrast to the real potential of their technology, Watt and particularly Boulton were focused on solving a contemporary problem with business potential - getting water out of flooded copper mines.

Product Innovation Toolbox: A Field Guide to Consumer Understanding and Research, Second Edition. Edited by Kannapon Lopetcharat, Dulce Paredes and Jacqueline H. Beckley.

Two lessons derive from the steam engine story. The first is that the skills of both participants were required. James Watt provided the technical knowledge and the motivation to create a technical improvement and Matthew Boulton encouraged and funded Watt through years of development and resolved a serious patent roadblock. By 1780, they had a commercial product. The second lesson is that many successful products or services extend far beyond the vision of their creators but they all must have at least one sustainable application to get them started, such as extending the life of flooded mines.

3.1.2 Nike: Bowerman and Knight (cf. Moore, 2006)

A modern example of the Watt-Bolton story is the development of Nike with the addition that the company branched out into adjacent categories. Bill Bowerman, coach of the University of Oregon track team, is known for producing numerous Olympic champions and world record holders and was also a cofounder of Nike. Bowerman liked to make running shoes for his athletes as he was dissatisfied with shoe design in the 1960s and early 1970s. He founded blue ribbon sports (BRS) with Phil Knight and they had Bowerman's designs for running shoes manufactured by Onitsuka in Japan. Bowerman, in search of a light shoe with traction for his track athletes, invented a "waffle" sole using a waffle iron in his home. This invention became the basis for the waffle trainer, the first really successful shoe sold by BRS before the company became Nike. When Frank Shorter won the Olympic gold medal for the marathon in Munich in 1972, a running boom was launched in the United States. With the running boom came a huge demand from the masses for comfortable shoes suitable for road running. Bowerman's designs were well suited to exploit that demand.

In this invention-innovation scenario, Phil Knight played the role of Matthew Boulton from the previous case and managed to successfully steer BRS through a contract dispute involving distribution and trademark issues with Onitsuka that could have destroyed the fledgling company. It is interesting to see once again the expression of the dual aptitudes required in a successful venture. The pattern of initial limited implementation followed by extensibility is seen again as Bowerman's interests were initially focused on the needs of high-performance athletes. The opportunity for mass marketing and expansion into adjacent businesses was successfully exploited by Nike, although they did not create the running boom.

3.1.3 The US Navy: Scott and Sims (cf. Morison, 1966)

Prior to 1898, gunnery accuracy at sea was dismal. In the space of six years, accuracy was increased by 3000% based on the ingenuity and doggedness of two men, Sir Percy Scott of the British Navy and William Sims, an American naval officer. Scott provided the basis for a process of continuous aim firing by adjusting guns on ships so that gunners could rapidly alter the positioning of guns to compensate for the roll of the ship. He also made changes to the telescopic sight so that gunners could continually sight their targets. Scott made these improvements in 1898 and began recording remarkable gunnery records. In 1900, Scott met Sims and showed him his new technique. Before long, Sims began to demonstrate similar improvements in gunnery accuracy to Scott. Then he set out to educate the US Navy who would surely welcome this new advance with open arms. On the contrary, they set out to prove that it was physically impossible to produce the results that Sims produced. He was dismissed and regarded as a falsifier of evidence. In a highly unusual move for a naval officer, Sims wrote to President Roosevelt to express his conviction about the value of continuous aim firing and in 1902 he was made Inspector of Target Practice. Scott's method was finally adopted by the US Navy over a period of about six years.

Continuous aim firing was a process made up of components brought together by Percy Scott, none of which he invented individually - guns, gears, telescopic sights - but he put them together in a highly successful way. William Sims, possessed of a desire to revolt against the rigidity of the status quo, provided the commitment and passion to bring Scott's process into use. This case not only illustrates again the dual aptitudes mentioned already but also demonstrates the role of chance in bringing innovative components together. We also see the resistance to change in any society where the people in it have limited identifications. In the Navy at that time, gunners were not influential due to the ineffectiveness of their craft, and others in that society, who focused on battle strategy, were not readily stepping forward to relinquish their power when gunners started to hit their targets. In a hopeful attempt to address the issue of limited identifications, Morison suggested:

> Any group might begin by defining for itself its grand object and see to it that everyone understands what it is.
>
> Morison (1966).

If one wants to create an innovative organization of a few people or of thousands, it is worthwhile to consider the aptitudes in staffing that would be required. Innovation is a messy, disruptive business often accompanied by personalities to match these qualities. In many companies, great ideas and concepts may be languishing for the attention of a Boulton, Knight, or Sims. Some inventions may not be seen as grand enough to warrant interest. This attitude misses the point that small-scale but profitable implementation may be all that is needed at first before the landslide of another industrial revolution, changes in warfare, or an exercise boom occurs. Like a forest fire, large-scale innovations may have very small beginnings.

3.2 Benefits

Inherent in the earlier definitions of invention and innovation is that an innovation provides a benefit to its user, one not obvious in current practice. An important consideration is that this benefit is consumer-perceived.

By the early 1970s, Philip Morris had acquired the Miller Brewing Company. By the middle of that decade, two products were introduced that had a major impact on their industries. One was Miller Lite and the other was a cigarette called Merit. These brands were based on remarkably similar benefits for consumers or, perhaps more accurately, perceived benefits. In the case of Miller Lite, a technical advance in brewing technology, the invention allowed the management teams at

Philip Morris and the Miller Brewing Company, the innovators, to introduce a product with extremely low carbohydrate content without sacrificing taste. Similarly, a novel advance in tobacco flavor technology provided the opportunity to introduce Merit and to promote a cigarette with flavor equal to full flavor rivals at half the tar. At the time, tar reduction implied a health benefit from the consumers' perspective. These two products contributed to new categories that became as important to their companies' revenues as the original categories.

The introduction of the Merit cigarette brand was not the first time that perceived health benefits drove fortunes in the tobacco industry. After Louis Pasteur connected disease to microbes in the 1860s, there was a lag until the general public became aware of the germ theory of disease in the 1880s. The spread of tuberculosis from sputum became a common concern and with it the health implications of chewing tobacco, the dominant form of tobacco use in the United States in the 19th century. Smoking forms, such as pipes and cigars, began to increase as chewing tobacco declined, and in 1910, the future of the tobacco industry looked firmly hooked on smoking, or pipe, tobacco. Then in 1913, the whole industry abruptly changed when R.J. Reynolds blended Bright and Burley tobacco to make a suitable inhalation form to create modern cigarettes. Lung absorption of nicotine and delivery to the bloodstream is far more efficient than buccal absorption as occurs with chewing tobacco. Ironically, a consumer health issue ignited consumers to turn to cigarettes and away from chewing tobacco, which they perceived to be an unhealthy alternative. This is the benefit that resonated with consumers and led to the creation of a multibillion dollar industry.

The Merit/Miller Lite scenario seemed ripe for a repeat after Philip Morris acquired General Foods in the 1980s. Technology for removing caffeine from coffee using a CO_2 extraction process seemed appropriate to take nicotine out of cigarettes to simulate what had been done with decaffeinated coffee. A product with little or no nicotine was test marketed in the late 1980s under the brand name "Next." This product was a failure and a valuable lesson because it underscored the importance of understanding what a company's products provide to consumers. Decaffeinated coffee is still a warm, good tasting beverage with consumer benefits. However, in removing nicotine from cigarettes, the main psychoactive substance that contributes to cigarette consumption was removed and there also may have been important sensory effects due to nicotine that disappeared on extraction.

Consumer-perceived benefits, whether justified or not, can drive major changes in the fortunes of companies and even create new businesses. Next, we will examine whether these ideas also apply in science and whether the simple dual roles discussed earlier still apply.

3.3 The invention–innovation paradigm in science

Contrary to popular opinion, and what is often reported in the media, science is not concerned with discovering the truth or facts about the world. Science is concerned with representations and testing them to choose the most useful model. Scientific models are inventions and, when they are broadly adopted for the benefits they provide, we can call them innovations. As an example, consumers of behavioral research in sensory and consumer science have certain basic needs and they judge the benefits provided by the methods and models proposed according to their ability to satisfy those needs. Broadly, there are two general interests. One interest is to scale features that differentiate among products and the other is to represent why people like or choose certain products or brands.

According to legend, Gustav Fechner, a physicist, lay late in bed on the morning of 22 October 1850, contemplating a log-law relationship between physical and mental quantities to explain known data. His conception that morning gave birth to the field of psychophysics in which theories concerning the relationship between the physical world and its mental representation are nurtured. 22 October has been traditionally celebrated as "Fechner Day" around the world by Fechnerian psychophysicists and the festivities include a special conference by the International Society of Psychophysicists. It is doubtful that their excitement will ever create a civil disturbance or compete with the Carnival in Rio de Janeiro. Nevertheless, for this small group of followers, Fechner made a scientific advance in psychological scaling that affected the thinking of all students of mental processes. Commercial applications of functions linking physical quantities, such as the time it takes for a gallon of water to exit a drain, to their mental representations, such as the perceived elapsed time (think of the immediacy of water leaving a clear drain compared to the seeming hours it takes for a partially clogged drain to clear), abound in consumer product categories. Although the ultimate judge of product quality is the consumer, physicochemical measures validated by psychophysical techniques reduce the cost and time of product development and improve the quality of consumer products. Fechner's conceptual invention from 1850 is today a successful innovation in the form of procedures for understanding the relationship between physical and chemical measures and subjective experience.

There are items for which a clear physicochemical correlate is not obvious, such as the beauty of art or handwriting specimens, and yet we still would like to produce relative scale values. In the consumer products area, an example might be the quality of a fine fragrance. In 1927, Louis L. Thurstone invented a basis for a "purely" psychological theory for scaling that met this need. His papers from this period led to a wealth of new models collectively referred to as "Thurstonian models" (Ennis, 2016). These models specify two basic ideas – the information and cognitive processes leading to decisions are probabilistic, and there is a definable decision rule that depends on task instructions. In the cases of invention and innovation discussed earlier, the separate roles of inventor and innovator were connected to individuals. Inevitably, this simplification may not describe the role of many other players in some innovations. In the case of Thurstonian scaling, it could be thought that Thurstone's invention was popularized and, in some cases, commercialized not by one innovator but by a community of scientists and programmers who contributed to the dissemination of useful tools. These tools were then used to bring Thurstonian scaling to those who would benefit from them.

Let us turn now from scaling intensities to models of hedonicity, including liking and preference. It is quite natural when thinking of a hedonic response to consider it to be based on a hedonic or utility continuum as we would a sensory variable such as sweetness. This idea is a direct extension of the previous section. Thinking of liking or preference responses as arising from judgments based on a hedonic or utility scale makes it possible to consider using Thurstonian models. Then to find explanatory variables for this hedonic or utility scale, one could use a linear combination of explanatory variables. An alternative to Thurstonian ideas for preferential choice is to consider using the logit, which can be expressed in closed form. Although the logit has well-recognized process limitations compared to its Thurstonian alternative, the logit's closed form propelled it to the status of a major innovation in a number of fields including economics, marketing, and public health. Contributors to choice models, such as the logit and its applications, include Daniel McFadden who was awarded the Nobel Prize in Economics in 2000, and R. Duncan Luce who was awarded the 2003 National Medal of Science for work he completed in 1959.

A more imaginative alternative to the idea of the hedonic or utility continuum, discussed in the previous section, is to consider that scaling emotional responses involves the possibility of individual internally generated points which are used to make liking or preference decisions. Sometimes these points are referred to as ideal points or motivation points. This conceptual invention is credited to Clyde Coombs (1950) and implementation of it leads to what is today called "unfolding." The result of an unfolding analysis is the generation of a spatial representation of the hedonic data in a low-dimensional space of liking or utility drivers. For half a century, unfolding did not become an innovation because it suffered from a deficiency that often led to degenerate solutions with limited or no interpretation. This problem was solved by combining a Thurstonian probabilistic similarity model (Ennis 2016, Chapter 7) with Coombs' insight and introduced in a method called Landscape Segmentation Analysis® (LSA[1]) in 2001. This method has many applications in the behavioral sciences which are discussed in Ennis and Rousseau (2020). A successful alternative based on a deterministic model was also introduced in 2005 (Busing et al., 2005.)

3.4 The time scale of innovations

The impact and even the source of innovations can take decades or even centuries to be identified. Watt conceived of the separate condensing vessel in about three hours but took fifteen years to implement the idea in an innovation. Coomb's ideal point concept required a half century and new analytic tools to reach successful implementation. Duncan Luce, who was seventy-nine when he received the National Medal of Science prize from President Bush, remarked: "This is a great honor for which I am most grateful . . . I'm also grateful for my genes, which have enabled me to live a long life and enjoy this honor."

3.5 Final remarks

For inventions to blossom into innovations, they usually benefit from the confluence of certain characteristics:

- They have an initial, possibly quite small, profitable application.
- They need to have consumer perceived benefits.

- They may coincide with a movement or trend, such as the running boom or a demographic shift.
- They may develop from seemingly unrelated fields.
- They require the skills of individual or community entrepreneurs who are often not the inventors.
- They should be easily implemented.
- They require significant amounts of vision, patience, and time.

Discussion questions

- What is needed to turn an invention into an innovation?
- Is commercial success required for innovation?

References

Busing, F.M.T.A., Groening, P.J.F. and Heiser, W. (2005) "Avoiding degeneracy in multidimensional unfolding by penalizing on the coefficient of variation". *Psychometrika*, 70(3), 587.

Coombs, C.H. (1950) "Psychological scaling without a unit of measurement". *Psychological Review*, 57(3), 145-158.

Ennis, D.M. (2016) *Thurstonian Models: Categorical Decision Making in the Presence of Noise*. Richmond, VA: The Institute for Perception.

Ennis, D.M. and Rousseau, B. (eds). (2020) *Tools and Applications of Sensory and Consumer Science*. Richmond, VA: The Institute for Perception.

Morison, E. (1966) *Men, Machines, and Modern Times*. Cambridge, MA: The MIT Press.

Moore, K. (2006) *Bowerman and the Men of Oregon: The Story of Oregon's Legendary Coach and Nike's Cofounder*. Emmaus, PA: Rodale.

Scherer, F. (1984) *Innovation and Growth: Schumpeterian Perspectives*. Cambridge, MA: MIT Press.

Schmookler, J. (1966). *Invention and Economic Growth*. Cambridge, MA: Harvard University Press.

Schumpeter, J.A. (1934) *The Theory of Economic Development*. Cambridge, MA: Harvard University Press.

Thurstone, L.L. (1927) "A law of comparative judgment". *Psychological Review*, 34(4), 273-286.

Note

1. Landscape Segmentation Analysis® (LSA) is a registered trademark of The Institute for Perception.

4

You cannot improvise unless you know exactly what you are doing

Christopher Walken

What this is about

The definition of two stages of consumer research is found in this chapter: (1) Upfront Innovation to look for consumer insights, and (2) New Product Development (NPD) that makes the insight ownable and validates the product opportunity. We set the stage for the different research tools for Consumer Explorers (CEs) by introducing the Iterative Experimentation Qualitative-Quantitative Research (IEQQR) model. We will discuss how to jumpstart the research process to engage the consumer from early concept to final product design.

Product Innovation Toolbox: A Field Guide to Consumer Understanding and Research, Second Edition. Edited by Kannapon Lopetcharat, Dulce Paredes and Jacqueline H. Beckley.

Chapter 4

Designing the Research Model

Kannapon Lopetcharat, Dulce Paredes, and Jacqueline H. Beckley

(updated from the original chapter by Kannapon Lopetcharat, Dulce Paredes and Jennifer Hanson)

4

Key learnings

✓ Factors influencing product innovation
✓ Setting up a successful product innovation research program
✓ Experimentation in practice
✓ Iterative Experimentation Qualitative Quantitative Research (IEQQR) process

4.1 Factors influencing product innovation

Product innovation comprises two major stages: (1) **Upfront Innovation** to identify product opportunities by understanding consumers' wants, needs, and pain points and (2) **New Product Development (NPD)** to refine, screen, and validate new product opportunities grounded on consumer insights. The latter is where you make the insight ownable. In practice, many companies lump the upfront innovation and NPD in one process, distinctly separate both stages from each other or somewhere in between. Regardless of the organizational structure, the success and efficiency of product innovation depends greatly on two factors: (1) Organizational, and (2) Execution.

4.1.1 Organizational factors

Organizational factors influence how different departments work with each other and, as a company, how the results from one stage will be utilized in the next step along the process. Organizational factors are dependent and unique to each company (a.k.a. company culture). Learning how the process works within your company and understanding how decisions are made along the process will enhance the efficiency of product innovation (e.g. reducing time,

Product Innovation Toolbox: A Field Guide to Consumer Understanding and Research, Second Edition. Edited by Kannapon Lopetcharat, Dulce Paredes and Jacqueline H. Beckley.

better communication, reducing cost). Here are four questions to answer to ensure efficiency of product innovation.

(1) **Do the goals of product innovation align with technology strategies and business strategies and goals?** The goals of Product innovation must align and support business strategies and goals; otherwise, the innovations will not be useful for the company. Also, technology strategies allow product innovation to realize its goals more effectively. This is where start-ups and big corporations differ. The main goal of start-ups is to find business models and innovation that allows "rapid expansion"; meanwhile, the main goal of big corporations is to find innovation that can be "executed on a large scale quickly." Many experienced professionals from big corporations fail when they try to manage start-ups and many successful start-ups fail to bring their start-ups to the next level because they do not align their mentalities and management skills to the real objective of their organizations.

(2) **Which projects should a company invest its money on?** Because there will always be a limited amount of money, senior management must invest adequate resources on the right innovation projects tactically and strategically. So it is critical for management to truly understand their goals.

(3) **Is your product innovation effective and sustainable?** Effective innovation means that the process is flexible and learns from idea to launch. The effectiveness of product innovation should not be defined by just the number of products launched. It should be defined as the ratio between the number of products that are launched and resource invested (i.e. money and time). With this criterion, developing one technology platform that needs five years for ten products will be more efficient than developing ten technologies that need ten years for ten products assuming both projects need the same amount of money.

(4) **Does your organization have a culture that nurtures innovation?** A cultivating culture that nurtures innovation requires commitments from senior management and true cross-functional teams to succeed. The signs of a successful innovation culture are the following:
- *seamless process from idea to launch,*
- *healthy discussion among the team members,*
- *experimentation is encouraged,*
- *fail early,*
- *iterative.*

Organization factors enhance the overall efficacy of product innovation. These factors should be guided by business strategies and goals and technology strategies within a culture that nurtures innovation. Execution factors are equally important because without it, the process cannot reach its full potential.

4.1.2 Execution factors

Execution factors are critical to product innovation because the factors influence directly and greatly on what will get "into" the NPD process. These factors have direct influences on the quality of the outcomes more than just reducing time and cost. Execution factors involve the people who actually conduct the research (e.g. which method they choose to use, how they use the results to persuade upper management). Therefore, the factors are not unique to a company or even product category. CEs are the ones who select research methods and execute studies; therefore, the ways that they utilize consumer research tools will have an impact on the outcome and ultimately the efficiency of product innovation. Throughout this book, we are providing tools, guidelines, pros and cons, limitations, and suggestions for conducting consumer research along the product innovation process that will allow CEs to select consumer research tools appropriately based on situations and goals. In addition, the success and efficiency of product innovation depends greatly on how well organization factors and execution factors work together.

4.2 Setting up a successful product innovation program

Business strategies and goals provide strategic focus to direct, define, and develop the strategies for product innovation which includes which product, market, and technology areas and technology strategies allow product innovation to realize its goals effectively. Consequently, NPD goals and roles in business are clearly defined and communicated. The best way to do this is developing product innovation strategies as "a part" of business strategies. With that in place, a company can start allocating resources accordingly and developing product roadmaps that ensure long-term success. There are four questions to answer before product innovation strategies become reality (or when a company starts spending money).

4.2.1 What are the goals and roles of new product innovation?

Defining and communicating these goals and roles clearly is very important. The goals and roles must impact overall business's goals such as "25% of business's sales will come from new products by the end of the year." This will create a common purpose for everyone involved in the process and will set a good environment for innovation.

4.2.2 What are the strategic boundaries and which directions do we want to pursue?

Defining which market, industry sector, applications, product types, or technologies on which your company will focus is very important because these will help your company to decide on which direction to pursue. Two factors are usually used to define the boundaries and directions: Attractiveness and Business strength.

- **Attractiveness** can be evaluated using many metrics such as growth opportunities (market size and value), competition, margin, and new product potential (technology maturity).
- **Business strength** is simply what your company can do better than other competitors in the boundaries of interest. This involves evaluating business core competency and strengths that can be leveraged.

4.2.3 How do you want to attack and enter?

Answering this question is very important for product innovation as it defines the types of innovation itself (disruptive, incremental, or cost-reduction) that dictate the speed of new product development. For example, "First to market" strategy demands more disruptive innovation than incremental innovation, "Fast me-too" demands incremental innovation and "Value for less" demands cost-reduction innovation.

4.2.4 Do you know the priorities of innovation projects?

This stage deals with the actual commitment and resource allocation to the project, and it is when strategies become reality as money and resources will be spent. Project priority should be based on business's strategic importance.

The priorities of innovation projects allow senior management to develop a successful strategic product portfolio. To execute this plan successfully and effectively, the management needs a set of criteria to decide which project should be continued or stopped and an effective product development process. The following are the requirements that each innovation project needs to fulfill at any time during the review:

- Does the project still fit with innovation strategies and business goals?
- Will the new product deliver a compelling value proposition to target consumers?
- Will the new product be easily differentiated among its competitions?
- Have the current market conditions (competitive landscape) and projection (volume, market share, and margins) changed from the last review? If so, do the changes impact the priority of the innovation?
- Will the company's leverage still be relevant and effective?
- What is the status of the technology feasibility to deliver the product?

4.3 Current approach to NPD

The questions above should be asked again and again during project review along the product innovation process where a product is developed from idea to actual commercialization. Generally, there are seven steps used by companies in their NPD and there are regular checking points built-in where Go/No Go decisions will be made to products under development (Figure 4.1). There are several NPD processes published and the most notable one is the Stage-Gate System proposed by Robert G. Cooper in 1990 (Moskowitz et al., 2006).

The seven-step process is a robust and flexible process that allows management to review and adjust its tactical planning to reach business goals through product innovation. Many successful companies differ from other companies at the "predevelopment" steps (Step 0 to Step 2) of the process as the successful companies invest time, money, and resources to "do their homework" by listening to their consumers to deliver differentiated products that deliver what consumers need. The purpose of this book is not to change the process but rather to enhance it by providing tools and methodologies to accelerate and increase

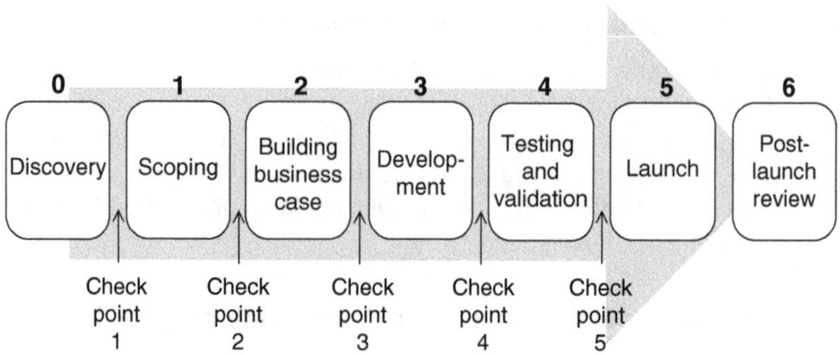

Figure 4.1 Seven general steps in NPD with five built-in check points. Source: Adapted from Moskowitz et al. (2006).

success rate in gathering consumer insights to build the pipeline that feeds to the NPD especially during Step 0 to Step 2 and Step 4.

Successful innovation companies reduce resources wasted on weak ideas, while the average company will spend more than 55% of their innovation budgets on ideas that should not have made it into their NPD in the first place (Cooper, 2001).

4.4 Experimentation in practice

Experimentation is a scientific principle where causal conclusions can be drawn with high confidence. In science, it is commonly known under the name of Design of Experiment (DoE). In a nutshell, an experiment is a structured study to test hypotheses or ideas and to discover insight(s) with the reason(s) (a.k.a. cause and effect) behind the insight(s). Around 2010, the concept of experimentation was introduced to the business world, and it has been a core competency for many successful organizations such as P&G, Tata group, Capitol One, Google, Amazon, etc. Anderson and Simester (2011) concluded that business experimentation was all about testing consumers'/customers' responses and thinking like a scientist. In this book, we implement the experimentation principle throughout especially in the tools section (Chapters 6–10) and encourage CEs to apply the principle of Iterative Experimentation Qualitative-Quantitative Research model (IEQQR).

Why is experimentation important? The true spirit of experimentation is "to learn" and "not to validate anything" (described as the pivot point in testing to learn vs. market testing to validate by Eric Ries, the author of Lean Startup in 2011 (Ries, 2011)). When applying experimentation in business, speed, and learning trumps meticulously planned large studies; therefore, experimentation is appropriate for early and middle stages of planning including product development to guide your actions to the right direction.

Having an agile learning approach goes hand in hand with a data driven organization that can quickly convert the insights to scaled impact.

The ability to draw insights from data using manual deep listening tools or sophisticated technologies such as artificial intelligence (AI) can enable companies to build innovation even during periods of uncertainty (Potter, 2020). Companies who put their ideas out for customer feedback and learn along the way builds resiliency in their innovation process (O'Reilly, 2021; Potter, 2020). A good learning experiment is not a study that will give you the results that you want but is a study that provides you a fair chance of failure and success with clear reason(s). Successful companies learn rapidly from failures through learning experiments and nurture the culture for experimentation to push boundaries for growth and/or create disruption. Learning experiments validate hypothesis and in the process, convert opinions into facts (based on its empirical result(s)), prove or disprove assumptions/ believes/myths, discover surprises about consumers/customers, help you make more informed decisions, and support claims and stories that you want to tell with confidence and believability.

Continuous learning from experimenting with incremental and innovative ideas allows organizations to optimize the consumer/customer journey with their products/services (Figure 4.2). Learning experimentation allows organizations to test new ideas in a low-risk environment that also provides empirical evidence for the success and failure of the tested ideas. Figure 4.2 shows the eight steps for conducting a learning experiment.

For CEs, these eight steps are high-level processes for guiding the creation of a learning experiment. In actual operation, certain circumstances will require expert assistance and detail-oriented activities; however, it depends on the nature of the actual stimuli in the experiment. It is worth noting that many business-level experimentation proponents are championing "rapid experimentation"; however, in actuality, the word "rapid" can be in the ranges of minutes to weeks depending on the nature of the business category and stimuli in

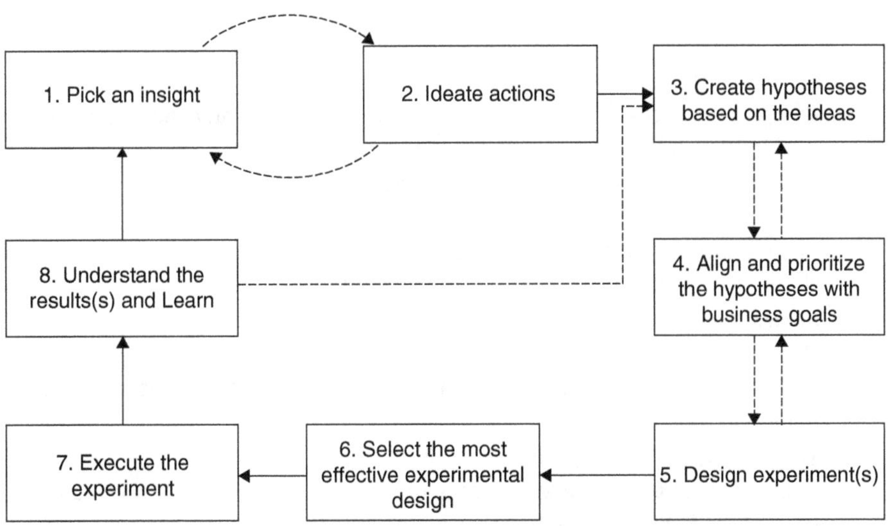

Figure 4.2 Eight cyclical steps for creating a learning experimentation.

the experiment. For example, if a manager wants to test an advertisement statement or concept online, it can be completed within hours or days. For actual product testing, creating the prototypes can take days to months depending on processing requirements and other considerations such as laws and regulations that the company has to follow. The meaning of "rapid experimentation" will depend largely on the nature of stimuli and circumstances around the actual introduction of the stimuli. Let us explain more about the eight steps in detail.

Step 1: Start with an insight: This step is very critical. Many books recommended researchers to pick an idea to create the hypothesis for testing; however, they forget to ask where the idea came from. If one has an insight ready, one can proceed to Step 2. An idea can come from new technology that requires consumer translation to create relevant products or from consumer inspiration that requires technology or new application for fruition. Ideas can come from the company's innovation pipeline or from internal and external data mining sources and knowledge-mapping activities. One can start with an idea for a new product/service/marketing message. Then one should think back, "will my target consumers/customers want this idea?" In short, WHY is this a good idea? The idea owner must come up with the reasons and that will lead to the fundamental insight(s) of the idea. Remember, learning experimentation is about "learning" and it is not for validating an idea. Therefore, the insight(s) behind an idea is critical for learning.

Step 2: Ideate actions: Once an insight is selected, the team can ideate potential actions around it. These could be product ideas that your company can do based on the insight. At this stage, team members should generate as many ideas as they can without any thought restrictions.

Step 3: Create hypotheses based on the ideas: Once all the ideas are generated, the team needs to convert the ideas to hypotheses that will be proven or disproven through experimentation. Hypotheses statements have these characteristics:

- Factors that you can control
- The differences between the factors must be significantly and/or perceivably different to the target consumers/customers
- Variable(s) that can be measured to signify success or failure.

The insight: Consumers want to buy <u>better</u> products.
Note: this is quite an oxymoron insight, everybody knows but acts differently.

Ideate: Claiming more benefits into the label.
Note: This is an action that your company can do. There are many ways to make consumers feel that they are buying better products though, this is just one of many.

Turning the idea to hypotheses
Hypothesis 1: Reduce the appearance of wrinkles by *50% within seven days*
Hypothesis 2: 80% of consumers said the appearance of wrinkles was reduced within seven days

This example is an online promotion campaign, one can see that with the same insight, the company can turn it into many actions (ideas) and each idea can be converted to many hypotheses. From these two hypotheses, which one is a better? Let us put the two hypotheses against the three characteristics:

	Hypothesis 1	**Hypothesis 2**
	Reduce the appearance of wrinkles by 50% within seven days	80% of consumers said the appearance of wrinkles was reduced within seven days
Factors that you can control	Yes, we had a clinical study done	Yes, we had claim substantiation study done
Significantly and/or perceivably different	Do not know yet because we have not selected any experimental design	
Measure of success or failure	Number of people who want to buy this product	Number of people who want to buy this product

Now we can see that the two hypotheses are fulfilling two out of three required characteristics for good hypotheses; however, the second characteristic is still undetermined. To determine the second characteristic, one must select an "experimental design" that we will discuss in Step 5. But before doing that, we need to go through Step 4 first.

Step 4: Align and prioritize the hypotheses with business goals: This step is very important as the business strategies/framework should drive the choices you make in research. However, in practice, after ideation and hypothesis formation, the team usually deviates from the original goal. Different hypotheses may generate different outcomes that serve the primary goal differently. Therefore, prioritizing the hypotheses again here is critical.

Step 5: Design experiment(s): This step is very important and things can turn very complicated. One rule to follow is "**keep the design as simple as possible.**" There are circumstances where complex design is needed especially in product formulation or product concept creation. Seek expert advice from a statistician for more powerful designs where you can look for main effects and interactions between ingredients. In this book, we introduce many powerful methods that simplify this step (refer to Chapter 5 for research design and Chapter 12 for applying statistics).

Business goal: To boost sales of current product.
The insight: Consumers want to buy a better product.
Idea: Our product can reduce wrinkles.
Turning the idea to hypotheses:
Hypothesis 1: Moisturizing and reduce the appearance of wrinkles by 50% within seven days.
Hypothesis 2: 80% of consumers said the product is moisturizing, and the appearance of wrinkles was reduced within seven days.
Current activity: Intense moisturizing.

By adding the current activity, as another condition for the claim, now, you can create many different learning experimentations. Let us start with *Current vs. Hypothesis 2.*

Current vs. Hypothesis 2
Current : Intense moisturizing
Hypothesis 2: 80% of consumers said the product is moisturizing and the appearance of wrinkles was reduced within seven days

By comparing the results between the Current and Hypothesis 2, we will learn the impact of the phrase "80% of consumers said the product is intensely moisturizing and the appearance of wrinkles was reduced within seven days" in the ads is positive or negative and by how much (the unit is dependent on the measurement selected). Now let us try another pair (*Current vs. Hypothesis 1*).

Current vs. Hypothesis 1
Current : Intense moisturizing
Hypothesis 1: Moisturizing and reducing the appearance of wrinkles by 50% within seven days

The same interpretation as the Current vs. Hypothesis 2 pair can be done but the conclusion is about adding "***clinical results, 50% wrinkle reduction in seven days***" in the claim.

Comparing two conditions as in *Current vs. Hypothesis 1* and *Current vs. Hypothesis 2* pairs is the most basic experimental design called two-sample design (a.k.a. A/B Testing in the business world). This design is one of the most-used designs in business experimentation. The two-sample design is simple yet powerful because it allows CEs to learn by comparing the "differences" between the two conditions, hence, giving them the ability to draw cause and effect conclusions. However, if we do not have enough budget to run two experiments for both pairs, how can we choose which hypothesis to test? This is the second characteristic (*The differences between the factors must be significantly and/or perceivably different to the target consumers/customers*). Comparing the Current against Hypothesis 1 and Hypothesis 2 separately as we are using two-sample design (A/B testing), which pair demonstrates more significant differences between each other? Visual and meaning differences are critical. It is quite clear that *Current vs. Hypothesis 2* pair demonstrates more differences in both visual and meaning impacts than that of *Current vs. Hypothesis 1* pair.

Current vs. Hypothesis 2
Current: Intense moisturizing
Hypothesis 2: 80% of consumers said the is moisturizing and the appearance of wrinkles was reduced within seven days

Current vs. Hypothesis 1
Current : Intense moisturizing
Hypothesis 1: Moisturizing and reduce the appearance of wrinkles by 50% within seven days

	Hypothesis 1	Hypothesis 2
	Moisturizing and **reduce the appearance of wrinkles by 50% within seven days**	80% **of consumers** said the product is moisturizing, and **the appearance of wrinkles was reduced within seven days**
Factors that you can control	Yes, we have clinical results	Yes, we have a consumer claim study
Significantly and/or perceivably different	Using Two-sample design (A/B testing) with Current claim (Intense moisturizing)	
	Clinical	Consumer
Measure of success or failure	Number of people who want to buy the skincare product	Number of people who want to buy the skincare product

Step 6: Select the most effective experimental design: Besides the two-sample design (A/B testing), there are several situations where comparing two samples are impossible or too costly. There is another alternative called Longitudinal Intervention Design (a.k.a. Pre-/Postexperiment). This design is not as refined as the two-sample design as it relies heavily on the assumption that "after introducing the factor (intervention), everything will be the same except the impact of the factor that has been introduced" that, in many cases, it does not hold true. The design relies on the perception and memory of the respondent at the beginning of the study and can be prone to innate biases (such as strong familiarity or affiliation with current product). Comparing the changes before and after the intervention is the key learning from this type of design. Even though the Longitudinal Intervention Design (Pre-/postexperiment) has a lot of bias and variations built in the design, it is one of the most practical used designs since it is easier and cheaper to conduct compared to the two-sample design (A/B testing). In product research, many layers of experimentations (factors) can be built-in one study to prevent potential biases and control unnecessary variations (or cover all the existing variation), more sophisticated designs are available for CEs such as Factorial Design, Central Composite Design. It is highly recommended to consult an expert in the field of experimental designs (e.g. statistician) to choose the most effective one.

Step 7: Execute the experiment: In Experimentation, the concept of "controlling variations" is highly valued because having control over the factors studied allows the CE to draw a true causal conclusion. Any variation besides the test factors need to be controlled as much as possible to create a valid experimental condition. This control concept may contradict the concept of "real-world" conditions where all variables are left as natural as they are. However, this is a misconception. To ensure that the testing condition is natural, one must have "control" on what they are doing so the factors that they are introducing do not make any significant variations (changes) in the actual environment. Conducting an experiment with consumers needs to ensure that the experiment itself will not alter any natural behavior especially between the two testing factors (in two-sample design) or before and after (in Longitudinal Intervention design). The length of time to conduct the experiment will need to be determined for the factors being tested to show any impact. This decision depends on the CE's experiences in each category. For example, if one wants to explore purchase behavior (e.g. repeat purchase), one needs to take into account the commercial cycle of the product/service. If one wants to study the impact of new technology on product efficacy (e.g. an anti-aging face serum), immediate response on single or short-term use may not be what you want to measure. Determining the length of testing time relies heavily on the objective and professional expertise.

Step 8: Understand the results(s) and learn: What differentiates an ordinary experiment from a learning experiment are the experimental design and interpretation of the results. Many books recommended relying on statistical criteria to ensure that the experiment conducted is a valid one. However, in this book, we take a different route. Our philosophy is about "what" we can learn and the confidence of the "what results" relies on experimentation (a.k.a. the planning of the study). Statistical significance provides guidance but is not absolute because, after learning, we will make many changes again and again. Therefore, we encourage CEs to pay more attention to the "whys" (or the reasons behind numbers or results) and any "perceivable differences" (not only statistical significance) especially in early stages of new product development. Pay more attention to "statistical validity" of the experiment in the late stages of new product development closer to product launch. Here are some recommendations for continuous learning through experimentation:

- Review and understand the metrics (your measurement) of success and failure in the experiment. Know the reasons why you chose a certain index and level as the threshold of success and failure.
- Pay attention to "surprises" and discuss why it happens because unexpected results indicate what you do not know about your consumers.
- Did your results prove or disprove your hypothesis? Any new questions and hypotheses that can be formed from the results?
- Pay attention to "new consumer insights." This usually happens when conducting the experiment and with special groups of consumers.
- Will the results make you change the idea? Any new ideas that can be formed by the results?
- Iterate and persevere, keep running additional learning experiments, take the learning to action, and continue to track and learn through time.

4.5 Iterative Experimentation Qualitative–Quantitative Research model

The Iterative Experimentation Qualitative-Quantitative Research process (IEQQR-process) covers upfront innovation and new product development in one seamless process (Figure 4.3). The IEQQR's goals are threefold: (1) enhance early identification of high potential consumer, market, and product insights early; (2) ensure effective resource allocation by increasing the probability of breakthrough product ideas grounded on consumer insight; and (3) build a

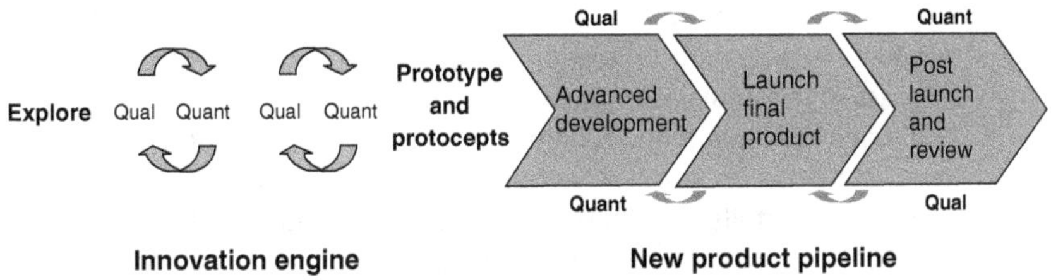

Figure 4.3 Iterative Experimentation Qualitative-Quantitative Research (IEQQR) process for deeper insights and meaningful metrics.

learning organization through a series of qualitative-quantitative iterative learning cycles.

Interactive innovation is the guiding principle for the IEQQR process and encourages an organization to:

(1) Experiment often;
(2) Fail early; and
(3) Learn from each iteration.

A series of qualitative-quantitative iterative learning cycles at the early steps of IEQQR process (equivalent to Step 0 to Step 2 in the traditional seven steps in NPD) (Figure 4.4) allows researchers to fail often and fail early to create prototypes or protocepts for screening quickly. Quickly coming up with prototypes

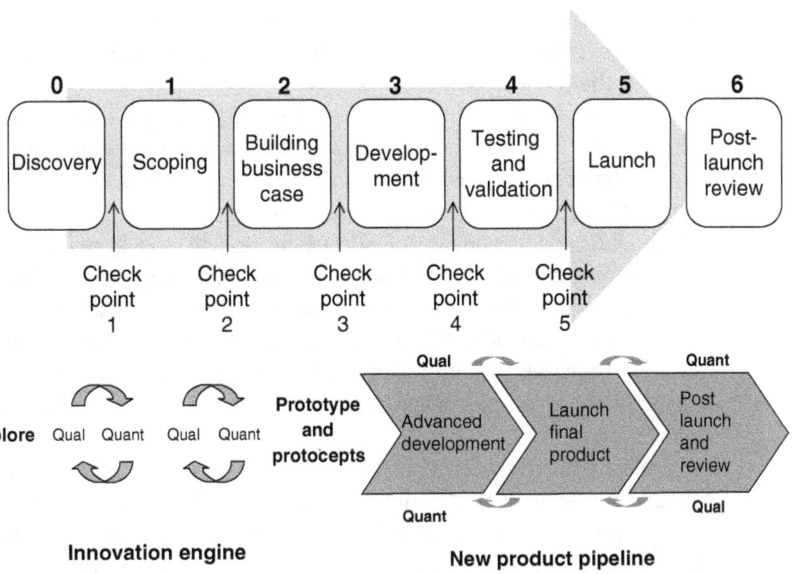

Figure 4.4 Comparing Iterative Experimentation Qualitative-Quantitative Research process to the traditional seven-step NPD. The IEQQR process encourages a series of iterative experimentation qualitative-quantitative learning throughout the process.

that clearly communicate the insights to consumers will significantly reduce product development time and money and will certainly increase probability of success in the final validation.

Instead of sequentially running small qualitative and large quantitative studies, the IEQQR process involves a series of iterative qualitative and quantitative work. This entails continuous and deep learning about consumer behavior through observations and asking questions to understand the *whys* and not just the *whats* from idea to final product form. Agile technologies from design to analysis provide both speed and quality output. In this book, we offer cutting edge tools that provide deeper understanding of consumers and allow the team to form appropriate questions, set relevant hypotheses, and build knowledge quickly as the springboard for new product ideas.

For example, the Consumer Explorer can conduct a series of small-scale qualitative research (e.g. in-context interviews and ethnographies) to identify insights on actual product usage and behavior upfront. The insights are then used to decide development paths, leading to new product ideas, target consumers, claims, and communication plans that are grounded on relevant consumer wants and needs in real-life situations. Then the product ideas, the claims, and the communication plans are developed and refined quickly through screening and rapid optimization studies, such as conjoint and discrete choice modeling. These steps result in the identification of an appropriate number of top candidates in final validation studies.

When you go through the IEQQR process, not only are the learnings grounded on the experiences of real people for a particular project, but you also gain additional knowledge for future innovation. The team can convert the additional knowledge to more insights for new product ideas that are then fed into the product development cycle quicker because the knowledge is already there ready for the team to pick up.

Our role as Consumer Explorers is to be strategic partners. We reduce risk and build confidence by guiding the team to navigate the unknowns and uncertainties that occur during front-end innovation. We design efficient research plans that lead to marketing and product development strategies that refine and validate product ideas for successful launch and in-market results. The advancement in digital technology (such as audio and video diaries, social networks, blogs and real-time testing) has enabled Consumer Explorers to get closer to what consumers think, say and do than ever before. In addition, new research tools like quantitative anthropology and network analysis allow Consumer Explorers to convert traditional qualitative information to quantifiable metrics.

In this book, the authors offer various new qualitative and quantitative consumer research tools that will provide detailed procedures, pros and cons, suggestions, and guidance that when readers finish reading each tool section, they can learn to apply the tool appropriately. Figure 4.5 shows suggested stages along the IEQQR-process that certain research tools will be effective and appropriate. The research tools described in this book along with new digital technology enable the Consumer Explorer to pay more attention to strategic questions rather than tactical problems (see Chapter 3 for details).

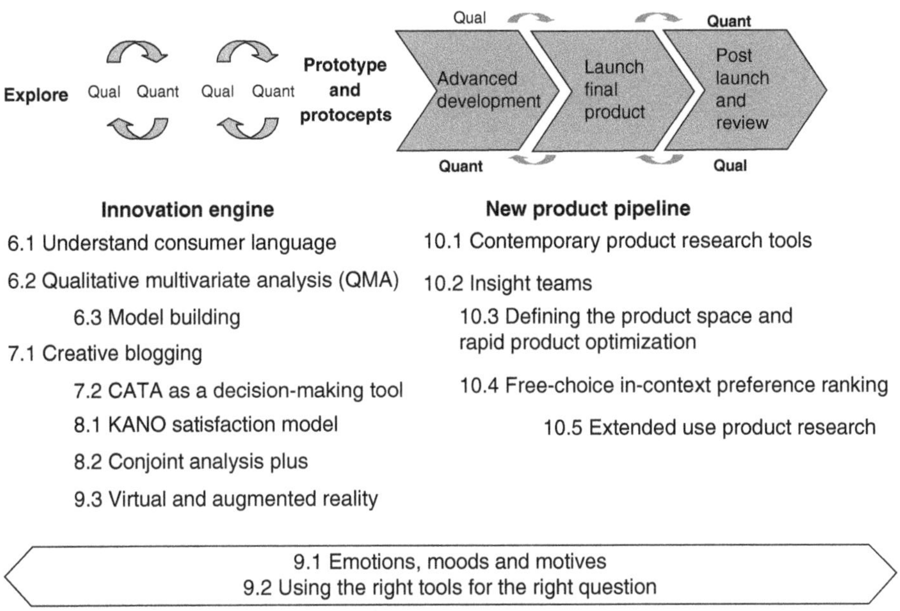

Figure 4.5 Suggested consumer research tools along IEQQR-process. The numbers indicate the chapter and section where detailed description of the tools can be found.

Discussion questions

- What makes the eight steps about raised about high-level processes for guiding the creation of a learning experiment important?
- How can you apply the concept of experimentation to your current NPD practice?
- What is to like/dislike about the IEQQR model given your current use of NPD practices?

References

Anderson, E.T. and Simester, D. (2011) A Step-by-Step Guide to Smart Business Experiments. *Harvard Business Review* (March). https://hbr.org/2011/03/a-step-by-step-guide-to-smart-business-experiments, accessed 24 May 2020

Cooper, R.G. (2001) *Winning at New Products: Accelerating the Process from Idea to Launch* (3rd edition). New York, NY: Perseus Publishing.

Moskowitz, H.R., Beckley, J.H. and Resurreccion, V.A.A. (2006) *Sensory and Consumer Research in Food Product Design and Development*. Ames, IA: Blackwell Publishing Professional.

O'Reilly, B. (2021, April 21) Unlearn Podcast [video]episode 12 Eric Ries The Lean Startup Pivot by Eric Ries. *YouTube*. https://www.youtube.com/watch?v=X7xOodGotdE

Potter, R. (2020, November 24) Data Drive Organisations: Making Data and Analytics the Top Business Priority. *Financial Times Live Global Digital Conference*. https://data-drivenorganisations.live.ft.com/ (accessed 06 March 2022).

Ries, E. (2011) *The Lean Startup: Use Continuous Innovation to Create Radically Successful Businesses*. New York: Crown Publishing Group.

5

Online, word of mouth changes. In the old world, someone might tell five people, but in the new world, you can tell 5,000. But be careful because it works both ways. If you make a customer happy, he can tell 5,000 people, too.

Jeff Bezos
American industrialist & founder of Amazon.com

What this is about

An insight is not a research finding but is derived from multiple interrelated observations put in context. It provides a view of what a person is feeling, doing, and dreaming. An ownable insight remains an information unless it is used for a specific purpose that impacts a desired strategic outcome.

Product Innovation Toolbox: A Field Guide to Consumer Understanding and Research, Second Edition. Edited by Kannapon Lopetcharat, Dulce Paredes and Jacqueline H. Beckley.

Chapter 5

What You Must Look For – Signs of High Potential Insights

Kannapon Lopetcharat, Dulce Paredes and Jacqueline H. Beckley

(updated from the original chapter by Kannapon Lopetcharat, Dulce Paredes and Jennifer Hanson)

5

Key learnings

- ✓ What is an insight?
- ✓ How to develop an insight
- ✓ Making an insight ownable

5.1 What is insight?

Many people have trouble defining an "insight." It is a term that rose in popularity among industry professionals once "market research" departments were renamed to "consumer insight" departments. This name change was designed to position market research departments as a more value-added department to the businesses they serve, in a time when market research was losing their "seat at the table" when it came to contributing to business decisions. This name change was to elevate the outcomes of research studies from delivering research findings based on facts to delivering insights. Unfortunately, instead of changing the outcomes, the term "insight" became synonymous with "research finding." Research departments and suppliers started to refer to themselves as "insights companies" or delivering "insights," when in fact they did not change what they delivered: research findings, merely facts that summarize data, such as, "s majority of Moms state oral care is extremely important for their children."

Product Innovation Toolbox: A Field Guide to Consumer Understanding and Research, Second Edition. Edited by Kannapon Lopetcharat, Dulce Paredes and Jacqueline H. Beckley.

As a Consumer Explorer (CE), if you cannot answer the question "Why?" with this statement, you have a research finding, not an insight.

An insight is not a research finding. An insight is a statement that connects the "whys" to consumer decisions. It delivers a very focused, yet deep understanding of consumer decisions. An insight must address the question "Why?" within the statement. For example, the research finding, "A majority of moms states that oral care is extremely important for their children" can be transformed into an insight by including the "why": "Moms focus on proper oral care so their children develop a great foundation for their adult lives."

Once communicated to key business stakeholders, the insight should be used for designing a new or better product and communication to go with it. Failure to activate an insight purposefully relegates it to mere data.

5.2 Good insights have the following characteristics:

- **Appear to be qualitative in nature** even though it can be based on quantified factual information and supported by knowledge gained from qualitative tools. Researchers sometimes refer to it as "you'll know it when you see it," a "hunch," an "intuition," and a "gut feeling." This feeling is a common revelation that you and others can see without talking to each other.
- **Allow you to see connections** across research findings. Scrutinizing information after information allows the CE to develop the experience to see connections and patterns among the ocean of data. This experience allows the CE to spot an insight. Insight is not the connections between different factual information, but it is an understanding of the wants and desires of consumers behind the data that enables CEs to connect the dots among the information.
- **Build hypotheses** for future exploration. After the CE gains deeper understanding of the consumers, it is natural to form hypotheses regarding the consumers and the products. This deeper understanding provokes more thoughts and helps researchers to see more subtle signals hidden among noises and obvious facts. Here are examples of thought starter questions:
  (a) Why do people behave that way?
  (b) Why do they feel this way?
  (c) What do they like to do after. . .?
  Good insights force the research team to think deeper about target consumers and form hypotheses about the consumers based on this new revelation.
- **Simple and easy to understand:** An insight is intuitive by itself. It is like a 30-second elevator pitch. If one cannot state an insight with one simple sentence, it is most likely that the statement is a series of facts laced together with justifications and therefore, is not an insight.
- **Seems to be "after the fact":** Because an insight is simple, it becomes obvious to audiences after it has been pointed out to them and increases their awareness of it.

5.3 What is an "ownable" insight?

On its own, an insight may not be unique. The insight stated above, "Moms focus on proper oral care so their children develop a great foundation for their adult lives" is not unique. It is something that any oral care product can claim as an insight. It is the Consumer Explorer's job to help make these insights unique. To do this, the CE needs to frame that insight in a way that is actionable to a particular product based on "how" consumers use the products they choose. If insights do not create the bridge between brands and consumers, they are considered category insights, and not ownable insights. It is the application to a product that makes an insight ownable. Only then can an insight drive marketing and product development plans that lead to sales and profitability.

A classic example of a good insight that drove business sales and profitability is when *Oral B*™ was looking to expand their manual toothbrush business. They hired IDEO™, a well-known consulting and designing firm. The IDEO team looked at the vast information about the manual toothbrush market. The team knew that their major consumers are adults and that young children do not like to brush their teeth. After many ethnography sessions, they found a great insight that brought *Oral B* millions of dollars. The insight was that "young children cannot manipulate the toothbrush well; hence, they do not like to brush their teeth" (Kelly and Littman, 2005).

This insight is clear and simple; it appears obvious and after the fact. The knowledge forces the project team to think deeper (more questions). Children cannot manipulate small objects because their motor skills have not fully developed. This led to the invention of a "big handle" toothbrush for kids that seemed to be counter intuitive at that time, but it was a big hit.

Another interesting and great example was Patagonia® dropping their clients' logos on its products in early 2021 (Hore-Thorburn, 2021). The famous US clothing company was known for its services for Wall Street corporates' garments (especially vest, tee shirt, jacket, etc.). Its corporate clothing gears usually had their clients' logos on them and the logos were unremovable. However, through many researches, Patagonia learnt about consumer's attitudes and habits, especially those toward clothing and branded garments.

- First, many corporate employees change jobs every few years.
- Second, the employees usually discarded their previous employers' clothing (especially those with previous employers' brands or logos on them).
- Third, people did not pass branded gears to their friends and family members.
- Fourth, people did not like to wear corporate-branded clothes during weekends or in their free time.

Consequently, many corporate-branded garments ended up in landfill prematurely (at least two years earlier than the expected life-time of the garments). These facts did not align with Patagonia's visions for sustainability. Patagonia knew that its decision of stopping this decades-long practice would cause the brand to lose many businesses. However, the brand hoped that many of its clients would realize the great cause and its commitment toward sustainability and would join the effort in the near future.

Patagonia's action connected many dots in many layers from product (the corporate-branded garment, product consumption life-cycle, product technical life-cycle), consumer/end user (as an individual and an employee along with consumers' attitudes, habits and behaviors), its own organization and brand (as producer, a member of the global community and solution provider) and to its corporate clients (as customers and members of global society). By connecting the insights together and deciding to drop its clients' corporate logos out from its gears, Patagonia made the insights their own and lasting impacts on society at large.

5.4 How to develop high potential insights

5

There are many tools that CEs can use to create insights. They range from observation to qualitative and quantitative research. It is the application of these tools that matter, more than the individual tool itself. The CE must always keep in mind the definition of an insight and potential business application when choosing research tools to deliver high-potential insights. Often, delivering insights will require the application of more than one research tool. The CE must gather these different inputs and build the insights from knowledge gained across all sources (Figure 5.1).

5.4.1 Behavior: the basis for all insights

Since an insight must provide an understanding of a consumer decision, it is important to focus on identifying the appropriate behavior to build the foundation to the insight. Therefore, the application of the tools the CE uses must result in an outcome of identifying a behavior that leads to a choice of product by consumers. Depending on the amount of existing research findings, the CE may be able to identify the behavior using previous research, rather than conduct new research studies. Behaviors can be identified from quantitative surveys or through observational techniques, such as ethnographies.

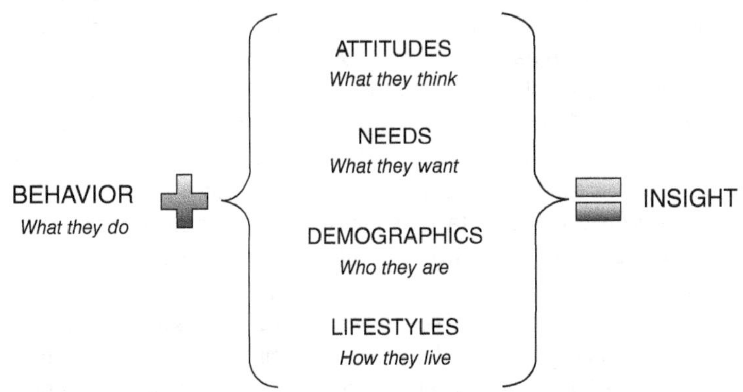

Figure 5.1 Insight is the result of understanding why consumers behave in certain ways.

5.4.2 Attitudes and needs: the explanation for behavior

Once the behavior is identified, the CE must develop a deep understanding of why the behavior exists. The tools chosen for this step must focus on the attitudes or beliefs the consumers have that lead to the behavior as well as the needs they are trying to satisfy. This aspect of insight development is usually gained through consumer discussions, whether one-on-one or in groups.

5.4.3 Demographics and lifestyles: the personal connection

Finally, the CE must consider the consumer demographics and lifestyles. Demographic information includes the age, gender, life-cycle stage, income, and occupation of consumers, while lifestyle refers to the attitudes, interests, and opinions of consumers, as well as the differences in behavior that relate to social values (ESOMAR glossary).

The insights must be relevant to the consumer target that exhibits the behavior. Understanding why consumers make decisions based on demographic characteristics such as life stage, ethnicity, or gender, will allow the CE to frame the insight to be relative to the behavior target. Many times, the CE can reference syndicated studies that provide a broad understanding of different demographic groups to obtain this knowledge.

In addition to general knowledge about these targets, the CE must keep in mind a consumer is not just one demographic, or lifestyle throughout the day. We coined the term **"multiple-selves"** to describe the many roles consumers play throughout the day or in many situations. For example, a woman can be a mother, a daughter, a girlfriend, a wife, a partner, and many other roles. Which role a person is in will influence the situation a consumer chooses to experience and *vice versa*. Situations are important as she has different product expectations and goals based on the situation. She can have a moisturizing lipstick for daily use at work, a shiny plumping lipstick for a night out and a lip balm for nightly treatment. Owning many products from the same category is a sign of multiple-selves phenomena. Bowman (2008) also referred to this phenomenon as "Split Personalities."

The complexities of multiple-selves have been increasing since the rise of the Internet and social network platforms. Physical boundary (a.k.a. geographical boundary) has been blurred and Internet societies rise. A person can have many social network accounts for different social circles. The word "Avatar" was prevalent especially in gaming societies. An 18-year-old man who is a high school student in a public school can also be a professional e-sport player (a.k.a. streamer) who earns US$ 30 000 a year by playing a game online. This example exemplifies many changes about consumers as we know them in the presocial network era. Traditionally, a person usually has few physical social circles, and now, a person can have a wide range of digital personas across multiple social circles. These different social circles can even have different cultures and subcultures. Consumers can have different decision rules in consuming products across their multiple social clubs. The rise of physical/digital hybrid space allows any consumer to become an influential figure (a.k.a. Influencer) in their digital social circles because they become an "expert user" in these circles. This leads to a disruption in marketing and mass communications (crowdsourcing reviews and influencers) and product-focus innovation. (see Chapter 13: Influencers' Impact on Innovation)

When a CE approaches the demographic and lifestyle understanding component of insight development, the CE must consider whether to treat the person as if they experience only one situation for their product, or many situations based on viewing a consumer as having "multiple-selves." A CE must consider physical and digital lifestyles and cultures equally. Making sure to consider all of these factors, especially among younger cohorts, is essential due to the self-finding that younger cohorts can exhibit 'multiple selves' as they move across social networks. With the outbreak of COVID-19 in 2020, many people adopted extensive online lifestyles very quickly and this behavioral shift is projected to be a fairly permanent.

5.5 Making insights ownable

5

It is easy to uncover an insight that is a category insight and not an ownable insight. The CE needs to provide ways to apply the insight to a product or a service. Only through this application, will an insight provide the foundation to sales, profitability, and evolved interaction. Many times, we refer to this as the "how" of insights:

- What are all the different ways that consumers use my products or services?
- Are there workarounds and trade-offs people make when using my product or services that can provide the team with ways to make it better fit with how consumers use it?
- How can I translate these specific behaviors to marketing messages that will deliver sales or enhanced satisfaction?

Since behavior is the foundation to all insights, it is important that CEs learn to look for the behavior-based signs that have the most opportunity to lead to an ownable insight. Here are examples of behavior-based signs to spot insights:

5.5.1 Routine and regular habits while consumers are using products

Every consumer has their own routines and rituals for using a product. Observing and understanding consumption routines and rituals allows the CE to spot where and how a new product can improve consumers' life and differentiate itself from the other products in a "seamless manner." The seamless manner is very important so consumers will not feel they must change their routine and habits or they have to trade off too much to achieve their goals. A new product that seems to fit with their life seamlessly and effortlessly is usually identified as more convenient. However, it is not only convenience that the product should deliver but also ensuring that a new product brings changes in habit and can fit with the consumer's routine or habit. Consequently, the product will differentiate itself from the rest of the market offerings. This kind of habit-changing or behavior-changing innovation mostly needs significant technical breakthroughs and, most of the time, the innovation creates a "new technology platform" as it is known as a "disruptive innovation." Examples of this are the launch of iPhone in 2007 and Beyond Meat® in 2012.

The iPhone was the first commercially available phone that did not have a dedicated keyboard or buttons as it used a "touch screen" technology that allowed the phone to have a new design (a rectangular shape) with large colorful screen (as opposed to black and white) and many functions at their fingertips (icons for apps). This combination disrupted the old mobile phone markets and created a new category called the "smartphone."

Beyond Meat (founded in 2009) introduced soy-based protein with hemoglobin that tasted like actual beef patty for hamburger in late 2012. This product reduced taste and flavor trade-off that flexitarians and anyone who wanted to eat less meat in their diets had to face as the beef patty tasted more similar to the actual beef patty than the gluten, grain, and mushroom-based meat analogues. In addition, it was marketed to be better for the environment as it used 99% less water, 93% less land, 90% fewer greenhouse gas emissions, and 46% less energy than those used in the actual production of a quarter-pound US beef burger (Beyond Meat, 2021). This product created a new category called "plant-based protein" globally and gained large market share from consumers who were not vegetarians or vegans but anyone who wanted to eat less meat (a.k.a. flexitarians) for health and other reasons (e.g., better for the environment, animal friendly, variety in the diet). The next technologies that are positioned to disrupt the meat category are cultured meat (lab-grown meat) and air meat (protein from the air using microbes).

These examples demonstrate that an insight that leads to habit and behavior changes needs a breakthrough technology to make it "ownable" and create a new market category and spawn technology growth.

5.5.2 Mistakes they make

Few consumers really read the instruction that accompanies a product that they purchase, and even fewer consumers follow the instruction provided. Studying the mistakes that consumers make fuels innovation through product adaptation. Product adaptation is a situation when consumers do not use a product as it is designed, and they devise a new way to use the product to reach their goals. Dismissing this sign misses an opportunity for improving current products to be more consumer friendly.

People make mistakes; however, they attribute the fault of making a mistake differently depending on the product category they used and that dictates different emotions toward the product and its brand. In the 1990s, many consumers made mistakes when they first assembled their personal computers. This insight led to the evolution of ready-to-plug-in personal computers. This category began with the introduction of the iMac series (starting with iMac G3 in 1998). Apple gained large market share from regular consumers (instead of its usual designer niche market) by focusing on consumer-friendly offerings (e.g., convenient two-step set-up, unique designs, and graphic display). The two-step set-up process was much easier than those of regular PCs from its rivals (e.g., IBM, Dell, Toshiba, and many other PCs) and, since then, Apple made its way to the general consumer personal computer market.

Skincare and cosmetic products are categories where consumers make many mistakes. Most of the time, consumers did not follow the use instructions properly and the consumers usually blame themselves for not taking care of their skin health properly or for lacking skill. One consumer skincare goal is to have naturally even looking facial skin. In 2019, P&G introduced the Opté Precision Skincare System that had the potential to disrupt skincare and cosmetic categories, and led the company to create new platforms in their corresponding markets. The system was a handheld inkjet printer for the face that combined image analysis algorithms and printing technology to spot and administer treatment and correct facial imperfections. The device scanned, decided the dose, and administered the treatment (both skincare and foundation combined) on the facial spot to give naturally even looking skin. This eliminated the lack of skill to apply cosmetic products correctly; hence, no more self-blaming and guilt. In addition, the device was claimed to be 70% less expensive and 30% faster than the leading technology. This product combined many different technologies (e.g., skincare, foundation, deposition, image analysis, inkjet) to create a breakthrough technology that could eliminate the mistakes made by consumers.

Another example that helps consumers eliminate mistakes, even though they do not know when and how they made the mistake, is the wireless charging technology. Many consumers had problems with broken charging cords after many uses with their electronic devices, especially their smartphones. Sometimes, they blame themselves for folding or bending the cord too tightly or blaming the cord for having poor quality. To eliminate this problem, manufacturers adopted a new technology called "inductive charging." Inductive charging was first proven in the late 19th century by electricity pioneer Nikola Tesla and was commercialized in 2008 in industrial sectors with the establishment of Qi standard. Inductive charging was commercialized in the consumer electronic sector in 2011 by Samsung and was known as "wireless charging." This technology allows consumers to charge their smartphones by just putting their phones on a charging pad without any cord. Later in 2012, it was installed in a Nokia® smartphone series followed by Samsung in 2014 (in Samsung Galaxy Note 4) and by Apple in 2017 (in iPhone 8). Now, wireless charging has been a standard function in high-end smart phones and many other electronic products. In 2014, Starbucks installed wireless charging stations in its 8000 coffee shops in the United States. In 2020, there were more than 50 car models offering Qi-based wireless charging in their cabins (Mearian, 2018).

Any insight that helps consumers to reduce or eliminate mistakes when they use a product is most likely to change the landscape by setting a new baseline (the must-have in KANO classification) (see Chapter 8.1 for applying KANO classification to create new products) for an existing category like what we have seen in personal computers, smartphones, and skin care.

5.5.3 Consumers combine products

This sign is usually found among avid users of a product category. They combine products because current products do not offer all the desired benefits and values, or the products do not offer benefits specifically for them. This behavior provides opportunities for creating incremental innovations (e.g., line extension), breakthrough innovations for unmet needs, or customizable solutions.

This type of insight influences the food industry more often than other sectors, and the drivers of this kind of innovation are usually product creators (such as culinary chefs, creative, and development teams). An example is Pretzel M&M's that was named the 2011 Product of the Year in the candy and snacks category by Product of the Year USA (MARS, 2011). Pretzel M&M's was created based on the insight that a combination of sweet and savory in snack is highly appealing to consumers (The Datamonitor Group, 2010). Another example is the Cronut created by the pastry chef Dominique Ansel of Dominique Ansel Bakery in 2013. A Cronut is a croissant-doughnut pastry started from a bakery and made it to mainstream when Dunkin Donut put it as a permanent menu item in 2015.

5.5.4 Home remedies

Understanding home remedies provides insights to consumers' beliefs about how certain things should work. It allows easy associations between product attributes and benefits that lead to many sustainable innovations. Line-extensions in personal care products and new flavors in food and drinks are good examples.

A limitation from using home remedies is that it is mostly regional or culturally specific. Turning a home remedy solution to a global product is difficult, but it is possible through thoughtful translation and communication of benefits and digital travel. A good example is the use of the term "super fruits" in combination with individual regional names (e.g., acai, goji, moringa, mangosteen) to convey strong antioxidant properties and the exotic nature of the ingredients.

Later in this book, we provide many methodologies that allow CEs to identify the links between product attributes (means) and aspiring goals (ends) in both qualitative and quantitative ways.

5.5.5 Consumers alter packages or use packages the way we do not expect

The product's package plays an important role in consumers' satisfaction, and it is very important to observe how consumers use your products in different real situations. This observation can yield insights for sustainable innovations (e.g., easy open cap, individual pack) to disruptive innovations (e.g., devices, applicators, and package-less).

The first thing to keep in mind when observing consumers, especially in the consumer-packaged goods sector, is that *the product is what consumers pay money for from the beginning to the end*. Assuming that the package is merely a container that houses items that consumers consume is a common mistake that researchers and marketers regularly make. Expanding the definition of packaging to be an integral feature of the product is a good start to gain more insights regarding packages.

Observing how consumers alter the original package provides insights for many innovations. Laundry detergent is an obvious product where consumers often alter the package (in this case, the bottle). Before 2006, most US consumers could only buy a 100 oz bottle of liquid laundry detergent (128 oz is a gallon), then the consumers had to lift the bottle and pour it into a cap provided to measure the amount of detergent every time they were doing laundry (this is the case for Americans who own a laundry machine). For those who do not own their own laundry machines, they had four choices: (1) carrying the big bottle to the laundromat, (2) transferring some of the detergent to a smaller container, (3) buying a more expensive detergent from the laundry mat, and (4) hiring someone else to do the laundry. The first two options were adopted by many Americans. Transporting and lifting a big gallon of liquid detergent is cumbersome for most consumers especially when they have to travel far from their homes to a laundromat and transferring liquid detergent from one big gallon to a smaller bottle is inconvenient and messy. It was obvious that American consumers needed a smaller size package and there were many, but the consumers had to trade-off between smaller size and spending more money to buy detergent more often. In 2006, Unilever® provided a solution to consumers by introducing a third smaller bottle (32 oz bottle) with 3x concentrated laundry detergent. This was hailed as a great sustainable innovation at that time because it benefited consumers, the environment, and the company. The smaller size with 3x concentrate liquid detergent solved consumers' problems of transporting, lifting and pouring, and delivered monetary value (providing the same number of laundry loads as the big bottle). It is good for the environment because it uses less resin to make, reduces pollution from transportation as the company can ship more products at a time, while using less paper for cartons. Unilever benefited by spending less in trucks and transportation, paper cartons, etc. The idea was so impactful that other companies followed, and it created a new subcategory called concentrated liquid detergent that took over the regular bottle (A side note to this, many users were familiar with delivering a certain amount of liquid detergent. So there was and is a certain amount of people who like the compact bottle but don't realize all of the benefits, since they misuse [pour too much] the product as packaged and directed).

Observing consumers' use behavior and habits and identifying pain points and patterns in the behavior and habit provide insights for great innovations.

5.5.6 How and where consumers store products

Storage of products tells CEs more than just location. It shows the relationships between consumers and their products. These relationships are formed by certain mindsets: The knowledge about products that consumers know; their expectations from products; and the relationships that manifest themselves in the form of nonverbal cues. Avon's Pro to Go one-handed lipstick (launched in 2008) was an innovative solution based on the observation that women store their lipsticks in many different places other than where they normally put on make-up and that they like to apply their lipstick while multitasking. The original light color palette was designed so it can be applied "mirror-free."

5.5.7 When consumers look for advice or help

It is important to identify situations in which consumers are seeking help to use a product or to understand whether a product works as hoped for in practice. This boundary provides opportunities for product and marketing innovations. For example, many personal care products use claims associated with salon or professional results. The role of the claim is to help the individual understand the product purpose.

Another example is single serve beverage machines that start with freshly brewed coffee products. These coffee machines provide high-quality coffee and are in a portion (single serving) to further suggest that the coffee is for a target of coffee consumers who are looking for convenient solutions and freshly brewed coffee. Another prominent example was image rendering technologies that were built-in for digital cameras or available for free to download. Having good and professional looking images was a very difficult task in the 20th century; however, with the advancement of artificial intelligence (AI) in the 21st century, image rendering and beautifying image technologies allow "all consumers" to take good-looking images with their smartphones without any professional help. Since it was first developed, image rendering technology has allowed selfies for a variety of apps to be even more enhanced or over-enhanced, thus raising questions regarding what is authentic in digital images.

5.5.8 Unachievable goals

This sign manifests itself in the form of complaints, pain points, disappointments, odd behaviors, customization attempts, and wishes.

Digital photography is a classic example of disruptive innovation that stems from this type of insight. In film photography, taking a good photograph requires skill, equipment, and practice that may cost a lot of money. In digital photography, consumers can make mistakes, take as many pictures as they want and choose the best photographs without spending too much money and waiting to develop the film. Again, in the 21st century, with the advancement of AI, all digital cameras were equipped with automatic image rendering systems that can give consumers very good and professional looking images in an instant especially self-images (a.k.a. selfie images). The AI systems eliminated the need for professional help to get satisfying outcomes almost instantly.

Another example are the services provided by third parties such as Airbnb, Uber, Grubhub. Calling taxis and making hotel reservations could take time and were not always easy until the introduction of these companies, the Internet, smartphones, and "apps" (applications that are designed to run with a computer device and the Internet). These on-demand companies began by owning no physical assets, but they provide "direct connections" between owners and consumers for desired services. This type of business model opened a new point-of-view for CE as the "C" encompasses not only the one who buys (*consumer*) but also the one who provides (*client*). For Airbnb, Uber, and Grubhub, the companies view the suppliers (room owners and car owners) as their clients or contractors and the ones who want rooms to stay or car services or food delivery as their consumers. In addition to connecting and perhaps providing "cheaper alternatives" to traditional products (e.g., hotel, taxi, limousine, food), these companies provide convenience as well. These on-demand companies adopted policies to build trust, safety, and confidence, e.g., upfront availability of providers to users and feedback tracking for both providers and users. These new business models also bring a raft of concerns such as social and legal that CEs may have to consider as part of their remit to deliver critical insights.

To develop disruptive innovation based on this insight, real advancement in technology or a highly creative application of existing technology is usually required and can result in a "new platform" for consumers.

5.5.9 Changes in moods and emotions

Emotional clues are everywhere yet many researchers miss them because the clues are subtle, mostly behavioral in nature and can be context-dependent.

Product categories like fragrances elicit strong emotional cues. An example of successful commercial innovation is Unilever's AXE fragrance brand that covers a wide range of personal care products. High school and college males are the target consumers based on the insight that sexual desire is a motivating factor for young men. The type and quality of fragrances were developed to mesh with these potential factors.

The best place to find emotional clues are the actual situations that consumers use products by studying the changes in moods people have before, during, and after product use. The behavioral signs of changes in emotions when observing, monitoring or interviewing consumers include change in tone of voice, talk about old experiences, length of time spent to discuss a particular product and experience, increase in engagement of a task, repeated touching of the product, changes in heart rate, and others.

Clorox's toilet cleaning wand is a good example for this case as it was developed through observing consumer habits and behaviors (Zink, 2008). A cross-functional team observed how consumers clean their own toilets. The team already knew that consumers were highly satisfied with the results of the current routine (Figure 5.2).

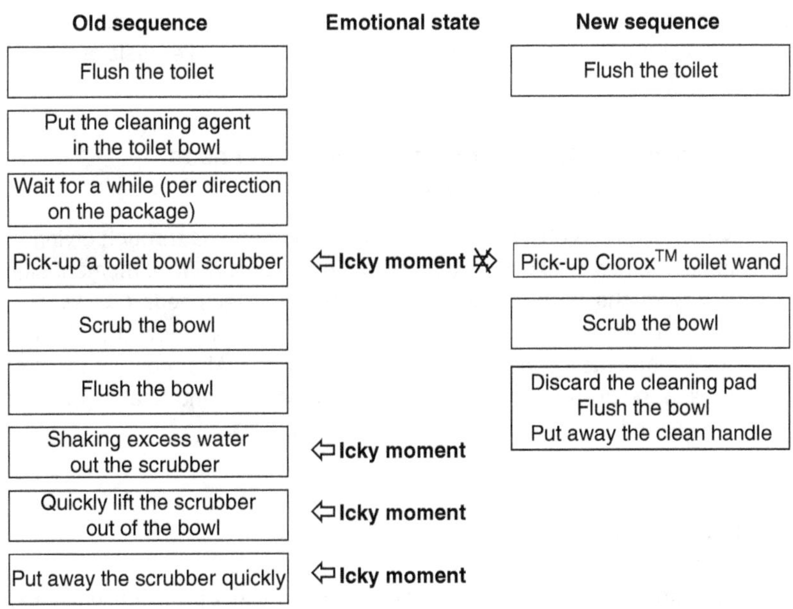

Figure 5.2 Comparison of toilet cleansing routine before and after the invention of Clorox® ToiletWand™ System. Toilet bowl cleaning behavior and emotional states that the Clorox team identified through observation of consumers' facial expression, habit, and behavior in the actual environment.

The team noticed an important behavior from this routine that was "every consumer made faces when they tried to put away the toilet scrubber." They named that behavior the "icky moment." Icky moment is very emotional as consumers want a clean toilet, but they do not really know how to clean a toilet scrubber after rinsing it with clean water in a newly cleaned toilet bowl by flushing. There was always a feeling that it is dirty and they do not want to store it, but there was no choice unless they really cleaned the scrubber thoroughly, which is another long process (and is an uncommon, and continues the ick). The team quickly brainstormed and came up with ideas of prototypes that eliminated this scrubber cleansing and storing steps. The prototypes allowed them to assess an optimal solution for the "icky moment" and the outcome was the Clorox® Toilet Wand with disposable sponge/scrubber. The wand revolutionized the toilet cleansing category as it changed consumers' behavior in a drastic and seamless way (Figure 5.2). Another similar innovation, P&G's Swiffer, can be analyzed in the same way. And yes, these innovations present other issues such as reuse/sustainability, which is another discussion topic.

Another great example of a product that made consumers say 'Wow' was the Tesla car, the first commercially available 100% electric car. Founded in 2003 and having released the first model in 2008, Elon Musk and his team managed to create buzz around every Tesla model and attained the most valuable car company in 2020 when its market value was higher than those of the top two incumbents (General Motors and Ford) combined. Tesla cars offered many features that made consumers say "Wow, this is cool". Being a 100% electric car was not unique but being a mass-produced, powerful 100% electric car with surprising features such as quick charging, long distance per charge, powerful engine, good design, programmable horn sound was 'Wow'. With an ingenious marketing campaign, Elon Musk managed to make Tesla a very attractive equity.

5.5.10 Wow! and Aha! moments are special cases to spot emotional cues

An easy way to observe Wow! and Aha! moments is through giving consumers opportunities to experience physical prototypes. A Wow! moment signals superiority of the product delivery for either unmet needs ("Wow, it works that well"), or unarticulated needs ("Wow, I don't even know that it can do that"). Aha! moments connote mostly unmet needs ("Aha, now I can do that too"). Presenting prototypes is effective in eliciting Wow and Aha! moments since they are elicited from spontaneous experiences.

5.6 Summary

The inputs to insights are everywhere, and it is up to Consumer Explorers to spot them. An insight is not a specific story or thing. It is an ability that enables researchers to connect the dots from pools of information and understand consumers better. Keen eyes and creativity are necessary for spotting high potential insights when observing actual consumers experiencing products in their normal environment. Using digital technologies such as portable cameras, cell phones, blogs, and social networks bring consumers closer to the research team in unprecedented ways, allowing CEs to inspire and influence their organizations to be empathic toward target consumers by showing proofs where consumers say, "at last someone understands me."

Discussion questions

- What differentiates an insight from a research finding?
- What are the different ways where you can generate compelling insights?
- What makes creating an ownable insight hard?

References

Beyond Meat (2021) Web. 3 March 2021. https://www.beyondmeat.com/about/#:~:text=Our%20Impact&text=A%20Beyond%20Burger%20uses%20significantly,GHGE)%20than%20a%20beef%20burger

Bowman, J. (2008) "Split personalities". *Research World: The Magazine for Marketing Intelligence & Decision Making*, 2(October), 48–51.

Hore-Thorburn, I. (2021) *Patagonia Is Dropping Corporate Logos From Its Products*. 6 May 2021. https://www.highsnobiety.com/p/patagonia-dropping-corporate-logos/?format=amp

Kelly, T. and Littman, J. (2005) *The Ten Faces of Innovation: IDEO's Strategies for Defeating the Devil's Advocate and Driving Creativity Throughout Your Organization*. NY, USA: Doubleday.

MARS (2011) *M&M'S® Pretzel Chocolate Candies Awarded Product of the Year*. MARS, Mars Incorporated. 07 July 2011. Web. 06 February 2011. http://www.mars.com/global/news-and-media/press-releases/news-releases.aspx?SiteId=94&Id=2811

Mearian, L. (2018) *Wireless Charging Explained: What Is it and how does it Work?* Computerworld. 28 March 2018. Web. 23 January 2021. https://www.computerworld.com/article/3235176/wireless-charging-explained-what-is-it-and-how-does-it-work.html#:~:text=Wireless%20charging%20has%20been%20around,a%20transmitter%20and%20a%20receiver

The Datamonitor Group (2010) *Successes and Failures in Consumer Packaged Goods Innovation and Marketing*. The Datamonitor Group. http://www.datamonitor.com/store/Product/successes_and_failures_in_consumer_packaged_goods_innovation_and_marketing?productid=CM00047-006, accessed 14 February 2011.

Zink, L. (2008) Sensory Science & Market Research: A Marriage of the Mind, Heart & Soul. *Presented in 7th Pangborn Sensory Science Symposium*. Minneapolis, MN, USA.

Part II

Gear Up for Your Journey

6

Listen for meaning, not necessarily the words.

Bill Moyers

What this is about

This chapter starts the "how to" discussion of research tools (both qualitative and quantitative) to identify new opportunities stemming from people's unmet needs and wants when combining these recommendations with the knowledge gained from the other chapters. This chapter will equip the Consumer Explorer with research methodologies and approaches that are cutting-edge, reliable, and proven to provide high-value and deep insights.

Product Innovation Toolbox: A Field Guide to Consumer Understanding and Research, Second Edition. Edited by Kannapon Lopetcharat, Dulce Paredes and Jacqueline H. Beckley.

Chapter 6

Tools for Upfront Research on Consumer Triggers and Barriers: Qualitative Tools

6.1

Understanding Consumer Language

Kannapon Lopetcharat and Jacqueline H. Beckley

6

Key learnings

- ✓ Intrinsic and extrinsic motivations
- ✓ How to select methods to understand consumers' perception of products
- ✓ Interview techniques
- ✓ Analysis of qualitative data

6.1.1 Consumers do not understand technical product language, so what should we say about our new products?

A product is valuable to consumers because it provides benefits or means that fulfill their needs and expectations. Therefore, benefits help constitute a product, differentiate a product from a set of close alternatives and define a product category. Defining attributes and their value used to belong to companies and their marketing groups; however, in the 21st century, there has been a shift with the users - consumers being in charge and having far greater influence due to rise in consumerism and social networking.

Product Innovation Toolbox: A Field Guide to Consumer Understanding and Research, Second Edition. Edited by Kannapon Lopetcharat, Dulce Paredes and Jacqueline H. Beckley.

It is very important to know how much consumers value a product's benefits and to identify which elements are more valuable than others. Two major factors determine a benefit's value: **Intrinsic motivation** and **extrinsic motivation** (Antonides, 1991). Intrinsic motivation is a consumers' perception of the extent to which a product's characteristic or attribute is believed to cause a consumer perceived product's benefit. Consequently, it is the reason consumers are satisfied and eventually love a product. On the contrary, the extrinsic motivation is a consumer's perception of the availability and acceptability of the characteristic in the market. Therefore, the intrinsic motivation is what consumers *demand* from products and the extrinsic motivation is what the market can *supply* to the consumers.

In this section, we will introduce a set of techniques that will allow you to identify consumer's intrinsic motivations (values, benefits, and attitudes) toward products, to understand how consumers differentiate new products from other products in the market (extrinsic motivation) and to translate product characteristics to individuals' benefits.

6.1.2 How to select a method?

Many techniques can be used to understand consumers' perception of products. It can be a simple product test with specifically designed questionnaires to understand consumer's choice and motivation. It can be a carefully designed experiment that extracts underlying important product characteristics such as Conjoint Analysis (detailed discussion can be found in Chapter 8.2). It can be a deliberately designed series of "why" questioning utilizing a laddering technique, the use of a carefully crafted interview guide in conjunction with a skillful moderator in a group interview (Bech-Larsen and Nielsen, 1999), or an expert interviewer implementing a skilled form of positive inquiry. However, each technique has its own limitations and optimal uses. The following discussion centers on a few techniques that have been proven in practice to provide consumer insights effectively.

The expectancy-value theory states that the best predictor of consumer's behavior is the belief or the most important association that the consumers have about their attitudes toward an object (Ajzen and Fishbein, 1980). Based on this framework, the outcomes of any techniques depend on three factors: (1) source of information, (2) task format, and (3) the use of stimuli in the task. For example, in Word Association, a Free Elicitation technique, the words that first come to the consumer's mind are associations, benefits, and some product characteristics that should be the most relevant for consumer behavior (e.g. choice, purchase). The consumers articulate the information to the interviewer as quickly as they can (task format) from their own memories (source of information) which depends greatly on familiarity with the product of interest (stimulus). Table 6.1.1 shows these three factors for several techniques: Free Elicitation or Zaltman metaphor elicitation technique (ZMET), laddering interview, Kelly's repertory grid, and Flash Profiling.

The source of information is virtually the most different and the other two factors, Stimulus use and Task format, seem to be more similar among the techniques. However, Stimulus use and Task format have equal or more influence on

Table 6.1.1 Source of information, stimulus use and task format, influence outcomes, and applicability of the outcomes for each technique.

Technique	Source of information	Stimulus use	Task format
Flash profiling	Product-driven	Multiple familiar products	Self-articulated and unstructured individual evaluation
Kelly's repertory grid	⇕	Multiple familiar products	Self-articulated and unstructured one-on-one interview
Laddering interview		Multiple familiar product/product class	Self-articulated and unstructured one-on-one interview
Free elicitation *or* ZMET	Need-driven	Single familiar product/ product class *or* No product evaluation	Self-articulated and unstructured one-on-one interview

6

outcomes and the applicability of the outcomes from each method than Source of information as the stimulus and the format influence the information source as well. Table 6.1.1 is organized based on the Source of information used in each technique from "need-driven" to "product-driven." The continuum between need-driven and product-driven methods is not distinct from one method to the next. ZMET, Free Elicitation, and laddering interview require only a concept about a product class but not the actual product; meanwhile, Kelly's repertory grid and Flash Profiling require actual products to be evaluated. The involvement of actual product and familiarity to product or product concept influence the level of consumers' responses from strictly differentiating product characteristics obtained from Flash Profiling to abstract responses such as benefits, value, and attitude from ZMET. These outcomes will impact the applicability in new product development (NPD) as shown in Table 6.1.2.

To select which method(s) to use, the Consumer Explorer (CE) must know how he/she wants to use the information as there is no one method that gives all

Table 6.1.2 Expected outcomes and applicability of the outcomes from the techniques in NPD.

Technique	Expected outcome	Target activity in NPD
Flash profiling	Differentiating product characteristics	Product development
Kelly's repertory grid	⇕	⇕
Laddering interview		
Free elicitation *or* ZMET	Values, believes, attitudes	Marketing

the answers. To get a more comprehensive picture about your consumers, the author recommends using two classes: Need-driven and product-driven methods.

6.1.3 Free Elicitation and Zaltman metaphor elicitation technique (ZMET)

Free Elicitation is a general class of techniques that is quick and convenient to explore consumers' perception. Free Elicitation includes many well-known techniques such as Word association, Direct elicitation, and others. In a nutshell, these techniques are based on the Expectancy-value theory (Ajzen and Fishbein, 1980) and the Memory Schemata concept (Collins and Loftus, 1975). The Expectancy-value theory states that the most important associations that a consumer has about his/her attitude toward an object is the best predictor for his/her behavior; hence, it will be revealed first when we ask the consumer. The Memory Schemata concept believes that a consumer has a preconceived knowledge (Memory) that is formed through their experience in life and the knowledge is organized in certain structures (Schemata) in which contents are interrelated. Therefore, Free Elicitation techniques rely on the consumer's memory about a product or product category (familiarity), consumer's ability to recall the memory during the interview and belief that the most important information is contained in the first responses. Regardless of the name, Free Elicitation, Word association, Direct elicitation, etc., the techniques share the same general eight steps (Figure 6.1.1).

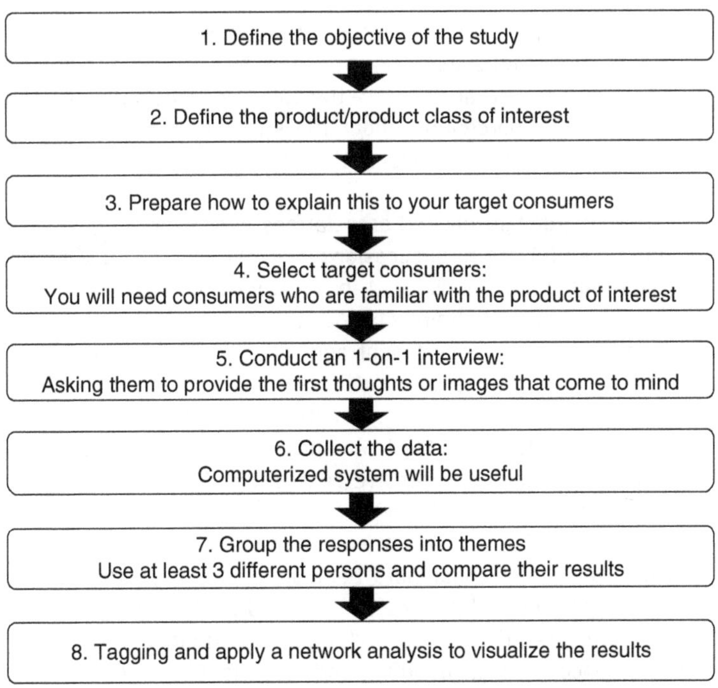

Figure 6.1.1 Seven steps to conducting a successful Free Elicitation.

The elicitation techniques require consumers to react quickly to the stimuli (a.k.a. a product or an idea of a product category). Therefore, it can be either a stimulus-dependent technique or a product-driven technique. If the stimulus is an actual product or a very specific product class, the outcomes will lean toward product attributes and some benefits that consumers think the product delivers (more information about extrinsic product differences). If the stimulus is an idea of a product class, the outcomes will lean toward benefits, values, and attitudes (more information about intrinsic motivation). In general, 16–25 consumers per market are sufficient to provide qualitative information and many related products or product concepts can be tested with the same group of consumers. However, if actual physical products are used, the influence of participant liking toward the products in conjunction with familiarity with the products will need to be considered.

Some enhancements can be applied to the Free Elicitation techniques to make it more discriminating, to enrich the qualitative information or even to quantify the information generated from the techniques. Below are few examples:

- **Adding negative and positive reactions or rating for each association (e.g. word or image):** Consumers can react or rate each word after writing the word down from hate to love or 1 (very negative/bad) to 5 (very positive/good). The reaction or rating provides the direction of the association and is usually analyzed after each response is grouped across all respondents.
- **Using ZMET**[1]: Consumers will select images from a set provided to them that they think reflect what they feel toward the stimulus in question. Therefore, ZMET is a projective technique. Usually, once the images are selected, the consumers are interviewed to explain why they chose the images using a regular Free Elicitation. Special care must be taken to select and prepare a set of images for a study. Relevancy to the products and objectives and quality of the images can influence the outcomes. The outcomes of ZMET are rich in high-level perception about the stimulus such as benefits, values, and attitudes and less in product attributes.

To select significant elicited associations (words or images), a cut-off of 10 is usually used for the total occasions of 400 or more (e.g.16 products × 25 consumers = 400 occasions or 1 product × 400 consumers = 400 occasions) (Guerrero et al., 2010; Roininen et al., 2006).

Free Elicitation is an efficient and rapid method to gain information about consumer perceptions of products. It allows researchers to identify high mental-level perceptions (e.g. affective, benefits, values, and attitudes) that are less conscious aspects of consumers' mind better than methods that rely on direct questioning, extensive evaluation of products, or highly structured interview techniques (Szalay and Deese, 1978). With good execution, Free Elicitation could provide a fast and convenient tool to identify and understand the motives behind consumers' behaviors; however, without good preparation, Free Elicitation could provide shallow and confusing outcomes that are difficult to interpret. If your research goal is to identify links between attributes, consequences, values, benefits, and attitude, Free Elicitation is not appropriate and other techniques that focus on identifying these links (e.g. laddering interview, qualitative multivariate analysis [QMA]) will be more appropriate.

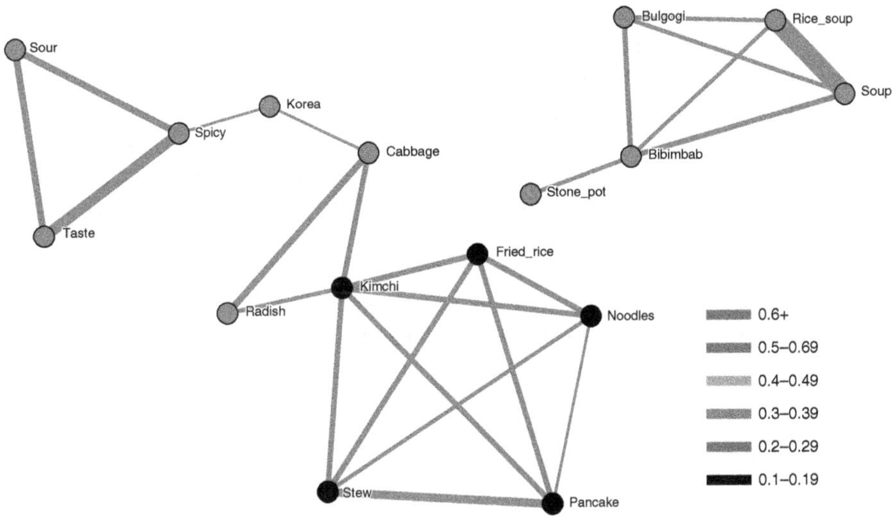

Figure 6.1.2 This is an example of applying a network analysis on free-word association technique. To gain a deeper understanding of words/terms consumers used on a topic. The links with different colors and thickness indicate different strengths of degree of affiliation between the words/terms used by the consumers.

6

6.1.3.1 Free Elicitation in practice

In 2017, an academic from a South Korean university (Dr. Mina Kim, an associate professor of department of Food Science and Human Nutrition, Jeonbuk National University) contacted one of the authors to help with her project on Gochujang. Gochujang (a type of fermented chili paste) is a staple in Korean cuisine and the Korean government wanted to promote the use of gochujang in China but did not know how to communicate it to Chinese consumers. One hundred Chinese consumers were interviewed and asked to give a minimum of three words when an image of gochujang was presented to them. The Chinese words were tallied, grouped, and analyzed using network analysis to visualize the relationships between words associated with gochujang among Chinese consumers (Figure 6.1.2). Spicy (49%), taste (45%), sour (23%), cabbage (19%), and kimchi (18%) were the top-five words among 79 unique word classes associated with gochujang (Lopetcharat, 2017).

Figure 6.1.2 shows deeper meanings of the words used by Chinese consumers. When they said spicy and anything about taste, they really meant spicy and sour (not spicy alone). In addition, when they talked about cabbage and kimchi, they related them to specific dishes such as fried rice, noodle, stew, and pancake. These word connections can be used to launch a communication campaign to promote an iconic Korean food in the Chinese market (Lopetcharat, 2017).

6.1.4 Laddering interview

If the question of interest is to identify the links between high-level perceptions about products (e.g., attitude, values, and benefits) and product attributes laddering interview can be an appropriate technique for the job. Based on the means-ends chain (MEC) theory (Gutman, 1982), laddering interview is designed to discover the links between concrete product attributes and intrinsic motivations (e.g. perceived benefits, values, and attitudes) and to explain why an attribute is more important than others due to the context of the linkage. The MEC states that the strength and numbers of cognitive links between a product attribute and its related consequences (perceived benefits and attitudes toward the product) and personal values determine the attribute's importance to consumers' decision or behavior. Therefore, a perceived benefit (an abstract attribute) that links directly to values and expectations of consumers will have more impact on final attitudes toward a product and consequently consumer's choice than physical product attributes that are mediated by an abstract attribute.

There are two general types of laddering interviews: Soft laddering interview and Hard laddering interview. There are two factors influencing the selection of laddering interviews: (1) Familiarity to products, and (2) Level of consumer involvement to the products (Grunert and Grunert, 1995). Soft laddering interview is appropriate when either factor is extreme (very low or very high). Hard laddering interview is appropriate for medium level for either factor (Table 6.1.3).

When there is no prior knowledge about any cognitive construct (links between product attributes, consequences, and values) of a product category, Soft laddering is recommended. This form of laddering is selected due to the exploratory/discovery nature of the research. Hard laddering can be recommended when there is substantial knowledge about the product of interest and therefore the interview is being used more for confirmation of understating, rather than discovery. Steps to conduct Soft laddering interviews are different from those of Hard laddering interviews. Let us start with Soft laddering interview steps (Figure 6.1.3).

The first four steps are common between Soft laddering (S) and Hard laddering (H) interviews. Since Soft laddering interview is an "unstructured interview," step S5–S8 will take some time to finish, and it is one of the limitations of Soft laddering interview that we will discuss later in this section (Figure 6.1.3).

Hard laddering interview begins with the first four steps mentioned in the steps for Soft laddering interview. Hard laddering interview is a "structured interview" that follows the next five steps (Figure 6.1.4).

Table 6.1.3 Guideline to select Laddering interview method.

		Consumer involvement level	
		Low involvement	**High involvement**
Consumer's familiarity to stimulus	**Low knowledge *or* Unfamiliar**	Soft laddering	Soft laddering
	Medium knowledge *or* Somewhat familiar	Hard laddering	Hard laddering
	High knowledge *or* Very familiar	Soft laddering	Soft laddering

1. Define the objective of the study: Do you want to know about consumers' preference or product perceived differences? This will dictate the activity of step 5.

2. Define the products/product classes/product concepts of interest

3. Prepare how to present the actual products/product classes/the product concepts (generally called stimuli) to your target consumers

4. Select target consumers:
You will need consumers who are familiar with the product of interest

S5. Elicit distinction between products from consumers: preference or similarity

S6. Ask why he/she selected the first choice (most like product or a group of products) compared to other products

S7. Laddering interview by asking "why this is reason important for you?"
Keep asking this question until the consumer cannot answer any further

S8. Continue steps 6 & 7 until you exhaust all the products or groups

Figure 6.1.3 Steps to conduct a Soft laddering interview. Letter "S" in front of the numbers indicates that these steps are unique to Soft laddering interviews.

H5. Collect knowledge about the cognitive constructs of the products and conduct preliminary works such as focus group interview, etc.

H6. Define a general cognitive constructs, select product attributes, consequences (function and psychological) and values (instrumental and terminal); ensure that the selected words are understandable by the target consumers

H7. Present the consumers with *Product attributes* list and ask the consumers to choose at least 1 and up to 3 attributes that are important to their preference or differentiating the products.

H8. Present the consumers with the words they have selected from Step H7 and ask "why it is important/differentiating for you?" for each word selected and provide them with the list of words from the *Consequence level* to select the reasons. The consumers must select at least 1 and up to 3 words.

H9. Present the consumers with the words they have selected from Step H8 and ask "why it is important/differentiating for you?" for each word selected and provide them with the list of words from the *Value level* to select the reasons. The consumers must select at least 1 and up to 3 words.

Figure 6.1.4 Steps to conduct a Hard laddering interview. Letter "H" in front of the numbers indicates that these steps are unique to Soft laddering interviews.

To conduct a Hard laddering interview, CEs **must** have prior knowledge about cognitive constructs of the product of interest in order to create lists of words that represent all the cognitive levels (product attributes, consequences, and values). By presenting target consumers with a "defined" list of words for each level, there is a possibility that CEs do not capture everything. We will discuss these limitations later in this section.

6.1.4.1 Analyzing results from laddering technique

Analyzing results from Soft laddering and Hard laddering interviews requires extensive coding and categorizing. Due to the discovery nature of the research, this is especially true for results from Soft laddering interviews. Figure 6.1.5 shows general eight steps necessary to analyze and present the data from the laddering interviews.

Step 1: Reviewing the data begins during the interview because the intent and meaning of every word generated from consumers should be confirmed by the consumers. Showing the data back to consumers at the end of interview sessions is recommended to make sure that all the elements, and the links are correctly recorded.

Step 2: Content analysis requires two to three researchers with experience in coding qualitative information, but they should not be intimately involved in the study. The information will need to be transcribed and transformed to elements (words or phrases that each consumer uses) and then arranged from the lowest level of abstraction (concrete product attributes) to the highest level of abstraction (terminal values). Figure 6.1.6 is a typical coding sheet of this step. Each researcher must perform these tasks separately. Once it is finished, the results from each decoder are compared. Any discrepancies

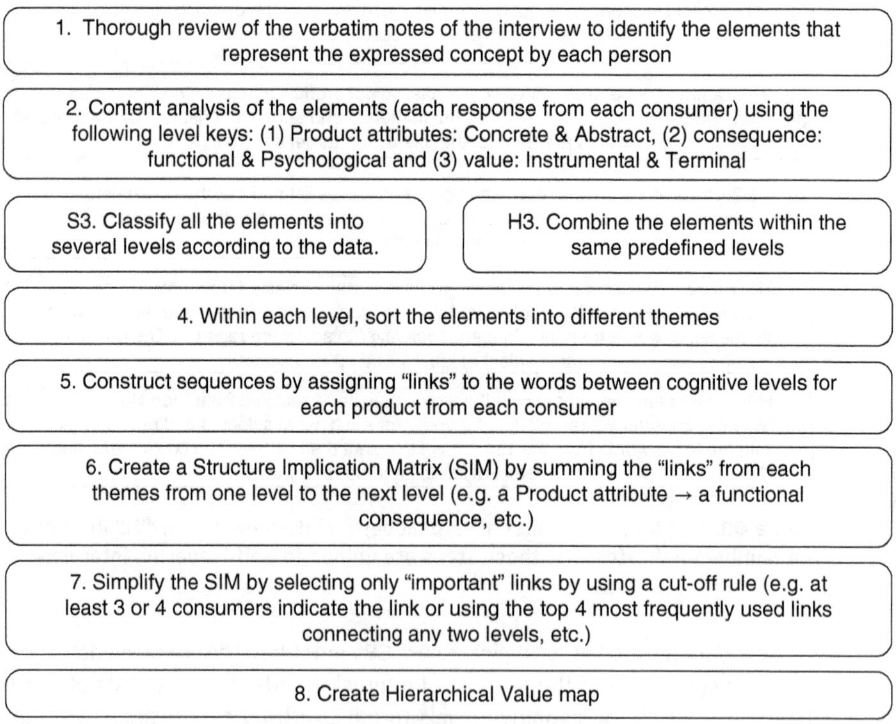

Figure 6.1.5 Steps to analyze the results from a Laddering interview. The Letters "S" and "H" in front of the numbers indicate that these steps are unique to Soft laddering interview and Hard laddering interview, respectively.

Consumer ID: ____________ Decoder: ____________
Date: ____________ Project: ____________

	Product 1: ________			
	Sequence 1	Sequence 2	Sequence 3	Sequence 4
Product attribute				
Physical	Word 1		Word 8	
Abstract				Word 9
Consequence				
Functional	Word 2			
Psychological		Word 6	Word 6	
Value				
Transitional	Word 3			Word 3
Terminal		Word 7		

Figure 6.1.6 A typical coding sheet with results after a content analysis. The results are generated from a consumer for a product by a decoder in a study.

among the researchers are usually resolved by discussion and consensus agreement. The discussion involves the assignment of the words to one of the three levels of cognitive structures (Product Attributes, Consequence, and Values) with two sublevels within each major level (Figure 6.1.6).

Step 3 (S3 & H3): (Note: this is the point where Soft laddering and Hard laddering interviews bifurcate). In Soft laddering interview (Step S3), the decoders assign any word to a cognitive level according to their understanding of the results. However, in Hard laddering interview (Step H3), the cognitive levels of words have been determined prior to the interview; therefore, the activity is combining the words within a preassigned level.

Step 4: This step is the same in both Soft laddering interview and Hard laddering interview. The coders sort the words into different themes within each sublevel (total six sublevels).

Step 5: Then the coders continue assigning "links" to the words between the three cognitive levels to construct a sequence (see the right column in Figure 6.1.6) of a product from each consumer. For example, Sequence 1 in Figure 6.1.6 comprises Word 1 (Physical Product Attribute)→Word 2 (Functional Consequence)→Word 3 (Transitional Value). The same word can be used as many times as it is dictated by consumers, for example, word 3 (a Transitional Value) is the end of both Sequence 1 and Sequence 4 or word 6 (a Psychological Consequence) is the beginning of Sequence 2 and the end of Sequence 3.

Step 6: Creating structural implication matrix (SIM) is a daunting task especially from Soft laddering interview, and it is the original reason for inventing

To (columns) / From (rows)

From \ To	T1	T2	T3	T4	T5	T6	T7	T8	T9
T1		1	0	0	0	0	0	0	0
T2	0		1	0	0	0	0	0	0
T3	0	0		0	0	0	0	0	0
T4	0	0	0		0	0	0	0	0
T5	0	0	0	0		0	0	0	0
T6	0	0	0	0	0		1	0	0
T7	0	0	0	0	0	0		0	0
T8	0	0	0	0	0	1	0		0
T9	0	0	1	0	0	0	0	0	

Figure 6.1.7 A SIM matrix from a consumer for a product based on the data in Figure 6.1.5 (assuming each Word is classified into a Theme after discussion and consensus agreement). The matrix is symmetrical in structure but not in the data as it contains consequential data (From Theme X to Theme Y). "T" stands for "Theme."

Hard laddering interview as it aids data collection and SIM creation. Figure 6.1.7 is an example of a SIM based on the results in Figure 6.1.6 by assuming that all nine words in Figure 6.1.7 are classified into nine different themes. For example:

Sequence 1: Word 1→Word 2→Word 3 that translates to Theme 1→Theme 2→Theme 3 that is represented in SIM (Figure 6.1.7) as "1" between row T1 and column T2 and "1" between row T2 and column T3

Sequence 2: Word 6→Word 7 that translates to Theme 6→Theme 7 that is represented in SIM (Figure 6.1.7) as "1" between row T6 and column T7

Sequence 3: Word 8→Word 6 that translates to Theme 8→Theme 6 that is represented in SIM (Figure 6.1.7) as "1" between row T8 and column T6

Once all SIMs from all target consumers are constructed, a final SIM is created by summing all the frequencies for each cell. This final SIM is the data to create the final hierarchical value map (HVM) for a product or target consumers.

Step 7: Simplify the final SIM by applying a cut-off rule below.

- **Rule 1: A cut-off level of 3- or 4-rule:** A cut-off of 3 or 4 means that the links are used at least three or at least four times by consumers. This rule is appropriate when the number of consumers within a target group is more than 30 (Lind, 2007).

- **Rule 2: The top-4 rule:** This rule was proposed by Russell et al (2004). The top-4 rule requires only the most frequently used 4 links that connect between two levels (e.g. Product attributes →, Consequences or Physical Product Attribute →, Psychological Product Attribute, etc.). This rule is easy to apply to Hard laddering interview data, but it is quite complicated when Soft interview data are used or with a big data set. Many network programs such as UCINET (Borgatti et al., 1999), or MecAnalyst software (Skymax Inc., Italy) can handle the SIM and provide many tools to aid the analysis.

Step 8: Creating the HVM: After a cut-off level is decided and applied to the SIMs, final HVMs can be created. Figure 6.1.8 is a simplified version of a typical HVM based on the data in Figure 6.1.7 with additional information. If the total frequency is 100, and an absolute cut-off level of three is chosen. Then every link in Figure 6.1.8 will be kept. If the top-4 rule is chosen, two links will be deleted from the data sets. The first link to be deleted is the link that connects Theme 1 to Theme 9 with 3% of the total frequency of the first cognitive level (denoted by rectangular) and the second link to be deleted is the link that connects Theme 2 to Theme 3 with 5% of the total frequency of the second cognitive level (denoted by triangle). The absolute number rule (three or four) is appropriate for Hard laddering interview because

6

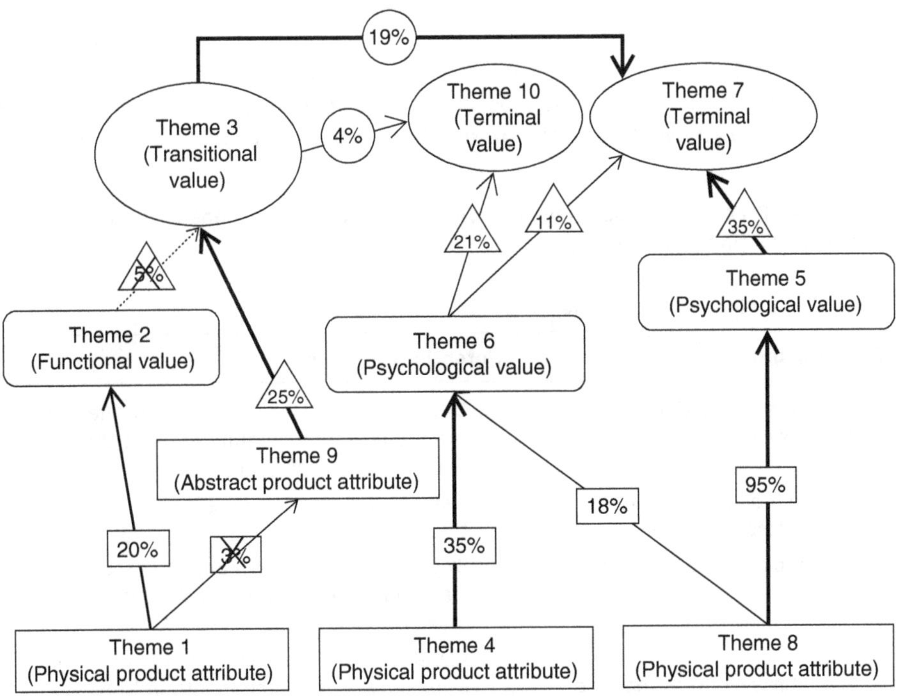

Figure 6.1.8 An example of HVMs from a Laddering interview based on the data in Figure 6.1.6 with some additional information. Links in the same level are noted with the same kind of geometric shapes. "X" denotes the links that are not the top-4 of their own level.

predetermined HVM is applied to every target consumer. The top-4 rule is appropriate for Soft laddering interview or highly complex HVMs because the rule normalizes selection criterion for all target groups by choosing only the most important four links for each association level.

6.1.4.2 Potential problems when applying laddering interview in practice

There are several potential issues when applying laddering interview in practice. Below are four most commonly encountered problems:

- Time-consuming and expensive
- Consumers make up their answers
- Biases from researchers
- Oversimplify the analyses and outcomes.

6

Since the laddering interview is an unstructured one-on-one interview, scheduling time for a project may be challenging. CEs do not want to rush the interview to just fit with the deadline and miss important insights. This issue must be considered. It is important to have highly skilled interviewers and consistencies especially for multiple site projects and international studies.

Table 6.1.4 shows potential problems and potential solutions for this time and cost issue. In addition to finding skilled interviewers and total time of a project, like with any research project, the interview can exhaust consumers, if the session lasts too long or lacks engaging structure. Therefore, making sure that consumers understand the procedure and time needed for the interview is the best way to avoid this issue. If an interviewee doesn't wish to continue, it is

Table 6.1.4 Problems and solutions for "Time-consuming and expensive" issue in applying Laddering interview in practice.

Time-consuming and expensive	
Potential problems	**Potential solutions**
1. Requires highly skilled interviewers	(a) Develop a database of skilled interviewers that you can rely on and alternate them from project to project
2. Will be too expensive for large-scale study	(b) Use computerized/on-line system to capture the interview
3. Takes too long for large-scale study	(c) Depending on research objectives, adapting Hard laddering interview in a group format may save some time (see QMA)
4. Consumers lost interest because the interview takes too long	(d) Explain clearly to consumers about (1) The procedure, and (2) the length of time needed for the interview (e) Respect consumers limitation and never force the consumers to answer any questions that they do not want to

Table 6.1.5 Problems and solutions for "Consumer made-up their answers" in applying Laddering interview in practice.

Consumers made up their answers	
Potential problems	**Potential solutions**
1. Consumers try to please the interviewer	(a) Use highly skilled interviewers with specific training for laddering interview
2. Consumers answer the question for the sake of answering the question	(b) Check for consistency of the answers from time to time during the interview
3. Consumers do not want to reveal their personal thoughts when the interview becomes too personal (happens at high abstract levels)	(c) Do not limit the answers to only words or phrases. Using story-telling or metaphor can enhance clarity of response (d) Use "pausing" to redirect the interview (e) Try "third person" probing to allow consumers to project their thought (f) Never interrupt the consumers. Ask follow-up question instead (g) Make sure that consumers feel comfortable about the their confidentiality and anonymity

6

better to let them go than coercing any responses from them. Anticipating the design before beginning the project is critical. Novice interviewers may typically design projects that lack accounting for time and attention concerns.

The second common issue, "Consumers make-up their answer," has many causes (e.g. consumers want to please the interviewer, consumers just want to get out as quickly as possible, and the interviewer asks bias or leading questions). The key to a successful interview is preventing this scenario from happening and noticing when it happens. Table 6.1.5 shows recommended potential solutions to this problem.

The third issue is from the researchers themselves as personal bias and expectation can blind the researchers from new insights that do not fit with their expectation. Even though it is human nature to see what they expect to see, preventing this bias from impacting the study is critical. Table 6.1.6 shows several potential solutions to curb this issue in practice.

The fourth issue, "Oversimplifying the results," is also rooted from the researchers, and it is due to the expectation and the complexity of data analysis for laddering interview. Turning the interview data into HVMs and insights can be viewed as a very laborious task, especially from Soft laddering interviews or inexperienced CDEs, and many researchers are intimidated by it. Table 6.1.7 shows few solutions for these problems.

Laddering interview is a powerful tool to uncover consumers' insights, especially the associations among product characteristics, consequences, and values that are perceived by consumers. Hard laddering and Soft laddering interviews provide consumer researchers some flexibility to cope with different objectives. Their ability to uncover different levels of consumer's product perception makes them suitable for a wide range of projects in NPD.

Table 6.1.6 Problems and solutions for "Biases from researchers" in applying Laddering interview in practice.

Biases from researchers	
Potential problems	**Potential solutions**
1. Personal biases influence the interview process and data analysis	(a) Audit the interpretation of the results in Step 5: Decoding the interview by peers will improve the quality of the findings (b) Accept that you have personal value system that will impact judgment and interpretation of the results and try to be neutral as much as possible
2. Researchers make the content analysis easier by over simplifying the reality	(c) Take advantage that consumers are there and let consumers validate your interpretation and ensure correct representation of their thoughts

Table 6.1.7 Problems and solutions for "Oversimplifying the analyses and outcomes" issue in applying Laddering interview in practice.

Oversimplify the analyses and outcomes	
Potential problems	**Potential solutions**
1. Interesting insights are missed because they do not agree with the preconceived patterns imposed by the technique	(a) If it possible, use more than one data collection method (e.g. computer aid, paper-and-pencil)
2. The questions are usually overtly positive attributes	(b) Compare the findings with existing literatures and evaluate them thoroughly and critically (c) Balance negative and positive questions. Converging in conclusions validate the results, discrepancy in conclusions provides more insights or indicate where you should conduct more research

6.1.5 Kelly's Repertory Grid and Flash Profiling

This section will discuss two methods, Kelly's Repertory Grid and Flash Profiling, as both methods share many characteristics:

(1) The methods are product-driven methods as both require multiple comparisons of products.
(2) Both require target consumers to be familiar with products of interest.
(3) Physical product characteristics are the results from both methods. Do not expect much about benefits or values from these methods as the purpose of these methods are to describe products.

(4) Both methods use multiple-comparison tasks to elicit the insights. Kelly's Repertory Grid uses triadic sorting and Flash Profiling uses ranking task.

6.1.5.1 Kelly's Repertory Grid technique

Kelly's Repertory Grid (also known as Kelly's Triadic sorting or, in short, Kelly's Grid) was developed by George Kelly in 1955 based on Personal Psychology Construct theory to map cognitive structures (Kelly, 1955). Applying the method to product research by using actual products as stimuli and asking consumers to describe what makes them different from each other uncovers the underlying perceptual constructs that describe the differences (Product Attributes). The Kelly's Grid procedure is designed to maximize the differences among products; hence, the method is not appropriate for detecting subtle differences and preference. Figure 6.1.9 describes general steps to conduct a successful Kelly's Grid.

The main objective of Kelly's Grid is to uncover product attributes that differentiate products in a choice set. Therefore, Step 1 is very important because the outcomes can be generalized only within the chosen product category. In Step 2, stimuli preparation and presentation must be done with care and consideration. Stimuli presentation can be a complex issue because of two commonly forgotten factors: (1) carry-over effect, and (2) use context. Products with long carry-over effects will need more recovery time (e.g. chili sauce, butter, lotion, and facial cream); therefore, the recovery period can impact consumers'

1. Define the products/product classes/product concepts of interest
2. Prepare how to present the actual products/product classes/the product concepts (generally called stimuli) to your target consumers
3. Select target consumers: you will need consumers who are familiar with the product of interest
4. For each individual consumer, present the consumer with a group of three products
5. Ask the consumer to select a product that is the most different from the other two products
6. Ask the consumer to explain what make the product selected in step 5 differ from the other two products
7. Record the descriptions elicited from the consumers
8. Continue steps 4 to 7 with different sets of three products
9. For each individual consumer, combine all product attributes elicited and ask the consumer to review the attributes and eliminate any redundant attributes from the list
10. Ask each individual consumer to evaluate the products and any additional products using the list created by each individual
11. Analyze the data using generalized procrustes analysis (GPA)

Figure 6.1.9 Eleven general steps to conduct Kelly's Repertory Grid elicitation.

memories. In addition, cleansing agents or procedures (e.g. water, cooking, and washing process) alter the physiology of modalities especially on the tongue (for food) and skin (for personal care products) to be different from the natural state under real consumption. The memory loss and modality alteration impact the interpretation and usefulness of the results. In Step 3, the target consumers must be familiar with the products of interest. If consumers are not familiar with the products, they will not be able to describe the product characteristics well and their results will be unusable due to the switching of evaluation criteria during use. This contradicts the underlying hypothesis of this method that consumers use their constructs (a set of systematic criteria) in their evaluation. Steps 4–8 describe the actual procedure of triadic elicitation. Warm-up sessions will be very useful to ensure that consumers understand the triadic elicitation. Steps 9–10 are the rating tasks where each consumer will use their own set of attributes to rate the products and any additional products of interest. Intensity scale (line or categorical) or Likert-type scale (or agreement scale) are commonly used for rating tasks. Intensity scale provides the differences in magnitude of attributes and the Likert scale provides the level of agreement that the attributes exist in the products. For product development, the author recommends to use intensity scale (e.g. a line scale or a categorical scale with anchors on both ends (None to Very Much)) over a Likert-type scale as the intensity scale provides more relevant information to product (the difference in magnitude of attributes) than the Likert scale. Step 11 is the analysis step, and we will discuss this in detail after Section 6.1.5.2 as both techniques use the same analysis, generalized procrustes analysis (GPA).

6.1.5.2 Flash Profiling

Flash Profiling was described by Professor Jean-Marc Sieffermann in late 2000 (Sieffermann, 2000). The method combines Free Elicitation (from another method called Free Choice Profiling) and a multiple comparison task (ranking) to uncover discriminating product attributes of a product set. Because of the use of ranking task, Flash Profiling has the same limitations as Kelly's grid, i.e. memory loss and vulnerability to carry-over effects. However, there are many remedies to these limitations as the author has applied this method to facial cream, soap bars, and many other high carry-over products. Figure 6.1.10 describes general steps to conduct Flash Profiling. The difference between Flash Profiling from Kelly's Grid technique starts at Step 4. In Flash Profiling, each consumer will be presented with <u>all</u> the products at once (not a set of three products). This provides consumers a holistic picture of the product set and helps them to generate the highly discriminating attributes. In Step 5, each consumer can spend as much time as he/she needs to evaluate the product set to come up with a list of attributes that he describes and differentiate the products from each other. It is important to instruct the consumers to generate "attributes that can discriminate the product set at least into 2 to 3 groups." Steps 4 and 5 differentiate Flash Profiling from Kelly's Grid elicitation. Flash Profiling provides attributes that can discriminate big differences (more obvious to consumers and most likely to be the first thing that they notice in the market); meanwhile, Kelly's Grid elicitation provides attributes that discriminate smaller

1. Define the products/product classes/product concepts of interest
2. Prepare how to present the actual products/product classes/the product concepts (generally called stimuli) to your target consumers
3. Select target consumers: You will need consumers who are familiar with the product of interest
4. For each individual consumer, present the consumer with a group of products of interest
5. Ask the consumer to generate a list of attributes that he/she thinks that describes the product set
6. Ask the consumer to explain the meaning and evaluation procedure of the attributes and combine redundant attributes, when it is possible.
7. Create customize ballot based on individual consumer's attribute list
8. Present each consumer with the same product set (additional product is permitted as long as the product is not drastically different from the set)
9. Ask the consumers to rank the product based on each attribute on his/her own ballot
10. Repeat steps 8 to 9 at least one more time to confirm the reproduce ability of each consumer
11. Analyze the data using generalized Procrustes Analysis (GPA)

Figure 6.1.10 Eleven steps to conducting a successful Flash Profiling.

differences (the results of triadic presentation) but the attributes many not be that obvious to consumers compared to those of Flash Profiling. However, both Flash Profiling and Kelly's Grid elicitation cannot discern very small differences between products (or confusable differences). In Step 6, the meaning and evaluation procedure that each consumer used to evaluate the sample for each attribute must be recorded. This is very important for the research team as it will be very helpful in explaining the results. Many consumers use the exact words to describe the same products; however, their rank orders are not the same or, in many cases, opposite from each other. This phenomenon is usually caused by using different procedures in product evaluation. Steps 7 and 8 are the actual product evaluation step, each consumer will use their own list of attributes to rank the products. Consumers are allowed to reevaluate the products as many times as they want and assign the same rank to products that they believe to have a similar level of an attribute. In addition, many improvements can be added to these steps to circumvent the memory loss and carry-over effects especially when many products are ranked. Dairou and Sieffermann (2002) conducted Flash Profiling with 14 jam samples and the author successfully used Flash Profiling for more than 20 body lotions. Below is the list of several improvements that have been used to improve Flash Profiling. Combining these improvements will enhance the ranking process for consumers.

- Using warm-up procedures allows consumers to familiarize themselves with the task. Also, the interviewer can allow them to rank the products according to their preference so the consumers do not feel the need to express their preference (which is usually the first attribute in the process) and can familiarize themselves with the ranking process.

- Allowing the consumer multiple sessions to rank helps prevent fatigue and memory loss. The consumer conducts partial ranking until he/she needs to rest to recover. Then the consumer comes back and continues the ranking process.
- Provide consumers enough physical space to arrange products: physical arrangement of products helps reduce memory loss because of visual presentation. It eases the ranking process by allowing consumers to place products physically away or close to each other according to their perceived relative intensity of each attribute.
- Use rank-rating procedure: This allows consumers to rank products and then assign ratings to all products. The rating works as a memory holder for consumers. It helps to avoid unnecessary comparisons between obviously different products for each attribute and works as a reminder for consumers to compare products that are close in intensity for each attribute.

The application of Flash Profiling in practice goes beyond just generating consumer languages. Its original purpose was to aid language generation in descriptive analysis; however, with the improvement mentioned above, direct application of Flash Profiling to naive consumers is possible. The information generated from Flash Profiling and also Kelly's Grid method is from each individual consumer. Therefore, the words and phrases generated by these methods are quite complex to understand and rich in insights about how consumers communicate their experiences with the products. A special statistical analysis technique called GPA is necessary to make sense from this rich data from Flash Profiling and Kelly's Grid method.

6.1.5.3 Analyzing and interpretation of Kelly's Grid and Flash Profiling methods

GPA is a special class of multidimensional analyses that its main goal is to "figure out a common multi-dimensional structure (revealing underlying constructs, concepts or meanings) among many multi-attributes data sets." GPA was first applied to data from Free-choice Profiling (Williams and Langron, 1984), and it has been used extensively to methods that generate this type of data (e.g. Kelly's Grid, Free-choice Profiling, or Flash Profiling) ever since. There are many statistical programs that can run a GPA such as Sensetool®, XLSTAT®, SAS®, SPSS®. These programs require different data arrangement formats; therefore, the author recommends any practitioner to consult the manual of any programs they wish to use. However, the results from different programs are quite similar, and there are two plots that are very important for consumer researchers: Loading plot and Correlation plot (Figure 6.1.11).

Loading plot (Figure 6.1.11a) shows the locations of the products in a multidimensional space (in this case, two dimensions). It is prudent to inspect the following: (1) the total variance explained and (2) the grouping of products.

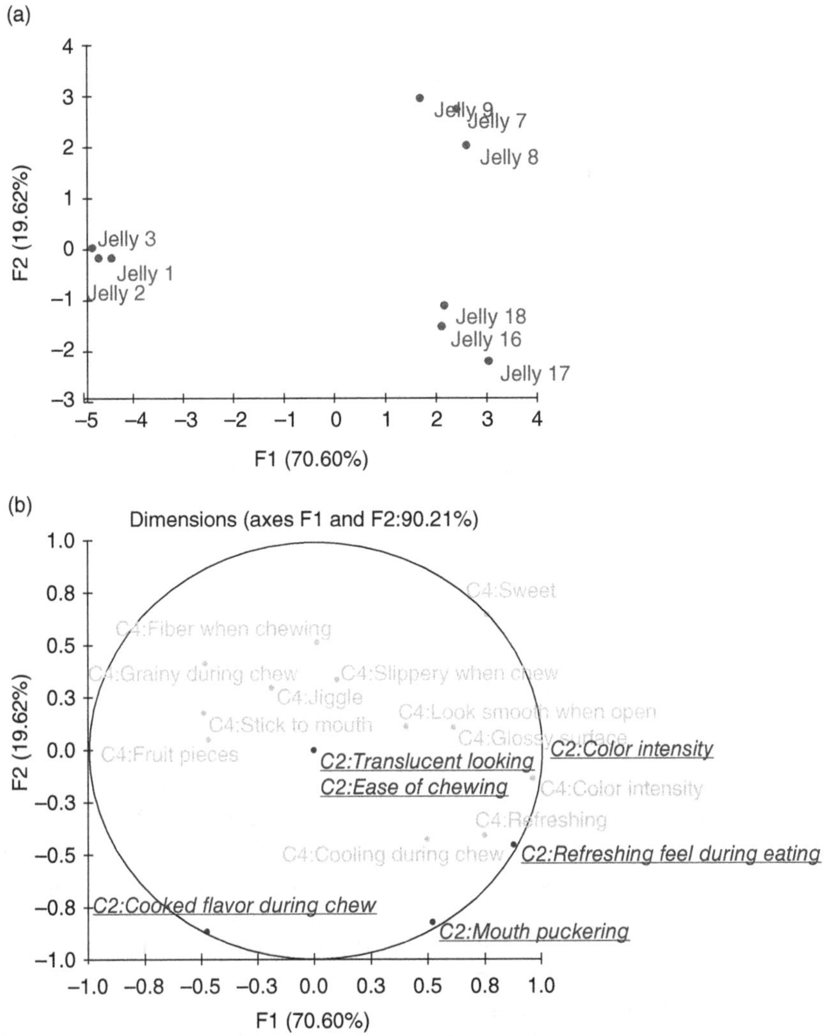

Figure 6.1.11 **Loading plot (a) and Correlation plot (b) from a GPA analysis. These two plots are important outcomes from the GPA as they provide both validation and the insights to consumer's language of products (in this case is Jelly).**

First, the total variance explained reveals the representativeness of GPA outcomes. The higher the variance is explained, the more representative is the outcome. Seventy percent or more is a good amount of variance. There is the usual question about the number of dimensions to select. Many programs have built-in algorithms to deal with this problem. In conjunction with the amount of total variance explained, inspecting a drastic shift in variance explained from one dimension to the next is also helpful. Usually, a shift of 10–15% is sufficient to use as a cut-off point to select the number of dimensions. In this case, the total variance explained is more than 90% by using two dimensions and inspecting the third dimension which contains only 6% (a shift of 13% from dimension 2 containing 19% of variance). This helps to confirm the two-dimension solution as an appropriate one. There is one cliché about these rules and techniques in selecting a number of dimensions that is "the meaning and perceived differences." If researchers know that there are certain subtle differences in product attributes that consumers can perceive consistently, but the intensity of those attributes are low, then using only "ranks" will help to avoid this problem, even though researchers may collect data using rank-rating technique or rating technique. This is the advantage of Flash Profiling over Kelly's Grid method.

Second is the grouping of the products. It is prudent for researchers to expect certain products to be grouped in the same or different group and visual inspection of the loading plot (Figure 6.1.11a) helps to confirm that. If the results do not align with the expectation, then there is a problem with the results. In Figure 6.1.11a, the author knew that there are three distinct groups of products within this Jelly category. The author intentionally put these three groups and expected the results to be distinct grouping as shown in Figure 6.1.11a. This validates the performance of the consumers. If unexpected outcomes are discovered, it is highly recommended to inspect the data carefully because there may be mistakes or very important insights hidden in the data. Checking for consumers who are very different from others is the first step as there may be segmentation among the consumers.

After inspecting the loading plot and validating the outcomes that there is no problem with the data, studying correlation plots will reveal the language that consumers use to describe their experience with products. Figure 6.1.11b shows only two consumers (C2 and C4) to simplify the plot. In this case, we can see that the two consumers are quite different in terms of what they use to describe the nine jellies. Consumer 4 uses texture (Fiber during chewing, Grainy during chewing, and Fruit pieces) and mouthfeel (Refreshing feel) to describe the jellies and consumer 2 uses mouthfeel (Refreshing feel and Mouth puckering) to describe the Jellies. It is interesting but expected to see that both consumers use the word "Refreshing" to describe the product in the same way. Also, consumer 2 reveals more that the refreshing is about the mouthfeel and not smell or flavor. Other interesting findings are that (1) consumer 4 uses the word "cooling" and consumer 2 uses the word "puckering." This insight suggests that the word "puckering feel" has some cooling aspect in it, and (2) the fact that one consumer uses texture more than the other consumer who uses mouthfeel to differentiate the nine products. If this happens when appropriate sample size is used (120 consumers or more), there is high probability of consumer segments in the market, and they pay attention to different cues (one texture and the other mouthfeel) for product differentiation.

Kelly's Repertory Grid and Flash Profiling are versatile methods for researching consumer reaction to products. Compared to other methods, Kelly's Grid and Flash Profiling may be viewed as having many advantages (Table 6.1.8). The methods are very open and dynamic yet with some structure (e.g. more numeration) to the information unlike many unstructured interviews such as Free Elicitation and Soft laddering interview, which can feel unanchored and frightening to CE's who like the feel of "numbers." In addition, both methods are applicable for both qualitative and quantitative study as the information generated is self-structured by each consumer and is captured using a computerized system with minimum or no impact on the outcomes. In conjunction with GPA,

Table 6.1.8 Advantages and limitations of Kelly's Repertory Grid and Flash Profiling compared to other consumer research methods for language discovery.

Advantages	Limitations
1. Open and dynamic	1. Too sensitive
2. Easy to conduct in a large scale	2. Prone to carry-over effect
3. Take in to account of individual differences	3. Low ecological validity in certain cases
4. Provide some statistical measures to guide qualitative interpretation	4. Does not provide the consequences of product attributes to perceived benefits and values
5. Allow hypothesis testing for product development	5. Not appropriate for disruptive innovation

both methods provide an overall and individual picture of product experiences. This allows researchers to anticipate segmentations in the market and to prepare for more costly quantitative studies. Because of the use of multiple products and GPA in these methods, statistical measures can be derived from a small number of consumers. These statistical measures, such as correlation coefficient (see Figure 6.1.11b Correlation plot) and *p*-values from Procrustes Analysis of Variance for scaling differences and structural differences among consumers, etc., increase the feeling of confidence in the interpretation of the results especially when a qualitative study is conducted. The use of multiple products has another benefit that is usually ignored by researchers that is allowing researchers to test hypotheses about products. Preparing products used in these studies in a systematic way provides invaluable opportunity to discover the links between product formulae or attributes and the impacts caused by these factors on perceived differences among the products. This insight is valuable for NPD as the methods are well suited for both qualitative exploration and quantitative confirmation.

Even though Kelly's Grid and Flash Profiling are versatile methods to capture consumer's language and elicit product experiences, the methods also have limitations. First of all, both methods use multiple comparison procedures (triadic elicitation and ranking) which are very sensitive and, sometimes, the procedures are too sensitive for actual situations. Many products are not used side-by-side or some are impossible to compare side-by-side in real life. For example, nobody has two bathtubs in the same bathroom therefore, trying to compare bubble bath samples side-by-side will be overkill. The use of multiple comparison procedures is prone to products with high carry-over effects as mentioned. Products such as chili paste, hot sauce, facial cream, or any high carry-over-effect products, require long recovery time and consumers will forget many details of their experiences. Therefore, certain adjustments in the procedure will need to be considered or considering alternative methods is recommended. This also brings up another issue with these methods: low ecological validity. If the main objective is to figure out product differences, these methods are appropriate. However, if the main objective is to figure out how consumers perceive product differences in real-use situations, these methods are not appropriate. Because Kelly's Grid and Flash Profiling require physical products and direct product comparison, the methods provide product-level information and guidance but will not provide the associations between product attributes and perceived benefits and values of using the products. Last, these two methods are not appropriate for disruptive innovation, since the methods require familiar products to make the methods work. This contradicts the definition of disruptive innovation, which is a truly new product.

6.1.5.4 Repertory grid in practice

A modified repertory grid with triadic elicitation approach was designed by Takasago International Corporation researchers (Rhodes et al., 2020) to identify how Americans would describe a Japanese taste expression called "koku." Koku is defined as the overall taste experience from the harmonious integration of aroma, chemesthesis (cooling, warming, tingling), and taste found in palatable foods and beverages (Kunieda and Mabuchi, 2016).

Four tomato juice prototypes (base [control], with taste modulator koku, with flavor, and with flavor and taste modulator koku) were presented to 23 consumers following a systematic rotation. Four warm-up samples, one of each kind, were evaluated by each consumer followed by four unique sets of triads where each prototype was seen three times. During the evaluation, panelists mentioned their likes and dislikes, placed samples on the attribute scales, and then were forced to pick the two samples that they thought were the most similar, explain why those were similar, and then explain what, if anything, stood out to them about the third sample. This approach identified sample similarities and differences and deep-seated consumer perceptions, attitudes, and feelings toward each prototype.

From the interviews, 138 constructs (or attributes) across appearance, aroma, taste, flavor, texture, mouthfeel, and aftertaste were identified and provided consumer language for product validation questionnaires.

Overall, differences between prototypes were relatively subtle and difficult to describe for some consumers. However, key differences were noted ranging in stepwise progression from the base which was viewed as plain, pleasant but unremarkable to the base with flavor and taste modulator koku which offered more "balance" and "complexity" with a strong tomato flavor and mouthwatering quality. This consumer language probe on a complex taste perception proved to be a fast and cost-effective approach with reproducible results as findings were later validated with trained sensory panels and consumer tests using language defined by the repertory grid interviews.

6.1.6 Summary and future

To have a successful new product, the product must be invented to (1) answer the needs of consumers, (2) be recognized by consumers that the product provides benefits for their needs, and (3) be different from other products in the market. To do so, identifying the needs, benefits, and values and the associations among these abstract meanings (psychological constructs) and product attributes is very critical. Free Elicitation is quick and easy to conduct for identifying these abstract meanings of products. Laddering interview provides the associations among product attributes, benefits, and values. And Kelly's Repertory Grid and Flash Profiling provide product attributes that consumers use to differentiate one product from others in the same category. Applying each method at the right time in NPD and combining the methods to generate comprehensive insights of product categories are important. Combining the advantages of these four methods in a single method is possible as qualitative multivariate analysis (QMA, see Chapter 6.2) show many traits of Free Elicitation and laddering interview with multiple product use in real-use situations. Using digital input devices will be easier and more consumer friendly as touch-screen technology is getting less expensive. New techniques based on check all that apply (CATA) tasks have been proposed, and it can generate comparable results to those of Flash Profiling and other rapid descriptive analysis techniques. Network analysis is highly recommended to deal with the data generated with these tools including the CATA tools (see Chapter 7.2). Incorporating ethnographic studies will enhance the sensitivity and resolution of the methods. Consequently, providing more enriched and more useful insights is possible with the advent of inexpensive video cameras, cell phones, blogging, and social networking technology.

Discussion questions

- How should one decide on which Language Probe approach to utilize for which stage of the NPD?
- What differentiates Soft vs. Hard Laddering?
- How might a novice CE become more experienced with consumer language?

Note

1. *Note: ZMET is a patented process. Any use of the process or elements (including neuroimaging) and other physiological techniques) is prohibited unless the user obtains a license or a written permission from the patent holder.*

References

Ajzen, I. and Fishbein, M. (1980) *Understanding Attitudes and Predicting Social Behavior*. Englewood Cliffs, NJ: Prentice-Hall.

Antonides, G. (1991) "Psychological variables in negotiation". *Kyklos*, Wiley Blackwell, 44(3), 347-362.

Bech-Larsen, T. and Nielsen, A. (1999) "A comparison of five elicitation techniques for elicitation of attributes of low involvement products". *Journal of Economic Psychology*, 20(3), 315-341

Borgatti, S.P., Everett, M.G. and Freeman, L.C. (1999) *UCINET 6.0 Version 1.00*. Natick: Analytic Technologies.

Collins, A.M. and Loftus, E.F. (1975) "A spreading-activation theory of semantic processing". *Psychological Review*, 82(6), 407-428. https://doi.org/10.1037/0033-295X.82.6.407.

Dairou, V. and Sieffermann, J.-M. (2002) "A comparison of 14 jams characterized by conventional profile and a quick original method, the flash profile". *Journal of Food Science*, 67(2), 826-834.

Eckert, M. (2020) Identifying Consumer Language and Ideals of Nuanced Samples Using Modified Repertory Grid and Triadic Elicitation. *EuroSense 2020: 9th European Conference on Sensory and Consumer Research*, 12-16 December. Live and On-Demand.

Grunert, K.G. and Grunert, S.C. (1995) "Measuring subjective meaning structures by the laddering method: theoretical considerations and methodological problems". *International Journal of Research in Marketing*, 12(3), 209-225.

Guerrero, L., Claret, A., Verbeke, W., Enderli, G., Zakowska-Biemans, S., Vanhonacker, F., Issanchou, S., Sajdakowska, M., Granli, B.S., Scalvedi, L., Contel, M. and Hersleth, M. (2010) "Perception of traditional food products in six European regions using free word association". *Food Quality and Preference*, 21(2010), 225-233.

Gutman, J. (1982)."A means-end chain model based on consumer categorization processes". *Journal of Marketing*, 46(2), 60-72. https://doi.org/10.2307/3203341.

Kelly, G.A. (1955. *The Psychology of Personal Constructs. Vol. 1. A Theory of Personality. Vol. 2. Clinical Diagnosis and Psychotherapy*. W. W. Norton.

Kunieda, S. and Mabuchi, N. (2016) The Development of "Koku" Providing Flavor. *17th International Symposium on Olfaction and Taste*, 5-9 June, Yokohama, Japan.

Lind, L.W. (2007) "Consumer involvement and perceived differentiation of different kinds of pork - a Means-End Chain analysis". *Food Quality and Preference*, 18, 690-700.

Lopetcharat, K. (2017) Deep Data Consumer Research for Deeper Understanding. *KoSFoST 2017 International Symposium and Annual Conference*, 21-23 June, Jeju, Republic of Korea.

Roininen, K., Arvola, A. and Lähteenmäki, L. (2006) "Exploring consumers' perceptions of local food with two different qualitative techniques: laddering and word association". *Food Quality and Preference*, 17(1-2), 20-30.

Russell, C.G., Flight, I., Leppard, P., van Lawick van Pabst, J. and Cox, D. (2004) "A comparison of three laddering techniques applied to an example of a complex food choice". *Food Quality and Preference*, 15, 569-583.

Sieffermann, J.-M. (2000) Le profil flash–un outil rapide et innovant d'évaluation sensorielle descriptive. *In AGORAL 2000, XIIèmes rencontres "L'innovation: de l'idée au success"*. Montpellier, France. pp. 335-340.

Szalay, L.B. and Deese, J. (1978). *Subjective Meaning and Culture: An Assessment Through Word Associations*. Hillsdale, NJ: L. Erlbaum Associates.

Williams, A.A. and Langron, S.P. (1984) "The use of free-choice profiling for the evaluation of commercial ports". *Journal of the Science of Food and Agriculture*, 35(5), 558-568.

6.2

Qualitative Multivariate Analysis

Kannapon Lopetcharat and Jacqueline H. Beckley

Key learnings

- ✓ Generating emphatic conversations with consumers
- ✓ Transforming a moderator to a facilitator
- ✓ Consumer hierarchy of needs and qualitative Kano map
- ✓ Value diagram

6

6.2.1 Consumers do not know what they want, really. Really?

Consumers do not know what they want.
Consumers have a hard time articulating what they need, want, or like.
Consumers are very fuzzy about what they want, specifically.

These are familiar expressions that we have often heard from our clients (product developers or marketing colleagues) or some of us (sensory professionals or designers who want more analytical responses) who did not hear what they had expected from a long day of focus group interview, one-on-one interview, a laddering study, or any conventional qualitative methods. However, *are these statements really true?*

It is very common for product developers or marketers to ask a Consumer Explorer (CE) to run a consumer study using X, Y, or Z method because the client is accustomed to those methods with which they have had familiar and feel positive about (through experience or positive outcomes). Yet, a good CE should not just execute a study because you know that your clients prefer a certain method over others. A good CE should contribute to his/her team by recommending appropriate methodologies and executing them flawlessly. Professor Mina McDaniel relayed the following to one of the authors (KL) when he was her student at Oregon State University:

> Every time clients call, they will tell you to run such and such test for them. It is our responsibility to listen and recommend to our clients the best or at least the most appropriate way to answer their needs.

Product Innovation Toolbox: A Field Guide to Consumer Understanding and Research, Second Edition. Edited by Kannapon Lopetcharat, Dulce Paredes and Jacqueline H. Beckley.

Let us come back to the three statements. There are two possible explanations for those statements:

- Did consumers really not know what they wanted? or
- Were the methods that the researchers used inappropriate?

Most of the time, it is the latter, when it comes to product innovation, especially disruptive innovation. Cammie Dunnaway, CMO at Yahoo! Inc. and Sergio Zyman, former CMO of Coca-Cola, thought that running focus groups was a waste of time as they never gained any useful information from them, especially for product innovation (Moskowitz and Gofman, 2007). Clancy and Krieg (2000) took aim at most aspects of classic market research in their book *Counterintuitive Marketing*, but in particular suggested that the traditional focus group was part of "death-wish research." These are yet a small sample of the evidence of using an inappropriate tool as the name focus group already hints that it needs *focus*. Therefore, if the client wants to know about unarticulated needs or subconscious insights, they will not find them by using a classic focus group format (Zaltman, 2003) or any method that utilizes interview guides in which it is assumed that the answer to the question is already known. The idea of getting people together and "forcing" them to think deeply around unarticulated and subconscious behaviors (i.e. no deep thinking) is rather crazy. How could that ever work? How can one bring out the unarticulated needs and subconscious insights from a consumer's mind and visually present them to the outside world without a lot of fuss and biases from researcher's and team's agendas? One answer lies in a protocol called qualitative multivariate analysis (QMA).

6.2.2 Introduction

Conventional consumer and marketing research processes can be quite robust. Many classic tools used (like quick Internet surveys via a range of vendors like Survey Monkey and Pollfish or many forms of central location test [CLTs]) can be lacking in some core implementation areas:

(1) Ability to reflect empathy toward target consumers
(2) Tools that appropriately unlock consumer's behavior and needs
(3) Use of relevant questions due to benchmarks/tradition or an incomplete picture of target consumers' views of product or situation.

These three important factors are interrelated (Figure 6.2.1) since one component is a basis of the others. To gain empathy toward target consumers, one must have the right insights that can be discovered through using appropriate consumer research tools. This depends on the objective/s of the study and on the research team's desired level of empathy toward the target consumers.

Solving these three issues early at the beginning of new product development (NPD) guides developers to innovate into the right direction quickly and to enhance the success of their innovations. QMA can provide solutions to these three critical issues with speed and accuracy.

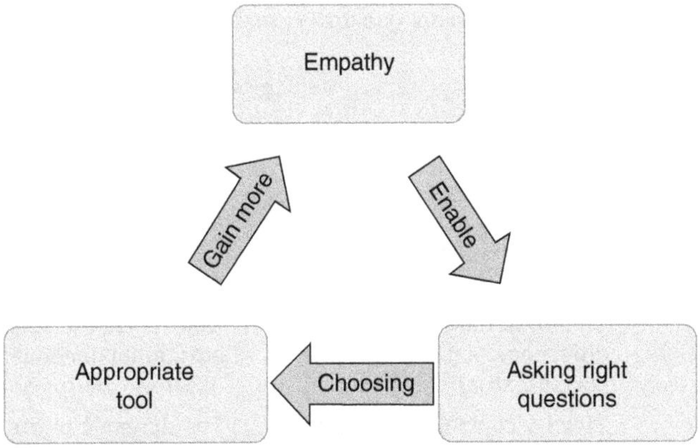

Figure 6.2.1 Empathy, asking right questions and choosing appropriate research tools are interrelated and support each other.

QMA consists of multiple steps, including contextual evaluations (home-use and more), a group discussion, and projective mapping (often referred to as tablecloth mapping or napping techniques) with a small number of individuals (10–15). Many published studies have applied some aspects of QMA, especially the projective mapping part successfully (Nestrud and Lawless, 2008; Perrin et al., 2008; Risvik et al., 1994). QMA incorporates all these steps purposefully to uncover different aspects of consumer insights.

6

6.2.3 Qualitative multivariate analysis in practice

QMA is an alternative consumer research protocol that was created to integrate unconscious behaviors, link them to more explicit insights from people and to discover the possible linkages between different and important values of products and use behavior (Drake et al., 2009). Jacqueline Beckley (one of the editors) from The Understanding & Insight Group invented QMA to take advantage of conventional qualitative tools (e.g. group discussions, one-on-one interview) while also avoiding many biases that can surround these methods. Practicing individual and group discussions without appropriate considerations for the situation and the individuals involved is believed by Beckley to be a large source of the lack of trust and reliability some find with "classic" focus group systems. Table 6.2.1 summarizes how QMA avoids or takes advantage of these biases.

Moderator's skills required for conventional research methods differ from those required for consumer-insight exploration research. The former is suitable for fine tuning assumptions (*known*), while the latter requires capability for exploration (*less known, unknown*). For example, skilled focus-group moderators are trained to probe and focus according to the objectives of the study (a priori determined assumptions); therefore, the moderators need to stay on task and create bases of understanding around what has been asked. This need for focus can be a barrier to hearing, seeing, and feeling what might be of actual consumer importance on a topic. For QMA, the protocol is designed to be open and

Table 6.2.1 Biases in conventional qualitative methods and how QMA avoids and takes advantage of the biases.

Biases	Focus group	QMA
Moderator's skill	Too structured and may be oblivious to actually important consumer's insights	Panelists lead the conversation and moderators are trained to detect insights, loosen command and control
Group composition	Cannot handle too much difference (e.g. age, income, social status)	QMA takes advantage of this emotional dialogue between different groups of panelists
Group dynamic	Strictly controlled by heavy probing and group composition; consequently, consumers will over rationalize their answers	The dynamic of the group is a part of the outcome as the topic of discussion is generated by the group, lead by stimuli and experience, not the leader
Personality of panelists	Outcomes are easily influenced by an alpha panelist or too many quiet ones	QMA makes sure that all panelists will have equal opportunities to express themselves through many activities
Moderating guide	Moderating guide limits panelists' true opinion as they will response to researchers' priori agenda	Panelists will guide the topics of discussion by themselves so no priori agenda to follow. Anchored in business-driven hypotheses

a QMA facilitator is trained to detect what are important to individuals through use of multiple cues (body, face, language, and tone) and stimuli and the ability to turn over apparent control of the discussion to the group of participants. The term *facilitator* will be used instead of moderator for the innovative qualitative tools for innovative consumer-insight exploration highlighted in this book. The rationale for the title of facilitator suggests that the group is driving the direction of the session which is enabled by stimuli and other tools, rather than modulating (moderating) a discussion.

Group composition is another potential problem for conventional methods, especially those that require group activity. The results of the study are easily influenced by the composition of the panelists (e.g. an alpha-panelist [a person that dominates the opinion of the group], a quiet majority [those who do not express their opinion], age differences [young vs. old panelists], social status differences [different income levels], etc.). Without control from moderators, the focus group discussion can turn into an argument or the participants may misrepresent their thoughts (lie) to save face, thereby leading to lost discussion opportunities around the topic at hand. This is not really a problem for QMA since it takes advantage of this emotional dialogue between different groups as these considerations are accepted as part of the results from QMA due to the grounded experiences of each panelist. Group composition also influences the group dynamic. To control the discussion, conventional methods such as focus group interviews rely heavily on probing activities that are assumed to be relevant, like frequency of use and typical market needs, such as consideration of choice in making a purchase. These specific questions may be detrimental to finding true and important consumers' needs and wants because consumers will overly rationalize their answers in an attempt to assist the moderator in achieving the discussion goal. In QMA, the dynamic of the group is a part of the outcome as the topic of discussion is generated by the group using subtle framing and positioning of a key discussion starter. Given how dynamic QMA sessions are, they clearly require the CE to have a good understanding of their personality and the personal interactions they may or may not be able to manage.

Personality of panelists is a very important aspect in group composition. As aforementioned, an alpha-subject can ruin the whole group outcomes. In the same token, too many quiet panelists can negatively impact the outcome as well. QMA was developed to make sure that every subject will remain engaged through the use of many activities (i.e. diary, round-robin discussion, note taking, voting, and map building) and prior/recent experience which establishes everyone to participate equally. Moderating guide or "the guide" influences the results of the study through moderators and their skills. Moderators have to follow the guide to deliver what their clients need (generally listed in detail in the guide); consequently, it does not leave much room for exploration and diverging from the guide when actual important insights are discovered. QMA does not really use any moderating guide. In fact, participants will guide the topics of discussion by themselves - creation of how to frame the research problem so as not to lead the group to what a "right answer" might be and understanding the three to five key business needs represent the "guide" created for QMA.

To avoid and take advantage of these shortcomings of conventional methods, QMA provides eight steps to real insight discoveries (Figure 6.2.2).

Step 1: Set up the objective(s) of the study

Great research starts with good objectives (a.k.a. right questions or first-order questions) as mentioned earlier in this book. In practice, especially during the insight discovery phase, this is actually hard to achieve because there are many objectives or requests from the research team and all of these needs appear to have equal importance. QMA accommodates this problem easily as the objectives for QMA study can be as simple as understanding consumer habits or as refined as understanding which package is good for a specific product in a specific format. QMA handles this through collecting all hypotheses from the team during the planning process. QMA answers the hypotheses through observing consumer behavior and, consequently, QMA provides not just answers but also reasons behind the answers. For any hypotheses, if consumers do not mention them, the topics are most likely not that important for them or they are not aware about those needs/wants/values. Because QMA provides both answers and reasons, it helps the team to identify any topics as opportunities (a.k.a. unarticulated needs) or insignificant needs.

Step 2: Identifying the product set

To discover insights about products, comparing many products is the best way for researchers to really understand what drives consumers' judgments

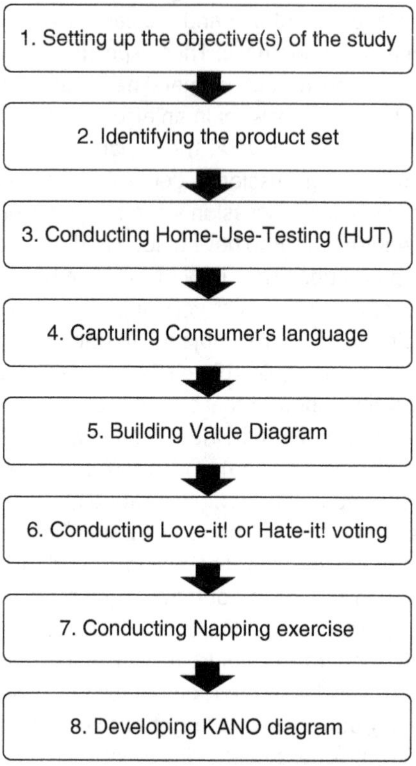

Figure 6.2.2 Eight steps in conducting a successful QMA study.

and provides the researchers true consumer insights (Moskowitz and Gofman, 2007). Testing one product with many questions (a.k.a. monadic testing) will not give one of those insights. QMA unlocks hidden insights through actual product evaluation in real-life situations. However, which products should we give consumers to use? In contrast to many other qualitative methods that utilize a small set of products that cover a small portion of a product category, QMA requires the use of products that cover the product category as much as possible (e.g. sensory properties, benefits, and packaging). Figure 6.2.3 shows the three criteria that should be considered to select products for any QMA study.

First, the most important objectives must define the aspect(s) of the products in your study. For example, if the main objective is to understand the impact of vanilla flavor in ice cream, then the products should cover a very broad range of vanilla ice creams. Different textures (even better if at the same flavor types as the flavor series) should be secondary in this example since the flavor issue has been positioned to be superior. Second, the selection should be systematic as much as possible. While not required by the method, in one of the authors' (KL) experiences, a profiling technique (e.g. descriptive analysis, Flash Profiling) plus experimental designs can be used to guide product selection to accommodate many objectives effectively and to represent any larger quantitative study such as category appraisal study.

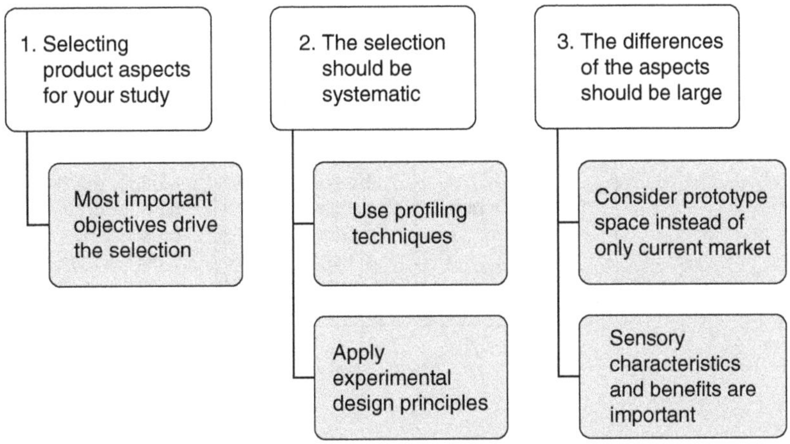

Figure 6.2.3 Three critical criteria to select samples for a successful QMA.

6

In addition, systematic selection allows researchers to interpret and understand the results easily. Having a mindful rationale for the inclusion or exclusion of each product sample facilitates the listening and understanding function within the session.

Third, the differences of the product aspects of the study should cover the whole category from an R&D perspective in addition to marketing/business requirements. In many cases, scent, taste, texture, color, and other sensory experiences with products (a.k.a. sensory aesthetics in personal care category or benefits) drive preference/choice behavior (both liking and disliking) and ultimately make consumers *love* or *hate* a product. Even though the main objective may not be about sensory aesthetics, it is recommended to consider these differences in product selection.

Step 3: Contextual experience (may be home-use-testing [HUT]) phase
This phase is the critical step as it influences the outcomes, study time-line, and cost. Unlike other conventional methods, QMA requires many products to be used by individuals in real situations for an appropriate period of time. Therefore, the length of the context phase depends on the objectives of the studies, the nature of the products, and the number of products selected from Step 2. To estimate the length of the context phase for a product category, researchers follow the steps in Figure 6.2.4. First, researchers have to define a use occasion. A use occasion is how a product is used and experienced regularly by people (this should be a realistic assumption and guided by the important objectives). This step is very critical as it defines *the length of time per use of a product.*

Some product use occasion and experience can be as short as a few minutes, such as eating an ice cream cone, or as long as a few hours, such as using a lipstick. Then the researchers multiply the use occasion and experience by three or more to get *the total length of time needed to use a product.* Finally, multiplying the total length of time to use a product with the number of products in the set provides *the total length of time needed for the whole*

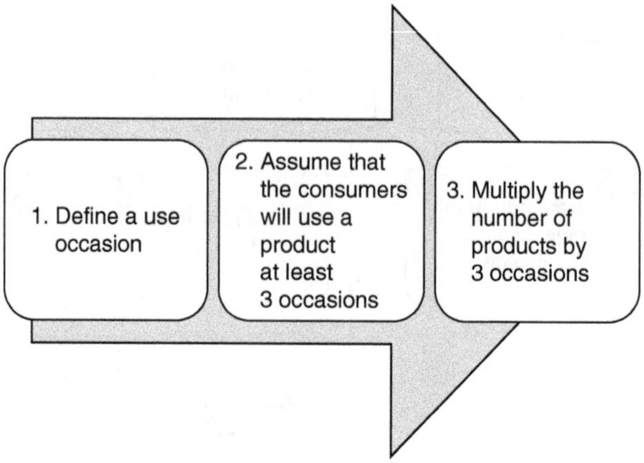

Figure 6.2.4 Steps in determining the length of HUT phase in QMA.

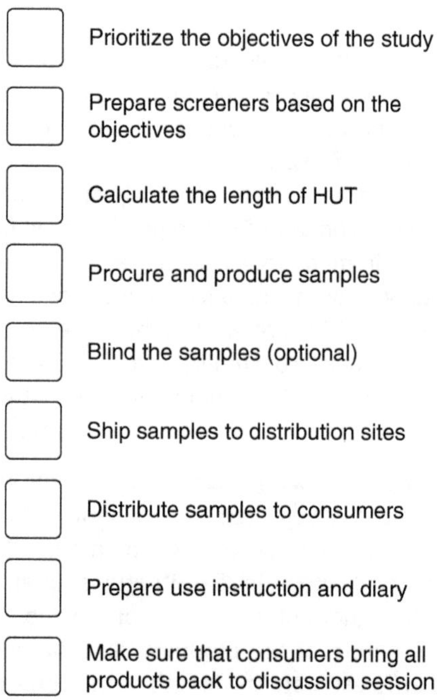

Figure 6.2.5 Important steps to conduct a successful HUT for a QMA study.

study. Behaviors that occur during this context phase assist in better understanding of both implicit and explicit experiences that the participants have at this phase.

To conduct a successful context experience/HUT for a QMA study, the following is the checklist for major steps in HUT (Figure 6.2.5). The details for managing and conducting a successful HUT will be further discussed in Chapter 11.

The nine steps in Figure 6.2.5 are interrelated, as each step will influence the others in terms of the length of the study and cost. The objectives will guide the screening of people needed for the study and the calculation of the length of this phase. The recruiting should start at least two weeks before the product pick-up date; therefore, the screener will need to be approved by the team/client about a week before recruiting starts. In addition, estimating the length of the HUT and the amount of product needed for the HUT is critical. The amount of the products must be in excess (at least twice of the estimated amount) because there always will be *other ways* to use products. Limited product may limit what is learned through the QMA. Procurement and production of samples usually takes about one month for all marketed samples and about two months for created samples with proper clearances. Proper sample packaging will need to be agreed upon by the team also. Blinding product is another issue as it impacts the packaging selection. This step is optional depending on the objectives of the study. Usually, when the study involves a prototype at a very early stage in NPD, the samples should be blinded; however, if the study is at a late stage in NPD and/or the package is part of the benefits of the product, then blinding the sample may be unnecessary or may impact the context of the experience. Shipping samples to the distribution/test site is very important for any product testing, especially any international projects and it will be covered separately in Chapter 12. Distributing samples to consumers and requiring them to sign a confidentiality disclosure agreement (CDA) are very important. Researchers must make sure that consumers sign a CDA before the consumers touch any sample. At this point, consumer orientation to the HUT and QMA should be done, including how to fill out the diary (Figure 6.2.6). The diary that will accompany the samples should have pictures of the samples with corresponding three-digit

> *"There are* XXX *samples in this bag. You have* YYY *days* to use these samples.
>
> Use these samples as you regularly use them. **You MUST use each sample at least once.**
>
> If you do not want to continue to use any sample after you use it once, please do so and explain the reasons why you do not want to continue using the sample in the diary provided to you in this bag.
>
> If you want to continue to use any samples more than once, please do so and describe your experience and the reasons why you want to continue using the sample.
>
> You will use the diary to discuss about your experience in a following group discussion, so please record as many details as possible.
>
> Again, **you MUST use each sample at least once.** Please feel free to start with any sample."

Figure 6.2.6 An example of a typical QMA HUT diary use instruction.

random codes and have plenty of space for consumers to write down their experience.

Usually, the authors prefer a blank page (half of an A4 paper) on which consumers write their experience, with the product picture and identifying code that are prominently displayed on the top-right corner of the page. Preference is to keep as limited a structure to the questionnaire as possible. If more information is needed (and we strongly recommend against this practice due to a desire to see how the consumer shapes their thinking and experience), a short list of questions (should be less than three questions in total) may be asked to provide a few structured pieces of information. During the orientation for the context experience/HUT, researchers need to make sure that consumers bring ALL samples back to the QMA discussion session. It is common that consumers transfer the samples from the test containers to their own containers before returning the samples back to the study coordinator. However, it is critical for QMA that consumers bring all the samples back to the discussion session as they will have to use the samples during the session to be reminded about their experience. On occasion, following the context experience, some people will not return for the QMA session. This can happen with items that are of high value or have long-lasting qualities (e.g. facial cream, electronics, and certain packaged foods). The most efficient way to handle this situation is linking the return of samples to payment and also including possible charges when the samples are not returned.

Step 4: Capturing consumer language

QMA captures words used to describe the product's characteristics and those that differentiate products that are loved vs. those that are hated in a very natural way; generally, the practice is group discussion without any discussion guide (a.k.a. live-conversation). The QMA leader/facilitator must *listen carefully, intently, intentionally* with a listening partner (at least one other person who is the silent leader who is tracking all conversation in real time [context integration]) must record all the details carefully. The discussion can be done in three easy steps (Figure 6.2.7). Start by telling participants to bring out their products since this step will start the people talking to each other. Usually, they will start to talk about their experience right away.

The facilitator should ensure that the participants do not influence each other's opinion at this point. It is helpful to have a small overview or a brief orientation for the QMA participants regarding what their job is and the job of the facilitator and the listening partner. Brief introductions and ice-

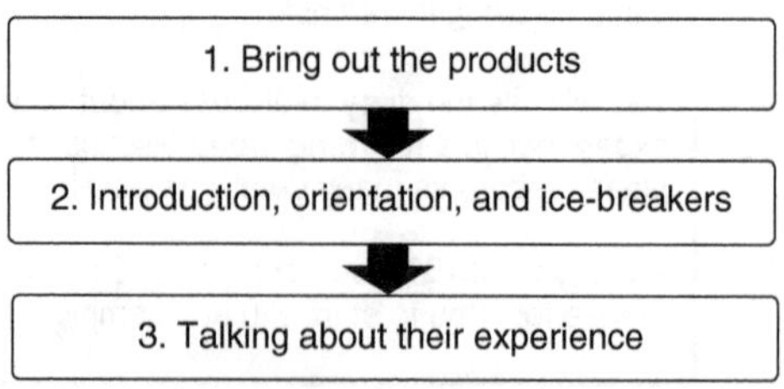

Figure 6.2.7 Three steps to start capturing consumer language in QMA.

breakers can be used since the participants will be together for a few hours and will eventually get to know each other. The goal of this step is to explain the objectives of the study, to ease any tension among the individuals and to ensure that their opinion counts and there will be no right or wrong answer. Also making sure that all participants are open-mined and respect other's opinions is very important and the facilitator must be very attentive to this throughout the rest of the session. It is common for consumers to think that their experience with a product is a shared experience and preparing them for the fact that we might experience the same product differently, is essential. Finally, the discussion about their experience with the products begins (generally no more than 15 minutes after starting the session). The diary is very useful here as the participants can use it as a reminder of their experience as there may be a length of time between the first product they used and the last product. Figure 6.2.8 shows key questions to be mindful of both as a facilitator and the listening team. Asking the questions and recording an individual's expression (e.g. language, phrasing, demeanor) are only half of the findings. Probing to understand the "relationship" among the expressions is the other important half of the findings. This leads to the next important activity, value diagram building.

Step 5: Building the value diagram

Value diagrams show the relationships among different levels of a consumer's values. It is the initiation of linking the casual language of the consumer to a hierarchical or laddered understanding of how they make sense of the product experiences they have had. The value diagram in QMA serves as the representation of the common experiences that everyone participating in the QMA has had. It will be referenced throughout the remainder of the session and helps to identify how many different individual experiences have been encountered with the products (called idiographs). Keep in mind, these values are discovered and captured through consumer's actual experience with product use in real situations. Consumers are asked to provide the tangible anchoring points for the product experience (called elements). These are the factors that are highly tangible yet general to the product experience and can be controlled by the client company.

- [ ] Why did they like a product and use a product again?
- [ ] Why didn't they want to use any product again?
- [ ] Which occasion did they use the product and how?
- [ ] What values did experiencing the product bring to them: functional, quality, monetary, emotional, safety, social, ethical, etc.?

Figure 6.2.8 Four questions that a QMA moderator and listening partner need to capture during the discussion.

People/consumers usually express their experiences through the use of sensory descriptors (i.e. taste, scent, sound, appearance, and feeling) and middle-level cognitive descriptors, phases or examples (e.g. fun, easy to use, pleasurable, remind me of . . ., this feature assures me that it is closed tightly). The defined words they use to describe their experience are usually the basic or top-of-mind level of their understanding and it is relevant to the actual product or situation. More complex descriptions (e.g. phases, examples, pictures, or additional products that they bring to the session) indicate higher level of experience for their context. The facilitator's function at this stage is to gather enough insights to ensure clear explanation of the relationships being described. Probing, if it needs to occur, keys off of the consumer conversations and *is not* driven by a prescribed list of questions which assume specific answers (Moskowitz et al., 2006).

Step 6: Love-it! or Hate-it! voting

Love-it! or Hate-it! voting forces the consumers to make decisions and brings structure to the broad QMA process. The voting is intentionally done right after value diagram building because it enables the consumers to articulate the reasons why they love particular products instead of getting a particular answer "I just love it." Remember the ultimate goal of QMA is to *understand the reasons behind the love* and *not only to find which sample is the most loved*. To conduct the voting, researchers, follow the simple steps (Figure 6.2.9).

Once the participants declare their preferences toward the products, they will know who agrees and disagrees with them. By asking them to explain the reasons behind their choices through group discussion for each group of individuals who love the same product and who hate the same product, the process focuses on identifying the trade-off of factors that ultimately lead to values.

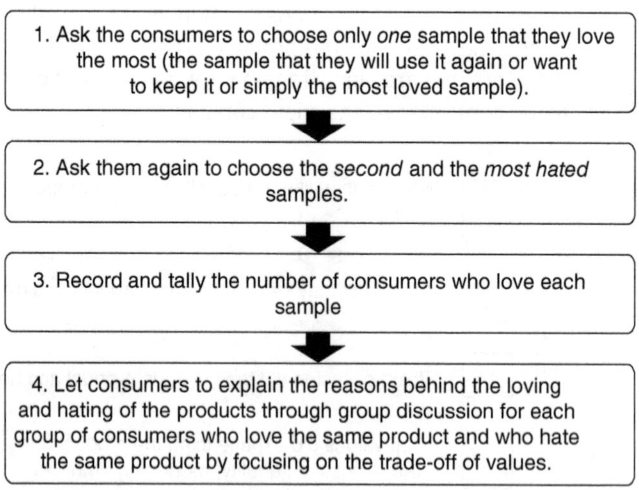

Figure 6.2.9 Four steps to conduct a Love it! Hate it! voting.

Step 7: Mapping/napping exercise

This activity is like the other side of the coin for Love-it! Hate-it! voting because it focuses on figuring out the order of samples along the most important values or characteristics (mainly the most two important values). While the QMA mapping method has been in place since the late 1990s, articulation of this approach was documented in 2003 by Professor Jérôme Pagès (2003, 2005) as an alternative means to other sample classification methods. Once the participants declare their preferences toward the products, they will know who agrees and disagrees with them. By asking them to explain the reasons behind their choices through group discussion for each group of individuals who love the same product and who hate the same product, the process focuses on identifying the trade-off of factors that ultimately lead to values. The whole sequence of factors linked to consequences and ultimately values provides prioritization of elements in the value chain for a product or service and ultimately leads to the sequencing of elements to provide a meaningful experience. This is what is required for the product designer or developer to link into the marketing promise and that which is understood most completely by the consumer of a given offer (Figure 6.2.10).

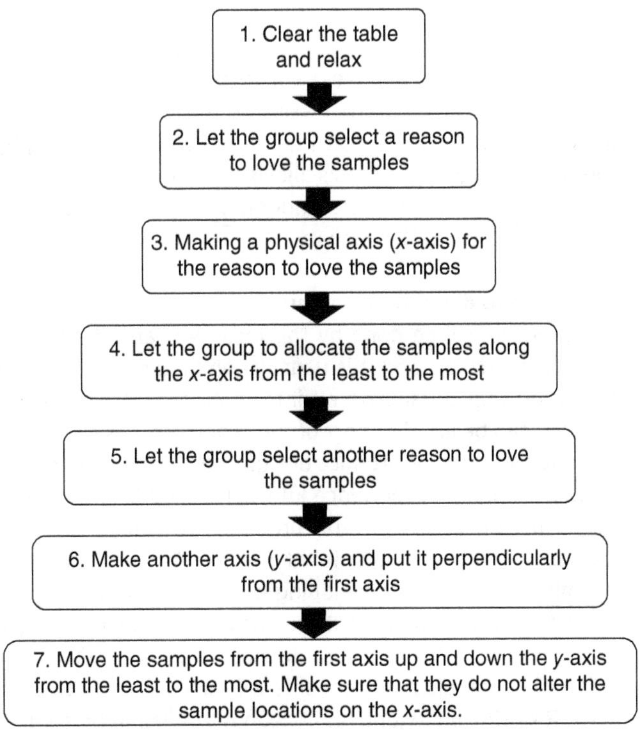

Figure 6.2.10 Seven steps to conduct mapping/napping exercise.

This reason can be anything such as a characteristic, a value, or a phase that describes what will make them love the products. The value diagram and Love-it! Hate-it! voting will be very useful at this stage as the individuals know how to articulate their opinion or at least give some examples. Then the facilitator will create a physical line (x-axis along one side of a table (masking tape is great for this purpose) and mark one end with the most *positive* side of the reason and the other end with the most *negative* side of the reason selected previously (Post It™ note is great for this purpose since the note is nonpermanent and highly flexible).

Next, the facilitator has the group allocate the locations of the samples along the x-axis from the least (close to the negative end) and the most (close to the positive end). Noticing and recording the interactions between the panelists is as important as the results. If the group has disagreement about the location of a particular product, the group can put the product in two or more places. This indicates a possibility of segmentation in the market. It is critical to *probe* the *why* of the differences of opinion and *observe* the interaction between the consumers as rationale is provided. If the interaction heats up, it means this is very important for them or at least a fraction of them (potential segmentation, potential factor which has high emotive consequences). The facilitator needs to make sure that the discussion will not turn into an argument, but rather keeps the dialog open to allow for understanding why the differences or the agreements occur. Once the group finishes allocating the samples along the x-axis, they continue with setting up the y-axis following the same steps (identify the *second* reason, make a line and allocate the samples). One restriction for this step is that the y-axis will be perpendicular to the x-axis and the group can only move the samples (all the samples) up and down the y-axis without altering the samples' locations on the x-axis. (Note: this allows the complex relationship of an intersection of two elements to be understandable to almost all people we have worked with during this activity.)

6

Step 8: Kano diagram development

This exercise will be done by the facilitator and the team who are present during the session, and it is usually conducted after finishing the consumer discussion and organizing all information collected. (Alternatively, if time allows, it can be another part of the group process and allows the consumers to apply the rules of product design that they have articulated throughout the understanding experience.) If the leadership approach is taken, consumers will not be involved in this process at this point. This should be done as soon as the session is completed as the memory is still fresh in the researchers' mind. The goal of developing a Kano diagram is to classify the product benefits, features, and attributes using participants' languages discovered throughout the QMA process into classes according to Kano's philosophy.

The Kano satisfaction model is a brainchild of Professor Noraki Kano of Tokyo Rika University and his colleagues in Japan that was introduced to the world in the 1980s (Kano et al., 1984; Berger et al., 1993). The Kano satisfaction survey was used as an alternative from an importance rating task (Berger et al., 1993), since it is difficult for consumers to assign relative importance to product attributes directly (Stelick et al., 2009). Professor Kano believed that the relationship between need fulfillment and consumer's satisfaction and dissatisfaction is not necessarily linear. The Kano satisfaction survey uses dual-questioning format with a predefined classification rule (a.k.a. Kano classification rule) to unlock the relationship. Kano survey format, its classification rule and more details can be found in Chapter 8.1.

There are three classes that are very important for product designers/developers:

(1) Must-have (that which defines the product and without which dissatisfaction occurs)
(2) Driver/optimizer (those factors that then follow the "more is better" philosophy)
(3) Delighter (the "wow!" component that can excite when present but can go unarticulated by people since its absence is only realized once it has been identified, e.g. the unexpressed want). To identify which product benefits/features/attributes (will use attribute from now on) fall into which types, the researcher can find the answer at different stages in the QMA process (Table 6.2.2).

- Must-have attributes are usually found during value diagram or Love-it! Hate-it! voting sessions by focusing on the reasons why the consumers dislike the least liked product and the reasons can be interpreted in-depth by understanding the value diagram.
- Driver/optimizer attributes are generally used for guiding sustainable innovation of a product category and the attributes can be found during the mapping/napping exercise.
- Mapping/napping exercise was designed to let consumers articulate the driver attributes by arranging the studied products along important dimensions. The driver/optimizer attributes are usually contained as parts of the axis descriptions.

6

Table 6.2.2 How to find KANO attributes from QMA and how to apply the leaning in NPD.

Type of attribute	What it is for?	Where to find it in QMA?
Must-have	The cost of entry of a product category	Love-it! Hate-it! voting: Focusing on the reasons for the most disliked product
Driver	Used for guiding sustainable innovation of a product category	Napping exercise The descriptions of the dimension are usually the Driver attributes
Delighter	Unarticulated needs Used to differentiate the products from others and hook consumers	Love-it! Hate-it! voting Napping exercise Focus on emotional reaction and dialogue among the panelists

- Delighter attributes can be found from value diagramming, Love-it! Hate-it! voting and the mapping/napping exercise. However, delighter attributes are hidden in behavior and emotional aspects of the results (this is the point where those with high levels of expertise in group dynamics/product design/and observation separate themselves from the rank and file "group moderator"). The most debated and discussed aspects during both exercises are most likely to be delighter attributes.

Using the information from the value diagram in conjunction with the observation in the Love-it! Hate-it! voting and the mapping/napping exercise, researchers will have a deeper understanding of why certain attributes are more important to consumers than other attributes, how they relate to each other and where the trade-offs intersect for different segments of individuals.

The qualitative Kano benefits or attributes discovered through QMA are a good starting point for a quantitative Kano survey that will be explained in detail in Chapter 8.1.

6.2.4 Qualitative Multivariate Analysis in practice: deeper understanding of cottage cheese consumption

At this point, the readers should get some idea about QMA and its process. In this section, we will discuss the application of QMA in understanding US consumer's consumption of cottage cheese in more detail.

In 2008, one of the authors (KL) received an e-mail from Professor Mary Ann Drake about a new project concerning cottage cheese. Cottage cheese is one of the profitable processed dairy products because of its short production time. However, the market for cottage cheese had been shrinking due to other dairy products, especially yogurt. In the beginning, Professor Drake had indicated that she wanted to conduct a focus group interview followed by a consumer taste test study; however, the author (KL) suggested trying a QMA approach instead. Since a primary objective of the research was to learn more about cottage cheese consumption behavior, the QMA approach could provide more insight into the behavior.

In the beginning of our research, we (KL, Professor Drake and her graduate students) brainstormed and discussed what was known/unknown about the consumption of cottage cheese. We concluded that consumers eat cottage cheese mainly for breakfast or snack by eating it as it is or with jam. We thought that the sensory characteristics of plain cottage cheese should be the most important things for consumers and the variety could be manipulated by providing different flavors of jams. We estimated that cottage cheese consumers will consume cottage cheese about twice a day maximum (for breakfast and snack). Hence, four regular cups were deemed adequate for the HUT portion of the QMA. Detailed information on this research can be found in Drake et al. (2009). These assumptions turned out to be quite far off from actual behavior observed through QMA as you will see later.

There are a few important issues about cottage cheese: (1) Cottage cheese flavor changes as the cheese ages and (2) It tastes best when eaten fresh.

Therefore, the research did not allow packing a big box of cottage cheese with 32 cups of cottage cheese inside and expecting that the cheese samples would not change during the HUT period. So the team decided to run the study locally where consumers could come and pick the samples up at our facility every two days. It is worth noting that, with perishable products, it is very critical to consider this aspect as it will influence the way the HUT will be tested and the screening of the participants. So the samples were made fresh on that day, and we gave them plenty of samples (much more than four cups that were estimated). With eight samples in the sample set and two days for evaluating each sample, this cottage cheese study HUT phase should last 16 days. Actually, it lasted 24 days as it covered weekends. Every time consumers returned to pick up a new sample, we were amazed that they actually used up most of the cottage cheese samples that they liked. This was unexpected as we provided them in excess of the amount from our calculation. We figured out the reasons during the discussion about their consumption habit that was not what we anticipated.

The value diagram in Figure 6.2.11 was an important outcome of the cottage cheese study. The hypothesis that cottage cheese was consumed mainly for breakfast or a snack that suggests that it was consumed with minimum alteration (e.g. adding jam or fruits) was disproved as consumers consumed cottage cheese as dessert, side dish, and salad dressing in addition to breakfast and snack. The QMA study revealed that convenience and health benefits are the major drivers of cottage cheese consumption (Figure 6.2.11).

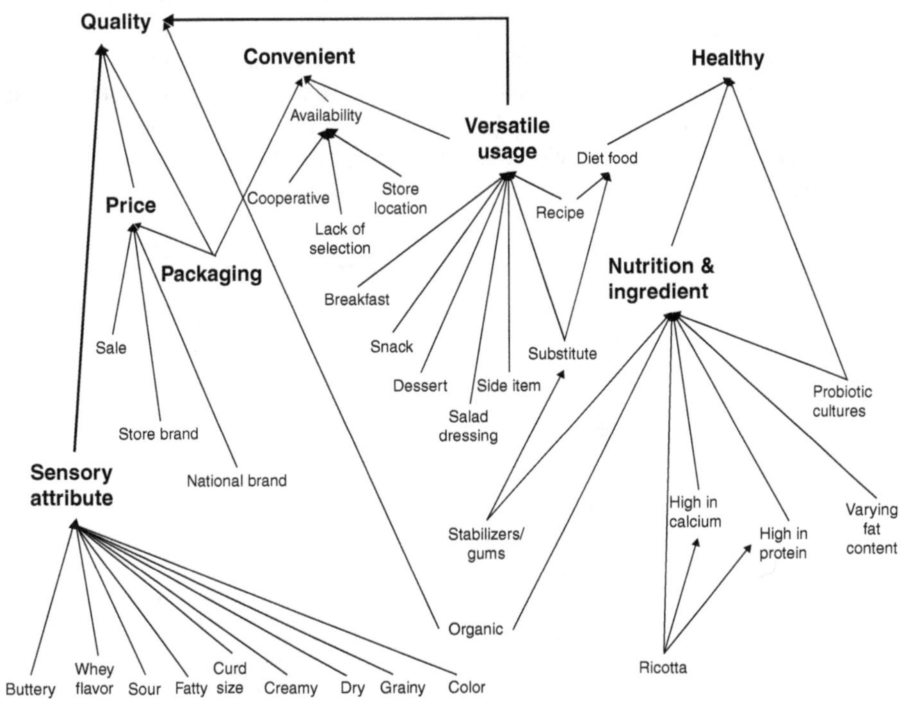

Figure 6.2.11 Cottage cheese value diagram. Source: Adapted from Drake et al. (2009).

However, many of the convenience aspects were hurdles that may cause cottage cheese consumers to switch to other dairy products as seen in the value diagram (Figure 6.2.11). There were two aspects of conveniences: Availability and versatile usage. Versatile usage was the main positive reason for cottage cheese consumers to keep consuming it, but low availability was the hurdle. There were few stores carrying cottage cheese and, if they did, there were not many choices to choose from compared to other dairy products.

In addition to usage, the impacts of the sensory characteristics (e.g. flavor, taste, and texture) were situational because cottage cheese's versatility of use was unexpected. For example, if consumers eat cottage cheese as it is, then creamy, small curd size, and not grainy texture are desired (a.k.a. good quality); however, if it is used as salad dressing, no flavor may be more important as it will get processed (e.g. blended or mixed) with other ingredients, or bigger curd size may be desirable as it contributes to the final appearance of the dressing. Figure 6.2.12 shows an important understanding of (1) Which values (e.g. health, convenience), (2) Product experiences (positive, negative, and neutral) and (3) Connections between the values and the experiences. These are very important and unique insights that can be discovered effectively by QMA but not through classic methods such as focus groups (in which discussion guides begin with questions you seek to confirm) or structured survey questionnaires (you get answers to questions you believe will reveal the important factors of the product design).

In this study, the team decided to modify the Love-it! Hate-it! voting exercise a little by asking the consumers to rank all the samples according to their collective liking (Figure 6.2.13). Also, the research team conducted a CLT for the same samples and the similarity in the rank order and the grouping of the products was striking (Figure 6.2.13). Further analyses of the CLT data reveal that the liking segmentation was not pronounced, and the ranking of the liking can be captured by a small sample size used in a qualitative study as the cottage cheese QMA (Drake et al., 2009). This shows an application of QMA as an effective screening tool.

Mapping/napping exercise was conducted, and the qualitative map shows that there are two important factors driving cottage cheese consumption:

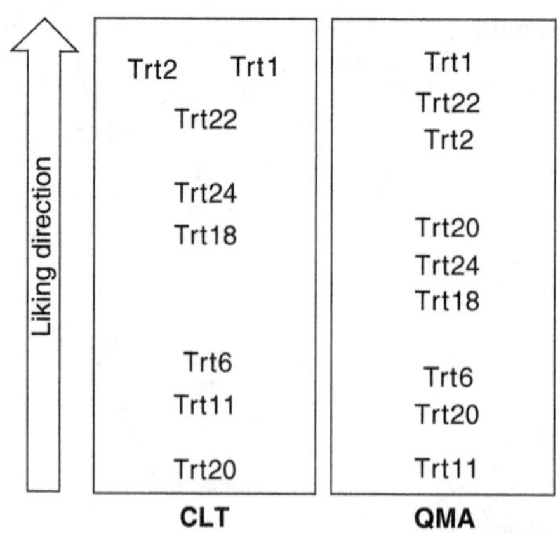

Figure 6.2.12 Without major segmentation, QMA provided almost the same results as those found in a subsequent quantitative study among cottage cheese consumers in North Carolina.

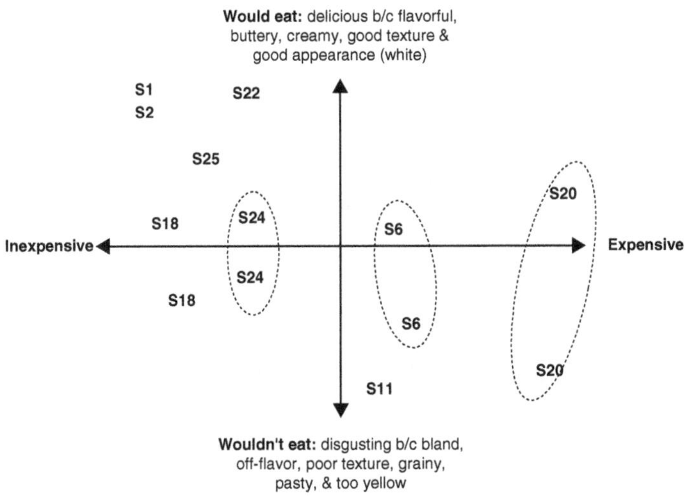

Figure 6.2.13 Preference map and locations of cottage cheese samples generated during a QMA session. Price and sensory properties were the two most important factors for the consumers to purchase and consume cottage cheese.

(1) Price (inexpensive vs. expensive) and (2) Edible (wouldn't eat vs. would eat) (Figure 6.2.13). There were few products that the consumers agreed to be both inexpensive and delicious (S1 and S2); meanwhile, expensive products seemed to be polarizing (as you can see two of S6, S20 and S24). QMA allowed the consumers to agree to disagree about the locations of samples. If there is disagreement about the final location(s) of a product, then it is recommended to *pay attention* to the reasons that the consumers used to persuade the others. This discussion reveals discriminating factors for consumer segmentation. In this case, it was not about price, but it was about individual personal liking of sensory characteristics: texture, appearance, and flavor.

As the results of the study were shared with the client, the sales of the client climbed to double digits in a few years and managed to compete against yogurt; meanwhile, its competitors' sales declined significantly.

6.2.5 Consumer perceived values

Value diagram is an important outcome of QMA; therefore, the authors would like to introduce the readers to the meaning and current understanding of *consumer's values*. The term *value* is a lasting organized set of preferential standards (a.k.a. belief) that privileges one mode of conduct or end-state of existence over another (Chryssohoidis and Krystallis, 2005). Values help a person to live and adapt in a society by helping that person to know and understand (a.k.a. internalize) how that society works (Grunert and Askegaard, 1997). There are five characteristics of values (Figure 6.2.14).

Values influence peoples' attitudes and behavior by serving as preferential standards in consumers' minds; hence, values are more closely related to behavior and attitudes than demographic measures (e.g. gender, age). Individuals

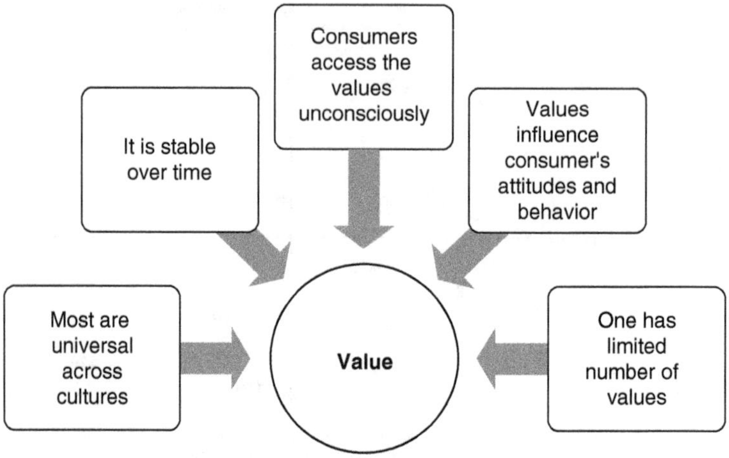

Figure 6.2.14 Five characteristics of a value.

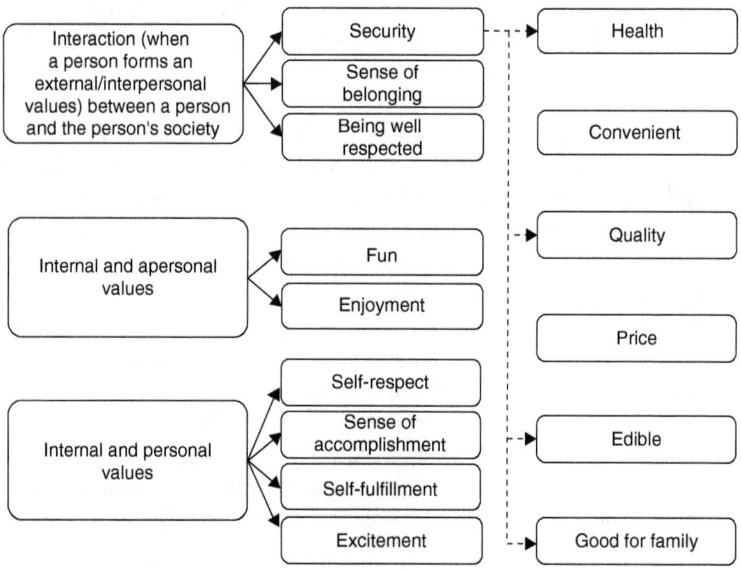

Figure 6.2.15 **The nine values based on the LOV of typology theory (Feather, 1984). These basic values are most likely to be the reasons behind the values uncovered by QMA. The dotted-line demonstrates an example of the linkages between security and few manifested values in Cottage cheese study. Source: Drake et al. (2009).**

tend to have limited numbers of values, and most values are universal across cultures. Values are stable over time and this characteristic differentiates values from attitudes that are not temporally stable. After values have been internalized, people *access the values unconsciously* (or behaving automatically), but the *consumers can articulate their values* as the reason behind their action and attitudes (Grunert and Askegaard, 1997).

Figure 6.2.15 shows nine types of values based on the list of value (LOV) of typology theory (Feather, 1984). The first three LOV values (security, sense of belonging, and being well respected) are relevant to the interaction (when a person forms an external/interpersonal value) between a person and the person's society. The other two values (fun and enjoyment) represent internal and personal values. The last four values (self-respect, sense of accomplishment, self-fulfillment, and excitement) are internal and personal values. QMA allows consumers to articulate their values through the use of multiple use products.

6.2.6 Summary and future of Qualitative Multivariate Analysis

QMA is a method that allows individuals to voice their needs and wants without any interference from researchers' preconceptions; consequently, it allows researchers to listen, observe, and discover real human insights. Value diagrams depicting the relationships between product characteristics, a person's experience and their values, a consumer's product perceptual map from mapping/napping exercise, and Kano classifications of product qualities are valuable for front-end research.

Advances in technology allow enhanced understanding of the contextual experience. Using both overt and passive monitoring with wearable tech, smartphone/handheld video devices, webcams, voice-enabled tools to augment classic diaries has created new understanding. Also, as the world's year-plus of pandemic showed us, high-speed Internet, less-expensive video conferencing, and interactive programs can be useful as another approach to the discussion session. While many believe that you cannot replace face-to-face research, virtual and interactive programs have their place.

Discussion questions

- When should you conduct a QMA study?
- How can you leverage a value diagram created by a QMA study?

References

Berger, C., Blauth, R., Boger, D., et al. (1993) "Kano's methods for understanding customer-defined quality". *Journal of Center for Quality Management*, 2(4), 3-36.

Chryssohoidis, G.M. and Krystallis, A. (2005) "Organic consumers' personal values research: testing and validating the list of values (LOV) scale and implementing a value-based segmentation task". *Food Quality and Preference*, 16(7), 585-599.

Clancy, K.J. and Krieg, P.C. (2000) *Counterintuitive Marketing: Achieve Great Results Using Uncommon Sense*. New York: The Free Press.

Drake, S.L., Lopetcharat, K. and Drake, M.A. (2009) "Comparison of two methods to explore consumer preferences for cottage cheese". *Journal of Dairy Science*, 92(12), 5883-5897.

Feather, N.T. (1984) "Protestant ethic, conservatism and values". *Journal of Personality and Social Psychology*, 46(5), 1132-1141.

Grunert, C.S. and Askegaard, S. (1997) "Seeing with the mind's eye: on the use of pictorial stimuli in values and lifestyle research". In L.R. Kahle and L. Chiagouris (eds). *Values, Lifestyles and Psychographics*. Mahwah, NJ: Lawrence Earlbaum Associates. pp. 161-181.

Kano, N., Nobuhiku, S., Fumio, T. and Shinichi, T.T. (1984) "Attractive quality and must-be quality (in Japanese)". *Journal of the Japanese Society for Quality Control*, 14(2), 39-48.

Moskowitz, H. and Gofman, A. (2007) *Selling Blue Elephants*. Upper Saddle River, NJ: Wharton School Publishing.

Moskowitz, H.R., Beckley, J.H. and Resurreccion, A.V.A. (2006) *Sensory and Consumer Research in Food Product Design and Development*. Ames, IA: Blackwell Publishing Professional.

Nestrud, M.A. and Lawless, H.T. (2008) "Perceptual mapping of citrus juices using projective mapping and profiling data from culinary professionals and consumers". *Food Quality and Preference*, 19(4), 431-438.

Pagès, J. (2003) "Recueil direct de distances sensorielles: application à l'évaluation de dix vins blancs du Val-de-Loire". *Sciences des Aliments*, 23(5-6), 679-688.

Pagès, J. (2005) "Collection and analysis of perceived product interdistances using multiple factor analysis: application to the study of ten white wines from the Loire Valley". *Food Quality and Preference*, 16(7), 642-649.

Perrin, L., Symoneaux, R., Maître, I., Asselin, C., Jourjon, F. and Pagès, J. (2008) "Napping® procedure: case of ten wines from Loire Valley". *Food Quality and Preference*, 19(1), 1-11.

Risvik, E., McEwan, J.A., Colwill, J.S., Rogers, R. and Lyon, D.H. (1994) "Projective mapping: a tool for sensory analysis and consumer research". *Food Quality and Preference*, 5(4), 263-269.

Stelick, A., Paredes, D., Moskowitz, H. and Beckley, J. (2009) Kano Satisfaction Model in Cosmetics. *Presented in Congress Cosmetic and Sensory: From Neuroscience to Marketing*. 24-26 June 2009, Tours, France.

Zaltman, G. (2003) *How Customers Think: Essential Insights into the Mind of the Market*. Boston, MA: Harvard Business School Press.

6.3

The Gameboard "Model Building"

Jennifer Vahalik, Melissa Jeltema, Jacqueline H. Beckley and Jeff Garza

(adapted from an original chapter on The Gameboard of "Model Building" by Cornelia Ramsey)

Key learnings

- ✓ In-context interviewing and model building to assist consumers in describing future desirable products
- ✓ Analysis and ensuring reliability of model outcomes
- ✓ Building on theory of mental models (or cognitive maps)

6

6.3.1 The problem – how to talk to consumers about new products that do not exist?

Developing new products that meet consumer needs and expectations involves engaging consumers in product research that explore these needs and expectations. Posing the question, "what do consumers want in new products?" captures the problem for product developers. We simply do not know what consumers want in new products. However, we can explore what they like and do not like in existing products by simply asking them. But how do we, as product developers and product researchers, take consumers with us into the future to talk about products and product attributes that do not yet exist? Beckley and Ramsey (2009) developed the technique of In-Context Interviewing using conversational approaches to framing the dialogue with consumers that place us both *in the situation* with people using current products. They are then able to describe their experiences to us in the context of product use. This discourse has proven to be valuable and insightful to product developers because consumers can refer to real-life experiences and concrete examples of product uses, benefits, and dislikes. However, the challenge of how to talk to consumers about **new** products remains – there is no "real" frame of reference for consumers OR product developers. To facilitate this conversation of new products, Beckley and Ramsey developed and tested new tools (based on empirical psychological and

Product Innovation Toolbox: A Field Guide to Consumer Understanding and Research, Second Edition. Edited by Kannapon Lopetcharat, Dulce Paredes and Jacqueline H. Beckley.

behavioral theory and practice) to use during interviews in order to assist consumers in describing the unknown, future, desirable products. Qualitative analyses of the outcomes of this process have been successfully used by research teams to craft new to the world products. But what about quantitative research? For some time, consumer and products researchers have felt that traditional survey research often falls short in providing the rich context consumers are capable of sharing, if only they are "asked."

All product developers know that consumers have different likes and dislikes based on personal preferences, individual experiences with products, culture and traditions, availability of products and personal characteristics (e.g. try new things vs. stick with what's known). For product developers, the key to designing successful new products requires a number of steps:

- Knowing the target audience(s) - groups people
- Precise screening strategies - demographic characteristics of interest
- Emphasizing development of *"consumer-centered"* products - tailoring products and attributes to meet consumer needs
- Engaging consumers in product conceptualization early in the innovation process using new methodology beyond focus groups
- Creative "thinking out of the box"
- Being open to whatever the consumer tells you!

In this chapter, we will introduce a new methodology, Gameboard "Model Building" that effectively solicits consumer participation early in new product development processes and define analyses that allow the researcher to apply this method to large-scale quantitative testing.

6.3.2 A new method: Gameboard strategy "Model Building"

This methodology is designed to explore how to actively engage research participants beyond the interviewing conversation. Experienced, qualitative researchers explore new techniques in working with consumers to design consumer-centered products. The model-building method demonstrates how people explore and guide the development of new, innovative products that would satisfy consumers on many levels including the sensory experience of the products. This new qualitative methodology is developed based upon the understanding of cognitive processes involved in learning, decision-making, and the theories of *schema*, mental models, *creativity,* and the *creative process.*

6.3.3 Construction: creative process model

There are two techniques developed using this methodology:

(1) a narrative technique and
(2) a graphic technique.

For each numbered slot, select from the corresponding list of choices.

A ___(1)___ that you ___(2)___ in your mouth. It is shaped like a ___(3)___ and tastes like ___(4)___. It lasts for ___(5)___ but the flavor lasts ___(6)___ after the product is gone. It is packaged in a ___(7)___ shaped like a ___(8)___. The main benefit is ___(9)___.

Choices for each slot

Slot 1. Gum, mint, paste, gel, liquid, mist
Slot 2. Chew, spray, savor, crunch, roll, dissolve
Slot 3. Oval, circle, rectangle, bead, capsule, strip, ball, toothpick
Slot 4. Mint, fruit, sweet, chocolate, spice, coffee, tea, cream
Slot 5. 30 seconds, 1 minute, 2 minutes, 3 minutes
Slot 6. 30 seconds, 1 minute, 2 minutes, 3 minutes 5 minutes
Slot 7. Tin, 6-pack, 12-pack, paper carton, canister
Slot 8. Pack of gum, book of matches, jar with lid
Slot 9. Breath freshening, wets mouth, whitens teeth, cleans teeth, wakes you up, relaxes you

Figure 6.3.1 An example of a "narrative technique" in Gameboard "Model Building" method. The writing in the box is called "Schema" that acts as the skeleton of the product concept of interest.

First, the *narrative technique* walks the participant through the process using words, cards, and a "Gameboard" requiring them to fill-in-the banks with their preferences (Figure 6.3.1). This narrative technique allows participants to talk through, with the interviewers, their preferred product attributes, talk about how these attributes work together for the total product experience, and thus combine the attributes into a complete product that they describe in narrative form. This technique works well with participants who are more verbal by nature and express themselves well with words.

For people who are more visual and comfortable expressing themselves graphically with pictures or physical items, the second technique works well. The *graphic technique* uses "game pieces" that visually represent the meaning of the attributes or benefits of interest and requires the participants to place these game pieces together on a Gameboard to represent their product preferences (Figure 6.3.2).

Although each technique accomplishes the same goals, the two offer the interviewer and participant their preferred way to achieve the goals. The *graphic technique* may provide more emotional information as visual stimuli are used, and works especially well for kinesthetic learners (i.e. people who learn through movement and experience - often those who choose hobbies or careers that allow them to work with their hands such as carpenters or knitters). However, preparing many options for each attribute is recommended as not everyone will interpret an image the same way, especially higher-level benefits (e.g. it is "cool" to use, romantic). From our experience, we found that engaging the participants in actions with the cards and game pieces through this novel process not only

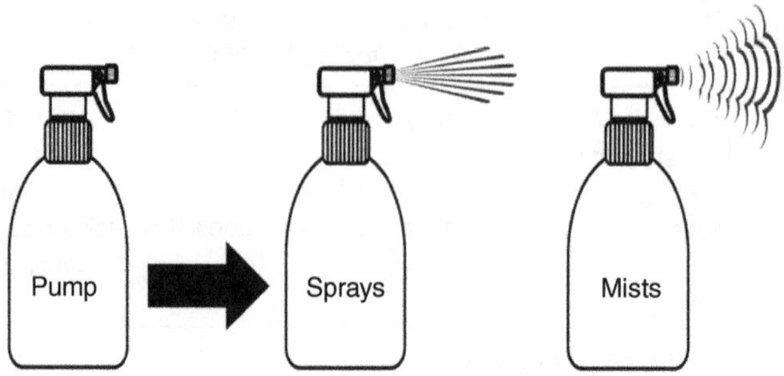

Figure 6.3.2 Examples of "game pieces" used in the graphic technique. In this case, two spray patterns were represented with two images in order to ensure the differences between "spray" and "mist" from a pump.

places the participants at ease but is also fun and allows the participant to "drive the bus." Thus, the outcome is a consumer-centered product.

There are six simple steps to successfully conducting the Gameboard "Model Building" (Figure 6.3.3).

6

First, the product development research team identifies and lists all product attributes for a new product of interest (referred as choices in Figures 6.3.1 and 6.3.2). Once all of the potential features are identified, a "story" is constructed to put them into the proper context. That is, sentences are constructed and "blanks" are created so that the choices can be filled in by the participant to complete the story. Each individual attribute/choice/image is then printed or pasted on a 5″×8″ card according to their category color. Different colors are assigned to each category as it will prevent confusion during the "modeling building." These cards comprise a "deck" that will be distributed to each person participating in the interviews. All cards representing potential shapes (e.g. square, oval, round) will be grouped together into a "shape" category and printed in the same color. *The color-coding of cards is important because the corresponding slot on the Gameboard is also color-coded.* That way, the consumer knows to put the green-colored cards in the green-colored slot on the Gameboard. Color coding removes a layer of complexity within the task for both the participant and the researcher, allowing the conversation about choices to flow more freely.

The *narrative technique* uses actual words to describe the attributes/benefits of interest while the *graphic technique* uses images or physical items with descriptions affixed to them rather than using individual words or descriptions alone. The process for the graphic exercise is the same as the narrative exercise; however, it uses more tangible game pieces that clearly demonstrate the product attributes in a highly visual way. Again, the product attributes are predetermined by the research team and are represented by game pieces of different shapes, sizes, and containers that correspond to the meaning of the attribute. For example (see Figure 6.3.1), potential product shapes will be represented by actual circles, squares, triangles, ovals, etc., also of different sizes. All game pieces will also be labeled with words (i.e. "circle," "square") for clarity to ensure participants have an understanding of the meaning of the pieces. For action attributes, such as "spray" or "mist" examples, prototype game pieces

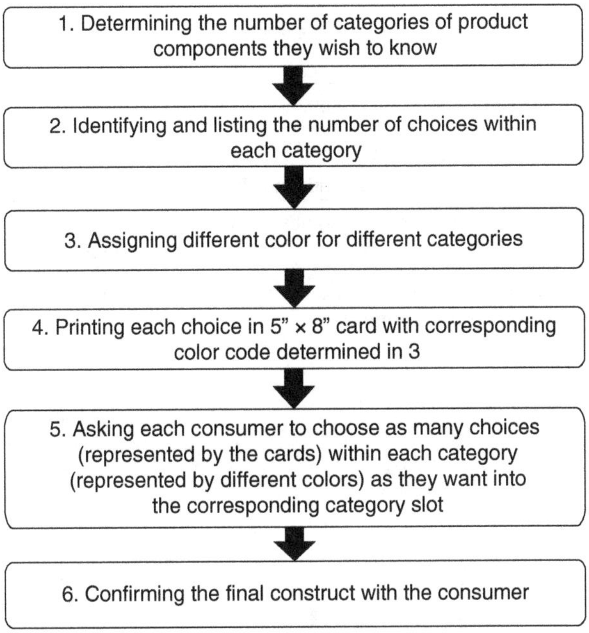

Figure 6.3.3 Six simple steps to conduct the Gameboard "Model Building" method.

that really spray or mist are used. These game pieces are called "action prototypes." Participants must try the action prototypes to make sure "spray" really represents what they prefer vs. a "mist" or "pump." As with the narrative exercise, they will be asked to select the game pieces that represent their preferences and place the pieces on the Gameboard slots. These pieces must be labeled as well. Using these materials, participants will be asked to describe the product that represents their experience, or the experience they would like to have, with a new product. Additional variable "pieces" should be available to participants to create and represent any attributes they identify that are not adequately represented by existing game pieces. These new game pieces can then be added to the decks for subsequent interviews.

For both techniques, participants are told they are not required to use all of the attribute pieces and that they can add other pieces. In addition, if they like more than one game piece for a particular slot on the Gameboard, they can use as many as they like. For example, the participant may like more than one flavor for a product or could define more than one purpose for a product. Through this Gameboard exercise, participants can identify product attributes, group attributes together in different combinations, prioritize the "must haves" from the "nice to haves," and talk with the interviewer about a consumer-centered product that meets their desires. The interview is essential to guide the participant through this novel approach but also to ensure the research team really understands what the consumer is saying, what they mean about each attribute and why they define their consumer-centered product the way they do. Figures 6.3.4 and 6.3.5 are sample outputs from a narrative and a graphic technique, respectively.

A (1) mist that you (2) spray in your mouth.
It is shaped like a (3) capsule and tastes
like (4) fruit and cream. It lasts for (5) 3 minutes but the flavor
lasts (6) 5 minutes after the product is gone. It is packaged in
a (7) Paper carton and Canister shaped like a (8) Jar with lid. The
main benefit is (9) Breath freshening.

Figure 6.3.4 An example of a completed Gameboard model by a consumers using the narrative technique.

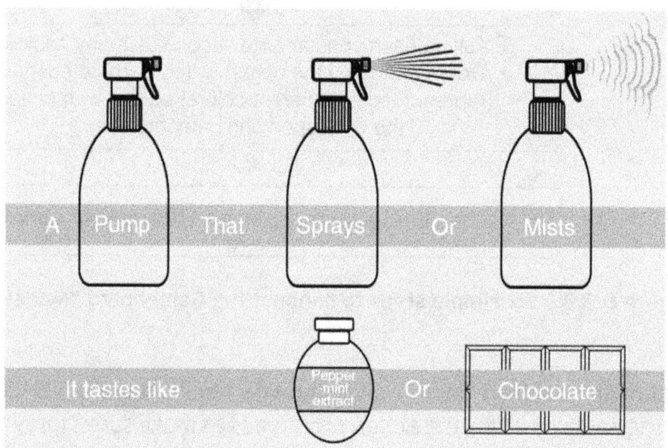

Figure 6.3.5 An example of a completed Gameboard model by a consumer using the graphic technique.

6.3.4 Interview guide for model construction methodology

The interview component is critical to guide the process and for product developers to understand the outcomes. It is the same for both the narrative and graphic technique. If there is one card/piece (attribute) that is more important to the participants than the others, it can be documented and explored in-depth via the discussion. Why is that attribute more important? In the same way, the "must have" attributes can be distinguished from the "nice to have" attributes via this discussion. These are important to know because the "must haves" can be dealbreakers and either encourage people to use the product or prohibit them from wanting the product.

In the same way, the common attributes that are included among most (or all) of the consumers' models can be recorded in the field notes and compared to those that are seen in fewer consumer models. These are recorded to explore later. For example, *which* people come up with new attributes and what are those attributes? Do certain segments of the population come up with similar attributes? Why? These attribute selections can then be used in the development of a quantitative version of this technique that is able to determine statistical attribute preferences among segments of the population (see Section 6.3.6).

Each interview should be directly related to the known attributes predetermined by the research team, and from existing consumer products as well as potential attributes that are not currently represented in the market but under exploration by the product development researchers. These attributes are, in turn, represented on the cards used in the construction of the narrative models or by the shapes or prototype containers used in the construction of the graphic models.

There are three critical phases in the interviewing process: (1) introduction, (2) model building, and (3) summary.

Phase one of the interview introduces the product attribute descriptions/images/actual objects to the participants through a series of open-ended questions. This is very important to ensure that all consumers understand the intended meaning of the attribute descriptions/images/objects in the same way and in the way that the researcher defined. Next, the facilitator who acts as the interviewer/researcher introduces the model-building concept to the participant. A Gameboard mat will be spread out on a table with the concept clearly printed with the open slots (Figure 6.3.1) and the facilitator will read the concept to the participant. For example,

A **blank** flavored oral product that **blanks**. It is **blank** and lasts **blank**

might be filled with the following attributes:

A ***peppermint*** flavored oral product that ***freshens breath*** and ***wakes you up.*** It is ***strong*** and lasts ***5 minutes.***

The participant will be asked to read the Gameboard concept out loud and explain in his/her own words the process of the game. The facilitator then reads each card out loud to the participant *OR* places the shaped game pieces or containers at each color-coded slot and talks about what each one represents in order to demonstrate the process. For lengthy or multipart Gameboards, it is helpful to cover parts of the mat, and only uncover sections as needed for discussion to minimize distraction.

Phase two of the interview is when the participants complete the Gameboard exercise. The participant is asked to describe in detail what they are thinking as they arrange the cards or pieces on the mat. Each participant will be given as much time as she/he needs to construct a model. Upon completion of the task, each participant will be asked to describe the overall product concept model. The facilitator will explain to each participant that her or his model will remain on the mat until the end of the interview, at which time participants are told that they will be allowed to review the model and change the model if desired.

Finally, in the *third phase*, the facilitator will ask semistructured summary questions to review the process with the participant, to confirm the selections of the participant and to give the participant the opportunity to think about the model that they constructed and make any changes they desire. If changes are made, the facilitator will query the participant to understand the reason for the changes.

6.3.5 Ensuring reliability of the outcomes

To ensure reliability of this creative methodology, author Ramsey searched for an established framework to guide the process. The *four-stage creative process model* presented by Lubart in 1994 provides a step-by-step example for model construction. Accordingly, the four stages of the creative process are the following:

- Preparation,
- Incubation,
- Illumination, *and*
- Verification (Lubart, 1994).

To further develop this qualitative process for consumer-centered product development research, Ramsey worked with Jacqueline Beckley, a senior executive and expert consumer product researcher, to develop the physical Gameboard and pieces for consumers to test.

Preparation is the initial acquisition of knowledge before a task is begun. The *preparation stage* was paralleled in this research by

6

- the initial interview questions and information exchange with the participant about the attributes in the model;
- the introduction of the model construction task and orientation to the game pieces to be used; and
- participants' initial attempts to build the model.

Incubation phase consists of successfully manipulating the task and discovering possible solutions. The *incubation phase* occurs when the participants arrange the pieces in different ways to determine the most accurate representation of the preferred product and combination of attributes.

Illumination phase occurs when the solution becomes evident. Illumination is the final arrangement of all of the pieces selected by the participant.

Verification occurs in the evaluation and refinement of the task. For example, the *verification* step can occur after model construction when summary questions are asked and the participant is given the opportunity to make any changes to their model.

It is important to emphasize that the participants are encouraged to "think out loud" *during the model construction process* so their thoughts are recorded. Researchers must pay attention to the participants' thinking processes in order to record any additional attributes and different definitions of existing attributes that participants include in the models. They should also document the definitions assigned to all attributes because these records will provide supplemental data for future interpretation of the results.

6.3.6 Analysis of the outcomes from Gameboard "Model Building"

As with other qualitative techniques, sample size is determined by saturation, meaning that when researchers begin hearing (or seeing) similar themes from the participants, and no new data is being discovered, adequate sample size is reached. This usually happens between 17 and 20 participants per site. The consumer researcher, also referred to as Consumer Explorers (CE) in this book, must also look across samples in different sites to see if themes are similar or different to determine if enough sites have been included. This type of sample size determination is appropriate because it is similar to other qualitative methodologies in that the research focuses on *concept development* and the goal of the research (Gameboard) is to develop consumer-centered product concepts that are *grounded in the consumers' real-life* events and circumstances (Sandelowski, 1995).

It is very important for the product development team to work closely with the facilitator and CE in the *preparation, duration,* and *debriefing* of this method. First, in the preparation, the meaning of each attribute under exploration must be clear to the team, the process of allowing the consumer to "drive the bus" is clear to product developers and the facilitator must be able to translate the intent product developers have for that attribute to the participant. In other words, the CE and facilitator must understand why the attributes were chosen and what the product developers think each attribute means. There must be communication *during* the interviews (in the form of "check-ins" after each stage of the process) to establish consensus among the team (e.g. facilitator, CE, and product developers viewing the interviews) about what the consumer/participant is saying, acting, expressing during the exercise, and to identify any confusion or contradictions the consumers may have because these contradictions often provide key elements of the product under exploration. For example, a contradiction may appear to be a consumer who wants a product flavor that lasts 10 minutes, but at the same time does not want a strong flavor in a product. Finally, the debriefing after each interview is critical to review key findings from the interview, review the combinations of attributes that consumers arranged on the Gameboard and discuss briefly how each interview contributes to the overall findings of previous interviews or if there are key differences. In a nutshell, total team communication and engagement are key.

The objective of the analysis of this method is to have a clearly defined consumer-centered product with specific attributes as defined by the consumers. The similarities and differences of models regarding the attribute selection, placements of attributes, relationships among attributes, and importance of attributes must be analyzed. Also, constructed models will be compared to each other for analysis.

A "triangulation" of data sources (actual models on Gameboard, interview records and field notes) help to construct and support reliability of findings. The three data sources allow cross-checking data for accuracy and established reliability of findings. In addition, consumer preference segmentation can be theorized at this time.

6.3.7 Analysis overview

The analysis of the models consists of (1) a preliminary descriptive analysis and (2) a secondary central theme analysis of the core elements of the product models. This analysis is iterative and data-driven, in other words, each analysis uncovers themes that are explored further in the next analysis.

(1) *Descriptive analysis* describes the process consumers followed and the characteristics of the outcomes. Four common things to observe from the data sources (actual models on Gameboard, interview records, and field notes) are listed below:
   (a) Which Gameboard pieces (attributes, words, descriptions, or images) are included? Which slots were filled first? What are the similarities, differences, and outliers?
   (b) Which new pieces are added?
   (c) Which pieces are omitted?
   (d) What is the organization of pieces between individual consumer's models - "must haves" versus "nice-to-haves"?

   Descriptive analysis answers *what happened* in the model construction.
(2) *Central theme analysis* describes why model construction happened in the way it did and helps researchers answer the following:
   (a) Why do participants select these particular sets of attributes per category/slot?
   (b) Why do they select the attributes in this order?
   (c) Why do they combine these attributes in this certain combination?

Any Theme Analysis is a search for relationships among the "parts" and relationships of the parts to the whole (Spradley, 1980). The completed Gameboard should be examined to:

(1) Determine which existing groups of attributes (e.g. those predetermined constructs) are included in the models and which grouping is most important, if any;
(2) Identify if new pieces (attributes) are added (*and* those pieces are recorded along with the definitions assigned to them by the participant); and
(3) Identify which attribute or attribute group is omitted, and why.

It is critical that detailed, contextual note-taking be incorporated into the interview process because this allows researchers to (1) reflect back to the participant, (2) review the selections that were made, and (3) provide quality assurance and rigor during the summary phase.

6.3.8 Consumer-centered products and Gameboard "Model Building"

We found Gameboard "Model Building" to be both *feasible* and *appropriate* for consumer-centric product innovation because the conceptual models of new products were consistently constructed by consumers directly. Although the actual model construction process was open-ended, the methodology was introduced consistently across all interviews using the following procedures:

- Each attribute was introduced through sequential interview questions;
- Game pieces were shaped like their meaning (circle represented round) and actively demonstrated the attribute relevant to the meaning of the attribute;

- Pieces were clearly labeled;
- Participants were encouraged to "think out loud" as they composed the models;
- All possibilities of attributes were encouraged; (there were no "wrong" answers)
- Models could be changed; and,
- Models were confidential.

From our experience, consumers found this Gameboard "Model Building" to be a "fun exercise" since there was no pressure to identify the "right" answers. This "fun exercise" increases individuals' comfort level and self-confidence in discussing their likes and dislikes in products via building a model. We found they had "relative ease" in constructing models which is important to observe because it provides more clues for further in-depth analysis. Moreover, consumers could invent even more complex features for the new products by combining attributes in new ways and assigning multiple attributes for each "slot" in the models. For example, consumers designed a product that could "mist" a light flavor on occasion but could also "squirt" a stronger flavor depending on their preference at the time. The stepwise process described here coupled with depth discussion allowed participants to have ownership of their individual models. In other words, the process of building the models, adding new constructs, or omitting constructs demonstrated an understanding of the definitions of the existing attributes, the purpose of the methodology and, again, ownership of the models. This method demonstrates the possibility of developing authentically consumer-centered products that meet consumer expectations from the onset of development.

6.3.9 Limitations

As with any method, there are limitations that must be considered:

- First, the research team guided by the CE must have comprehensive knowledge of existing attributes under exploration and must have clear definitions of new, innovative attributes being explored and be able to transfer the knowledge effectively to participants so that participants understand the new, nonexisting attributes, and can use them consistently across models.
- The CE must have a facilitator knowledgeable in this process to assist and guide but *not influence the responses* of the participant.
- The facilitator must have clear understanding and have pieces, cards, and Gameboards that accurately reflect attributes and make sense across participants.
- The CE must educate other research team members to know what they are observing and to extract the most information out of this process.
- This process takes interview time to orient the participants to the process.

With work upfront, these limitations can be managed. First, *preparation* is key. Devoting time and energy to the up-front work (e.g. preparing the inputs of the Gameboard as well as physical Gameboard and pieces) is essential. As with any other method, pilot testing with consumers is essential to ensure definitions are clear and the process is understandable.

6.3.10 Quantitative Gameboard

6.3.10.1 Overview of quantitative Gameboard

There are times when a research team may require a larger sample size to (1) test their hypotheses around mindset segments that emerged from the qualitative Gameboard sessions or (2) validate the learnings among a broad population prior to committing to further research and development efforts on a new idea. To meet these needs, a quantitative version of Gameboard, called Quessence®, can be conducted as a follow-up to qualitative sessions.

For Quessence, the Gameboard "mat" and "pieces" are created digitally by partnering with an online survey provider that can provide program customized screen layouts. The game pieces need not be color coded in this instance, as each "blank" has a drop down menu containing the words gathered during the qualitative learning stage. The participant simply scrolls through the options and picks the best choice, for them. Just as in the qualitative version, participants may create their own game piece if none of the supplied words fit their needs. The list is either alphabetized for all participants or randomized across participants depending on complexity of the topic or objectives of the research. If randomization is chosen, a small-scale pilot should be run, even if only among the research team, to get a feel for the level of difficulty in making choices that this option can present. Once again, a complete story with multiple blanks and sentences can be shown at once, or parts of the story can be grouped to be shown together on separate screens for better clarity. Again, depending on the objective, there are times when one rationale will make more sense than the other.

Once the participants have completed their "story" in digital format (whether in sections or as a whole), they are presented with the entire story to evaluate whether or not it truly represents their thoughts, feelings, or experiences. If it does not, they are then given an opportunity to change any part of the story that they wish. Finally, they are asked to once again read through their story and rate how it makes them feel using an emotion word bank.

6.3.10.2 Analysis of quantitative Gameboard

The collected data from the Gameboard exercise naturally leads to frequencies, percentages, and other standard approaches. Some common standard text analysis presentation tools, such as word clouds, may also be beneficial. These approaches along with simple percentage comparisons are useful and informational, but they leave much to be desired in a multivariate problem setting (such as in the Gameboard setting). More interestingly, some different classes of multivariate techniques may also aid in uncovering patterns within the data. principal component analysis (PCA) may be the most obvious method, but other related techniques can also be applied.

PCA allows a multivariate data matrix to be summarized by a few meaningful dimensions. This allows patterns to be uncovered and presented on a map in order to simplify the interpretation of the results. The technique is flexible, requires few assumptions, and can be tailored to the situation at hand. Although other (related) methods are available and, sometimes, preferable, PCA is very robust, and it is the recommended starting point for data of this structure.

Jeltema et al. (2019) showed that consolidating the Gameboard stories and using PCA provides a useful and robust method for uncovering patterns even in traditionally difficult areas to study such as Mouth Behavior. Differences between products are easily visualized and understood using a simple mapping approach. The impact of specific factors of interest, such as time (Jeltema et al., 2019), can also be represented in order to form conclusions and derive hypotheses for future research.

Future uses of the Gameboard methodology, especially when linked to product performance measures (consumer survey oriented, in market performance, etc.), may require more advanced predictive modeling methods. These hold immense promise, but require future research.

6.3.10.3 Limitations of quantitative Gameboard

It is critical to avoid the human tendency toward confirmation bias by going from idea origination straight to quantitative validation without first testing your hypotheses about what you think you know. Therefore, Quessence is not recommended as a standalone methodology without at least some minimal amount of qualitative input, even if (or especially if!) the team feels like the subject is well understood.

6.3.11 Theoretical background of model construction methodology

6.3.11.1 Cognitive processes

Interviewing consumers about new product ideas creates opportunities for new ways of thinking for both the CE and the consumer and requires techniques for putting existing knowledge together in new ways to produce new concepts.

All of us have existing knowledge that we have learned through day-to-day living, formal education and informal information acquisition such as observations. We have existing knowledge and ways of thinking based on our beliefs and experiences. These pieces of existing knowledge give us *mental models* of how life is and what we know. The mental models are cognitive maps that represent real or hypothetical situations and give us starting points in how to learn new information (Goel et al., 2010; Johnson-Laird, 1980). This is important to understand because when we interview consumers, we know they are responding to our cues based upon their existing knowledge and mental models. But this also tells us that because they (consumers) have existing mental models of knowledge based on their product experiences, with guided conversation and tools via interviews, they are able to make inferences regarding new information and think about new products and new product attributes that they have not used but that they may be interested in if those new products existed. So how do we help our consumers draw on their existing mental models of products to help them articulate the product attributes that they would want in new products? The answer is that we learn together (researcher and participant) how to identify new attributes and how to put those attributes together to make a *consumer-centered* product. By guided in-context interviews and using tools we have developed to facilitate the conversation, we, as researchers, are able to take consumers with us into the future. Through in-context interviewing (Beckley and Ramsey, 2009), we frame the conversation so that the participant is placed in the experience and in the situation (context) of the product. We do not just talk about the product and its attributes, but we describe the situation, environment, feelings, aromas, thoughts, people, and surroundings that were happening to and with the consumer when she/he was using the product. We augment this interviewing with concrete tools (e.g. Gameboard and pieces) that participants use to identify and define what attributes and combinations of attributes they would like to have in new products.

6.3.11.2 Schema theory

A *model* is considered an individual's mental picture or graphic that represents specific events or situations. Frederic Charles Bartlett (1886–1969) developed the theory of "schema" to explain how individuals mentally represent knowledge (Brewer, 2000).

"A schema is a knowledge structure that captures regularities of objects and events" (McNamara, 1994). This structure is composed of "slots" in which events or categories of things can be specified (McNamara, 1994). These slots in schemata represent *". . . typical properties of an object or event and contain default values"*(McNamara, 1994). For our research, these properties are attributes of products as shown in Figure 6.3.1.

6

6.3.11.3 Theory of creativity

The theory of creativity is critical for researchers to understand when developing new products and exploring product attributes that do not yet exist. Creativity is the ability to produce work that is both "novel" and "appropriate" (Lubart, 1994). "Novel" means it is original and not predictable. "Appropriate" means it fulfills a need and/or is useful. This research proposed a "novel" and "appropriate" approach to understanding new consumer products by requiring participants to construct a graphic representation of their favorite product attributes into new combinations of products while engaging in descriptive conversation with the interviewer.

6.3.12 Summary and future

The Gameboard model building technique enables consumers to identify and graphically present the key elements and combinations of elements (attributes) for new products. By identifying and defining the key elements via the model construction procedures of the methodology, consumers are able to examine their current product experiences in new ways because they deconstruct the product by individual attribute and via the interview talk about each attribute – existing or new. It is plausible that this methodology would be applicable to other research. For example, organizations could conduct "internal inventories" of existing products by constructing models of the existing products and asking consumers to deconstruct, then reconstruct the product with their individual preferences and priorities for attributes and different combinations of attributes.

The model construction methodology could also be used to identify which attributes are most important to consumers and which attributes they could manipulate in new products. The Gameboard model building technique engages the people and once engaged, their true preferences are brought out via the interview. This process adds quality and depth beyond conversation. It engages the participants. It is fun and when people have fun, they are relaxed and more open to "out of the box" thinking which is essential for new product development research with consumers. There are no wrong answers and consumers can design whatever they desire. As one consumer commented upon completing a Gameboard exercise, "This is so cool! I'd for sure buy a product that was designed to do what I wanted it to do!"

This method offers a concrete picture of consumer preferences in a cost-efficient and time-effective manner by using cards or prototypes early in the consumer research process without having to commit research and development time and money into many different complex prototypes with predetermined yet untested attributes that may not meet consumer needs or interests.

Discussion questions

- What is the problem that Gameboard is trying to solve?
- Why might this approach to "model building" work?
- Why would one select a quantitative Gameboard over a qualitative Gameboard?

References

Beckley, J.H. and Ramsey, C.A. (2009) "Observing the consumer in context". In H.R. Moskowitz, I. Saguy and T. Straus (eds). *An Integrated Approach to New Food Product Development*. Boca Raton: CRC Press. pp. 233–245.

Brewer, W.F. (2000) "Bartlett's concept of the schema and its impact in theories of knowledge in contemporary cognitive psychology". In A. Saito (ed.) *Bartlett, Culture and Cognition*. Guildford: Psychology Press UK. pp. 69–89.

Goel L., Johnson, N., Junglas, I. and Ives B. (2010) "Situated learning: conceptualization and measurement". *Decision Sciences Journal of Innovative Education*, 8(1), 215–240.

Jeltema M, Beckley J., Vahalik J and Garza J. (2019) "Consumer textural food perception over time based on Mouth Behavior". *Journal of Texture Studies*. 1–10. https://doi.org/10.1111/jtxs.12479

Johnson-Laird, P.N. (1980) "Mental models in cognitive science". *Cognitive Science: A Multidisciplinary Journal*, 4(1), Jan–Mar 1980, 71–115.

Lubart, T.I. (1994) "Creativity". In R.J. Sternberg (ed.) *Thinking and Problem Solving*. San Diego, CA: Academic Press. pp. 290–332.

McNamara, T.P. (1994) "Knowledge representation". In R.J. Sternberg (ed.) *Thinking and Problem Solving* (2nd edition). San Diego, CA: Academic Press. pp. 81–117.

Sandelowski, M. (1995) "Sample size in qualitative research". *Research in Nursing and Health*, 18, 179–183.

Spradley, J.P. (1980) *Participant Observation*. University of Minnesota: Holt, Rinehart and Winston.

7

If we want to change the world, we have to keep renewing ourselves.

Jil Sander, a world-famous clothing designer

What this is about

Two versatile tools, Creative Blogging and CATA (Check All That Apply), can be used throughout the iterative product research process. These approaches provide both language and more structured responses to people's perceptions.

Chapter 7

Tools for Upfront Research on Consumer Triggers and Barriers: Qualitative-Quantitative Tools

7.1

Creative Blogging

Kannapon Lopetcharat and Dulce Paredes

7

Key learnings

- ✓ Questionnaire fallacies
- ✓ Designed conversation
- ✓ Blogging and experimentation
- ✓ Blogging platforms

"Read between the lines" is a popular idiom that originated in the mid-19th century when people were using invisible inks in the military to communicate. The idiom reflects that when people explicitly express their opinions, those explicit opinions do not really contain all their thoughts, feelings, and other intended meanings. In 1970, Prof. Albert Mehrabian of the University of California in Los Angeles proposed that all intended meanings (feelings, attitudes, and beliefs) in our personal communication should be deciphered from spoken words (7%), tone of voice (38%), and body language (55%)). Even though this partition seems to oversimplify the complexities of all types of communication, he cautioned that:

Product Innovation Toolbox: A Field Guide to Consumer Understanding and Research, Second Edition. Edited by Kannapon Lopetcharat, Dulce Paredes and Jacqueline H. Beckley.

Total Liking = 7% Verbal Liking + 38% Vocal Liking + 55% Facial Liking. Please note that this and other equations regarding relative importance of verbal and nonverbal messages were derived from experiments dealing with communications of feelings and attitudes (i.e., like–dislike). Unless a communicator is talking about their feelings or attitudes, these equations are not applicable.

Extend this line of thinking to traditional consumer research methodologies that heavily rely on consumers' explicit expressions of their opinions, attitudes, judgments, and feelings through the use of question-and-answer type tasks. These methods include answering questionnaires (online or offline) and structured interviews (focus group interview [FGI] or in-person one-on-one/in-depth interview). The latter has advantages over the former because nonverbal cues can be observed and studied due to the "in-person" component of the method. In many instances, a one-on-one interview with a structured questionnaire is used for data collection and researchers miss all the subtle and nonverbal clues.

Social networking has been rising since the 1970s when the PLATO (programmed logic for automatic teaching operations) system that allows people to chat in a forum (a.k.a. chatroom) to current social networking platforms (e.g. WhatsApp®, Line®, WeChat®, Twitter®) that allow consumers to send message instantly (instant messaging), leave messages (mail and blog), chat, send photos, and send video clips (normal or 3D). These technologies enable researchers to "interact" with people directly with ease and to express themselves in many more ways than before. As these social network tools turn to be go-to channels for people to interact and express themselves, how can Consumer Explorers (CEs) take advantage and adapt to this new environment to interact with their target consumers more naturally? The answer lies in a new protocol called *Creative Blogging*.

7

7.1.1 Introduction

Asking a question accompanied by a scale is one of the most common methods used in consumer, market, and product research, and it also has an apt name called "instrument" in psychology. In practice, besides the battery of questions with scales (the instrument), open-ended questions are added to provide additional learnings beyond the numbers gathered from the instrument. Legacy tools of summarizing learnings from open ends include subjective reading and/or creating word clouds based on frequency counts. Contemporary tools include text analytics tools to transform unstructured text data to structured data for quantitative analysis (see Chapter 16 on for examples of using text analytics on unstructured data sets). The questionnaire structure is essentially fixed, especially the instruments with predesigned questions and scales. Consequently, consumers who answer the questionnaires can only "react" to the questions and to the scales chosen for them. Therefore, the questionnaire may not be appropriate for consumers to express their full thoughts and attitudes naturally because all the choices are chosen for them in advance based on the researcher's point of view. Many researchers add the open-ended questions to allow consumers to freely express their opinions; however, it is still reactive to the questions asked. Methods that allow people to express their thoughts "proactively" are live interviews (individual or group), but the methods have three major limitations: risk of deviating from objectives, memory bias, and risk of exposure to disease. Deviation from research objectives happens frequently in fully unstructured interviews where people lead the conversation based on topics they mention without prompting. It is very suitable for in-depth understanding but can be extremely time-consuming. A structured interview will end up not much better than using a questionnaire. Memory bias is another limitation where the person must rely heavily on their memories that may not be accurate unless the memories are very important to them. The latest issue is fear of infection as we have seen in COVID-19 where people are asked to distance themselves to prevent potential infection. Many researchers have used online interviews successfully to eliminate these biases, but they require a seasoned interviewer to steer the conversation adeptly.

7.1.1.1 Without context vs. with context

Context during questioning is very important as expectations, moods, and emotions are heavily dependent on it. Survey is usually done in a secluded context and rarely done within the actual context of the experience. Context of use is very important especially for product research when actual product use influences the expectation of outcomes, satisfaction of experiences, and performance of products. Mobile technology helps researchers overcome the context issue to a certain degree as in-context online questionnaires can be executed easily using mobile phones. However, full-length questionnaires especially with a battery of questions still cannot be done in a short period of time during actual product use or right after product use as many products' efficacy/performance and consumers' memories are temporal dependent. Therefore, an in-context short questionnaire is more effective than a full-length questionnaire because of the temporal and memory effect.

7.1.1.2 Noninteractive vs. interactive

In addition to the reactive nature of questionnaires, a consumer answering a series of questions on a screen or paper is totally noninteractive, which induces boredom and reduces information quality. That is why live interviews (individual or group) can provide more in-depth insights and the synergistic interactions between group members and interviewer/moderator/facilitator.

A good questionnaire is better used for testing hypotheses and quantifying how people react to the product but is not suitable for generating creative insights. Live interviews and open-ended questions provide room for people to express their thoughts and opinions fully; however, memory bias and out-of-context of actual use can hinder the method's full potential. Mobile technologies can help overcome some of these issues when programmed to collect input in real-time and in context.

Creative Blogging was created using mobile and Internet technologies in conjunction with new consumer behavior of chatting online and blogging.

Nowadays, chatting is a natural way of communication that has transcended phone calls. Consumers converse with their peers, family members, colleagues, etc., through various applications. Creative Blogging combines the unstructured nature of chatting and a semistructured interview protocol with a facilitator. With Creative Blogging, CEs can turn most people to be creative bloggers through a chatting platform.

7.1.2 The rise of blogging platforms enables new mode of data collection

7.1.2.1 Blogging genesis

Before we talk about blogging, we need to know the meaning and the origin of the word "blog."

> Blog (noun): a website where an individual person, or people representing an organization, write regularly about recent events or topics that interest them, usually with photos and links to other websites that they find interesting.
>
> *Oxford dictionary*

Blog is a new word born from the shortening of the word "web log" (Figure 7.1.1). Like its original words, a blog was where people "log" their lives on the "web," in short, it was an Internet version of a paper diary. The evolution of internet and mobile technologies ushered "blogging" to new levels since its inception in the 1970s. The changes include video blogging, micro-blogging, and eventually sharing the blogs and, consequently, social networking was born in the 1990s dominated by popular platforms such as Facebook®, Twitter, Instagram®, Line, WeChat, Whatsapp.

7.1.2.2 Types of blogging

With the advancement of Internet and mobile technologies, blog, a one-way communication tool, has changed from a place one can deposit one's experiences in written language (a.k.a. text only) to an interactive communication platform that allows people to share, discuss, or broadcast one's experiences. Popular platforms found themselves to be a verb such as Twitter where "twittering" means "micro-blogging or writing a very short blog." Interactions between the owner of the blog and the blog readers made blogging a very useful research tool for the 21st century.

There are six types of blogging techniques based on two critical criteria: (1) content creators and (2) editing power of the content creators and platform

Web log ➡ Weblog ➡ Weblog

Figure 7.1.1 The origin of the word "Blog."

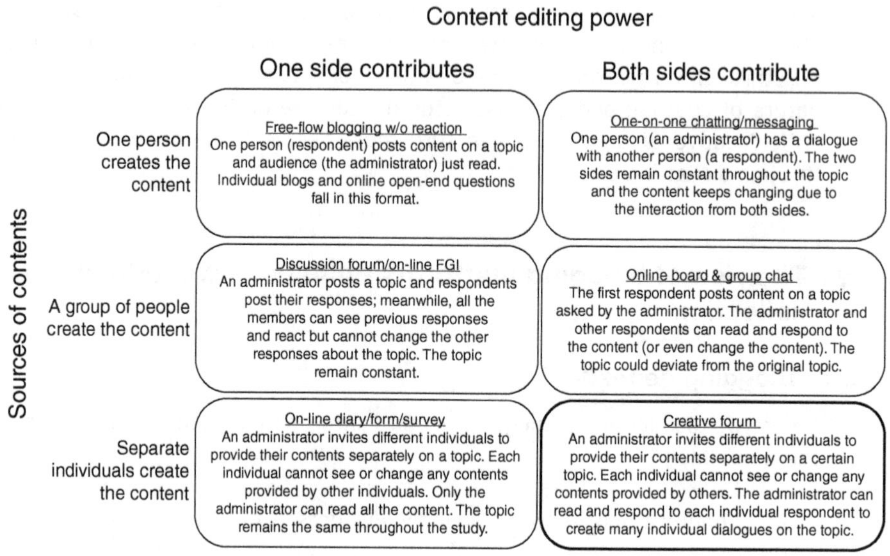

Figure 7.1.2 Six types of blogging techniques.

administrators (Figure 7.1.2). This chapter will discuss in-depth the Creative Blogging technique that is an application of Creative forum type blogging. There are several words or phrases that readers need to agree before continuing with this tool.

(1) **Content:** Contents of a blog comprise topics, stimuli, and responses.

(2) **Topic:** This is the theme of the content and is usually controlled by platform administrator(s) who created the study. In traditional Internet search studies, researchers set up a topic, and search through already created contents online. For example, a researcher is interested in "egg" and "its usage," then the researcher scrapes data from purchased or public datasets, or tracks posts for a certain time period. The topic is kept constant by the researcher (who is also called administrator here).

(3) **Content creators:** Content creators can both be the administrator(s) and respondent(s) using blogs as a dynamic research tool. Anyone who contributes to the "conversation or dialogue" between both sides and whose contribution can significantly impact the meaning of the content can be considered a content creator.

Our inspiration comes from Facebook. Like a Facebook post, an administrator posts a statement (a.k.a. stimuli) under a topic of interest, and then many respondents comment on the statement. Everyone (both administrator and respondents) see each other's posts and respond to the original statement or a specific post after the statement. Taking advantage of these dynamic interactions, the administrator can "control" the flow of the dialogues to "stay strictly" within or "deviate" from the original topic. Other respondents have this power as well. We see this naturally in Facebook conversations where one starts with a topic and their friends change the topic later within the same post. On the other hand, in online FGI, a group of respondents are invited to an online forum with a moderator. The role of this moderator is to ask the questions prepared from the discussion guide and to manage the conversation so all respondents can express their opinions equally. In this case, the content of the online FGI is generated from the respondent side only and the moderator does not significantly contribute to the content.

In summary, content creators are anyone who contributes significantly to the contents created throughout the length of a dialogue for a certain topic for a period of time in a study.

(4) **Administrators:** In this chapter, an administrator is responsible for controlling or moderating the flow of the dialogue, facilitating the interaction between the respondents and/or interviewing the respondents depending on the method used. For example, in online FGI, a group of respondents are invited to an online forum with a moderator. This moderator is considered the administrator for this online forum.

As with any interviewing platform, it is important for the administrators to be seasoned practitioners who study how to listen before asking the question.

(5) **Interactions between content creators:** The dynamics between content creators are dictated by methods and platforms used in a study and this results in the six types of blogging (Figure 7.1.2). Different blogging platforms have different degrees of flexibility and dynamics. Some are rigid, and some are very fluid. Choosing a platform depends on the objective(s) of the study. For example, if one wants to conduct an online survey for 10 000 consumers, one may use Google form®, or other online survey tools rather than choosing Twitter. The interactions between content creators are the criterion that classifies types of blogging because of the contribution of the contents called content editing power.

(6) **Content editing power:** There are two types: (1) One side contributes and (2) Both sides contribute.

One side contributes type is when the methods used allow only one side (either administrator or respondent) to contribute significantly to the content. Online diaries, forms, and surveys are examples because only the respondents can contribute to the answers in the questionnaire or form designed for the study. The administrator's role is to design the form and distribute it to the respondents. The real contents will be created by the respondents.

Both sides contribute type when the methods used allow anyone participating in the study to change the contents significantly. For example, one-on-one chat or a post in a blogging platform is considered this type of blogging because both ends of the dialogue can contribute significantly to the content of the blog. You and your friend in a one-on-one chat can put a significant amount of information in your conversation or even change the topic of your conversations in the chat. Therefore, knowing the differences between "One side contributes" and "Both sides contribute" types allow CEs to choose appropriate blogging platforms and techniques for their research objectives.

(7) **Sources of content** is not about who generates the content but how much and how the content is collected from "respondents" exclusively. In Figure 7.1.2, there are three types of Sources of contents:

1. **One person creates the contents:** This content source is the most found online and can range from blogs from influencers, members of discussion groups to web novels that can be readily mined by topic. The author, in this case, contributes almost 100% to the content.
2. **A group of people create the content:** This type of content source has been used in research for a while. There are many tools and platforms under this type of content sources such as online FGI, forums, discussion boards and wiki-ish sites (which we will refer to as *nucleated discussion forums*). What is common among these tools and platforms is that the contents are generated by a group of people either from respondents alone or from both respondents and administrators. The interaction among respondents and administrator and the allowable outcome of the study differentiate methods that employ these tools and platforms. For example, an online FGI where a group of respondents discuss a topic posted by a moderator (an administrator); meanwhile, all group members can see each other's responses and can agree or disagree with the responses. The final content of this activity will be mostly from the discussion among the respondents with all the verbatims. On the other hand, a nucleated discussion forum allows an administrator to start posting content and then let other people read and make any changes including altering the original structure of the content. The people and the administrator see only the latest change and make changes based on the latest response. This type of platform is very useful for quick concept refining. Both respondents and administrators can contribute significantly to the final content.
3. **Separate individuals create the content:** This type of content source is equivalent to the first type multiplied by the number of individuals (respondents) involved in the study. It is also one of the most used research tools as online surveys and blogs fall in this category as the method collects contents (the answers) from the respondents using a designed set of probing questions. Therefore, online survey type study gets their contents from one side only (from the respondents). Creative Blogging also falls into this category but opposite from the survey, this method uses the one-on-one chatting tool to create separate conversations with many respondents simultaneously. Therefore, Creative Blogging gathers contents from both sides (respondents and administrator) as administrator (the interviewer) can contribute significantly to the outcome of the final content at individual and overall levels.

7.1.3 Creative Blogging

Creative Blogging is created based on "experimentation" and "freedom of creative expression" principles, which differentiates it from other traditional online protocols if one considers how contents are flowing among administrator/moderator and respondents (Figure 7.1.3).

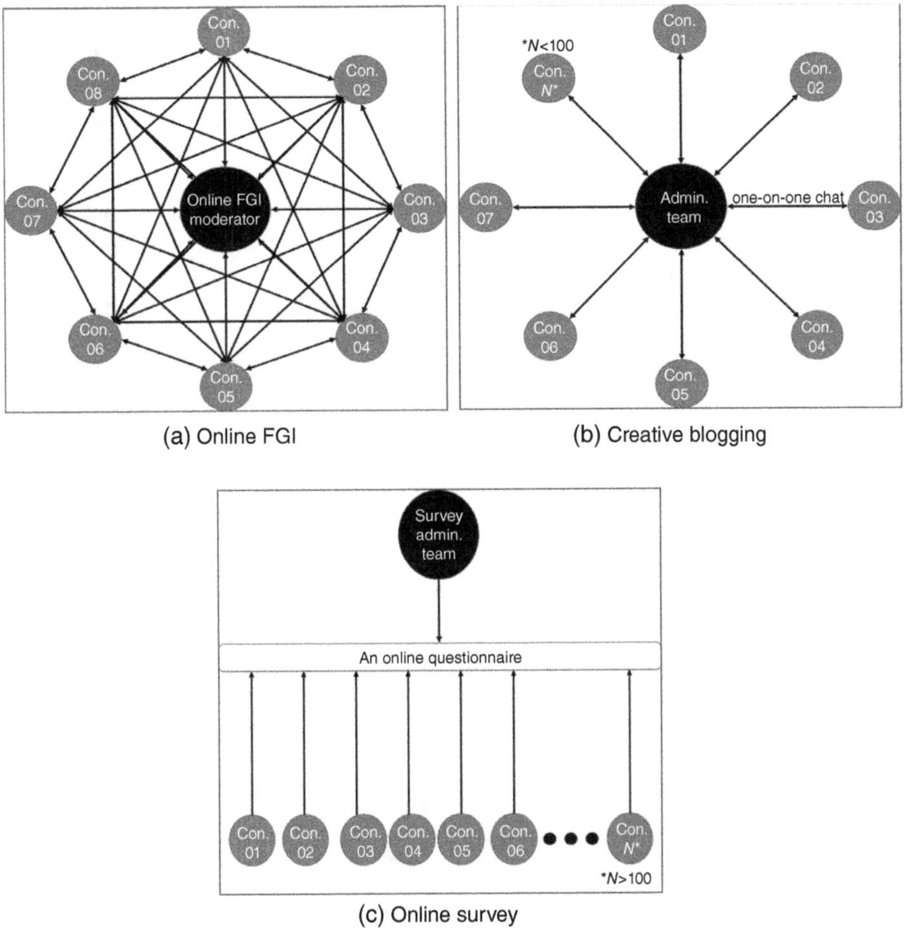

Figure 7.1.3 Comparisons of content flows between Online FGI (a), Creative Blogging (b) and Online survey (c).

For online FGI type protocols (Figure 7.1.3a), the contents are generated from every respondent to everyone including the moderator (in true FGI, the moderator's role on content is minimal and restricted to the questions and giving all respondents equal chance to speak out). With the contents flowing everywhere in both directions continuously within the FGI, the group size is usually limited to not more than eight to ensure moderator's control on the discussion topics.

In a traditional online survey, where researchers and moderator team (administrators) create a questionnaire and post it online to target respondents (Figure 7.1.3c). The contents are mainly generated by the respondents, and the contents flow in one direction toward the administrators. Because of this orderly content flow, online surveys can be executed with many respondents (100 and more is a norm) with ease.

For Creative Blogging, the content flow is restricted to be between respondents and facilitators in both directions by employing a one-on-one chatting platform (Figure 7.1.3b). The interaction between the facilitator and the respondents is dynamic since there is no "peer pressure" in a one-on-one chatting platform.

In addition, the conversations between the facilitator and the respondents stay focused on topics and provide in-depth discussions because of the use of semistructured interview protocol. Because of the simpler content flows compared to online FGI protocols, "experimentation" can be implemented easily in Creative Blogging protocol by designing the flow of conversations to answer specific hypotheses about the research stimuli or systematically presenting the stimuli following an experimental design. In this area, the rule of thumb for participants vs. facilitators is 30-70 respondents will need about two to three facilitators (Figure 7.1.4).

Appropriate and minimum number of respondents needed to execute a protocol to give a meaningful result directly impacts the research budget. Creative Blogging can give a meaningful result from 1 person (like in clinical case study) or 70 respondents (for in-depth understanding of consumers). FGI requires at least three and not more than eight people (within a target subset) to execute it properly. Online surveys and Internet-content studies require larger sample

	Creative blogging	Online FGI	Internet survey	Internet content search
Minimum number of respondents /execution/topics	1	3	60	100
Recommended number of respondents per execution	30–70	8	100+	400+
Content flow	Both directions	Both directions	One direction	One direction
Ease to control the content flow	Easy	Difficult	Very easy	Very easy
Room for respondent's creativity	High	Moderate	Limited	Limited
Major task for respondents	Having private dialogue/ conversation	Discussion in group	Answering questions	n/a*
Potential for primary in–depth understanding	High	Moderate	Limited	Limited
Application of the Experimentation principle	Very easy and built-in in every step	Difficult due to time constraint	Only in questionnaire design and may be in product testing	n/a*
Qualitative or Quantitative	Can do both	Qualitative	Quantitative	Quantitative

*n/a = not applicable

Figure 7.1.4 Comparison of some technical specifications between Creative Blogging, Online FGI, Online survey, and Internet content search.

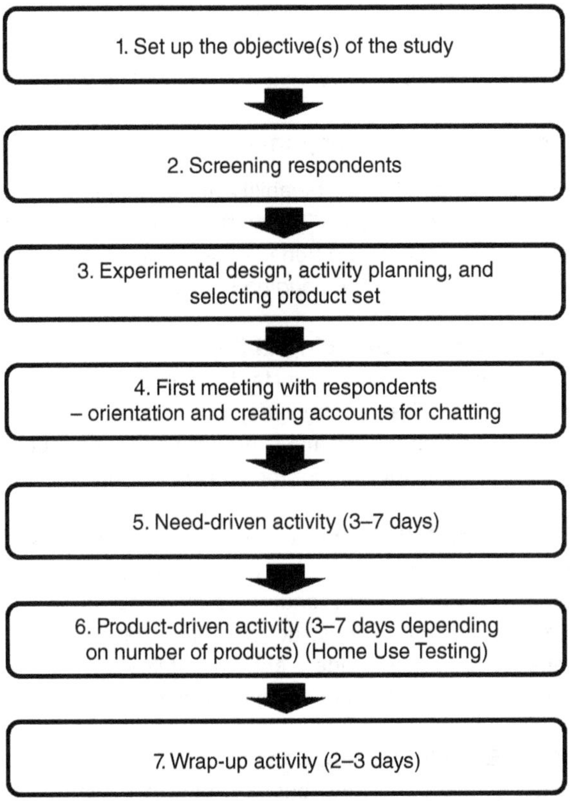

Figure 7.1.5 Seven steps for successful execution of Creative Blogging.

sizes for statistical quantification. Because of the dynamic nature of Creative Blogging, this protocol can provide in-depth understanding of a person as well as a group of people through activities and experimentation. On the contrary, an Internet-content study where researchers search for target consumers who already blogged or talked about the topic of interest for further analysis does not allow for experimentation due to its retrospective nature.

Figure 7.1.5 shows the seven steps to follow to take full advantage of the flexibility of the Creative Blogging protocol.

7.1.3.1 Step 1: Set up the objective(s) of the study

A good and successful research starts with a clear objective. In practice, especially for multifunctional projects and at the early development stage, identifying a single objective is quite a challenge. However, it is critical for the exploration team to prioritize the objectives and focus on the few top ones. Creative Blogging is suitable for broad-meaning objectives as it is quite powerful to explore and understand consumer behaviors and expectations. With experimentation mindset and product-driven activities (including product testing), Creative Blogging is also appropriate for screening prototypes, and identifying formulation directions. Another tool that can be used in conjunction with Creative

Blogging is qualitative multivariate analysis (QMA). QMA can accommodate many objectives usually found during the insight discovery phase. Chapter 6.2 of this book provides step-by-step explanations and case examples for QMA.

7.1.3.2 Step 2: Screening respondents

Creative Blogging relies on the ability of the respondents to express their opinions and perceptions readily. In addition to consumer profiles targeted by the team, there are three additional criteria to be considered when screening respondents for a Creative Blogging study.

7.1.3.2.1 Step 2.1: Blogging ability

Blogging ability is the most important among the three criteria. It includes both "willingness to use" and "familiarity with the blogging platform." Fortunately, online chatrooms and social media platforms are very popular and have become mainstream forums for most of the consumers including many baby-boomers.

7.1.3.2.2 Step 2.2: Articulation and expression ability

Articulation and expression ability to describe experiences using words and images is important but secondary to blogging ability because of the use of experimentation and semistructured interview. With experimental design, preprepared interview guides, and the ability to search for images on the Internet alleviate the need of using lengthy words or writing skill.

7.1.3.2.3 Step 2.3: Discipline

Discipline is very important, but it is one of the hardest to screen. Creative Blogging is quite lengthy and intense with different activities every day throughout the study. Therefore, respondents must follow the two following rules closely: (1) Timely response to probing questions and (2) Timely use of products according to schedule. From our experience, about 10–15% of respondents are not disciplined enough to go through a Creative Blogging study. Hence, over-recruiting of respondents by 10–15% is recommended.

7.1.3.3 Step 3: Experimental design, activity planning, and selecting product set

This step can be done simultaneously with Step 2 and right after Step 1.

What is experimentation? Science-trained CEs know this in its more extensive form called "design of experiment" or "DOE." However, applying full-fledged DOE principle in this business context may be unmanageable for some. Therefore, only the core principle of DOE has been adopted, and the word "Experimentation" was born. There are seven steps in applying "Experimentation" in consumer research including Creative Blogging studies (Figure 7.1.6).

Applying these seven steps for successful experimentation in Creative Blogging is critical. Designing need-driven and product-driven activities in Creative Blogging are based on experimentation principle and this practice allows CEs to draw "true causal conclusions." The authors will explain how to apply experimentation principle at different stages in Creative Blogging in more detail in Steps 5 and 6.

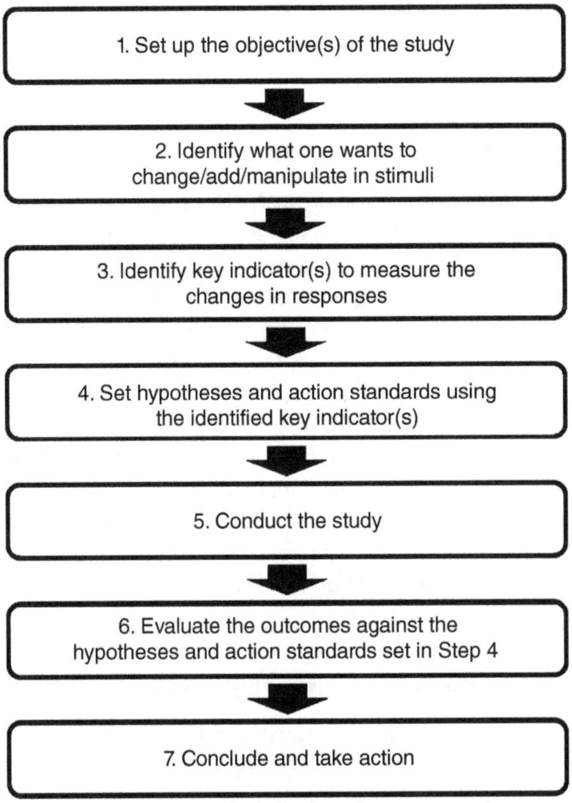

Figure 7.1.6 Seven steps in applying "Experimentation" for a successful research project.

7.1.3.4 Step 4: First meeting with respondents – orientation and creating accounts for chatting

Putting the respondents at ease with the platform and the activities is key. What needs to be done in this step are the following:

7.1.3.4.1 Step 4.1: Explain the objectives and activities planned for respondents

Any questions, especially the meaning of words/phrases used in the study questionnaire should be clarified.

7.1.3.4.2 Step 4.2: Form chatroom between facilitator(s) and each respondent

Ensuring connectivity between both parties is a must especially when a close-platform is chosen over an open platform.

7.1.3.4.3 Step 4.3: Answer any concerns

This step helps put the respondents at ease particularly how the data will be used and shared by the team and maintaining privacy for the respondents.

7.1.3.4.4 Step 4.4: Obtaining all required documents

Respondents should sign confidentiality (standard) and media release (if used for testimonials) forms.

7.1.3.5 Step 5: Need-driven activity

7.1.3.5.1 Applying experimentation in need-driven activities

Need-driven activities are assigned before product-driven activities to study people's beliefs, attitudes, moods, and expectations of a product category and not clearly written product concepts. Due to its "broad nature," applying experimentation in these activities can be done upfront through a "series of questions." For example, first, CEs can ask respondents about their "regular lives with a product" of interest (e.g. habits, attitudes, beliefs, and expectations) and then, ask the respondents again to imagine their lives "without the product" and how they will feel about not having the product. By contrasting the responses from both questions can yield many insights. Another way to apply experimentation at this stage is by asking respondents to think about a certain broad concept (e.g. moisturization) and, then, ask them to provide examples of their experiences or any products that give them a "moisturizing feeling" and describe why they chose those products.

7.1.3.6 Step 6: Product-driven activity Home Use Testing

7.1.3.6.1 Applying experimentation in product-driven activities

Product-driven activities in Creative Blogging are assigned "after" need-driven activities to prevent biases due to product experiences. Depending on the stage of innovation and the focus of the study, the order of test product prototypes and well-written concepts are very important. Testing a well-written full concept before product prototypes will impact how consumers evaluate the prototypes because their expectations have been set. Usually, if a product does not fit with the concept, the consumers will evaluate the product badly, even though, in blind testing, the consumers liked this product.

Experimentation principle is usually applied in "product formulation level" or "benchmarking activities." The former will give a more concrete conclusion as the changes will be controlled by controllable and measurable factors such as ingredients, processes, or sensory characteristics that can be traced back to product formulation and processing. This practice is recommended for product formulation-led projects. If formulating prototypes is not applicable, benchmarking with reliable measurements is encouraged. Reliable measurements such as physical, chemical, or sensory measures need to be identified and the changes in these measures from actual products (prototypes and benchmarks) must be reliably identified, perceived, and measured. In short, the CEs will need to know exactly what and by how much the products differ from each other so the consumer reactions can be linked back to the product design.

There are many activities and tasks for consumers and blogging facilitator(s) to do in this step. Figure 7.1.7 shows an example of a probing sequence for a product-focus study.

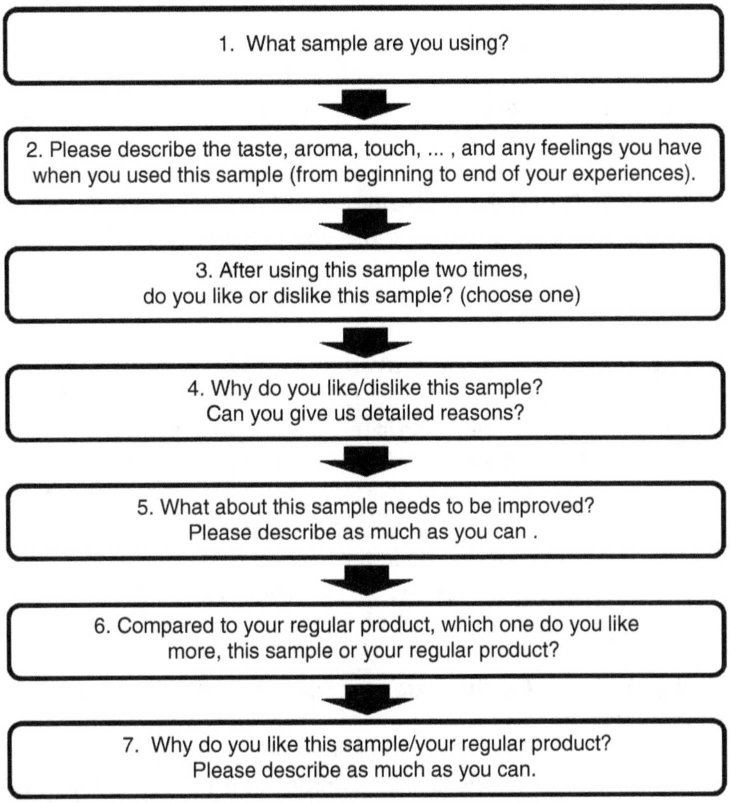

Figure 7.1.7 Seven step probing for a product-focus study in Product-driven activity Home Use Testing.

Step 6.1: Ask to confirm and identify the actual sample they evaluated

This step is critical especially, if the consumers will have all the test samples with them at once (due to budget constraints), and they can use whatever sample they want (free choice). Product sample presentation order can be provided to ensure balancing the order effect; however, people can still make mistakes. Asking for verification is still necessary. Beside typing in the number, a photo of the actual sample can be taken and submitted as well.

Step 6.2: Probe for their experience

This section can be varied to make the blogging more fun. For example, one can play "the three word game" (e.g. give me three words that come to your mind after using this sample) or ask them to find images representing their feelings after using the sample. Beside open-ended type questions, many categories (e.g. flavors and perfumes) can benefit from using choose-all-that-apply task (CATA) listing a range of potential descriptors (curated by the project team and/or based on hypotheses). Using the CATA task in this step is suitable when one wants to know whether the consumer's perception of the sample matches with what researchers have in mind or not (details of CATA can be found in Chapter 7.2).

Step 6.3: Ask for their judgments

Hedonic judgement such as preference is a usual key indicator so liking rating can be asked here. Asking them to answer a short questionnaire online is also possible at this stage. This step makes Creative Blogging encompass Home-use testing as well.

Step 6.4: Probe for the reasons behind the judgments

This step is critical and the reasons obtained are usually beyond sensory characteristics. It is also sensitive to their expectations and can be impacted by context and testing order.

Step 6.5: Probe for negatives and possible solutions

This probing question could be asked using a projective technique such as "if you want to improve anything in this sample, what would that be?" It is important to ensure that this answer in this step is dependent on the answer in Steps 3 and 4. In some cases, the sample is liked so much, and they did not want to make any change to the sample or the sample is disliked very much that there are too many things to change. The latter phenomenon usually ends up with answers such as "changing everything" or "no change required, just throw it away." In this case, skillful probing further is required.

Step 6.6: Ask for their judgments compared to their internal benchmarks

This step provides key evaluation of the sample such as why the sample fails or wins over the regularly consumed product. If the sample is well liked but still cannot beat the regularly consumed product, other reasons beyond sensory characteristics are the usual suspects.

Step 6.7: Probe for reasons behind winning or losing to the benchmarks

This step aims to probe the reasons behind their judgment in Step 6. Being liked very much does not mean a sample will beat the regularly consumed product. From the authors' experience, there were several incidents where the product was liked very much but failed to win over the current product because the sample was "the same or similar to" the current products. Most winning incidents were found when the samples were somewhat different, yet they still feel familiar about it. The degree of differences and familiarity depended on product categories. In Creative Blogging, the answers to this question can be found in Step 2, 4, and 5.

Other tasks beyond what have been described here can also be added from steps 2 to 7, depending on the objectives of the study. Keep in mind tasks that require immediate or instant answers are not suitable for this platform.

7.1.3.7 Step 7: Wrap-up activity

This step usually lasts for one to two days for the team to wrap-up or ask additional questions. Concept-Product fit can be asked here as well as evaluation of well-written concepts. Do not try to add too many activities in this step as the consumers will be exhausted and unreliable outcomes can occur.

7.1.4 Creative Blogging in practice: a case example in Thailand

A few years ago, the authors conducted an interesting research to gain deep understanding about "refreshing feeling" among Thai consumers. Results in many surveys revealed that "refreshing" and "overall liking" were highly and positively correlated. However, from many regular interviews, we knew that they were different. Therefore, the second objective of this study was to understand the differences between "overall liking" and "refreshing" as well. To accomplish these objectives, the team decided to use the Creative Blogging method.

The project was designed in two major sections. The first section was the need-driven step (Step 5), where our experimentation was how Thai consumers reacted to the word "refreshing." What was interesting about this activity was that we did not ask the respondents to describe the meaning of the word refreshing but instead instructed them to talk about and/or buy (within set budget) whatever products they think were refreshing to them and blog about the experience. After the first week, we switched to the product-driven step (Step 6) where we gave them a set of preselected beverage products that we know have different attributes that can be described as refreshing. We asked them to try the products one at a time and chat with us. The process took two weeks to finish.

Our expectation was that Thai people would buy many "cold beverages" as Thailand is a very hot and humid country. However, what we got was a surprise for all. First, surprise was that the beverage did not need to be cold to be refreshing. Hot beverages could be refreshing as well. The second surprise was food can be refreshing as well as beverages, and the third surprise was there were at least four variations of refreshing feelings among Thai consumers. Let us take you through some details. The first week resulted in 31 product categories that they thought were refreshing, and these 31 product categories covered hot beverages to food items. There were more than 1000 words used to describe their feeling "refreshed." Among the thousand-plus words, 61% were positive, 28% were neutral, and 11% were negative. Analyzing the conversations about each product category (from 31 categories) revealed that there were several key decision steps and feelings based on context that Thai consumers considered to describe refreshing beverages. In a certain context, a cold bubbly drink would not be considered refreshing, but a hot drink would be. Thirst quenching was a requirement for a beverage to be refreshing, but was not the most fundamental property for all foods. Figure 7.1.8 shows a partial network of a beverage that Thai consumers thought was refreshing.

Network analysis of a Creative Blogging on a refreshing beverage revealed that, for this beverage, being refreshing meant that they were energized and cold was the primary product characteristic followed by having good identifiable flavor, thirst quenching and revitalizing. This answered our first objective. In addition, from "refreshing" through "cold," Thai consumers "like-it." These two-links revealed the relationships between "refreshing and liking" that answered our second objective. For this particular product, if the beverage was consumed when it was cold, it was considered "refreshing" and also "liked" but liking was

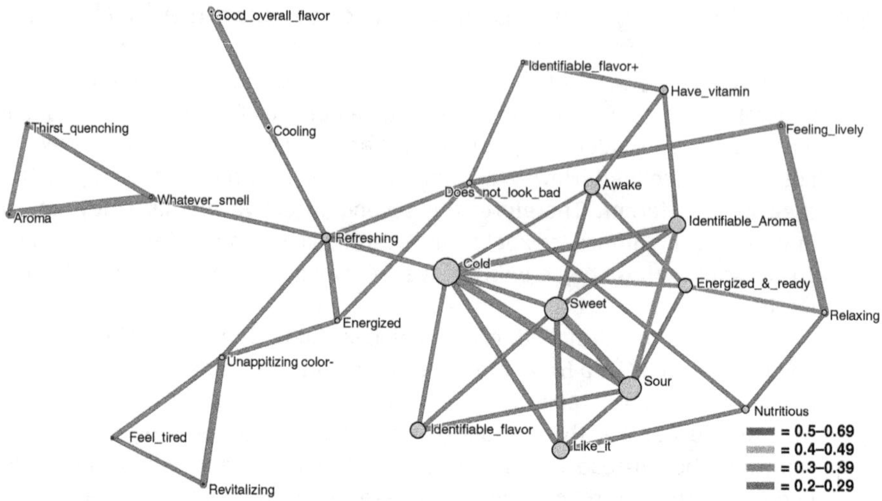

Figure 7.1.8 The refreshing network of a beverage from Thai consumers. Size of the bubbles represents the contribution of each word to the total variance. The strengths of the links were Jaccard similarity coefficients.

not directly influencing refreshing like we had thought from other methods through correlation analysis.

7

7.1.5 Choosing the platform: Close- or Open-platform

There are many chatroom applications available for Creative Blogging activities. They generally fall into two types: Open and Close.

Open platforms are chatroom programs or applications that research teams do not really have a control over. These open platforms are readily available and most consumers are familiar with them, e.g. WhatsApp, Line, Band®, Kakao Talk®, and WeChat.

Close platforms are proprietary chat room programs or applications that an agency or even your company owns, and the CEs have all the controls over the program. Primary advantages of using close platforms are security and data collection, e.g., ensure the discussion will not be hijacked by bots or trolls. Figure 7.1.9 compares certain operative advantages and disadvantages between open and close platforms.

A close platform is recommended if the stage of innovation is where information security is a priority. If the project is about exploration, using an open platform is recommended as it will not disrupt consumers' lives that much, and they are already familiar with the application.

7.1.6 Read between the lines: dialogue with consumers

Open-ended questions, one-on-one interviews, and live discussion provide very rich information because the methods are interactive. Creative Blogging takes advantage of this interactive aspect from writing and having two-way

Operative features	Platform: Open	Platform: Close
Consumer familiarity (they can use it with minimal learning)	✓	-
Information security (no spam or unwanted party access)	-	✓
Data collection & transfer for analysis (automation)	-	✓
Disruption to consumers' lives (they have to use 1 more application)	-	✓
System stability and bugs	✓	-

Figure 7.1.9 Comparison between close- and open-chat room platforms on operative features.

communication, and combines them with digitalization, gamification, and mobile technologies to enhance the dialogues with consumers. These enhancements can help consumers to be more expressive, bring out their creativity, and facilitate capturing information. To ensure that the objectives of your study will be met, three issues need to be decided when designing a Creative Blogging study:

- **Interview rigidity:** Structured, semistructured, and unstructured interviews: Creative Blogging employs all three types of interviews.
  - Structured interview and rigid-type questioning are very useful for getting uniform answers from consumers because all consumers will be asked the same question in the same way. Survey is the most structured interview (if an interview is used as a data collection method) with the most rigid questioning because the whole sequence cannot be changed and survey questions are mostly accompanied by predesigned scales where consumers must answer using the scales. However, its rigid nature prohibits creative expression and probing. Consumer explorers can use structured interview and rigid-type questioning where you want to get quite uniform answers. There are two levels in Creative Blogging that apply the structured interview and rigid-type questioning principles: (1) during designing the whole sequences of the conversation, and (2) when asking for a particular question (e.g. liking or purchase intent) and using the answer as foundation for further probing. We discussed applying experimentation principles in Steps 5 & 6 of designing a Creative Blogging study. These two steps are where we can apply the structured interview principle to a Creative Blogging study to ensure that all the objectives will be answered by designing the whole sequence of conversation in advance. Another place to apply rigid questioning principles is when designing the questions within the conversation sequence. We can ask consumers to evaluate a

product and ask them their liking level and then probe it with less rigid type questioning like an open-ended question. Example below shows such a sequence of interviewing.

FACILITATOR: After trying the product number 379, What is your overall opinion about this product? (choose one of the choices here: dislike it extremely, dislike very much, dislike moderately, dislike it slightly, like it slightly, like it moderately, like it very much and like it extremely).

CONSUMER: I dislike it slightly.

FACILITATOR: You said you dislike this product 379 slightly, can you tell us why you dislike it and why it was only slightly?

- Semistructured interview is very useful and essential for Creative Blogging. Semistructured interviewing technique is applied in Creative Blogging at a higher level than applying structured interview. A structured interview is applied on a question or wording and choices of questions. However, semistructured interview principle is applied when we arrange the flow of the interview sequence including when to ask a question and when to probe for deeper insights. This activity happens impromptu during the study, and it takes a facilitator's judgement to do so. The example above shows the use of a rigid-type questioning and then it is followed by an open-ended type questioning (much less rigid than the question with choices). Here, in Creative Blogging, we encourage facilitator(s) to "deviate to probe for insights." For example, following the example above, a consumer answered

7

CONSUMER: I disliked it because it tasted like licorice, but it was not that bad. It was just not expected.

Here, we know that the consumer perceived the "licorice" flavor, and it was ok. However, the consumer expected something else and the facilitator is encouraged to probe further to get to "what the consumer expected." Once it is done, the facilitator must "revisit" the predesigned conversion sequence. The "deviation from" and "revisit" the predesigned conversation sequence is the semistructured principle.

- Unstructured interview is very useful for an in-depth interview where the consumer's expression will lead the whole conversation (no predesigned interview structure). The facilitator needs to spot an "interesting" topic of the conversation and probe it further. Therefore, the topic of the conversation in an unstructured interview can change from the beginning due to its impromptu nature. In Creative Blogging, we use the unstructured interview principle in a limited fashion by using it for deep probing and asking open-ended type questions. Let us continue the above example. The facilitator(s) should spot "*it was just not expected*" as an important point to probe. The facilitator can ask the following with another question that is usually an open-ended question.

FACILITATOR: Oh, what did you expect from this sample? Tell me more.

Here, one can imagine that the conversation can get deeper and deeper. The next issue is how deep we should probe the consumer.

- **Probing depth, how deep should you probe them:** In Creative Blogging, the facilitator has room to decide and change the probing questions. In a live interview, if the interviewer decides to "stop" probing, then they will not have a chance to come back and probe it again. However, in Creative Blogging, the depth of probing is determined by two factors: (1) time constraint, and (2) consumer's reaction. Even though the time frame in Creative Blogging for a topic is much greater than that of interviewing (many hours vs. minutes), there are also many activities planned in advance for consumers to do (the structured nature of the method). Therefore, from the authors' experience, two to three rounds of questionings should be enough for an articulated need or want; otherwise, the facilitator needs to change tactics in probing from direct questioning to something else (e.g. gaming, collaging). Another cue that the facilitator can take is how the consumer responds to the latest question. From our experience, consumers "stop" answering altogether after two rounds of questioning. Another cue is that consumers just answer with a "holistic response." For example, if we ask them about what to improve in the product and the consumer says, "just change the whole thing" or "there is nothing to improve, it's just bad." These are the signs to "stop probing" as the way you ask is ineffective.
- **Beyond words - images, gamification, and other media:** In addition to asking questions, there are other techniques that CEs can use to gain insights beyond words. Instead of asking consumers to describe their feelings toward a concept or a product sample, a facilitator can ask the consumers to submit images that represent their understanding and feelings. Besides the topic of the images submitted, colors, and tones of the images are very useful for marketing and packaging design.

Gamification is another way to create a fun environment for blogging. For example, a three-word game is very useful to probe for top-of-mind and immediate important thoughts about product experience or concepts.

Example questions for a three-word game:

- After tasting the product, can you give me "three words" that come to your mind.
- Thinking about brand xxx, can you give me "three words" that you think this brand makes you feel.
- After using the product for a while, which are the top three words from the list provided that represent your opinion about this product?

Another game that is very easy to implement is "fill in the blank." This technique is very useful when one wants to construct a short concept of products or services. Chapter 6.3 describes this gamification approach in detail.

7.1.7 Future of Creative Blogging

There are many ways to use Creative Blogging forums during the product research cycle. The method will continue to evolve with new digital technologies, i.e. artificial intelligence (AI), robotics, conversation platforms, data capture, natural language processing, advanced data processing, analytics, and visualization. What keeps it unique from other blog platforms is the application of the experimentation concept where Consumer Explorers can learn how people react to products through hypothesis testing and design of experiments as backbone to systematic learning. The challenge remains on training and finding facilitators who can listen, probe, and uncover in-depth insights while balancing time requirements. Additional support such as creating a conversation guide where facilitators can anticipate possible answers and prepare questions in advance can be built into the program. A double-edged future projection and challenge is the use of chatbots and digital assistants in interviews. This is probably okay for simple interviewing, but seasoned facilitators must be adept in active listening and skillful conversation to capture true reactions to complex scenarios.

Discussion questions

- What are advantages and disadvantages of Creative Blogging for new product development (NPD) over other traditional methods such as focus group interview, in-context research, or home-visit interviewing?
- How much structure and metrics are appropriate for Creative Blogging activities?
- How might the experimentation concept in Creative Blogging take form? What are the advantages/disadvantages for bringing experimentation into a blogging study?

7.2

CATA as a Decision-Making Tool

Kannapon Lopetcharat and Dulce Paredes

Key learnings

- ✓ Big data and data science
- ✓ Choose-All-That-Apply
- ✓ Total Unduplicated Reach of Frequency (TURF)
- ✓ Network analyses

7

7.2.1 Introduction

Check all that apply or CATA task has been used in market research for many years as a part of a survey questionnaire. CATA has been gaining popularity among sensory scientists since the presentation at the 7th Rose-Marie Pangborn Sensory Science Symposium by researchers from Frito-Lay (Adams et al., 2007). The task allows product researchers to (1) differentiate products, (2) characterize product differences, (3) study correlations among attributes, (4) identify drivers of liking, and (5) help in product formulation/creation efforts (Meyners et al., 2013).

CATA task is very attractive for product researchers because it is easy to use (not-much to no-training required), intuitive and relatively quick to execute. Varela and Ares (2012) and Buck and Kemp (2018) comprehensively reviewed the applications of CATA for descriptive analysis; while Meyners et al. (2013) provided a guideline on analyzing data obtained from CATA tasks for descriptive and product formulation purposes.

For this chapter, we will not focus on-product characterization alone but on how to create a benefit proposition using a CATA task with accompanying data analysis techniques: total unduplicated reach of frequency (TURF) and Network analyses.

TURF analysis had its origin from media campaign planning, and it has been adopted by market research for product line planning (Moskowitz, 1996).

Product Innovation Toolbox: A Field Guide to Consumer Understanding and Research, Second Edition. Edited by Kannapon Lopetcharat, Dulce Paredes and Jacqueline H. Beckley.

Its original purpose was to find a combination of products (a.k.a. product line) that can satisfy the most unique consumers (a.k.a. reach) in a market. Extending this line of thought, TURF can help narrow down the number of benefits to create a concept with many benefit propositions that can attract the most consumers in a targeted market. However, TURF assists only in "identifying" benefits, but "not understanding," how the benefits are perceived by consumers who want them. To gain this in-depth understanding, a Network Analysis is needed.

Network Analysis is an application of Graph Theory where the focus of the analysis is on "type connection and its strength" between things and presenting the connections on a two-dimensional picture unlike other conventional visualization methods such as principal component analysis (PCA), correspondence analysis (CA), or multi-factor analysis (MFA), where the "connections" are presented in a multidimensional picture.

In Network Analysis, researchers know and measure a type connection (e.g. friendship, occurrence, employment, or even a physical distance from one place to another) and convert the connection to see how the objects of interest "connect" to each other. If there is a connection between two objects, a line (called Link or Edge) will be drawn. Finally, after all connections among the objects are identified (via actual research), then a full network can be elucidated. The notion that researchers know about the "connection" and "collect data on that particular connection" fundamentally differentiates Network Analysis from the other traditional methods (e.g. PCA, CA, MFA). Because, in applying the traditional methods, researchers "do not know the connection," and they are "looking for the type of connections" and then visualize "the derived connection" on a multidimensional space; even though mathematical calculations may be similar in principle such as the Chi-square distance used in CA.

Figure 7.2.1 shows a comparison between a network and a multidimensional map. For a multidimensional map like a result from PCA (Figure 7.2.1b), there are two axes with arrow heads. Each axis has a meaning; hence, the arrow head (can be double or single arrow heads). The arrangement of any object (i.e. 1, 2, 3, 4, 5, 6) needs to be interpreted along each axis.

The construction of each axis gives the meaning to the "relationship" among the six objects. In this example, the researcher can conclude that object 1, 2, and 3 are more similar to each other and object 6 is the most different. There usually is a problem interpreting object 4 as it falls on (0,0) co-ordinate of both axes. Object 4 can be the most different as it needs the 3rd axis or it actually is on (0,0) coordinate.

Now, let us look at Figure 7.2.1a, which is a network representing the same data in Figure 7.2.1b; however, there is no need for any axis. The connections and their strength between the objects are in the links (thinner = weaker and thicker = stronger). In Figure 7.2.1b, the interpretation of objects 1, 2, and 3 as more similar to each other is the same as Figure 7.2.1a, but we see the underlying reasons why. Objects 1, 2, and 3 in Figure 7.2.1a are connected by three links (two thick lines linking Object 1 to Objects 2 and 3 and one thin line linking Objects 2 and 3). This supports the interpretation of Figure 7.2.1b. In addition, two thick links connecting Object 1 to Object 2 and Object 1 to Object 3 reveal the reasons why the three objects are similar. Furthermore, there is a thin link between Objects 2 and 3 that enhance the similarity.

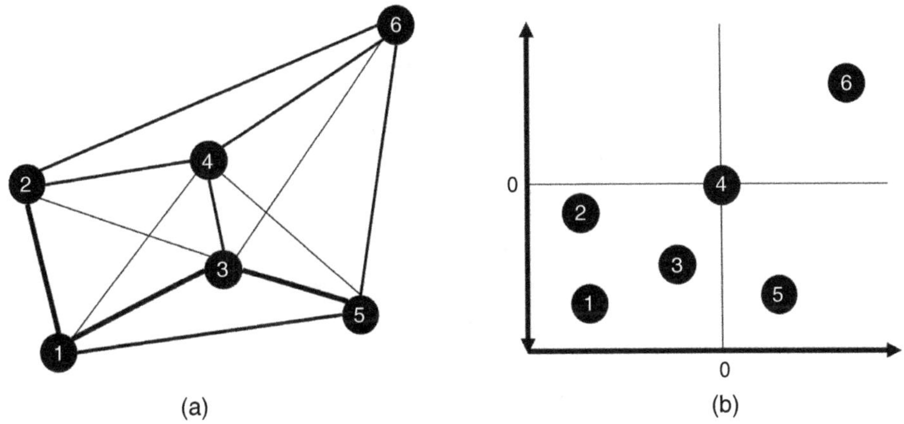

Figure 7.2.1 An illustration comparing a network (a) and a multi-dimensional map (b).

Therefore, instead of providing a bunch of points floating in an empty space, Network Analysis produces visual images that have actual links connecting the objects with quantified strengths. So Consumer Explorers (CEs) can actually "see all the connections" and "make decisions" with confidence.

The following discussion lays out best practices for applying and conducting CATA tasks with a case study in identifying and selecting benefits for creating a product/marketing concept for a first in market product launch.

7.2.2 Check All That Apply (CATA) task in practice

CATA task is a "reactive" type task where consumers react to a question by using a set of choices. There are three critical steps that CEs need to pay attention to when they want to apply a CATA task in their research, namely: (1) Framing the question, (2) Creating a choice set, and (3) Meaning alignment with consumers. These three critical steps need to be completed properly before conducting any study involving the CATA task. Figure 7.2.2 outlines seven steps for CATA best practice.

Step 1: Set up the objective(s) of the study: This is the most important step for the whole study (not the CATA task section) and will dictate the design of the CATA task as well as the overall direction of the study. For example, if the CEs decide to use qualitative multivariate analysis (QMA: Chapter 6.2) because we want to understand fundamental functions, benefits, and values of a product category, generate consumer friendly language for future concept creation, the CATA task can be applied in several sections in QMA procedure especially right after the Value Diagram building (see Chapter 6.2 for more details).

Step 2: Frame CATA question: This is the first critical step in CATA task creation - the objectives. There are many variations in asking the CATA questions based on the objectives. We will use a skincare study to illustrate the steps.

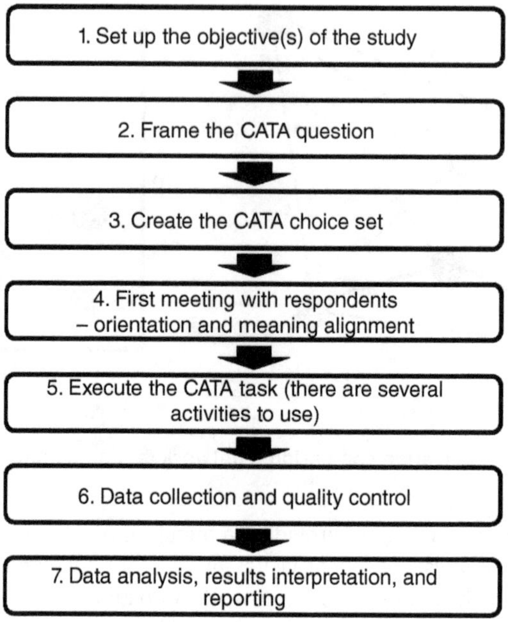

Figure 7.2.2 Seven steps for a best practice in implementing CATA task in any research technique.

Objective: Identify the **most-wanted** benefits for a skin cleansing product.

Examples of CATA question:
From the list of skin cleansing benefits provided,

- **Version 1:** Please choose any benefit(s) that you wish to have in your skin-cleansing product. You can choose as many as you want.
- **Version 2:** Please choose three benefits that you want to have in your skin-cleansing product.
- **Version 3:** Please rank all the benefits from what you want to have the most in your skin-cleansing product to what you want to have the least.
- **Version 4:** Please choose five most-wanted benefits that you want to have in your skin-cleansing product and then rank the five selected benefits from what you want to have the most in your skin-cleansing product to what you want to have the least.

First, we need to focus on the objective that the examples here are focusing on "want." This is the key word to use in the actual phrasing of the CATA questions. If the objective is to characterize or describe products, then the CE needs to change the word "want" to another more appropriate word such as "prominent."

Version 1 is the most commonly used phrasing. However, if you have too many choices (such as 30–50 choices) to choose from or have a small number of respondents (such as 1–30), the chance of getting "not important selections or nice-to-have selections" is high. This will interfere with interpretation and the final conclusion; therefore, other versions are more appropriate. However, the interpretation of the results will be a little different. This version is impacted by the presentation order of choices as well (Buck and Kemp, 2018).

Version 2 asks the respondent to choose the top three most-wanted benefits from the list. This version is suitable for situations where the CEs know that they will work on a limited number of choices from the beginning so the respondents will not "casually" choose the choice and "focus" more on what they really want. The choice-presentation order effect has not been studied for this type of task. Randomization of choices is still recommended.

Version 3 asks respondents to rank all the choices. This is actually a ranking task (not CATA task); however, one can convert ranking data to CATA-type data by applying a decision rule. For example, if there are 10 choices, the team can select only the top 3 choices to be "the chosen ones," and the other 7 choices are considered not chosen. This ranking and then converting to CATA-type information equivalent to Version 2 with the flexibility to change using only three choices to using four or more choices in subsequent analysis. However, this kind of CATA-task should not be used with more than 10 choices as it will be extremely time consuming. Ties should not be allowed as well. This version of questioning is resistant to the presentation order effect as the respondents must consider all the choices before ranking.

Version 4 is the hybrid between Version 2 and Version 3 (choosing a smaller set of choices and then ranking them). It is suitable when the choice set is quite large, such as 10–30 choices. The top-most choices should be between three and five choices as too many choices to rank will be time-consuming. The presentation order effect has not been formally studied for this version. From our experience, this version is quite resistant to presentation order effect and the respondents seem to consider all the choices properly with the orientation section.

Step 3: Creating CATA choice set: This is the second critical step in preparing CATA activities and the most important step. The action ability of CATA studies depends greatly on the list of choices. There are several guidelines for CEs to follow:

- **First guideline:** The choice list must "follow the primary objective of the study." If one wants to identify words/terms for market launch, only positive words/terms should be used as negative words/terms will have no use for the next step. If one wants to characterize or describe products, then the list should contain words/terms that balance each other and not use the word "not" in front of another word. For example, if researchers want to use the word "hard," then the other opposite word can be soft, supple, etc., but the word "not hard" should not be used as its meaning is too ambiguous.
- **Second guideline:** The number of choices/terms in a list should not exceed 25 (Buck and Kemp, 2018; Jaeger et al., 2015). We did not state the lower limit here as it mainly depends on the primary objective and the actual task performed. As low as five can be used, if you are interested only in five choices. From our experiences, 30–35 choices can be used with Creative Blogging technique (Chapter 7.1). For online questionnaires, we had successfully used 5–30 choices with repeated exposure to the choice set.

7

Step 4: First meeting orientation and meaning alignment: This is the third critical step. It can elevate the restriction of the number of choices in a list, increase focus of respondents, and reduce misunderstanding the meanings of the choices. Even though a CATA task is very easy to perform, it relies on the understanding of the choices (words/terms) used in the list. This step is very critical for choosing benefits for concept writing and testing, where the words/terms will be generated by the research team and not directly from consumers. Other applications where understanding the meaning of words/terms is important are studying consumers' perception of products, especially smell, flavor, and texture terms. For example, the word "white flower" is commonly used in fragrance. This term in the United States can connote the smell from white lily, peony, gardenia, and other common white flowers in the United States. However, white flower can be interpreted to different flowers in other countries such as Thailand where the term white flower mainly connotes jasmine. Therefore, an orientation to go through all the words/terms in the list before conducting the CATA task is very important. Going through and discussing the meanings at least once with the respondents will make them focus on the list, reduce the chance of "casual" selection, and minimize presentation order effect.

Step 5: Execute the CATA task: CATA task can be administered through many modes such as one-on-one interview, group interview, and online survey. The presentation of the list should be randomized and balanced across respondents as much as possible; however, each respondent should get the same order for all his/her exposure to the list within a study (Ares and Jaeger, 2013; Ares et al., 2014).

Step 6: Data collection and quality control: Digital collection and data arrangement of the CATA task are very easy and mostly error free. In situations where a CATA task is done using a paper ballot or a group interview setting like in QMA (see case study below), care must be taken in data transfer. The data structure of CATA study usually has respondents as a row and each choice as a column. The number in each cell is 1 or 0, where 1 means selected and 0 means not selected. If the ranking task is used (as in Version 2 and Version 4), after a cut-off rule is set, the ranks will be converted to 1 or 0 according to the cut-off rule.

Step 7: Data analysis, results interpretation, and reporting: Depending on the primary objectives, data analysis strategy varies. Meyners et al. (2013) described many statistical procedures for product characterization and optimization. For fundamental understanding in early phase of new product development and creative use of results such as selecting benefits for concept creations in late phase of new product development like the example in this chapter, the procedure involving TURF and Network analysis described in this chapter is more appropriate because they will provide both the identified benefit bundle and underlying reasons as mentioned in Figure 7.2.1. An additional example will be described in detail later in this chapter.

7.2.3 Selecting benefit propositions for a new product: a case study of a cleansing product using CATA

In 2016, one of the authors was contacted by a multinational personal care product company that wanted to launch a product from North America in Asia. They wanted to evaluate a new "American" facial cleansing product. Technically, the cleansing product was designed to perform a very specific function and provided specific benefits that were very new to the market globally. They used "new English words" to communicate the new function and benefits. When the product was launched in the United States, it performed quite well. Few months later, they decided to launch this particular product in Asia.

With new benefits communicated using a new set of vocabularies in English, the Asian team was unsure that the United States concepts would work in the Asian market especially after translation. The author conducted Qualitative Multivariate Analysis (QMA - Chapter 6.2) with additional activities using CATA task to investigate which fundamental values, benefits, and functions for skin care and cleansing products to lay the groundwork for concept creation. Figure 7.2.3 shows an example of the list of benefits generated and used in two QMA sessions. The list was created by 13 respondents from the two sessions during the Value Diagram building step. The list was a consensus from the first QMA session and the second QMA session also agreed that the list was comprehensive for them (Figure 7.2.3).

The respondents were asked to select three most important benefits for their ideal skincare and cleansing product from the list of 16 benefits (Figure 7.2.3). This task is equivalent to a CATA task where the respondents had to choose all benefits that are "applicable" for being "ideal skincare and cleansing products with three proposed benefits." This modification is suitable for situations, where the lists are all "positive" like the one in this study. Otherwise, the respondents would choose all of the benefits on the list.

In business communication and industry formulation activities, practitioners usually deal with good to better products; therefore, balancing of negative and positive attributes will not be appropriate as the negative attributes will not be of any use for them. On the other hand, in fundamental research such as product characterization and new technology screening, balancing between negative and positive attributes is highly recommended.

Hydration	Smooth texture
No blemish	Soft
Bright	Balance oil content
Supple	Firm
Acne free	Glow
Clean	Normal
Not oily	Small pores
Not dull	White / fair

Figure 7.2.3 An example of a list of benefits from a country in the Asian market. Note: This is an English translated version of the benefits that were generated in native language.

7

As expected, the benefits listed in Figure 7.2.3 did not contain any of the new benefits in the new product concept from the United States. Therefore, the team decided to create concepts that contain these fundamental benefits obtained from QMA and introduced the new benefits along with the fundamental benefits found from the study. A three-step analysis scheme was used to define the number of benefits and benefit combinations to use in the concepts.

7.2.3.1 Step 1: Select benefit bundles using total unduplicated reach frequencies (TURF)

Back to the study, after finishing this CATA task, the respondents continued other activities in the QMA workshops. The information obtained were arranged in a regular "attribute matrix," where each column represents an attribute (a.k.a. variable and in this case, a benefit) and each row represents a case (in this case, a respondent) where the number within each cell (where a column and a row intersect) represents the value of that variable that belong to that row. In this study, the value in each cell was nominal or ordinal at best as it indicated that each respondent wanted to have which benefit in their ideal skincare and cleansing product (yes/no) (Figure 7.2.4).

TURF analysis was done on the attribute matrix in Figure 7.2.4 to identify how many consumers can be reached by different benefit combinations. The TURF analysis discovered 20 benefit combinations that could reach at least 80% of the respondents. There were 5 pairs of benefits and 15 benefit triads. Figure 7.2.5 shows the Reach (%) of each benefit that indicated how many respondents from all respondents wanted each benefit.

Figure 7.2.5 shows a normal format for CATA-type or many choice-type tasks. Hydration was the most desired followed by bright and supple. Acne free was the fourth choice. From Figure 7.2.5, the team could have chosen hydration, bright, and supple for communication for their new product. This was where a mistake happened. The mistake was in the efficacy of the top four benefits they selected. How many people were actually reached by the top four benefits? There were six unique pairs and four unique triads of benefits could be selected from the top four benefits. How many people were reached by the pairs and the triads? This was where TURF analysis helped the researcher to solve the dilemma. Figure 7.2.6 shows the Reach (%) of some pairs and triads of the 16 benefits, and the team used this to select their final benefit bundles for the new product.

TURF identified only five pairs of benefits that could reach 80% of the respondents and four of them were from the top four benefits. One can see that hydration was the most prominent benefit to reach 83% of respondents, but it was not absolutely necessary as a supple-bright pair could reach 83% of respondents as well. In the case of benefit triads, again, hydration was the prominent benefit, but it was not absolute. Bright-supple-no blemish triad actually reached 100% of respondents without the help of Hydration.

At this stage, the team could use the results (Figure 7.2.6) to guide their selection of benefits (how many and which ones) in conjunction with other information such as technical performance and claim substantiation. The team could have chosen many benefit bundles, but they did not know if the benefit bundle "makes sense" for the consumers or not. To answer this issue, a Network Analysis was needed.

ID	Hydration	No blemish	Bright	Supple	Acne free	Clean	Not oily	Not dull	Smooth texture	Soft	Balance oil content	Firm	Glow	Normal	Small pores	White/fair
R01	1	1	0	1	1	0	1	1	0	1	0	0	0	0	1	0
R02	1	1	0	1	1	0	1	0	0	0	0	0	0	0	0	0
R03	1	1	1	0	1	1	0	0	0	1	0	0	0	1	0	0
•	•	•	•	•	•	•	•	•	•	•	•	•	•	•	•	•
•	•	•	•	•	•	•	•	•	•	•	•	•	•	•	•	•
•	•	•	•	•	•	•	•	•	•	•	•	•	•	•	•	•
•	•	•	•	•	•	•	•	•	•	•	•	•	•	•	•	•
•	•	•	•	•	•	•	•	•	•	•	•	•	•	•	•	•
•	•	•	•	•	•	•	•	•	•	•	•	•	•	•	•	•
R11	0	0	0	1	0	1	0	0	0	0	0	0	1	0	0	0
R12	1	1	0	0	0	0	0	0	0	0	0	1	0	0	0	0
R13	3	0	1	1	0	0	0	1	0	0	0	0	0	0	0	0
SUM	7	4	2	4	3	2	2	2	0	2	0	1	1	1	1	0

Figure 7.2.4 An example of an attribute matrix from the CATA study. Each row represents a respondent and each column represents a benefit (an attribute).

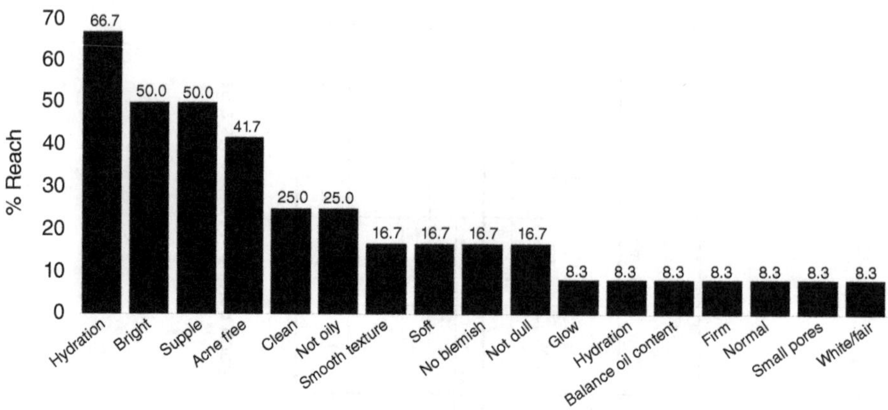

Figure 7.2.5 Reach (%) of individual benefit.

Benefit combination	Reach (%)	Benefit 1	Benefit 2	Benefit 3
2	83	Hydration	Supple	
2	83	Hydration	Bright	
2	83	Hydration	Acne free	
2	83	Hydration	Clean	
2	83	Supple	Bright	
3	100	Bright	Supple	No blemish
3	92	Bright	Supple	Acne free
3	92	Hydration	Acne free	Clean
3	92	Hydration	Acne free	Supple
3	92	Hydration	Acne free	Bright
3	92	Hydration	Supple	Bright
3	92	Hydration	Supple	Smooth texture
3	92	Hydration	Supple	No blemish

Figure 7.2.6 An example of benefit pairs and triads with Reach (%) identified by TURF.

7.2.3.2 Step 2: Visualize the conversation through CATA and Network Analysis

To visualize the information from CATA and TURF results, the rectangular attribute matrix in Figure 7.2.3 was converted to a new square matrix. From a rectangular matrix, two square matrices could be generated: (1) attribute-by-attribute matrix and (2) respondent-by-respondent matrix. To choose which matrix to use depends on the objective of your study. In this study, we wanted to visualize the "benefits" and their "co-occurrences." Therefore, we used attribute-by-attribute matrix to create a new square matrix called "co-occurrence matrix" (Figure 7.2.7).

The matrix was used to create a Benefit Network (Figure 7.2.8). This network visualized how often the benefits were "mentioned together (a.k.a. co-occurred)"

	Hydration	No blemish	Bright	Supple	Acne free	Clean	Not oily	Not dull	Smooth texture	Soft	Balance oil content	Firm	Glow	Normal	Small pores	White/ fair
Hydration	1.0	0.6	0.4	0.4	0.3	0.1	0.4	0.3	0.1	0.3	0.1	0.1	0.0	0.1	0.1	0.0
No_blemish	0.6	1.0	0.2	0.2	0.6	0.1	0.3	0.1	0.0	0.3	0.0	0.2	0.0	0.2	0.2	0.0
Bright	0.4	0.2	1.0	0.2	0.2	0.3	0.1	0.1	0.3	0.1	0.2	0.0	0.0	0.2	0.0	0.2
Supple	0.4	0.2	0.2	1.0	0.4	0.1	0.3	0.3	0.0	0.1	0.0	0.0	0.2	0.0	0.2	0.0
Acne free	0.3	0.6	0.3	0.4	1.0	0.1	0.3	0.2	0.0	0.4	0.0	0.0	0.0	0.2	0.2	0.0
Clean	0.1	0.1	0.1	0.1	0.1	1.0	0.0	0.0	0.3	0.3	0.0	0.0	0.3	0.3	0.0	0.3
Not oily	0.4	0.3	0.1	0.3	0.3	0.0	1.0	0.3	0.3	0.3	0.3	0.0	0.0	0.0	0.3	0.0
Not dull	0.3	0.1	0.3	0.3	0.2	0.0	0.3	1.0	0.0	0.3	0.0	0.0	0.0	0.0	0.5	0.0
Smooth texture	0.1	0.0	0.1	0.0	0.0	0.3	0.3	0.0	1.0	0.0	0.5	0.0	0.0	0.0	0.0	0.5
Soft	0.3	0.3	0.2	0.1	0.4	0.3	0.3	0.3	0.0	1.0	0.0	0.0	0.0	0.5	0.5	0.0
Balance oil content	0.1	0.0	0.0	0.0	0.0	0.0	0.3	0.0	0.5	0.0	1.0	0.0	0.0	0.0	0.0	0.0
Firm	0.1	0.2	0.0	0.0	0.0	0.0	0.0	0.0	0.0	0.0	0.0	1.0	0.0	0.0	0.0	0.0
Glow	0.0	0.0	0.0	0.2	0.0	0.3	0.0	0.0	0.0	0.0	0.0	0.0	1.0	0.0	0.0	0.0
Normal	0.1	0.2	0.2	0.0	0.2	0.3	0.0	0.0	0.0	0.5	0.0	0.0	0.0	1.0	0.0	0.0
Small pores	0.1	0.2	0.0	0.2	0.2	0.0	0.3	0.5	0.0	0.5	0.0	0.0	0.0	0.0	1.0	0.0
White/fair	0.0	0.0	0.2	0.0	0.0	0.3	0.0	0.0	0.5	0.0	0.0	0.0	0.0	0.0	0.0	1.0

Figure 7.2.7 An example of co-occurrence matrix from the CATA study's attribute matrix in Figure 7.2.4. Numbers in each cell are Jaccard similarity coefficients.

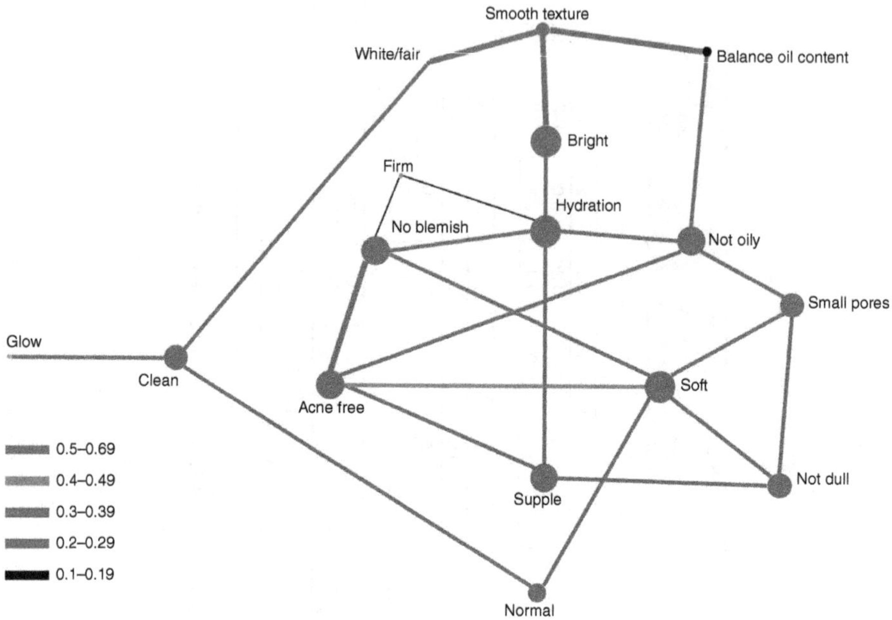

Figure 7.2.8 Benefit network for skincare and cleansing products for an Asian country generated from two QMA sessions. Different colors represent different strengths of the links based on Jaccard similarity coefficients.

through the use of an index called Jaccard similarity coefficient[1] (Borgatti et al., 2002). Figure 7.2.8 has two important graphical elements that help the interpretation of the network: the size of the circles and the color of the links.

In Figure 7.2.8, there are 16 circles representing the 16 benefits. The circles are different in sizes representing different Eigenvalues[2] of the benefits. ***The larger the circles***, the higher contributions of the benefits to total variations in the data or, in short, ***more people selected those benefits***. For example, hydration is much larger than firm, glow, and white/fair.

This means there were more people who selected hydration than the three benefits (Figure 7.2.5). In fact, in Figure 7.2.4, there were seven respondents who selected hydration, while only one-person selected firm, glow, and white/fair. From the size of the circle alone, we can see the benefits respondents wanted (the same way as Figure 7.2.5). In this case, the Asian consumers were interested in hydration, not oily, bright, no blemish, and the other big red circles than other smaller circles (benefits).

Now let us pay attention to the links in the network that contain very valuable information (Figure 7.2.8). Any direct link (a direct connection between two benefits) indicates that those two connected benefits "co-occurred" (or the two benefits were selected by the same person). The more consumers chose the same pair of benefits, the "stronger" the link or the "higher" the Jaccard coefficient. For example, Figure 7.2.8 shows firm links with no blemish and hydration with very thin black line (Jaccard coefficient between 0.1 and 0.19) and blemish links together with a thick pink line (Jaccard coefficient between 0.5 and 0.59).

If we look back to the original data (Figure 7.2.4), there was only one person (R12) who chose the benefits (firm, no blemish, and hydration); and, at least four respondents chose the hydration-no blemish pair. Figure 7.2.7 shows the actual values of Jaccard similarity coefficients of the links. Therefore, Figure 7.2.8 can guide how the marketing team to choose a benefit (the larger circle), a pair of benefits (the thicker link), a triplet of benefits, and so on.

When three or more benefits are of interest, there is an important aspect to consider: ***the structure of the whole connections of interest***. If the links connect and form a **close structure**, it indicates that there is a group of people (like a segment of consumers) who want the same set of benefits. If the links connect but branch out ***without forming a close structure***, it indicates that those benefits can "reach" more unique consumers, and they may share some benefits but not all of the benefits. For example, in Figure 7.2.8, hydration formed a close structure with firm and no blemish. This close structure suggests that the team can form a concept around hydration, firm and no blemish for a particular group of consumers who wanted a combination of these benefits.

To reach more consumers, the team could choose hydration, bright, and no blemish instead. However, it is worth mentioning that the open structure of hydration, bright, and no blemish where hydration is in the middle carries a particular meaning. For consumers who wanted hydration, some wanted either bright or no blemish and "not enough respondents" wanted bright and no blemish. Put it simply, the triad of hydration, bright, and no blemish were targeting two groups of people who wanted hydration-bright and hydration-no blemish because there were not enough consumers who wanted bright-no blemish pair (no direct link between the two benefits).

7.2.3.3 Step 3: Making decision using the results from TURF and Network Analysis

The team used the information from both TURF (Figure 7.2.6) and Network Analysis (Figure 7.2.8) in their decision. Figure 7.2.6 guided the team on which benefits and how many should they use and Figure 7.2.8 helped to screen those selections to make the story coherent based on respondents' perception. For example, if the team chose to use Bright, Supple, and No blemish together because this triad reached 100% of consumers, could the team use only these three benefits? Figure 7.2.8 suggests that without hydration, this triad would not really satisfy the respondents because "no one" wanted these three benefits together (no direct links connected the three benefits), but "everyone" wanted each benefit as a part of three pairs hydration-bright, hydration-supple, and hydration-no blemish (all three benefits connected through hydration). Therefore, omitting hydration risked not satisfying everyone to the fullest. What if the team wanted only three benefits including one "new benefit" not included in the list? In this case, based on Figures 7.2.6 and 7.2.8, hydration-bright and hydration-supple were the prime candidates.

This study with full information was used to create concepts to promote the new product with new benefits in two Asian countries and turned out to be the best launches of the year for the company.

7.2.4 Summary and future of CATA in product research

The CATA task is simple and easy for consumers to understand and perform correctly without much guidance from researchers. It has some resemblance to real situations where consumers have to choose a product or products from a set of choices. TURF analysis effectively helps heighten the benefit combinations and the benefit network provides additional insights to guide decisions. Benefit network analysis contains fundamental insights and can help CEs, developers, and marketers to develop new products with new benefits to reach more consumers or to target a niche group. Benefits without closed-loop links should be used cautiously because consumers may get confused and frustrated as they have not formed any values for those unconnected benefits (especially brand-new benefits) in the product. To help reduce the set of benefits, the results from TURF should be used in conjunction with network analysis to select connected benefits that can reach and effectively satisfy most of the consumers.

Discussion questions

- Why do you think CATA has become popular?
- How can you apply this CATA tool in product development?
- What are the benefits of using network analysis in conjunction with TURF analysis over traditional methods like multiple correspondence analysis (MCA) for CATA responses? What are the limitations?

Notes

1. The Jaccard similarity coefficient (ranges from 0% to 100%) compares members for two sets of data to see which members are shared and which are distinct. The higher the percentage, the more similar the two populations.
2. An index that measures the influence of a node in a network. A high eigenvector score means that a node is connected to many nodes who themselves have high scores for a relationship (e.g. co-occurrence).

References

Adams, J., Williams, A., Lancaster, B. and Foley, M. (2007) Advantages and Uses of Check-All-That-Apply Response Compared to Traditional Scaling of Attributes. *7th Rose-Marie Pangborn Sensory Science Symposium*. Minneapolis, MN, USA.

Ares, G. and Jaeger, S.R. (2013) "Check-all-that-apply (CATA) questions: influence of attribute order on sensory product characterization". *Food Quality and Preference*, 28(1), 141-153.

Ares, G., Etchemendy, E., Antúnez, L., Vidal, L., Giménez, A. and Jaeger, S.R. (2014) "Visual attention by consumers to check-all-that-apply questions: insights to support methodological development". *Food Quality and Preference*, 32(Part C), 210-220.

Borgatti, S.P., Everett, M.G. and Freeman, L.C. 2002 *Ucinet for Windows: Software for Social Network Analysis*.

Buck, D. and Kemp, S.E. (2018) "Check-all-that-apply and free choice description". In S.E. Kemp, J. Hort and T. Hollowood (eds). *Descriptive Analysis in Sensory Evaluation*. West Sussex, UK: John Wiley & Son Ltd. pp. 579-607. https://doi.org/10.1002/9781118991657.ch17

Jaeger, S.R., Beresford, M.K., Paisley, A.G., Antúnez, L., Vidal, L., Cadena, R.S., Giménez, A. and Ares, G. (2015) "Check-all-that-apply (CATA) questions for sensory product characterization by consumers: investigations into the number of terms used in CATA questions". *Food Quality and Preference*, 42, 154-164. https://doi.org/10.1016/j.foodqual.2015.02.003

Meyners, M., Castura, J.C. and Carr, B.T. (2013) "Existing and new approaches for the analysis of CATA data". *Food Quality and Preference*, 30(2), 309-319. https://doi.org/10.1016/j.foodqual.2013.06.010.

Moskowitz, H.R. (1996) *Consumer Testing and Evaluation of Personal Care Product*. New York: Marcel-Dekker.

Varela, P. and Ares, G. (2012) "Sensory profiling, the blurred line between sensory and consumer science: a review of novel methods for product characterization". *Food Research International*, 48(2), 893-908. https://doi.org/10.1016/j.foodres.2012.06.037

7

8

People think focus means saying yes to the thing you've got to focus on. But that's not what it means at all. It means saying no to the hundred other good ideas.

Steve Jobs, Apple

What this is about

This chapter covers quantitative research tools to understand reasons behind consumers' satisfaction, to select the best ideas from good ones and to prioritize consumers' desired benefits and their motivations through understanding their hierarchies. These tools are used to validate the insights with consumers.

Product Innovation Toolbox: A Field Guide to Consumer Understanding and Research, Second Edition. Edited by Kannapon Lopetcharat, Dulce Paredes and Jacqueline H. Beckley.

Chapter 8

Tools for Up-Front Research on Understanding Consumer Values

8.1

KANO Consumer Product Satisfaction Model

Alina Stelick, Kannapon Lopetcharat, and Dulce Paredes

Key learnings

✓ KANO classification system
✓ KANO satisfaction survey
✓ Understanding true meaning of consumer voice

8

8.1.1 What consumer satisfaction can do to your business

Many consumer goods companies continue to find themselves in a tough position. The competition is as fierce as ever. Cost of production and supply chain continues to climb. Shareholders and Wall Street demand profits. Shopping behaviors have changed with more digital options and people looking for experiences beyond the products. Consumers, squeezed from rising cost of living, lost incomes, and fear for their health are less reluctant to part with their hard-earned money unless the product offered is an absolute necessity, something they believe in or something they love, love, love, and cannot live without. So what is the key to keeping consumers buying our products? How do we make sure that our customers keep coming back to us time after time for their favorites? How do we make our products stay current, exciting, and new?

Product Innovation Toolbox: A Field Guide to Consumer Understanding and Research, Second Edition. Edited by Kannapon Lopetcharat, Dulce Paredes and Jacqueline H. Beckley.

One way to do this is to understand the fundamentals of consumer satisfaction with our products and continually build upon it.

This is how Apple and the like are bringing their customers and new folks to their stores year after year.

The KANO Product Satisfaction Model is one of the methodologies that allows the Consumer Explorer to gather insights, determine product fundamentals and prioritize its attributes to optimize the product development process. The KANO was originally developed to gauge customer satisfaction with a product or a service (Kano, 1979; Kano et al., 1984). However, this chapter will focus on the product optimization and insights cataloging applications of the KANO methodology.

The KANO model differentiates the reasons behind consumer satisfaction into four main classes of attributes: must-have, driver, delighter, and indifferent. Must-have attributes are product features that require robust quality control and assurance. Driver attributes highlight desired performance, that is, the more the better. Delighter attributes are new features that are not currently in the product but are highly desired if present. Indifferent attributes are those not missed when absent and can be traded off for more desired features.

Here are some of the areas where the KANO can be useful:

- Product innovation and development:
  - o What features should we "innovate"?
  - o What technologies should we seek out?
  - o How do I know that a new benefit will be "relevant" to my consumers?
- Marketing:
  - o What claims/product features should we be highlighting in the advertisement?
  - o What is motivating product purchase?
- Quality control:
  - o What product attributes must be delivered?
  - o When developing cost-reduced products within the current product category, what attributes can be "trade-off" and what must be maintained?

The first area that is a natural application of the KANO method is in Product Innovation and Development. The results from the KANO allows product developers and managers to decide which area in their product portfolio they should "innovate" and to evaluate that they develop "right" products because the KANO method brings in consumers' perspective about benefits that the products are expected to deliver. In addition, KANO results enable product developers to evaluate their capability to deliver the benefits so they can develop their technology strategy accordingly. In marketing, KANO method brings in consumer voices and language to the process. It enables marketers to strategize their offerings and communication that have the most impact on consumers. Using KANO methodology in conjunction with other process management tools (e.g. Quality Deployment and Six Sigma) allows product developers and engineers to develop their quality assurance and quality control programs beyond the basic safety parameters and identify other parameters that impact consumers' perception of quality. This is very important especially for cost-reduction innovation to ensure perceived "must have" features remain intact.

8.1.2 Philosophy behind KANO's consumer satisfaction model

Until 1984, many researchers believed that consumer satisfaction or liking (collectively called consumer hedonics) was a one-dimensional construct that is driven by the increase or decrease of perceived product attributes (also known as requirements or qualities). However, in 1984, Professor Noraki Kano of Tokyo Rika University and his colleagues in Japan proposed a new philosophy, the KANO satisfaction model, which has changed the way consumer researchers understand the meaning behind consumer hedonic responses (Kano et al., 1984).

The KANO satisfaction model challenges the notion that consumer satisfaction is only driven by an increase or a decrease in a product attribute (one-dimensional attribute); however, the model introduces three additional reasons behind consumer's satisfaction: Must-have attributes, Delighter attributes, and Indifferent attributes (Figure 8.1.1).

One-dimensional attributes are attributes that fit with the conventional meaning of drivers of likings as the higher the attribute, the higher the liking and *vice versa*. General characteristics of one-dimensional attributes are the following:

(1) Most of the time, consumers can articulate these attributes.
(2) These attributes are quite specific for consumers.
(3) The linear relationship with consumer's liking makes it easy to measure.
(4) Most of the time, there are technical definitions of these attributes.

8

Must-have or Must-be attributes are product qualities or requirements that consumers find to be very important to them. Usually, Must-have attributes confound with one-dimensional attributes in any Important Rating study because the study does not allow consumers to differentiate these two types of attributes. General characteristics of Must-have attributes are the following:

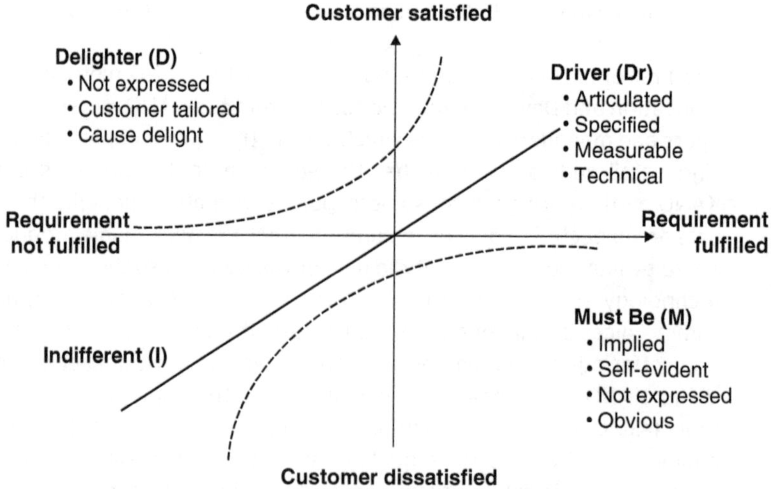

Figure 8.1.1 Original KANO philosophy to classify consumer's satisfaction into Delighter, Driver, Must-be, and Indifferent as reasons behind the satisfaction.

(1) They are needed by consumers.
(2) Because these attributes are needs, consumers usually have a difficult time to articulate any specifics about the attributes, but they usually describe the attributes indirectly (e.g. giving examples of situations).
(3) These attributes are usually obvious to consumers. It is so obvious to consumers that they usually forget to mention Must-have attributes when they are asked.
(4) Must-have attributes are self-evident. If the attributes are not there, consumers will believe that the product will not work.

Delighter attributes are unimportant attributes to consumers; however, they will be very satisfied when there is one in their product. For example, Westin Hotel chain introduced Heavenly Beds in the late 1990s and engaged people in a greater discussion around the importance of a comfortable bed and quality linen. General characteristics of Delighter attributes are the following:

(1) Consumers cannot articulate Delighter attributes readily.
(2) Delighter attributes are tailored for specific needs and wants for a particular group of consumers.
(3) Consumer's response when they encounter Delighter attributes in a product is delight.

Indifferent attributes require special treatment in practice. First, the data must be checked for consumer segmentation (Berger et al., 1993). If there is no major segmentation in the consumer sample, then the second interpretation, which is that consumers do not really care about these attributes, is warranted.

The next two classes of attributes, Reverse attributes and Questionable attributes, do not have theoretical definition as the attributes can be found only in practice.

Reverse attributes are found in practice when the results of certain attributes are opposite from what usually expected from the attributes. Reverse attributes are usually a product of misunderstanding of the questions by consumers. Therefore, the best way to prevent it from happening is piloting the questionnaire with target consumers. If the questionnaires have been confirmed to be easy to understand by target consumers, but there are still Reverse attributes in the results, it is possible that there are segments of consumers. This is not a bad finding as researchers learn more that there is at least a group of consumers who do not interpret things the same way as the researchers do.

Questionable attributes are indicators of misunderstanding of the questions by consumers, and we recommend discarding these attributes from further interpretation.

KANO satisfaction survey allows Consumer Explorers (CEs) to understand reasons behind numbers of important attributes by indirectly deriving from asking questions in two different forms: positive/functional and negative/dysfunctional forms. By doing so, CEs can recommend appropriate actions to the team.

KANO satisfaction survey is a valuable tool to understand consumer's needs, but it is not a tool to discover the needs itself. The exploratory qualitative methods described in Chapters 6 and 7 are better tools to discover consumers' needs that can be verified by a quantitative KANO satisfaction survey.

8.1.3 KANO survey step by step

There are six simple steps in conducting a successful KANO satisfaction survey (Figure 8.1.2).

8.1.3.1 Step 1: Identification and generation of consumer requirements

This is the single most important step that can determine the success of your KANO research, and it may be the most time-consuming step of the whole process. To ensure success, first:

(1) **Clearly define the scope of the products of interest:** For example, if you were doing a study about toothbrushes, you should be asking "Are vibrating electric toothbrushes my category? Or am I interested only in the basic toothbrush? Or both?" To help you answer these types of questions, the following are recommended steps:
   - **Review all the previous research you have available in the product category:** Historical sensory panels, ethnographies, surveys, focus groups, syndicated data, etc.
   - **Competitive review:** Go to the store or any retail channel and see what the consumer is seeing on the shelves and in yours and competitive ads.
   - **Internet review by visiting competitor's websites, online product reviews, consumer blogs, and social network sites:** These days anyone with a Facebook® page or Twitter® account can give your product a review which can be eye-opening.
   - **Catalogue and translate technical terms and industry jargons into consumer-friendly:** For example, the word "unctuous" are used commonly among cosmetic product developers to describe a particular product feel on the lips. We conducted a study to figure out its meaning

8

Step	Main objective	Tips
Step 1: Identification and generation of consumer requirements	Gather languages to communicate benefits and describe product characteristics	Define the scope of your products clearly and do not use technical words or jargons in your study
Step 2: Refine the element list	Edit for exclusivity between questions or each question on should be actionable for the team	30–40 questions are recommended and make sure that translation to other languages is done appropriately
Step 3: Create KANO questionnaire	Make sure that the questions are easily understandable in the same way among target consumers	Pilot tests for KANO's scale & questions are recommended to ensure clear interpretation of the scale
Step 4: Define your target consumers	Create inclusion and exclusion criteria to screen consumers for your product category	Recommend 200 consumers per target group
Step 5: Field the study	Properly administer KANO questionnaire to consumers	One-on-one interviews and Internet surveys are standard Presentation of questions has to be randomized
Step 6: Analysis and interpretation	Convert the frequencies of each questions into indices that easily understandable	Presentation should be simple – table or chart

Figure 8.1.2 Six steps to conduct a successful KANO satisfaction survey.

among consumers and discovered that about half of the respondents, all native English speakers, could correctly interpret that word as "oily," "creamy," "rich" feel in the context of a lipstick. The rest said that either they did not know what it meant or provided incorrect interpretation (my favorite - "an obnoxious person").

- **(Optional) Conduct preresearch:** This step provides two levels of benefits: (1) identifying unarticulated needs and (2) aligning new team members. For the first, identifying unarticulated needs, this step is highly recommended because past research and current existing information provide only the already "articulated" needs and wants. In order to exceed consumers' current expectation, analyzing consumer's problems, values, and behavior is critical. Here are the three methodologies you can use to identify unarticulated needs and generate product descriptors/elements quickly and efficiently:
  - o Free Choice or Flash Profiling or equivalent technique (Chapter 6.1)
  - o Qualitative Multidimensional Analysis (QMA) (Chapter 6.2)
  - o Ethnography

(2) **Add new product attributes and/or determine "where" in your company's NPD is going to take your product in:** This list gives a few recommendations on how to identify new product attributes:

- Review current consumer habits and attitudes (H&A) and future trends data.
- Talk with your innovation or research organization about new and "upcoming" technologies they are interested in.
- (Optional) Conduct preresearch: This step provides two levels of benefits: (1) identifying unarticulated needs and (2) aligning new team members. For the first, identifying unarticulated needs, this step is highly recommended because past research and current existing information provide only the already "articulated" needs and wants. Here are the three methodologies you can use to identify unarticulated needs and generate product descriptors/elements quickly and efficiently:
  - o Free Choice or Flash Profiling or equivalent technique (Chapter 6.1)
  - o Qualitative Multidimensional Analysis (QMA) (Chapter 6.2)
  - o Ethnography

For the second, aligning new team members, this step will help the whole team to be on the same footing quickly and reduce any confusion in the future.

8.1.3.2 Step 2: Refine the list of consumer's requirements

Keep it short, simple, and current. At this step, you are still editing the list of consumer requirements. It is highly recommended to keep the number of requirements below 35 to 40. This will limit respondent fatigue and boredom while answering the questionnaire as each requirement will be asked twice, that is, in functional and dysfunctional forms (total number of questions be 70 to 80). Pick and choose how to word attributes in consumer speak. Review it with a typical category user and if they don't understand what a word means, a regular person on the street won't either.

Lastly, when running global research, ensure that the words/phrases you choose are easily translated. There are many English words that cannot be translated directly and easily, so providing some options (words/phases) will be very helpful. Verify translation before fielding by "back translation," so that the meaning of each attribute is communicated correctly and consistently.

8.1.3.3 Step 3: Create the KANO questionnaire

Once the requirement list is finalized, word each requirement in two ways: (1) as if it is present in the product (this is called "functional form), and (2) as if it is absent in the product (this is called dysfunctional form) (Figure 8.1.3).

For example, a lipstick "has a pleasant taste on the lips" ("pleasant taste" is present in the product or functional form), and a lipstick "does not have a pleasant taste on the lips" ("pleasant taste" is absent in the product or dysfunctional form). Group all the "present"/"positive" attribute wordings into one list and "absent"/"negative" wordings into another. It is critical to make sure that the "meanings" of the wordings of the functional form/present and the dysfunctional form/absent of a requirement are "diametrically opposed." Adding "no" or "not" in the front may not give this property, and it is very critical to check the diametrically opposed property after translation. At this step, you should have two final lists of the requirements: (1) functional form one, and (2) dysfunction form one.

Another aspect of the KANO questionnaire that needs some attention is the KANO scale. The KANO scale is critical in defining KANO classification of requirements, and there are different versions derived since the original paper published in the 1980s (in Japanese). The classic 5-point KANO scale (in English) is as follows:

- I like it that way
- It must be that way
- I am neutral to this feature
- I can live with it that way
- I dislike it that way

Functional form of question How would you feel if the lipstick has a pleasant taste?	✠ I like it this way ❑ It must be this way ❑ I am neutral ❑ I can live with it this way ❑ I dislike it this way
Dysfunctional form of question How would you feel if the lipstick does not have a pleasant taste?	❑ I like it this way ❑ It must be this way ❑ I am neutral ❑ I can live with it this way ✠ I dislike it this way

Figure 8.1.3 Function and Dysfunctional forms of each consumer requirement (in this case of a lipstick). The functional form and dysfunctional form should be "diametrically opposite" in meanings. The right side for both forms shows the KANO's scale.

Original	New variations
I like it that way	• I enjoy it this way • This would be very helpful for me • I like it
It must be that way	• I expect it this way or it is the basic necessity • This is a basic requirement for me • I expect it
I am neutral to this feature	• I'm neutral to this feature • This would not affect me • I'm neutral
I can live with it that way	• I dislike it but I can live with it this way • This would be a minor inconvenience • I can tolerate it
I dislike it that way	• I dislike it and cannot accept it • This would be a major problem for me • I dislike it

Figure 8.1.4 Examples of new variations of KANO scale in English.

Having been translated from Japanese into English, some feel it is not as easy to interpret by the public as it should be. Figure 8.1.4 shows some other new variations of the English KANO scale. It is up to the researchers who know their audiences to choose the most appropriate scale wording.

It is also very important to test the scale after translation. One of the authors tried to use the direct translation of the original English scale in Thai and the consumers rejected the scale completely as the translation did not make any sense to them. This problem can be avoided by testing with a small group of consumers (especially those who speak "only" the native language) as those consumers will have no bias toward the translation.

Some have argued that for some attributes the whole scale wording does not make sense and recommended leaving off those scale choices. We, however, feel that it is up to the person answering the survey to make the decision. And rather than leaving off one of the scale options, we recommend instructing the respondents to read the question and scale choices carefully and choose the most appropriate answer from those available that does apply to the question. That is, the same scale choices should be available for both positive and negative lists.

8.1.3.4 Step 4: Define your target consumers

Your study population should reflect your real consumers and their experiences in the marketplace within their particular brand consideration sets. The respondents should be product category users or consumers that have used several versions of the same product type, for example from different brands, forms, and situations. You may also consider a quota of respondents that are more "early adopters" who tend to buy the newest products available on the market. This would be especially important if the attribute set in your KANO questionnaire includes many innovative product features.

When it comes to the population size, use at least n = 200 per test market. This would allow you to conduct segmentation analysis on the study results *post hoc*. Also, include a few demographic and psychographic questions to be answered by each respondent after the main KANO questionnaire is completed. This will help describe your segments.

8.1.3.5 Step 5: Field the study

This step is very simple and straightforward. There are several agencies that can provide fielding and data analysis services for this type of study. What used to be a painful "paper and pencil" process of cross tabulating results about 5–10 years ago is now as easy as a simple "push of a button." However, we must caution you to screen your fielding agency to make sure that they have all the appropriate randomization and programming capabilities and they understand the methodology before committing your research monies (see Chapter 11). For fielding a KANO questionnaire, there are two special requirements:

(1) For each respondent, a unique randomization scheme must be created separately for the "present"/positive attribute list and "absent"/negative attribute list.
(2) The "present"/positive attribute list must always be presented first to each respondent; only then can the "absent"/negative attribute list is to be presented.

8

It is also important that the agency has capability to include additional or "classification" questions after the main KANO questionnaire (Stelick et al., 2009).

8.1.3.6 Step 6: Analysis and interpretation

Data analysis of KANO satisfaction surveys requires the pairing of the responses from the functional and dysfunctional forms of a requirement. For each consumer, each response-pair will be assigned a KANO class according to Figure 8.1.5.

For example, if a consumer answers that the consumer "likes" the requirement when it was asked in Functional form (Present) and "Dislike" when it was asked in Dysfunctional form (Absent), then for that particular consumer, the requirement is a "Driver." Once the KANO classes were assigned to each requirement from all consumers, the frequencies of each requirement fall into each KANO class (Questionable, Driver, Delighter, Must-have, Indifferent, and Reverse) will be tallied. This step researchers should have a KANO data set (Row = requirement and Column = KANO classification) (Figure 8.1.6). At this stage, many advancements have been made in analyzing KANO data.

However, based on our experience, the simple frequency counting is sufficient and easy for communicating the results to other team members. Before moving

An requirement in Absent/Dysfunctional form

An requirement in Present/Functional form	Like	Must Be	Neutral	Live with	Dislike
Like	Q	D	D	D	Dr
Must Be	R	I	I	I	M
Neutral	R	I	I	I	M
Live with	R	I	I	I	M
Dislike	R	R	R	R	Q

Figure 8.1.5 KANO classification of a requirement based on a consumer's responses from Functional form and Dysfunctional form using the classic 5-point KANO scale. Note: *Q*: Questionable, *D*: Delighter, *Dr*: Driver or One-dimensional, *M*: Must-have or Must-be, *I*: Indifferent, and *R*: Reverse.

	Delighter (D)	Driver (Dr)	Must-have (M)	Indifferent (I)	Questionable (Q)	Reverse (R)
Requirement 1	35%	24%	15%	15%	10%	1%
Requirement 2	8%	22%	23%	20%	15%	22%
Requirement 3	38%	17%	3%	35%	5%	2%
	•	•	•	•	•	•
	•	•	•	•	•	•
	•	•	•	•	•	•
Requirement n	...	...	...	...	...	...

Figure 8.1.6 An example of the KANO data sheet. Consumer's requirement (row) and KANO classification (column) form table of frequencies where each cell contains frequencies of a requirement to be assigned to different KANO classes.

forward with any analysis, conclusion, and interpretation, the following two steps must be performed to ensure correct interpretation and conclusion.

(1) **Check the frequency of "Reverse" and "Questionable" categories for each requirement:** If the frequencies of a requirement for these classes are more than 20% (in combination or alone) (Figure 8.1.6-Requirement 2), it is

recommended to discarding all the data for this particular requirement as consumers do not understand what you are trying to say.

(2) **Check the distribution of the frequency of each requirement:** this step is usually missed when an automated process is used. Requirement 3 in Figure 8.1.6 is an example. Using an automatic process, usually it will conclude that this requirement is "Delighter" as it gets the highest percentage; however, inspecting the distribution of the frequencies, we can see that it is also an "Indifferent." This is a classic sign of segmentation in the target consumers. Therefore, both "Delighter" and "Indifferent" should be the conclusion for Requirement 3. The same rule applies to other classes (except Questionable and Reverse).

Once these steps are completed, you may present your data in a graphical format to reflect the original KANO philosophy or show it in a table. The following formula is one of many developed to determine their coordinates (Figure 8.1.7) and a graph can be constructed from the coordinates (Figure 8.1.8).

Degree of satisfaction (Y axis) = (D +Dr)/(D + Dr + M + I)

Degree of requirement (X axis) = (Dr + M)/(D + Dr + M + I)

Figure 8.1.7 Formula to create an (X,Y) coordinate for a requirement. Source: Berger et al. (1993).

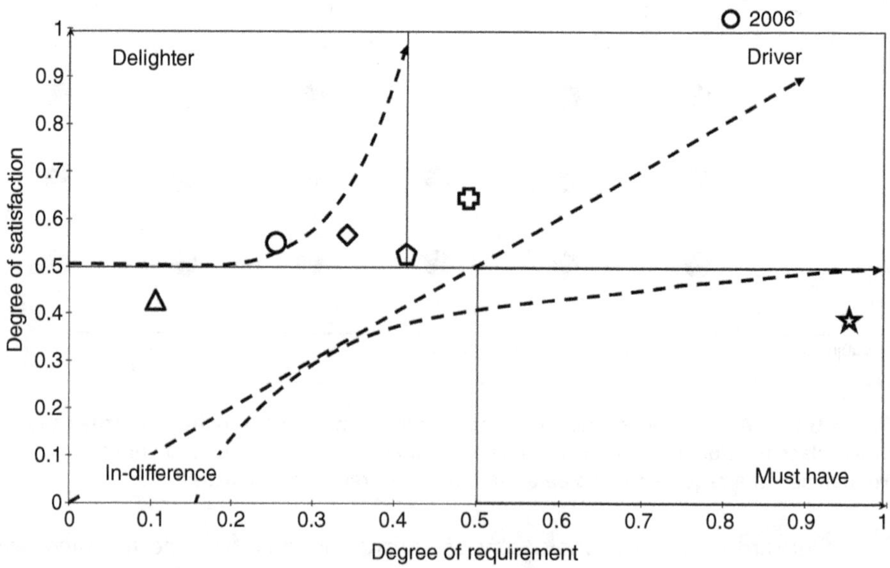

Figure 8.1.8 A typical result from a KANO satisfaction survey after transforming the KANO data into Degree of Satisfaction (Y-axis) and Degree of Requirement (X-axis) using the formula in Figure 8.1.7. Each symbol represents a benefit or a product feature. The dotted line shows the hypothetical lines based on the original KANO philosophy. The boxes and their boundaries were based on a 2006 study for a specific product category.

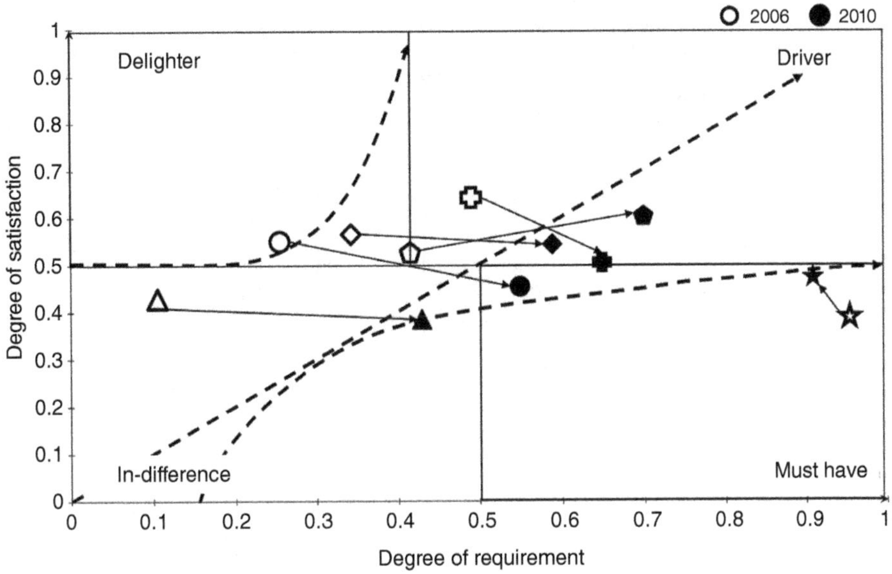

Figure 8.1.9 The movements of KANO classes of benefits or product features from 2006 (empty symbols) to 2010 (filled symbols). The dotted line shows the hypothetical lines based on the original KANO philosophy. The boxes and their boundaries were based on the 2006 study for a specific product category.

The graphical presentation is useful when you need to compare the attribute classification, for example across time. Figure 8.1.9 shows how attributes (different symbols) migrate through time. Delighter attributes can migrate as performance drivers and "must haves" once discovered by consumers and/or matched by competitors.

These coordinates can be used to present the results visually in the two-dimensional space that reflects the original KANO's philosophy (Figure 8.1.1). The graphical presentation is useful when overall assessment and comparative assessment for the requirements are of interest. However, we have found that our clients can digest the information better when the results are presented in a table format (Figure 8.1.10), especially when you are trying to organize attributes from different product aspects.

8.1.4 Case Study: Lipstick KANO survey

Objective: Identify what attributes contribute to consumer satisfaction and/or dissatisfaction in a Lipstick

- **Study Set up**
  - Generate attributes/elements using past research and knowledge of category
  - Execute using KANO questionnaire structure/scale
  - Automated data collection using online program

	Product Aspect 1 (e.g. Visual Appeal)	Product Aspect 2 (e.g. Sweet)	Product Aspect 3	Product Aspect 4
Must Have		Requirement 5 Requirement 6	Requirement 12 Requirement 13 Requirement 14 Requirement 15	Requirement 25
Driver	Requirement 1 Requirement 2	Requirement 4 Requirement 7 Requirement 8 Requirement 9 Requirement 10	Requirement 11 Requirement 21	
Delighter		Requirement 16	Requirement 17 Requirement 18 Requirement 19 Requirement 20 Requirement 22 Requirement 23	Requirement 3

Figure 8.1.10 A suggested table format to present the conclusion from a KANO satisfaction survey. Product requirements are grouped into different aspects and reported as which requirement is Delighter, Driver, or Must-have for consumers (Prokopchuk et al., 2007).

- Analysis
  - Based on total number of consumers
  - Segmentation based on response pattern or predetermined (e.g. demographic variables)

8

- **Data Collection**
  - The research fielded online
    - Consumers were invited via e-mail to participate in the online interview.
  - Respondents (n = 200) were prescreened to be women, lipstick users.
  - Each respondent saw a randomized list of Elements (in functional and dysfunctional form).
  - Figure 8.1.11 shows partial results from the study.

8.1.5 Comparison with degree of importance surveys

In the past, we used to routinely run a Degree of Importance Surveys to determine what attributes were important in a product. For that, a list of product attributes was compiled and presented to respondents, product category users, with a simple Importance Scale. As follows. . .

"Thinking about this widget, would you say having this widget be easy to use is. . .

- Very important to me
- Moderately important to me

	Application	Appearance	Feel on lips	Scent & Taste	Wear	Care
Must Haves	Attr. 3	Attr. 5 Attr. 6	Attr. 12 Attr. 13 Attr. 14 Attr. 15 Attr. 16	Attr. 25		Attr. 33 Attr. 34
One Dimensional	Attr. 3 Attr. 1 Attr. 2	Attr. 4 Attr. 7 Attr. 8 Attr. 9 Attr. 10	Attr. 16 Attr. 11 Attr. 21		Attr. 26	
Attractive			Attr. 17 Attr. 18 Attr. 19 Attr. 20 Attr. 22 Attr. 23		Attr. 27 Attr. 28 Attr. 29 Attr. 32	Attr. 35
Indifferent	Attr. 24			Attr. 30		Attr. 31

Figure 8.1.11 KANO classification of lipstick requirements.

- Somewhat important to me
- Not very important to me
- Not at all important to me"

The respondent would answer this and all other questions. The Top 2 Box counts would be calculated to determine whether a significant majority of respondents (≥95% LOC) found this particular attribute "important." Based on the Top 2 Ratings it is possible to "tier" the attributes into different levels of importance.

As a result, the researcher would typically get a "laundry list" of attributes all or 99.9% of which would be considered "important" by the criteria described above. Although this looks like a great result and something that's very easy to do, it complicates things for the product developer. Because now you're telling a product developer that "everything" in their product is "important" to consumers, there is no way to prioritize. To remedy this, that is why we turned to the KANO satisfaction survey. Because the KANO method allows us to not only determine what's important and what's not, but also to prioritize within the important and unimportant categories (Figure 8.1.12).

KANO satisfaction survey	Degree of importance survey
Provides ability to classify attributes quantitatively and differentiate between "Must Have" and "One Dimensional" with attributes that are "important" to the consumers	Are not as restricted in number of attributes asked as Kano, therefore, attributes do not have to be as targeted/hypothesis-driven as in Kano.
Provides ability to identify new opportunities ("Attractive" attributes) where as they may be classified as "not important" in the "Degree of Importance" Surveys	Provides opportunity to test hypotheses with target population

Figure 8.1.12 **Comparison of advantages provided by KANO satisfaction survey and Degree of Importance survey.**

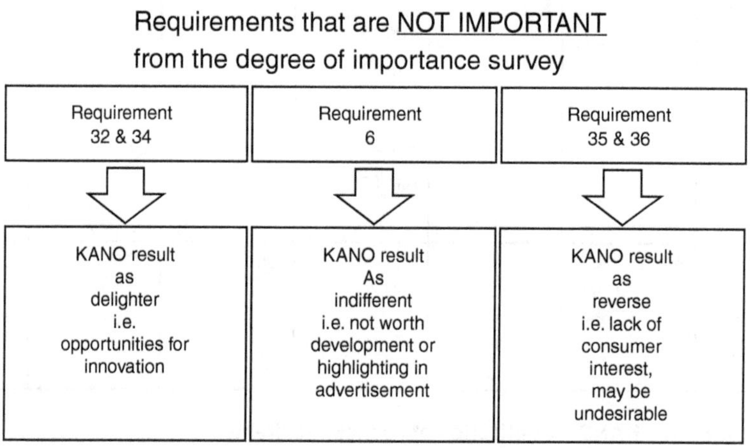

Figure 8.1.13 **An example when Degree of Importance Survey fails to identify opportunities or guide product innovation and development compared to KANO Satisfaction Survey.**

As an example, we run the same attribute list in a KANO study and in a Degree of Importance Survey with similar groups of women, *n* = 200 regular users of lipstick. What we found was that in the Importance Survey, 20 attributes were found "important" in the same importance tier, whereas the KANO survey was able to "pluck" out 7 out of these 20 as "must have" attributes and the rest it classified as "drivers." This provided a clear usable hierarchy of attributes. At the same time, several attributes that ended up on the "not at all" important list were classified as either "Delighter" or "Indifferent/Reverse" in the KANO study (Figure 8.1.13). The KANO satisfaction survey was able to tease out what's less important and identify opportunities for innovation and development via attractive attributes.

In many cases, the Degree of Importance Survey is built into NPD as key parameters in the process, it is highly recommended to use KANO Satisfaction Survey to supplement and explain the reasons behind the ratings of the Degree of Importance Survey. For example, KANO results can be used in conjunction with Quality Improvement Index (QI) (Griffin and Hauser, 1993) to guide product development and create Quality Function Deployment programs (Matzler and Hinterhuber, 1998).

8.1.6 Future of KANO satisfaction survey

The KANO methodology has been widely accepted across multiple scientific disciplines and industries, since its debut in 1984. The major reason for this popularity is that unlike other quality models (e.g. Gap model, PZB, 1985), it assumes that there is asymmetry and nonlinearity between a particular attribute level of performance and overall satisfaction with the product and sorts product attributes into different categories that could deliver incremental improvement in its performance (i.e. one-dimensional) or take the product to the stratospheric heights of success (i.e. delighter). Thus, recognizing that not all product attributes are created equal, some are more important than others to the product user. This is the piece that seems to ring true across many products and services, as evident after running a quick search on Google Scholar and within a millisecond examining a short selection of recent scientific articles that used KANO satisfaction to evaluate consumer/customer/user/business partner/citizen satisfaction with home appliances in Slovenia (Južnik Rotar and Kozar, 2017), computers in Italy (Violante and Vezzetti, 2017), presidential election outcomes in Iran (Shahin et al., 2017), electronic library service in China (Zhang et al., 2017), wine festivals (Velikova et al., 2017), and patient healthcare quality in the USA (Materla et al., 2019), interdisciplinary college program in South Korea (Seo and Um, 2019), and pharmaceutical logistics in Taiwan (Chen et al., 2019).

8

Over the years, researchers have tried to improve on the classic KANO Satisfaction model by proposing more simplified questionnaires (rather than paired functional and dysfunctional attribute list), modified scales or more complex mathematical models to extract the attribute classification more precisely. Mikulic (2007) and Rashid (2010) reviewed many applications of the KANO classification system and concluded that other modified versions did not provide significantly better results.

However, in 2006, MacDonald and others presented a new way to use KANO classification system on product design where different levels of features can be evaluated instead of using only presence (functional form) and absence (dysfunctional form). The method utilized two questions (1) satisfaction (from "unsatisfied" to "satisfied") and (2) usefulness (from "not at all useful" to "very useful") along with an updated version of KANO classification system (MacDonald et al., 2006). This method allows product developers to fine-tune product features and optimize the offer beyond just only identifying the feature important to consumers or not and eliminate the need of asking the questions twice in function and dysfunction forms (Audiverth et al., 2021).

With the rise of big data analytics, some, like Bi et al. (2019), are starting to extract online customer reviews and develop a model to measure the customer product satisfaction based on sentiments expressed by using natural language processing tools.

Adapting the KANO classification system to filter the content from social media may be a quick way to help identify new directions for product features and understand the importance of new product features launched in the market.

Discussion Questions

- How can you apply the and KANO concept to incremental innovation where there is no negative or absent function, only good and better offerings are available?
- Do you know which product features and attributes belong to which class in KANO classification for your product category?

References

8

Audiverth, H.L.F., Garcia, A., Ormenese, R. and Pacheco, M.T.B. (2021) "Use of the Kano model and sensory evaluation in the development of an iron supplement for women". *Journal of Sensory Studies*, 36(4), e12655. https://doi.org/10.1111/joss.12655

Berger, C., Blauth, R., Boger, D., Bolster, C., Burchill, D., Du-Mouchel, W., Pouliot, F., Richter, R., Rubinoff, A., Sken, D., Tinko, M. and Walden, D. (1993) "Kano's methods for understanding customer-defined quality". *Journal of Center for Quality Management*, 2(4), 3-36.

Bi, J.-W., Liu, Y., Fan, Z.-P. and Cambria, E. (2019) "Modelling customer satisfaction from online reviews using ensemble neural network and effect-based Kano model". *International Journal of Production Research*, 57(22), 7068-7088, https://doi.org/10.1080/00207543.2019.1574989

Chen, M.-C., Hsu, C.-L. and Lee, L.-H. (2019) "Service quality and customer satisfaction in pharmaceutical logistics: an analysis based on Kano model and importance-satisfaction model". *International Journal of Environmental Research and Public Health*, 16(21), 4091.

Griffin A. and Hauser, J.R. (1993) "The voice of the customer". *Marketing Science* (Winter), 12(1), 1-27.

Južnik Rotar, L. and Kozar, M. (2017) "The use of the Kano model to enhance customer satisfaction". *Organizacija*, 50(4), 339-351, https://doi.org/10.1515/orga-2017-0025

Kano, N. (1979) On M-H property of quality. Nippon QC Gakka, 9th Annual Presentation Meeting, Japan, pp. 21-26.

Kano, N., Seraku, N., Takahashi, F. and Tsuji, S. (1984) "Attractive quality and must-be quality". *Journal of the Japanese Society for Quality Control*, 39-48.

MacDonald, E., Backsell, M., Gonzalez, R. and Papalambros, P. (2006) The Kano method's imperfections, and implications in product decision theory. *Proceedings of the 2006 International Design Research Symposium*, October (pp. 1-12).

Materla, T., Cudney, E.A. and Hopen, D. (2019) "Evaluating factors affecting patient satisfaction using the Kano model". *International Journal of Health Care Quality Assurance*, 32(1), 137-151, https://doi.org/10.1108/IJHCQA-02-2018-0056

Matzler, K. and Hinterhuber, H.H. (1998) "How to make product development projects more successful by integrating Kano's model of customer satisfaction into quality function deployment". *Technovation*, 18(1), 25-38.

Mikulić, J. (September 2007) The Kano model-a review of its application in marketing research from 1984 to 2006. In *Proceedings of the 1st International Conference Marketing Theory Challenges in Transitional Societies* (pp. 87-96).

Prokopchuk, A., Pereira, B., Paredes, D., Katz, R. and Moskowitz, H. (2007) Lipstick: Kano Satisfaction Study Presentation. *Presented in 7th Pangborn Sensory Science Symposium*, 12-16 August, Minneapolis, USA.

Rashid, M. (2010) "A review of state-of-art on Kano model for research direction". *International Journal of Engineering Science & Technology*, 2(12), 7481-7490.

Seo, Y.-J. and Um, K.-H. (2019) "The asymmetric effect of fairness and quality dimensions on satisfaction and dissatisfaction: an application of the Kano model to the interdisciplinary college program evaluation". *Studies in Educational Evaluation*, 61, 183-195, https://doi.org/10.1016/j.stueduc.2019.04.007

Shahin, A., Mohammadi, S., Harsij, H. and Rahbar Qazi, M.R. (2017) Revising satisfaction and dissatisfaction indexes of the Kano model by reclassifying indifference requirements: a case study of the presidential elections. *The TQM Journal*, 29(1), 37-54. https://doi.org/10.1108/TQM-05-2015-0059

Stelick (Prokopchuk), A., Paredes, D., Moskowitz, H. and Beckley, J. (2009) Kano Satisfaction Model in Cosmetics. *Presented in Congress Cosmetic and Sensory: From Neuroscience to Marketing*. 24-26 June, Tours, France.

Velikova, N., Slevitch, L. and Mathe-Soulek, K. (2017) "Application of Kano model to identification of wine festival satisfaction drivers". *International Journal of Contemporary Hospitality Management*, 29(10), 2708-2726. https://doi.org/10.1108/IJCHM-03-2016-0177

Violante, M.G., Vezzetti, E. (2017) "Kano qualitative vs quantitative approaches: an assessment framework for products attributes analysis". *Computers in Industry*, 86, 15-25, https://doi.org/10.1016/j.compind.2016.12.007

Zhang, Y.-l., Bi, R.-F., Xiao, M. (2017) "Studying on enhancing readers' satisfaction model of electronic service quality in library based on LibQUAL+ and Kano". *Procedia Engineering*, 174, 260-266, https://doi.org/10.1016/j.proeng.2017.01.132

8.2

Systematics of Communication: Conjoint Measurement, Emotions, Cognitive Economics, and Consumer Mind-sets

Howard Moskowitz and David Moskowitz

Key learnings

- ✓ Systematic research through design-based experimentation
- ✓ Building study elements
- ✓ Identifying appropriate scales and metrics
- ✓ Data analysis, interpretation, and reporting
- ✓ Mind genomics or science of the mind

8.2.1 The issue

During the past 50 years, marketers have grappled with the issue of how they communicate product ideas for the development of new products. During the early days, the notion of the golden tongue was popular, namely that someone in the corporation or its advertising agency would "know" the mind of the consumer, and be able to specific the features of the product, and the language by which to communicate the product so that customers would willingly buy the product.

The era of the golden tongue and the mad men (and women) of advertising eventually gave way to simple measurement. For the development of new products, the initial studies were called needs gap analysis, the needs people had, the gap between the products available and the expressed needs. This would lead to the selection of new ideas about the features of the product, and then the evaluation of such ideas. There were different names for this process, but in the end, the objective was to rank order the different features of the product as those features were attractive, and solved consumer problems. These processes were

Product Innovation Toolbox: A Field Guide to Consumer Understanding and Research, Second Edition. Edited by Kannapon Lopetcharat, Dulce Paredes and Jacqueline H. Beckley.

so fundamental that they did not lead to standardized methods or to norms. The question was simple about the "promise" of the feature to solve a problem, hence, the name "promise testing."

Beyond the promise test was the description of the product by a concept, a set of sentences, with these sentences attempting to spell out the customer need, the feature of the product, and how the feature itself or the product as a whole solved the problem. There were various ways that the exercise could be done, that the concept could be constructed, such as "problem-solution," and so forth. The idea was to move beyond the specific problems or features, and instead present the new product (or service) in its entirety, not so much to sell the product as to describe it, and determine the degree to which the product would be interesting for the consumers enough to purchase it.

8.2.2 Consumer research: experimentation vs. testing

In consumer research, the notion of experimentation has evolved into two distinct forms: (1) Testing and (2) Experimentation. Testing uses the responses of the consumer to judge the performance of test stimuli on a number of characteristics. From the ratings, the Consumer Explorers (CE) discovers the attributes that perform well, and the attributes that perform poorly

When testing several unconnected stimuli, the CE must perform additional analyses to discover the specific characteristics that are responsible for, or which drive, the strong performances vs. the poor performances, respectively. The bottom line is that whether testing single stimuli or an array of unconnected test stimuli, the CE generates a report card about performance, and may even discover some possible causes of acceptance. The key thing to keep in mind is that the discovery of the drivers is left to the intelligence and experience of the CE, rather than being built into the system so that anyone can discover those drivers.

In experimentation, the goal is to make these drivers easy to discover, through systematic up-front work. Experimental psychologists and statisticians have recognized the value of systematic variation of the test stimuli to understand how to create better products, and how we make decisions. In science, fundamental ideas have a way of migrating from one domain to another. The powerful methods of experimental design, so popularized by companies such as DuPont, would find their way into psychology. Researchers such as Norman Anderson would recognize the value of mixing ideas together, in a discipline he called functional measurement. Through statistical methods such as regression, Anderson would disentangle the different variables, discovering their individual-level contributions to the mixture.

Around the same time, mathematical psychologist R. Duncan Luce, a formidable intellectual giant in the field was exploring the issue, but from a formal mathematical approach, one he and his colleague, John Tukey, would call Conjoint Measurement. Unlike Anderson, who was interested in the applications of design to psychological processes, Luce was interested in axiomatic measurement theory, the formalization of the measurement into axioms. One of these topics was conjoint measurement, the measurement of individual variables from the measurement of their interaction with each other to drive a response.

In practice, the CE systematically varies the factors that can be controlled. Experimental design lays out the specific combinations of the variables. Each respondent evaluates one, several, or even all of these combinations, rating each combination on one or more scales. From the ratings, the CE uses statistical tools, especially least-squares regression, to estimate the specific contribution attributable to each element.

The scientific discipline of experimental design has made it popular among development scientists in industry, as well as consumer researchers who want to apply it to realms such as the combination of ideas (Box et al., 1978). Furthermore, the widespread use of shrink-wrapped software, for example SYSTAT (2008), has made experimental design accessible to everyone who works with a personal computer. The bottom line is that drivers or causes of responses can be isolated when the CE takes the time to create, test, and analyze responses to systematic variations.

8.2.3 Conjoint analysis (aka conjoint measurement)

Conjoint analysis (CA) quantifies the response to mixtures. From that response, CA attempts to identify the underlying impact or utility of each element in the mixture. At its very core, CA is the implementation of experimental design. It has come to mean systematic exploration of combinations of independent variables which are discrete, either appearing or not appearing in a particular combination.

The notion of conjoint measurement moved in two directions, resulting from deep philosophical points of view. On the one hand were those methods based on choosing between pairs of vignettes, with the vignettes being known to systematically create blends of ideas from experimental design. The research comprised a limited number of vignettes. Indeed, it was difficult to do more than 5–10 vignettes, simply because of the onerousness of the task, reading two vignettes, and deciding between them. The mind-set (MS) of these studies could be traced to the antecedent test methods using so-called "Thurstonian scaling," where it was not the percentage of times one concept was chosen over another, but rather a mathematical transformation of the percentage of choice of one concept over another. The goal was to deconstruct the concept into the contributions of its components. The approach could be called measurement by error or measurement by inconsistency. It was absolutely necessary for the vignettes to be confused so that sometimes one of the vignettes was chosen, whereas other times the other concept would be chosen.

A major breakthrough occurred when the researcher opted to substitute rating scales in place of direct choice, so that the job of the respondent was to read a complex combination of methods and score that concept on a scale. By itself, the approach generated a rating for the concept. With systematically varied vignettes, one could create a model showing how each element drove the rating, viz., contributed to the rating. In the end, both methods would flourish, but over time, the paired comparison system would appeal more to the larger-scale one-off studies, whereas the rating system would appeal to the faster, simpler, smaller, and affordable studies.

Experimental design, in general, and conjoint analysis, in particular, operate under the assumption that people learn the most about independent variables when they are presented in combination, the way they would be experienced in real life. Indeed, people often cannot evaluate single elements (ideas) in a valid way. For example, brands of products are important, and would be rated important when evaluated singly. Yet, when evaluated in combination with product features and health benefits, quite often brands do not perform as well. Conjoint analysis pulls out the effect of the individual elements when they are evaluated in the "ecologically-valid" mixtures (vignettes) that constitute the experience of daily life.

Conjoint analysis is a branch of statistical design. The stimuli or elements are typically discrete options, present or absent in the vignettes. These stimuli or elements are often cognitively "rich," having meaning in and of themselves, such as brand names, benefits, and prices.

8.2.4 Doing the basic conjoint analysis experiment

There are seven steps to creating a successful study using conjoint analysis (CA).

8.2.4.1 Step 1: Set the objective(s) of a study

This step is the most critical as it dictates the conditions, details, and results of the study. In practice, it is rare to run a study with a single objective. Conjoint analysis studies are no different. Therefore, prioritizing is the key to success in this step.

Conjoint analysis is one of the best methods for assessing the magnitude of interest or impact of product attributes in order to select the most suitable combinations of product offerings that suit both consumers' needs and business strategy. However, CA is not an appropriate tool for classifying attributes into different kinds of measuring the impact of product attributes that are functionally critical but not emotionally delightful. Thus, as typically implemented by consumer researchers, CA works with stimuli that can be understood when the respondent reads a product or service description.

8.2.4.2 Step 2: Create and/or select elements (attributes) and classify the elements into relevant categories

The rationale for this second step is that categories and their elements are the crucial building blocks of the CA study. They are fundamental building blocks of experimental design, and the phrases that will be presented in the test stimuli. Category is a collection of elements that are similar. Technically, category is also known as "variable" in experimental design language. In practice, a category does not always contain the same kind of elements and the consequence is not really severe. However, having the same kind of elements in a category allows researchers to assess the overall impact or importance of that category compared to other categories in the same study.

An element is an individual attribute or a product offering to which researchers want consumers to respond. An element can be as short as "red color" or as long as "made from organic white tea exclusively grown in China with a fair-trade agreement," It can be abstract as "creamy, rich and silky-smooth chocolate for decadent sensation in your mouth" or as factual as US$8.99 per pound.

The key for a successful Conjoint Analysis (CA) study is that the elements must fit with the objectives for which the team has an action plan and standard. For example, when the innovation team would like to figure out which product offerings are the best for a long-term project, then the action standard should be to identify highly motivating elements that have the highest impact regardless of feasibility. However, when the objective is short term, the action standard should be to identify those particular elements which have high impact and can be easily claimed and/or produced.

8.2.4.3 Step 3: Select an experimental design

The experimental design governs the specific combinations of elements that the respondent will see and evaluate. The experimental design can be likened to a recipe book with many recipes. In turn, each test combination constitutes one recipe in this book. The experimental design is a critical part of CA studies, but often the complexity behind the design makes the design a hindrance.

It is common for CEs to have many questions. The technically appropriate experimental design is too costly. Therefore, the prioritization of study objectives allows researchers to select the best and most practical designs to use.

When the CE is interested only in main effects or the impact of interaction terms are negligible, then many screening methods can be used such as Plackett-Burkman designs, Simplex designs, 2^k factorial designs, fraction of the 2^k series, for example, (Montgomery 1991; Moskowitz et al., 2006).

If the objective is to study interaction effects between the elements from different categories, high-resolution experimental designs such as full factorial designs, D-optimal designs, some fractional factorial designs, central composite designs (for second-order effect), Box-Behken designs (for third-order effect), may be used.

Many statistical packages have experimental design modules that enable users to design experiments easily. There are many service providers that integrate experimental design into their conjoint analysis services.

One of the more popular methods, IdeaMap®.Net, a do-it-yourself conjoint package, the experimental design is set up to provide a specific set of combinations of elements in an efficient way. With IdeaMap.Net, the user doesn't have to know the principles of experimental design but instead has to select a specific design, with a fixed number of categories and equal number of elements in each category. The popular designs (categories X elements) are the large 4 X 9 and 6 X 6 designs. In 2014, the authors created a simplified 4 X 4 design under BimiLeap® (details on Section 8.2.16.4). The 4 X 4 template has enough elements to get good data and insight that even high school kids can do it using an inexpensive, uncomplicated program with automatic reporting, mind-set segmentation analysis, and timed consumer response as added measure.

8.2.4.4 Step 4: Develop the appropriate scale by which to obtain consumer feedback

The scale provides a great deal of information about the consumer's response. It is important to develop the appropriate scale. It often looks quite easy and developing a scale without considering what one might learn is, in fact, easy. Yet a bit of thinking will yield other types of scales that will tell much more.

The most typical scales are fixed-point category scales that "evaluate" the test concepts. Examples include the traditional nine-point hedonic scale first proposed by Peryam and Pilgrim (1957), as well as the five-point purchase intent scale (1 = definitely not purchase → 5 = definitely purchase).

Purchase intent scales are often confused with liking scales but are different. The respondent evaluates the likelihood of buying the product. Buying may call into play other dimensions, such as price and efficacy not addressed by liking.

Appropriateness for a specific end use is also often used. When the respondent evaluates a concept using appropriateness for a specific occasion or end use, the decision calls into play both acceptance of the concept and the context of use. A product can be liked and yet be inappropriate for a specific end use.

Generally, these scales are treated in one of two ways. One way is to use the actual numerical rating as the dependent variable when one "models" the data to develop utility values. Thus, the actual rating itself is used in statistical analysis. A second way is to convert the rating scale to a binary value, and then use that value as the dependent variable. Thus, in a 9-point scale, one might convert ratings of 1–6 to 0, and ratings of 7–9 to 100. One would then use regression analysis with the binary values as dependent variables. Other evaluative scales can be used, such as a 7-point scale, or indeed any other type of evaluative scale. The rating scale need not be limited to evaluation. Ratings of price and selection of feelings/emotions are also valid rating scales.

When it comes to price, the rating scale can present the consumer with different price points. The question would be phrased along these lines: "Select the one price that you feel fits the concept (or vignette) that you have just read." The rating scale would then present the consumer with a graded series of prices. The prices have to be monotonic (increasing from low to high), but do not have to be equally spaced. There should be a price point for US$0 (would not pay). Best practice dictates each scale point to have only one price, and the instruction to be around the consumer picking the scale point whose price is closest to the appropriate price one would pay for the product or service described by the concept.

More recently, CEs have begun to explore the inner experience of emotion and feeling. These terms don't intend to limit the focus to classically defined emotions but call attention to attributes that the person feels applies to himself or herself, rather than to the stimulus (H. Meiselman, personal communication, 2010). The field of emotion and feeling is increasingly being seen as the "new frontier" in consumer research and has been incorporated successfully into the IdeaMap.Net system.

Figure 8.2.1 Example of a concept (package design) with an emotion question. Source: Moskowitz, 2009/with permission of Academic and Business Research Institute.

Emotion has to be treated differently from other scales. We are accustomed to working with conjoint analysis where we rate our feelings when we read the test concept. Typically, our ratings fall along some sort of a criterion variable, such as purchase interest or liking, or even price. What happens, however, when we select one of several qualitatively different responses? For example, the respondent can be instructed to read the test concept and select the feeling that he most strongly experiences. Figure 8.2.1 shows an example of this type of question, for a package created according to an experimental design.

The analysis for such a study differs. When dealing with a set of different emotions, for example five emotions, we assume the respondent feels that emotion and does not feel the other four emotions. We code the selected emotion 100, and the remaining emotions 0. Across a large number of concepts, we relate the presence/absence of the elements to the emotions felt. Typically, in such cases, we run five ordinary least squares regressions, one per emotion. We do not use the additive constant when computing the regression model. The result of the analysis shows the conditional probability of a respondent selecting an emotion when a specific element is presented in a test concept.

8.2.4.5 Step 5: Analyzing the data

We are accustomed to one-at-a-time analyses where the respondents evaluate one stimulus or a set of related stimuli. We then look for patterns using regression modeling, or do tests using inferential statistics to determine whether or not two or more stimuli differ from each other. The focus of conjoint analysis is more toward the side of recognizing patterns and relations, rather than inferential statistics, although there is room for inferential statistics when one wants to establish the degree of confidence that two stimuli differ from each other.

Any conjoint study begins with an underlying experimental design or recipe. A conjoint study comprises many of these recipes, not just one. Furthermore, the elements in the recipe, that is, in the test concepts or vignettes, appear several times, against different backgrounds. Finally, the elements are statistically independent of each other (by the nature of the experimental design).

Since the elements are statistically independent of each other, the experimental design leads to regression analysis. The regression analysis can either be logistic (which is appropriate when the independent variables are binary), or the regression analysis can be ordinary least squares (OLS). Ordinary least squares are a little less appropriate, statistically, but with it, one can create a powerful science and communicate the results in a way management can easily understand. In this book, we will use OLS, fully recognizing there is a somewhat more powerful statistical method, but one that is not easily understood, hard to communicate, and does not lead to a science. It's a matter of value judgment here.

Analyzing overall rating data is done in two ways: first, by using the actual ratings as the dependent variable, and second, by transforming the ratings to a binary scale. When we use the actual ratings at the dependent variable, OLS tells us the number of rating points contributed by each element. When we convert the ratings to a binary scale (e.g. 1 to 6→0; 7 to 9→100), we are changing our focus from studying how elements "drive intensity of feeling" to "how elements drive membership in a group" (don't accept the concept vs. accept the concept). The two dependent variables sound the same, and in fact get to the same end result, namely which individual elements in the conjoint analysis perform well. However, the focus differs. The former, using the actual rating as the dependent variable, focuses on intensity of feelings, and is typically employed by those interested in the respondents, that is, how they feel. The latter, using the binary transformed data, focuses on membership in a group and comes from a long history of sociology and consumer research where the interest is in external, measured behavior that is typically "no/yes." Typically, the OLS analysis uses the additive constant, a measure of basic interest in the product or service. The individual coefficients are the utilities and show the impact of each element.

When the dependent variable is dollars or some other economic indicator that has a numerical value (e.g. years that one would use), the OLS model begins by transforming the ratings to the actual metric. For instance, when the rating scale is a 1-9 scale, with each scale point corresponding to a dollar value, we substitute the dollar value first, and then run the OLS regression. We do not use the additive constant. The result is a dollar and cents value for each element.

When the dependent value is emotion or another type of "feeling" or "end use," the respondent is instructed to select the appropriate emotion or end use from a list. For example, there may be five different emotions. The one-of-five response is converted into five new dummy variables, one dummy variable for each possible response. For each test concept, the respondent has had to select one of the five choices. The dummy variable corresponding to that choice is given a value of 100. The four remaining dummy variables, corresponding to the choices not selected, are each given the value 0. Then, one runs the OLS regression for each newly created dummy variable. The additive constant is not used in the OLS regression. The result is a conditional probability or percentage for each concept element, showing the proportion of times in 3-4 element concepts that one can trace a particular emotion to that element. Summing all the coefficients or percentages across an element for the five emotions, for example, will generate a total of around 27%. The total is not 100% because the OLS is working with three to four elements per concept, and the OLS is estimating the assignment of emotions to elements in a full concept.

8.2.4.6 Step 6: Interpret the results (what the utilities tell you and what the insights are)

Regardless of the method of eliciting consumers' responses, the utilities from Conjoint Analysis (CA) allow Consumer Explorers (CEs) to know the impact of product attributes on consumers' decision-making on liking (or believability, purchase intent, uniqueness perception, etc.). Consequently, the results enable the research team to create ideas and know their potential by summing the attribute utilities. Knowing the potential of product ideas allows the research team to select, prioritize, and manage NPD processes more efficiently.

We will now discuss the specific meaning of the utilities that emerge from the IdeaMap process. We will talk about the results that emerge when the dependent variable is the binary 0/100 rating, with ratings of 1-6 first transformed to 0, and ratings of 7-9 transformed to 100. Afterwards, OLS is run. The output is coefficients or utilities. These are averages across all the respondents in a key subgroup, for example total panel, males, users of a specific product, and so forth. The typical regression model is expressed as follows:

8

$$\text{Binary result(0/100)} = k_0 + k_1(\text{element 1}) + k_2(\text{element 2})\ldots + k_n(\text{element } n)$$

First, what do the numbers mean, are they additive constants? So how big is an impactful utility? There are two answers to this question:

(1) Using benchmark(s) and
(2) Comparing against historical data.

Using benchmarks allows researchers to ground the findings with a reference product idea(s) that will be exposed to the same conditions as new product ideas generated later. Benchmark(s) also assures researchers that the results of the study are valid; however, using benchmarks(s) means having less space for new attributes. The other way is comparing utility values against a norm. Using norms is a common practice in concept testing. However, it takes time and money to develop enough data to generate norms for each category. Some companies have generic norms for clients to compare to; however, since companies face different market conditions, the norms are rarely comparable.

8.2.4.7 Step 7: Report and learn

This step is the most important as it moves business forward in an efficient fashion. There is no specific right way to report the results; however, this list offers some advice generally useful in reporting CA results:

(1) **Report data, not just conclusions**: There's a tendency by many to abstract the results, to identify "what happened" and then to report the key factoids. As tempting as it is to tell a story, skipping the data because there are only so many numbers one can handle, should be avoided. Those who commission CA studies enjoy looking at the numbers, and coming to their own conclusions.
(2) **Highlight strong performing elements (high-positive utilities), and strong performing disutilities (high- negative utilities):** Typically, when we look at the model with the binary (0/100) as the dependent variable, we highlight elements whose utilities exceed +8 on the positive side, or exceed −5 on the negative side. At these levels, previous experience has shown that there is some correlating behavior in the external world, for example the marketplace.

As tempting as it is to work with inferential statistics, the real information comes from the size of the utility vs. the cut-off points of +8 and -5, and not from differences between pairs of elements based on the utility values. In other words, comparisons using statistics may be valid, from a technical sense (i.e. inferential statistics are valid), but there is little to gain from such analysis.

(3) **Wherever possible, make the highlighting tell the story:** The story becomes even clearer when one sorts the elements by utility value. The sort, along with the highlighting of strong performing elements, allows whoever is looking at the data to formulate a hypothesis. The elements are cognitively "rich," meaning that each element tells a story. Look for what is common, that is, for the story or theme, among the winning elements, and among the losing elements, respectively.
(4) **Revisit the objectives:** Ensuring that the stories you are going to tell fit with the objectives and are still the same (both content and priority).
(5) **Know the audience and what you expect from the audience:** When the meeting is a sharing and working session, full details of the results should be discussed in order to gain consensual conclusions. When the meeting is an informing or decision-making session, a short background of the study, objectives, results, and recommendations should be presented concisely. The most common mistakes are presenting too much information and assuming that the audience has the same level of understanding as the presenter.

8

8.2.5 The raw material of CA

This chapter will classify the stimuli into silos and elements. Silos (also known as variables, buckets, categories) are general groups of related elements. Elements (also known as levels) are the specific phrases that will be combined. Table 8.2.1

Table 8.2.1 Example of silos and elements appropriate for conjoint measurement. The elements would be those used for an "unspecified food" label. The elements are "cognitively rich." Each element can be evaluated by itself, or in a mixture, and be meaningful in either format.

Silo 1: Calories and sweeteners	
A1	0 calories per serving
A2	60 calories per serving
A3	160 calories per serving
A4	Sweetened with sucrose
A5	Sweetened with high fructose corn syrup
A6	Sweetened with fruit juice concentrate
A7	Sweetened with Equal (aspartame)
A8	Sweetened with Splenda (sucralose)
A9	Sweetened with Truvia (Stevia extract)
Silo 2: Ingredients for weight and health	
B1	Contains caffeine
B2	Contains green tea
B3	Contains soluble fiber
B4	Contains aloe vera
B5	Contains guarana extract
B6	Contains natural polyphenols
B7	Contains antioxidants
B8	Contains plant phytonutrients
B9	A good source of folate
Silo 3: Vitamins and minerals	
C1	Contains vitamin C
C2	A good source of vitamin C
C3	An excellent source of vitamin C
C4	Contains L-taurine
C5	A good source of calcium
C6	A good source of potassium
C7	A good source of magnesium
C8	With vitamin B12
C9	With beta carotene
Silo 4: Consumer issues: Taste, price, satiety	
D1	Keeps you feeling fuller longer
D2	Prevents food cravings
D3	Can stop snack attacks
D4	Great refreshing taste
D5	Energizing taste
D6	Optimized taste
D7	All the nutrients for one low price
D8	A low-cost source of key nutrients
D9	Affordable nutrient rich choice

Source: © IT! Ventures LLC, a limited partnership between the Understanding and Insight Group and Moskowitz Jacobs Inc.

8

shows an example of silos and elements for an "unspecified food" label. The same organizing principle, of the underlying structure comprising silos and elements, holds whether the topic is a description of a food, a professional service, or even a topic in public policy such as health care and nutritional compliance. The elements can be short or long, simple words or complete sentences. Ideally, the elements should constitute single ideas.

8.2.6 Experimental design

The key to conjoint analysis is the experimental design. The proper experimental design has the following two statistical characteristics:

(1) The elements are statistically independent of each other, so that regression analysis can be used to pull out the effect of each element.
(2) Each element appears against many different backgrounds, preventing pairs of elements from always appearing together so that their separate contributions cannot be estimated. This is called confounding, which often happens when the researcher doesn't use a design, but rather "guesses," or puts together so-called "rifle shots."

Experimental designs come in many types. Statisticians may use off-the-shelf statistical programs to create the design so that there are relatively few combinations to be tested. Of course, there must be more combinations to be tested (observations) than there are elements, in order for the modeling to work. Interested readers are referred to classic texts such as Box et al. (1978) which deal with the issue of design.

8.2.7 Building models

The essence of the analysis creates a model or equation that relates the presence/absence of the elements to the rating assigned by the respondent.

(1) **Using ordinary least square modeling:** Earlier forms of conjoint analysis used logistical regression, requiring a great deal of computer time. Although technically a slightly better form of regression, logistical regression does not produce results that are user-friendly. The user and, in turn, management commissioning the study cannot easily understand the output of logistical regression. They cannot determine what the elements mean, nor can they combine the elements mentally to come up with new combinations. In CA, the models can be made much more intuitive and user-friendly by using conventional regression analysis, often called OLS, or ordinary least-squares modeling.
(2) **The best modeling generates individual-level equation:** One doesn't have to worry about sample balancing, because the data from each respondent generates a model for that particular respondent. The approach to creat-

ing individual models and ensuring that each individual evaluated a unique set of combinations was solved in the early 1990s, and optimized for the Internet about ten years later (Moskowitz et al., 2001). The approach, called IdeaMap, began with a basic experimental design structure.

(3) **Each concept was created within that basic structure:** The key difference between this basic structure and so-called "fixed design" was that the structure itself sampled elements from the full set of available elements, checking that there were no previously defined pairs of incompatible elements, and then presenting the test concepts to the respondent. The entire process was done in real time. The key change was that every respondent now evaluated unique combinations.

(4) **Every respondent evaluates a different set of test concepts:** Traditionally in CA, the researcher would create a set of combinations. The test stimuli would comprise only that set of combinations. Many respondents would evaluate the different combinations in a randomized order. A high base evaluating the same set of combinations coupled with randomized order, reduced bias. In the IdeaMap approach presented here, each respondent evaluates a unique set of combinations, rather than having everyone evaluate the same set of combinations. The unique set of combinations totally eliminates the potential of bias, caused by testing a limited number of the same combination. The unique set of combinations for each respondent also means that the data can be analyzed in terms of both main effects and interactions among elements (Gofman and Moskowitz, 2010; Moskowitz and Gofman, 2004).

(5) **Using dummy variables in the input data for modeling:** The input data for the modeling is typically in the form of a data matrix that looks like Table 8.2.2. The matrix is, of course, much larger. Each respondent in the study on food labels generates his own matrix, comprising 36 independent variables, the elements, and a dependent variable, the ratings. In the manner of consumer researchers, we convert the nine-point rating to a binary scale, with ratings of 1–6 converted to 0, and ratings 7–9 converted to 100. The rationale behind this transformation is the history of CA as a form of consumer research, wherein focus is placed on membership in a category (e.g. acceptor of the concept), rather than focus being placed on intensity of feeling. The former is sociological in nature, membership in groups, the latter is psychological, intensity of personal experience.

(6) **The equations can be run either at the individual respondent level, or at the total respondent level:** When the data is run at the individual level, the focus is typically limited to the parameters of the model. These parameters are averaged across all of the respondents to generate sets of averaged coefficients, including the additive constant and values for the different elements. We see the average coefficients (utilities) in Table 8.2.3, for the total panel and for some key subgroups defined by the classification, such as gender, age.

(7) **The utility values that we compute have the property of being "absolute" rather than relative:** This is important for the creation of a science using CA. The experimental design comprises incomplete concepts. Rather than forcing every concept to comprise exactly one element from each silo, the concept comprises a limited number of elements (3–5 to make reading easier). In some concepts, a silo is completely absent. Working with absent

Table 8.2.2 Example of data used as input from one respondent for modeling. The table shows partial data (12 of 48 test concepts), the presence/absence of 14 of 36 elements, and the two dependent variables: the nine-point rating, and the binary expansion of that nine-point scale. A "1" signifies that the element was present in the test concept; a "0" signifies that the element was absent from the test concept. The column titled "binary" shows the result of transforming the nine-point rating to a binary 0/100 scale.

Conc	A1	A2	A3	A4	A5	A6	B1	E6	F1	F2	F3	F4	F5	F6	Rating	Binary
1	0	0	0	1	0	0	0	1	0	0	0	0	0	0	7	100
2	0	0	0	0	0	1	0	0	0	0	0	0	0	0	6	0
3	0	0	0	0	0	1	0	0	0	0	0	1	0	0	1	0
4	0	1	0	0	0	0	0	0	0	0	0	1	0	0	3	0
5	0	1	0	0	0	0	0	0	0	0	0	0	0	0	9	100
6	0	0	1	0	0	0	0	0	0	1	0	0	0	0	5	0
7	0	0	0	0	0	0	0	0	0	0	0	0	0	0	6	0
8	0	0	0	0	0	1	0	0	0	1	0	0	0	0	7	100
9	0	0	0	0	1	0	0	0	0	1	0	0	0	0	5	0
10	0	0	0	1	0	0	0	0	0	0	0	0	0	0	3	0
11	0	0	1	0	0	0	0	0	1	0	0	0	0	0	5	0
12	0	0	0	0	0	0	0	0	1	0	0	0	0	0	7	100

silos and incomplete concepts allows one to estimate the absolute utility value of each element (vs. only compatible within the silo). That absolute value, in turn, leads to comparability across studies. It encourages databasing, the foundation of a true science rather than just an analytical method.

8

8.2.8 Presenting the result – numbers, text, data, talk, move to steps

Even before we deal with the actual data, it's good to deal with what we're going to do with the data. Most discussions of research methods focus on what the methods provide, how they are implemented, and, of course, the appropriate statistics. The truth of the matter is that the methods are not only as good as the application but also only as good as the presentation. Sometimes, especially in the world of consumer research, one hears about reams of perfectly valid data sitting in warehouses, unused. It's just that the researcher has not been able to present the results in a sufficiently cogent way. It ought to be that management has the wisdom, but all too often that connection is missed, even though the work has been done.

How, then, do we ensure that the results of conjoint analysis reach the eyes of management, and more importantly, drive decisions? What are we to do? Here are a few pointers, some of the Realpolitik of the situation:

- **Management likes numbers:** It's tempting to fall into the trap of giving management a verbal explanation of what you find and keeping the numbers to

Table 8.2.3 Sample output from the study on nutritional labeling, showing the results from the total panel, and from pairs of complementary subgroups (males vs. females; younger versus older respondents). Each respondent generated his own model, relating the presence/absence of the elements to the binary 0/100 rating of interest. The table shows the average of the parameters across the respondents. The 36 elements are sorted in descending order, based on the utilities from the total panel.

		Total	Male	Female	Age under 45	Age 45 +
	Base size	320	71	249	107	213
	Additive constant	22	35	18	25	21
A1	This product is a good source of protein	17	13	18	15	17
A2	This product is an excellent source of protein	16	6	18	14	16
B2	This product is high in fiber	14	15	14	15	14
C4	This product has no saturated fat	12	12	11	11	12
D6	This product is without sodium	12	2	15	14	12
A4	This product is an excellent source of vitamin C	11	7	13	14	10
B4	This product is rich in calcium	11	6	12	7	13
C6	This product has zero cholesterol	11	9	12	9	13
C2	This product is fat free	10	7	11	6	12
D4	This product is free of added sugar	10	4	12	5	12
A3	This product is a good source of vitamin C	9	0	11	11	7
A6	This product is high in vitamin A	9	1	11	9	9
B1	This product contains fiber	9	3	11	7	11
D2	This product is sugar free	9	10	8	6	10
D5	This product is a low source of sodium	9	3	11	10	9
B3	This product contains calcium	8	5	9	5	10
D1	This product is low in total sugar	8	2	9	8	8
C1	This product is low in total fat	7	2	8	6	8

C5	This product is a low source of cholesterol	7	9	6	7	7
E6	Puts more nutrient power on your plate	7	6	7	10	5
F5	Be at your best. Enjoy good taste and good health	7	5	8	11	5
A5	This product provides vitamin A	6	1	8	9	5
E2	A good way to balance your diet to keep it nutrient rich	6	2	7	7	5
E4	Naturally packed with nutrients for better health	6	-1	8	6	6
B6	This product is rich in iron	5	2	6	1	8
E5	Meets your daily nutrient needs without too many calories	5	1	6	3	6
F4	You and your family can eat right – for life	5	2	6	6	5
B5	This product provides iron	4	-2	6	2	5
D3	This product is low in added sugar	4	-1	5	5	3
E1	Wholesome food that gives you more nutrition per bite	4	0	5	4	4
E3	A total nutrient package with more nutrients than calories	4	0	6	3	5
F1	A great way to enjoy your healthy lifestyle	4	-1	6	7	2
F2	Takes the stress out of healthful eating	4	0	5	5	3
C3	This product contains little saturated fat	3	-6	5	2	3
F3	Let's you eat well to live well, starting today	3	-3	4	2	3
F6	You can trust the nutrition label to guide smart eating	3	-1	4	4	3

yourself. We've all done that. Indeed, one of the earliest lessons was a presentation author H. Moskowitz gave to a packaged goods company in the Midwest. The instructions were to talk about the results in text. Yet, during the course of the presentation, some numbers and tables "snuck in." Marketing management gravitated to those simple tables of results, not the presenter's verbose explanation. It was a good lesson (see discussion in Chapter 1 on "winstonizing" data).

- **Don't bore with statistics:** Give a rule of thumb, for example utilities of five or more are significant, utilities of eight or more are worth attending to, and then stop. Conjoint analysis may have an intellectual history in academic and science, but to make sure people use it, give them simple rules. They'll make the data their own.
- **Finally, tell a story:** With lots of elements as stimuli you are dealing with a cognitively rich environment. You have two things going for you; take advantage of them. The first is that you are dealing with elements that have meaning, whether these are brands, benefits, ingredients, and so forth. That type of stuff is interesting, and second, you have numbers which reflect how the mind responds to these cognitively rich elements. The numbers fascinate. The combination is a winner. The combination is a peek into the customer's mind. Management cannot resist that. The peek always fascinates.

8.2.9 Using the results – what do the numbers tell us?

8

In its early business use, conjoint analysis for messaging never really achieved the wide distribution it deserved, for the reason that many "creatives" in advertising agencies believed strongly that messaging was an art, and that it was virtually impossible to explore messaging by rational, systematic means. The argument was often made that using conjoint analysis to understand messaging and create optimal messaging was akin to "monkeys typing Shakespeare." The criticism was wrong, of course, but effective in holding back the use of conjoint analysis in the work of messaging.

However, the force of business is such that improved methods eventually find their "place in the sun," despite the best efforts to dismiss them. Conjoint analysis began to find wider acceptance after exposure to business needs forced it to deal with more mundane, realistic problems of an everyday sort.

With the foregoing in mind, let's look at the results in Table 8.2.3, to see what a researcher might extract from these results.

(1) **The additive constant, a measure of readiness to up-rate the concept:** The additive constant tells us the proportion or conditional probability that a person will rate the concept 7–9, in the absence of elements. The additive constant, a purely estimated parameter, nonetheless provides us with a sense of the baseline. Here the additive constant for the total panel is 22, meaning that without elements, we expect 22% or about one person in five to rate the concept 7–9. The elements have to do the rest of the work. The additive constant is higher for males (35) and lower for females (18), suggesting a gender difference.

(2) **We've sorted the utilities from high to low, based on the utilities from the total panel:** We did this for a simple reason. It's the elements themselves,

and not the silos or categories, that are important. The six silos are, in reality, just bookkeeping devices. We learn a lot from the elements that perform well. We should look for patterns. The pattern that we see is that familiarity is important. The elements that score highest are those that we recognize.

(3) **By putting the elements in descending order, we quickly get a sense of the range of utilities (+17 to +3):** We know what elements work and what don't. For our data, all of the elements drive interest in the label.

(4) **There are very few negatives:** A negative utility means that the element drives away consumer respondents and diminishes the number of interested individuals. In this particular study, almost none of the elements are negative. When we work with new products; however, rather than with conventional label information, we're likely to see more elements showing negative utilities.

8.2.10 Beyond individual groups to segments – finding mind-sets using conjoint analysis

Conjoint analysis is a perfect example of the evolution of a research tool when it is applied to the practical problems of industry. In this section, we deal with three of these advances: discovering synergisms among pairs of elements, applying CA to graphics, and finally using CA to mind-type consumers (so-called "Mind Genomics™").

People differ from each other. Table 8.2.3 showed us some of these differences when we divide the respondents in a CA study by the conventional ways of dividing people; who they are, what they do, and even what they say about a topic. The history of CA is also one of dividing people by the pattern of their utilities. Individuals showing different utilities can be considered to differ from each other in the ways their minds work. The typically narrow focus of a CA study, generally dealing with one topic, makes the assessment of individual differences a very instructive exercise. One learns just how people differ in a granular aspect of their lives.

We get a sense of this "mind-segmentation" for labels from Table 8.2.4. The specifics of the segmentation are not important; they can be found in most texts on clustering, and in many off-the-shelf or "shrink-wrapped" statistical programs. What's important is that the people in a specific segment show similar response patterns in the CA study. The centroids of the three segments are quite different from each other. Whether the people in the three segments will cluster together when it comes to other topics, such as vitamins or healthy lifestyle, is not known. The granularity of the CA study means that the segmentation is limited to the particular topic area being investigated.

Table 8.2.4 shows the data from the 172 adults who evaluated the cheesecake, first showing the total panel, then showing the three emergent mind-sets.

Mind Set 1 - Imaginers - word picture driven, not food category driven, they need help to crave

Mind Set 2 - Elaborates - crave the category plus other stuff

Mind Set 3 - Classics - Crave the category in and of itself . . . don't mess with it . . .they like the flavors and textures as they are and have been for years

Table 8.2.4 Results from the adult cheesecake study, showing the performance of elements from the Total Panel, and from the three emergent mind-sets (MS) (highlights indicate important utilities of elements).

		Total	MS1	MS2	MS3
Base Size		172	36	88	48
Constant		41	55	32	45
Mind-set 1 - "Imaginers"					
B1	When its cold outside, cheesecake is cozy and inviting	3	11	6	-8
A3	Rich cheesecake flavor with a silky texture	13	9	11	17
A5	Cheesecake so rich it melts slowly to release delicate, intense flavor	17	8	17	25
C2	When you think about it, you have to have it . . . and after you have it, you can't stop eating it	4	8	5	1
D7	Made fresh . . . especially for you . . . by you	1	8	4	-7
Mind-Set 2 - "Elaborates"					
A8	Cheesecake with swirls of raspberry, chunks of white chocolate, baked in a crunchy crust, and garnished with pecans	19	-38	42	20
A7	Dense cheesecake swirled together with ribbons of chocolate chips in a chocolate crust . . . served with a raspberry sauce drizzled across your plate	18	-20	35	17
A9	Cheesecake wedges with thick filling, gooey sauce, chopped nuts, and slices of fresh fruit	14	-27	35	5
B7	With all the toppings you want . . . chocolate sauce, berries, caramelized topping, chopped nuts, . . . whatever	14	1	22	9
A2	Cheesecake with a smooth appearance, light flavor, fluffy creamy texture and a graham cracker crust	16	-3	15	32
A6	Heavy texture with complex flavors and a velvet appearance	10	-2	13	15
Mind-Set 3 - "Classics"					
A4	Real cheesecake made with ingredients like eggs, cream cheese, sugar, vanilla, and lemon juice	11	4	7	23
A1	A smooth, dense slice of cheesecake	8	7	7	11
B9	So rich and creamy . . . you practically have to lick your lips twice after each bite	7	6	6	10
C8	Pure ecstasy	5	7	2	9
Not Strong in any mind-set					
B4	Premium quality . . . that great classic taste, like it used to be	6	7	6	7
B6	100% natural . . . and new choices every month to keep you tantalized	4	5	3	5

8

Table 8.2.4 (*Continued*)

		Total	MS1	MS2	MS3
C9	It feeds the HUNGER	2	1	1	5
C6	A joy for your senses . . . seeing, smelling, tasting	4	5	3	4
C7	An outrageous experience . . . shared with family and friends	2	1	2	4
D5	From the Cheesecake Factory	3	3	4	2
D3	From Sara Lee	1	0	1	2
C5	Now you can escape the routine . . . a way to celebrate special occasions	4	2	6	1
C3	Fills that empty spot in you . . . just when you want it	2	1	3	1
D8	Simply the best cheesecake in the whole wide world	3	2	5	0
C1	Quick and fun . . . eating alone doesn't have to be ordinary	2	2	4	-1
B5	You can just savor it when you think about it during work and school	2	4	3	-1
B3	In bite size pieces, stored in the freezer for a fast taste	0	2	1	-1
D2	From Philadelphia Brand	2	-3	6	-3
D6	From a fine local pastry shop	1	1	2	-3
B8	You can imagine the taste as you walk in the door	-1	-3	2	-4
C4	When you're sad, it makes you glad	-2	-5	1	-4
D4	Made by Eli's	-5	3	-6	-7
B2	With a hot cup of coffee, tea, hot cocoa. . .or carbonated beverage	-3	3	0	-13
D9	With the safety, care and cleanliness that makes you trust it & love it all the more	-4	1	0	-14
D1	Available by mail order	-11	-9	-8	-17

8

Once the mind-sets were identified, Consumer Explorers can inspect the distribution of mind-set membership among other demographic groups such as adult vs. teen. Compared to Adults, Teens have typically larger-sized Elaborate Segments and smaller-sized Imaginer Segments (Table 8.2.5)

8.2.11 Scenario analysis – discovering synergisms and suppressions (interactions) among elements in a conjoint analysis study

One of the traditional aspects of concept research has been the claim by some creatives in advertising and marketing that they are privy to the different ways people process ideas, and "know" from their years of experience what elements work together, and what elements do not work together. When put to the test, however, most individuals cannot predict what elements synergize,

Table 8.2.5 Distribution of mind-sets for three products (cheesecake; hamburger; steak).

	Total		Mind-set 1 Imaginer		Mind-set 2 Elaborate		Mind-set 3 Classic	
	Teen	Adult	Teen	Adult	Teen	Adult	Teen	Adult
				Cheesecake				
Base Size	89	172	24	48	37	88	28	36
Percent	100%	100%	27%	44%	42%	19%	31%	37%
Constant	38	41	40	45	27	32	51	55
				Hamburger				
Base Size	77	150	21	69	31	25	25	56
Percent	100%	100%	27%	46%	40%	17%	32%	37%
Constant	18	30	38	7	31	49	18	51
				Steak				
Base Size	83	168	24	101	36	13	23	54
Percent	100%	100%	29%	60%	43%	8%	28%	32%
Constant	30	41	47	26	39	54	51	67

and what elements suppress each other. The notion of interactions among elements has continued to fascinate researchers. Among the creative professionals, copywriters, and graphic artists, there is a belief that developing good concepts and packages is partly due to talent; namely, one's ability to spot synergies among words in language or in visual stimuli. Accordingly, it is important to be able to spot and quantify synergisms and suppressions among pairs of elements.

Such capability has typically been beyond the ability of standard conjoint procedures, simply because there are too many possible combinations to test directly. For instance, a conjoint study comprising three silos, each with eight elements, has a possibility of 64 interactions for each pair of silos. This makes 192 combinations to test. From the practical point of view, it's just too much to test 192 combinations simply to find interactions that may or may not even exist.

Traditionally, CA has focused on "main effects," wherein each of the elements has been treated as an independent entity. In a CA study, the individual element might appear against several different backgrounds. Nonetheless, when it comes time for the analysis, the question is often asked whether or not two elements "interact" with each other, so that the combination is far higher or lower than one might expect, or the performance of an element (e.g. promise or product feature) in the presence of another specific element (e.g. brand) is otherwise strongly affected.

The Mind Genomics approach identifies interactions by dividing the data into a set of no-overlapping strata, based upon the four elements in the up-front design, and the condition of no element. The interaction becomes the measurement of the performance of one element in those vignettes where another type of element (e.g. Brand) is not represented, or is represented by any of four options. This is so-called "scenario analysis," shown in Table 8.2.6. Scenario analysis simply looks at the pattern of coefficients for a single element when the element is put into the same vignette with one of the four elements of another question, and with no answer. "The method is straightforward. One way to discover interactions forces in the linear terms and observes how much variance is explained. Then, one can assess the additional amount of variance in the data. This reveals which pairs of elements may synergize, over and above their individual ability to explain the variability" (Gofman, 2006).

In scenario analysis, one identifies a specific silo, such as brand name, whose elements, in turn, define the different strata. A stratum is thus defined by the particular brand name (or another criterion) in the test concept. Operationally, the researcher first sorts the full set of test combinations into the different strata. By design, each concept comprises only one element from the key variable defining a stratum (e.g. brand name), or, as a result of the design, the stratum may comprise those concepts because the design called for that stratum to be absent in any elements from the silo. Each stratum ends up having only one constant element (e.g. brand name). We see the results from the analysis in Table 8.2.6. The table shows the utility values for the model without interactions (the last column "Not Present"), and the utility values of those elements in a silo chosen for studying the interaction effect (the other column headers), as well as the utility values of the other elements from the other silos in the study (rows).

Let's use the first row (element E16: Farm-to-table fresh produce. . . crisp fruit, vegetables, and eggs), as an example, to examine synergism and suppression (interaction effects) between this element and the interested silo (in this case, it was the certification such as non-GMO, USDA organic). One can see that without these certifications, the utility of element E16 was only 2.75. However, with each of the certification silo's element, the utilities were changed (higher when paired with Non-GMO verification, Vegetarian, Certified OU & Kosher, Locally Grown and Certified Sustainable and lower when paired with All Natural, Gluten free, International Halal and USDA organic). For element 34 (E34), pairing "Produced by eco-conscientious farmers" with "Gluten free" had suppression effect with each other as the pair's utility was decreased from 2.75 to 1.96. By conducting a scenario analysis, CE can identify interaction effects (synergism, neutral, or suppression), quantify the effect sizes and guide concept creation.

8

8.2.12 Dealing with prices

Pricing presents us with an extensive set of topics, often left to economists and marketers, but very addressable by conjoint analysis. Indeed, conjoint analysis presents the pricing issue in terms of the newly emerging area of behavioral economics. With conjoint analysis, one can deal with pricing either as an element in an offering, or a response to an offering.

Table 8.2.6 Scenario analysis for discovering synergisms (highlighted cells) and suppressions (white cells) between elements from a silo (the column titles) and other elements from the other silos (rows).

		Non-GMO verification	Certified Sustainable	USDA Organic	Locally Grown	Vegetarian	Certified OU & Kosher	All Natural	Gluten free	International Halal	Not Present
	Average	2.64	2.60	2.47	2.46	2.39	2.37	2.35	2.20	2.05	1.83
E16	Farm-to-table fresh produce . . . crisp fruit, vegetables and eggs	3.65	2.84	2.73	3.33	3.64	3.60	2.37	2.46	2.46	2.75
E34	Produced by eco-conscientious farmers	2.76	2.07	3.12	2.15	2.49	2.64	2.56	1.96	2.02	2.75
E17	Gourmet, hand-made chocolates	2.88	3.00	3.69	2.92	2.05	2.40	2.47	3.05	2.91	2.69
E12	Freshly baked goods . . . artisanal breads, cakes, pies and cookies	3.04	3.50	3.04	3.73	3.32	2.80	3.49	2.12	2.61	2.46
E10	Packaged meat . . . ready to cook	3.19	3.39	3.83	2.31	1.90	2.18	1.84	1.51	2.71	2.39
E36	Recommended by health professionals (e.g., doctors, dietitians)	2.20	2.53	1.01	2.20	3.33	3.11	2.64	3.09	2.01	2.29
E31	Regularly inspected by U.S. Department of Agriculture and religious authorities	2.69	1.84	2.73	3.36	1.93	1.82	2.57	2.69	1.55	2.21
E18	Freshly made pasta . . . just boil and add your favorite sauce	2.44	2.55	1.78	3.04	2.75	2.95	2.91	1.90	2.04	2.21

E30	Ingredient list reflects claim on the label	1.76	2.46	2.06	2.74	2.19	2.85	1.40	1.99	2.12	2.13
E29	Endorsed by an environmental watchdog organization	2.37	2.19	2.63	2.21	2.59	1.74	2.59	2.12	2.37	2.06
E32	Label is prominently displayed . . . easy to read	2.34	3.07	2.31	2.54	3.24	2.27	2.82	1.91	1.89	2.05
E35	Follows ecological/ environmental guidelines	3.39	3.39	2.51	2.14	2.25	1.14	2.31	1.31	1.98	2.02
E28	Traveling . . . perfect time to try new foods	2.92	2.20	1.41	1.55	2.11	2.36	2.10	2.37	2.07	2.01
E13	Spices, herbs and other flavor enhancers	2.75	2.56	1.80	1.59	1.43	2.19	2.24	2.67	1.32	1.98
E14	Packaged pre-washed ready-to-serve salad . . . just add dressing and toppings of your choice	2.13	2.10	1.87	2.59	2.05	1.99	1.80	1.95	2.04	1.90
E21	Manufactured by a well-known, national food company	2.84	2.04	2.51	1.64	1.64	2.00	2.39	2.18	2.42	1.90
E33	Healthful and permitted by religious teachings . . . a winning combination	1.44	2.44	2.97	1.87	1.78	2.19	1.42	2.18	0.29	1.86
E19	Invited to celebrate a holiday with family . . . with a feast no doubt	2.29	2.63	2.74	2.10	3.06	1.95	2.19	1.10	2.02	1.81

(Continued)

Table 8.2.6 (*Continued*)

		Non-GMO verification	Certified Sustainable	USDA Organic	Locally Grown	Vegetarian	Certified OU & Kosher	All Natural	Gluten free	International Halal	Not Present
E11	Ready-to-eat soup . . . just heat and serve	1.52	0.71	1.49	1.74	1.59	2.12	2.85	1.60	2.37	1.60
E15	Bottled juice	2.69	1.68	2.87	1.66	1.68	1.59	2.19	2.02	1.02	1.59
E23	Prepared and packaged for sale right in the same place . . . convenient	1.83	2.94	2.39	3.10	2.28	2.22	1.87	2.51	2.18	1.44
E24	Prepared by people who share your cultural and religious values	2.28	2.68	2.38	1.83	1.57	2.51	2.16	2.61	1.50	1.36
E22	Look for it in your neighborhood grocery store	1.67	2.21	2.17	2.60	1.74	2.63	2.63	2.57	2.17	1.31
E27	Based on sacred teachings . . . not your faith, but shares your values	1.61	1.53	1.62	2.36	1.54	1.94	1.83	1.86	1.27	0.95
E20	From a local street vendor	2.13	2.44	2.03	1.66	1.91	1.37	1.94	1.99	2.17	0.92
E26	In line with values in the Qur'an	0.27	1.68	1.39	0.59	1.42	0.70	2.05	0.93	1.64	0.45
E25	In line with values in the Torah	0.74	0.94	1.45	1.40	1.34	0.30	1.09	1.52	1.39	0.43

Today, you will help us understand people's behavior and feelings regarding TYPES OF FOOD AND THEIR LABELS.
It will take you about 25 minutes to complete this survey.

On the following screens you will be presented with 60 sets of 2 to 4 statements about TYPES OF FOOD AND THEIR LABELS and related information
Although the screens may seem similar. please note that each set of statements is UNIQUE.
You will respond to the same two questions for each set of statements:

(1) If a package of this food typically costs $6.79, what would you be willing to pay for it.

1 = $7.35, 2 = $2.95, 3 = $9.09, 4 = $6.79, 5 = $9.99, 6 = $8.19, 7 = won't buy, 8 = $4.99, 9 = $3.54

(2) How do you feel about the food as described here?

1 = Distrusting, 2 = Suspicious/Concerned, 3 = No feeling/Indifferent, 4 = Curious, 5 = Enthusiastic/Delighted

Please rate each screen as a single unit.

Following the 60 sets of statements, you will be asked 50 short questions.

You can track your progress in completing the study by looking at the numbers at the top right of each screen.

Figure 8.2.2 An example of an introduction page for using price and emotion as a rating instead of liking or interest rating.

The former, price as an element in the mix, creates a set of elements dealing with price or even a whole silo of elements with price. Price is then embedded in the study, as a set of elements to be tested. The impact or utility values for specific items with prices attached show the marginal or part-worth contributions of price for that item. The second approach uses price as a dependent variable (Figure 8.2.2). The respondent reads the test concept and picks a price that would be appropriate for the concept. The ratings are replaced by the dollar value selected, and the regression analyses are done as before, but without an additive constant.

From the analysis, one determines the dollar value of each individual element. The results show the dollar value of each element. Table 8.2.7 shows an example of this approach.

Which approach is better, price as an element or price as a rating, remains for this particular experiment. The results of many studies with price evaluated in each way suggest that respondents have little or no problem with price, either as the stimulus or as a response. H. Moskowitz and others published a jointly authored book on pricing, with a large number of examples showing how conjoint analysis can be used for pricing (Galanter et al., 2011). The introduction of price as a dependent variable, with a scale, brings up the interesting problem about the willingness of a respondent to pay more for a product which is liked more. The answer is both yes and no, depending upon the particular product. The answer is yes when the product is eaten for pleasure. The answer is a definite "maybe" if the product is purchased for reasons other than sensory pleasure.

The final aspect of price is the results from clustering and segmenting respondents on the basis of their liking or interest in buying (homo emotionalis) vs. segmenting respondents on the basis of the price they are willing to pay. As one might expect, the mind-sets are more different when the dependent variable, the rating scale, deals with liking or willingness to buy, and more similar, less pronounced when the dependent variable is price.

Table 8.2.7 Results from using price (in USD) and five emotions as ratings when see a vignette (calculated without intercept/additive constant) highlights indicate important utilities of elements).

	n = 165	Price	Distrust	Suspicious/Concern	No Feel/Indifferent	Curious	Enthusiastic/Delight
B7	Farm-to-table fresh produce . . . crisp fruit, vegetables and eggs	US$2.28	-3	-1	10	8	18
A5	USDA Organic	US$2.20	-1	4	10	11	10
B8	Gourmet, hand-made chocolates	US$2.15	-1	-1	6	13	14
B3	Freshly baked goods . . . artisanal breads, cakes, pies and cookies	US$2.12	-1	-2	13	11	13
A9	Non-GMO verification	US$2.04	2	0	10	13	9
B1	Packaged meat . . . ready to cook	US$2.03	1	5	11	9	4
A7	Certified Sustainable	US$1.98	0	4	11	12	7
D7	Produced by eco-conscientious farmers	US$1.97	-1	2	8	14	9
D9	Recommended by health professionals (e.g., doctors, dietitians)	US$1.92	1	-2	8	13	12
A1	All Natural	US$1.90	1	8	8	8	11
C1	Invited to celebrate a holiday with family . . . with a feast no doubt	US$1.90	2	3	10	16	2
B9	Freshly made pasta . . . just boil and add your favorite sauce	US$1.82	-1	1	13	11	8
A8	Locally Grown	US$1.80	2	5	10	4	9

C3	Manufactured by a well-known, national food company	US$1.71	2	3	8	11	7
D3	Ingredient list reflects claim on the label	US$1.71	4	4	11	10	1
D8	Follows ecological/environmental guidelines	US$1.71	1	1	12	13	5
D5	Label is prominently displayed . . . easy to read	US$1.66	1	3	14	8	3
C5	Prepared and packaged for sale right in the same place . . . convenient	US$1.64	1	7	6	11	7
A2	Certified OU & Kosher	US$1.62	4	5	14	8	2
B4	Spices, herbs and other flavor enhancers	US$1.62	1	0	15	13	5
C4	Look for it in your neighborhood grocery store	US$1.58	3	4	13	14	1
D4	Regularly inspected by U.S. Department of Agriculture and religious authorities	US$1.58	3	3	12	6	9
D1	Traveling . . . perfect time to try new foods	US$1.57	0	5	16	8	3
A6	Vegetarian	US$1.54	2	3	7	9	8
D2	Endorsed by an environmental watchdog organization	US$1.53	2	5	13	4	9
A3	Gluten free	US$1.52	3	9	12	3	6
C6	Prepared by people who share your cultural and religious values	US$1.51	6	5	12	12	0

(Continued)

Table 8.2.7 (*Continued*)

	n = 165	Price	Distrust	Suspicious/Concern	No Feel/Indifferent	Curious	Enthusiastic/Delight
B2	Ready-to-eat soup . . . just heat and serve	US$1.30	2	2	14	9	6
B5	Packaged pre-washed ready-to-serve salad . . . just add dressing and toppings of your choice	US$1.26	2	0	14	3	9
C2	From a local street vendor	US$1.26	10	11	-1	13	1
B6	Bottled juice	US$1.23	2	4	19	5	5
D6	Healthful and permitted by religious teachings . . . a winning combination	US$1.23	4	9	6	8	2
C9	Based on sacred teachings . . . not your faith, but shares your values	US$1.17	10	12	12	1	-1
A4	International Halal	US$1.02	7	7	7	7	3
C8	In line with values in the Qur'an	US$0.51	14	13	10	1	-5
C7	In line with values in the Torah	US$0.38	13	11	14	2	-5

8.2.13 Linking elements to emotions

The topic of feelings/emotions is an increasingly important one, as shown by the numbers of papers in the academic literature. It has long been known that emotions are linked to food and drink. Indeed, some of the earliest papers were on the emotions dealing with aromas, on the pleasantness and unpleasantness of tastes, and on the general issue of hedonics in food. A great deal of the world of consumer research is to ensure that foods are liked, indeed liked more than they were before, under the assumption that the more liked foods will enjoy a greater market share.

Mind Genomics provides an interesting technology by which to measure the linkage between feelings/emotions and foods, or ideas about foods. Recent developments have linked emotions in three ways.

(1) **Make emotions and feelings one of the questions (categories):** In the It! Studies, one of the silos did comprise feelings and emotions (silo #). The data showed the degree to which the stated emotion drove acceptance. These early efforts represent first attempts, efforts that would be supplanted by far more powerful approaches. We can say that incorporating emotions and feelings into vignettes is not particularly productive. One cannot transmit the feeling by making it part of the stimulus.

(2) **Make emotions and feelings the rating scale:**. Rather than having the respondent rate acceptability or price, the respondent can be instructed to select one of a limited number of feelings/emotions after reading a vignette. Let us assume that there are five such emotions, but there could be more, up to a reasonable number, around nine. We will work with the five emotions, recognizing that they are hardly exhaustive (Figure 8.2.2. and Table 8.2.7).

(3) **Make emotions the set of elements that are combined and presented as vignettes:** The respondent reads a vignette (product or sales concept about the food), or eats the food, and afterwards evaluates the fit of each of the 24 vignettes to what was just read or eaten. Each feeling/emotion is a separate element. The pattern of coefficients provides an easy-to-create, nonintellectualized emotional signature.

8

8.2.14 Measuring response time

The opportunity to measure the responses to individual vignettes virtually automatically by computer acquisition of responses meant that one could also measure response time, the time between the presentation of the stimulus and the response to the stimulus. Response time (also known as reaction time) plays a big role in the history of experimental psychology and in current day marketing. The notion is that longer response times mean that the respondent takes longer to process the information, either because the information is complicated, or engaging, or both.

The BimiLeap app (an evolution of conjoint analysis technology, we will talk about this tool later in this chapter) returns with an automatic analysis of response time for each element. The equation used to estimate the individual response times for elements is almost the same as the equation relating the Top-3-Box to the presence/absence of the elements. The only difference is the absence of an

Table 8.2.8 Estimated reaction time (RT) for each element calculated from a model without additive constant for Total panel and different mind-set (MS) solutions (highlights indicate important reaction time of the elements).

Group (RT in seconds)		Total	MS 1 of 2	MS 2 of 2	MS 1 of 3	MS 2 of 3	MS 3 Of 3
Base Size		51	25	26	16	21	14
A1	Formulated to contain healthier ingredients	1.0	1.0	1.0	1.4	0.8	0.7
A2	Good for you and the environment	0.9	0.9	1.0	0.8	0.8	1.2
A3	Contains natural ingredients	0.8	0.7	1.0	0.9	1.0	0.6
A4	Produced with sustainable ingredients	1.1	1.0	1.1	0.9	1.2	1.1
B1	Helps you to lose weight	1.2	1.1	1.3	1.2	1.4	0.8
B2	Plant-based foods increase fiber intake. . . strengthens your digestive system	1.5	1.2	1.7	1.5	1.8	1.1
B3	Plant-based foods are rich in vitamins	1.4	1.1	1.7	1.2	1.5	1.5
B4	Plant-based foods reduce the risk of type-II diabetes and cardiovascular diseases	1.1	0.9	1.3	0.9	1.4	0.9
C1	Appearance: exactly like meat	1.0	1.0	0.9	1.4	0.9	0.7
C2	Aroma: exactly like meat	1.0	1.1	1.0	1.5	1.2	0.2
C3	Texture: exactly like meat	1.0	1.0	1.1	1.0	1.1	0.9
C4	Flavor: exactly like meat	0.8	0.6	1.1	1.0	1.2	0.0
D1	Use: pastry, bread & meat pie	1.0	1.1	1.0	1.0	1.0	1.2
D2	When: at a business lunch	0.9	1.0	0.9	0.9	0.5	1.6
D3	When: at home with the family	1.1	1.2	0.9	1.2	0.6	1.6
D4	Form: meat slices and patties	1.2	1.5	0.9	1.5	0.6	1.7

8

additive constant, the rationale being that the response time is zero if there are no elements. Table 8.2.8 shows the results for a study on plant-based food with the estimated response time (seconds) for each element within a different mind-set. One can expect that the patterns governing response time will be different from the pattern governing interest (e.g. liking rating or purchase interest rating).

8.2.15 Discovering the "new" through conjoint analysis – creating an innovation machine

About 30 years ago, in the early 1990s, the opportunity arose to work on a new form of dental cleaning. Jonathan Kalan, then director of market research of the Oral B Company, now part of Gillette and P&G, presented a challenge to create new, more powerful methods to clean teeth. The field of alternatives was open. Kalan had previously used conjoint analysis for positioning and advertising, but this time the question on the table was whether one could throw together ideas from different worlds, into a "mixmaster" and by doing so come up with new-to-the-world technologies for oral care.

It became clear from those early experiments that the conjoint analysis silos need not be the simple, well-defined variables of features for a toothbrush or a paste. Rather, as part of the effort, the group took ideas from different types of fields, including industrial abrasives, industrial cleaning machinery, and oral care. The elements were edited and then incorporated into concepts according to experimental design. Consumers reacting to the test stimuli had no idea that the elements in their test concepts came from such diverse worlds. The concepts seemed "reasonable," at least at a superficial level, even if they weren't immediately recognizable.

The outcome, discussed in detail in a book written in 1995 (Moskowitz, 1995), showed how one could use conjoint analysis to create an entirely new category, now called power brushing. Whereas today the notion seems so obvious, it was not in the 1990s. The ability of conjoint analysis to act as this "mixmaster" became a hallmark of innovation in many other product studies afterwards. Furthermore, with the advent of the Internet as an almost infinite source of information, conjoint analysis becomes an ever more powerful tool for innovation by recombination.

8.2.16 Mind Genomics™: a new "science of the mind" based upon conjoint analysis

8.2.16.1 Mind-set segmentation, Mind Genomics™ and Addressable Minds™

The notion that the world of consumers can be divided into various groups is hardly new. A century ago, merchants had already recognized that customers were divided into those who wanted upper scale products and those who wanted economy products having fewer features but at clearly lower cost. One need only look at the evolution of the automobile industry to recognize that manufacturers intuitively recognized these different strata of desire (Wells, 1975).

The key problem for marketers is not the existence of these segments, but how to understand the nature of the segments, what to offer the segments and, most importantly, how to communicate to these segments. The problem of assigning a new person to a segment, called "scoring the person," is equally important.

One can segment consumer respondents by any number of criteria. One of these is division into groups, based upon who the respondent "is" (e.g. male vs. female, older vs. younger). Another way to segment analyzes their patterns of behavior. Respondents exhibiting a specific pattern of behavior might fall into one segment, whereas respondents exhibiting another pattern of behavior might fall into a different segment.

There is no end to the ways that one can segment individuals. Whether done on the basis of who the respondent is, or on the basis of what the respondent does; however, segmentation will not easily predict the specific stimuli to which the respondent will react. For example, when the segmentation is for a credit card, individuals in the same segment, whether defined by spending pattern, payment pattern or geo-demographics, will respond in different ways to the same messages. That is, a segment may be "homogeneous," but generally people in the same segment respond in different ways to messaging. What is necessary is segmentation based on response patterns to messages.

Fortunately, conjoint analysis produces segments which are homogeneous with respect to the tested messages. The conjoint experiment generates a set of utilities for each respondent, showing how that person reacts to the different elements. By segmenting the individuals based upon the patterns of their utilities, it becomes easy to develop segments or homogeneous clusters of individuals showing similar response patterns to the tested messages.

8.2.16.2 Systematized conjoint analysis generates databases and produces a substantive science

Up to now, we have been talking about methodological issues and the change of conjoint analysis from an arcane, difficult-to-understand rubric into a tool that can be used to solve simple problems at a reasonable cost. One of the newer approaches in conjoint analysis is the creation of databases of information about the way people react to their external world. It is an example of a tool by which we can simulate the complexity of nature and understand the algebra of the consumer mind.

For many years, conjoint analysis was used primarily for high- profile projects. In early 2001, author H. Moskowitz and his colleague Jacqueline Beckley at the Understanding and Insight Group were approached by Maryanne Gillette of McCormick & Company to create a system by which to understand craveability. At that time, in late 2000 and early 2001, the notion of craveability for food and drink was just being promoted as a prospective new way to understand foods (Beckley and Moskowitz, 2002).

As part of the solution to the issue of craveability, Beckley and Moskowitz used conjoint analysis to create a set of 30 studies. Each study comprised four silos, with each silo in turn comprising nine elements. The silos were the same and elements were either the same or very similar across a number of studies; therefore, it was possible to compare one element across different products and generate a set of test concepts (see Figure 8.2.3). Thus, across studies of different foods the data comprised responses to 1080 elements. Each element of the 36 had a specific function, whether as a simple description of the food, a more complex description, an emotional reassurance, or a brand.

The craveability studies used conjoint analysis to identify the contribution of the different elements. The results generated a database that retains its usefulness almost a decade later. One can return to the data to

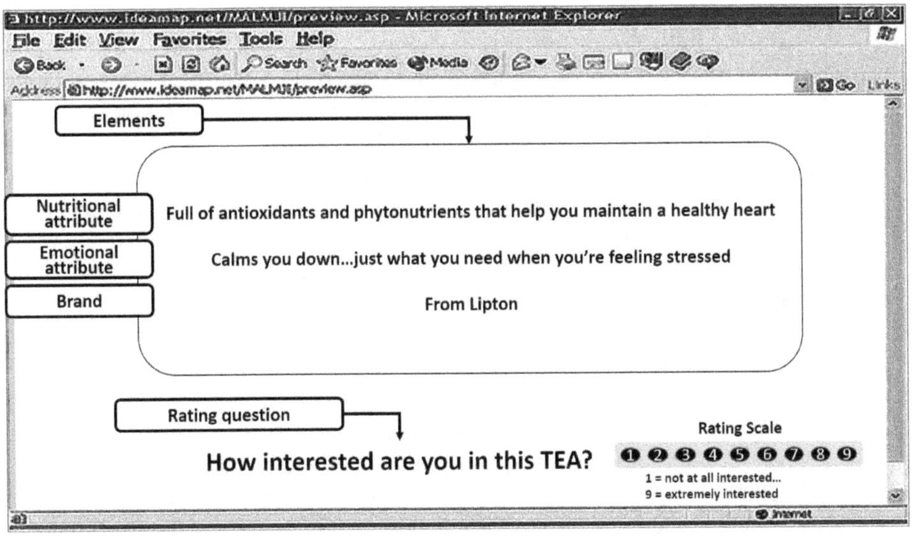

Figure 8.2.3 Structure of an It! Concept. Source: IT! Ventures, LLC, a limited partnership between the Understanding and Insight Group and Moskowitz Jacobs Inc.

discover how a specific term fared across all of the different foods or how different elements fared in a single food.

8.2.16.3 Beyond conjoint analysis and databases to the emerging science of Mind Genomics™

The application of conjoint analysis as the foundation of a new science is called Mind Genomics. The term was first used by Moskowitz et al. (2005) paper and then used extensively by Moskowitz and Gofman (2007) in their book Selling Blue Elephants.

The notion of Mind Genomics is patterned on the science of genomics. Very simply, we live in a world of concrete experiences that can be divided up into meaningful wholes, of everyday life, such as shopping for a specific item. In turn, the experience can be dimensionalized into various components by extracting its different aspects, putting them into silos, and then creating alternative elements to represent different parts of the experience. Thus, shopping for a car can be deconstructed into different aspects, ranging from the nature of the store, to that of the interaction with salespeople, to the nature of the purchase. These aspects are the actual experiences and the feelings about the experience.

Once the experience is structured, with multiple elements for each silo, the apparatus of conjoint analysis comes into play, presenting different vignettes to people, acquiring the relevant rating, such as how closely does the vignette match the ideal experience, and then deconstructing the vignettes into the part-worth contribution of each element.

In this first step of Mind Genomics, the work product is a granular "science" of the specific experience. This first exercise produces a database of the mind in terms of that experience, couched in terms of the world of the everyday (Moskowitz et al., 2006). We don't live in abstractions, but rather in the concrete aspects of everyday experience. To reiterate, it is the science of that concrete experience that is the focus of Mind Genomics.

In the second step in the Mind Genomics science, people are segmented based upon the patterns of their responses to these granular aspects of everyday life. This second exercise with the data reveals how different people "think about" the specific experience. Often, there may be as few as two or as many as 5–10 mind-set segments. The segments represent individuals with different points of view about the same everyday experience, as well as represent theoretical states that the mind could take (e.g. price orientation vs. convenience orientation). The fewer segments the better when it comes time to apply the science for both understanding and commercial application.

Traditional segmentations provide the basis for interesting stories about the way people divide, whether the topic area is foods, health concerns, cars, and so forth. Based upon the segmentation, the researcher creates a persona, a small picture, and story about the individual. The personas and stories, called heuristics, are interesting and make intuitive sense. However, these heuristics soon take over. Users of the segmentation data begin to believe that the persona must be real, despite the fact that originally the persona was a way to make the segmentation alive, and, in turn, when an individual who looks like the persona shows up, the natural response is to assume that the individual is a member of the segment.

In the third step of Mind Genomics, people are assigned to the different mind-sets, based upon a typing tool. In most cases, it will be virtually impossible to assign a good proportion of people into these mind-set segments when the only information available is previous purchase behavior and/or geodemographics. That is, the assignment is not based upon who the person is, but rather the way the person thinks. The assignment is done by means of a short "intervention test", comprising three to five questions.

The rationale for the typing test is the analogy to the way medicine is practiced. Today's medical community uses "intervention" tests, such as blood tests and urine tests, to identify medical conditions. A productive, perhaps "best" way to identify membership is through a short intervention, in the same way that the doctor takes a person's blood, sends it to a testing laboratory, and receives a report. The doctor may take down a detailed family history, but the information about the patient's physiological condition is best obtained by the different tests. The key here is that the patient provides the information through the interaction. To identify a person as a member of a mind-set segment requires the same type of thinking, a short intervention, or typing test. From the pattern of responses to the test, one easily identifies segment membership. The identification isn't perfect, being usually 60–70% correct. However, that percentage is far higher than guessing. Figure 8.2.4 shows an example of the typing tool to identify a mind-set segment.

Figure 8.2.5 shows the output of the typing tool, what to say to the segment, as well as what to avoid.

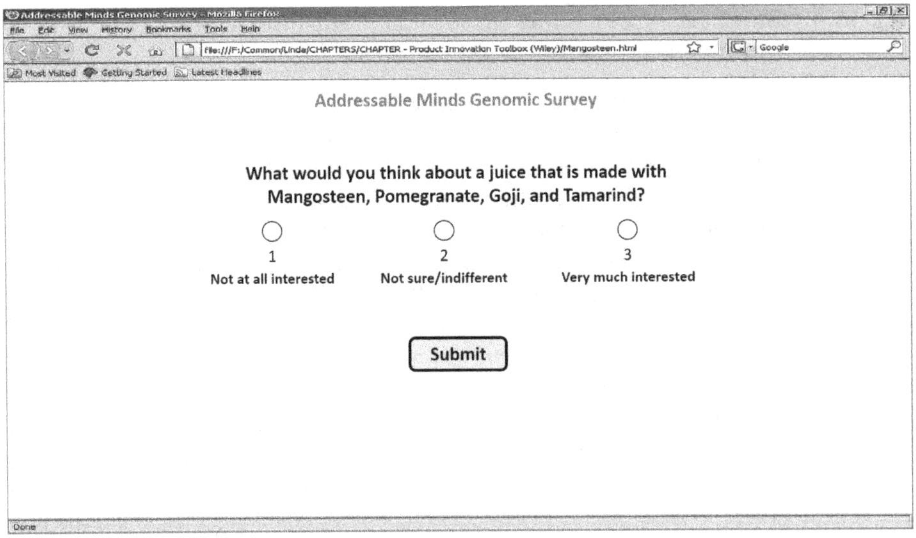

Figure 8.2.4 Example of a typing tool for mind-set segmentation. Source: IT! Ventures LLC, a limited partnership between the Understanding and Insight Group and Moskowitz Jacobs Inc.

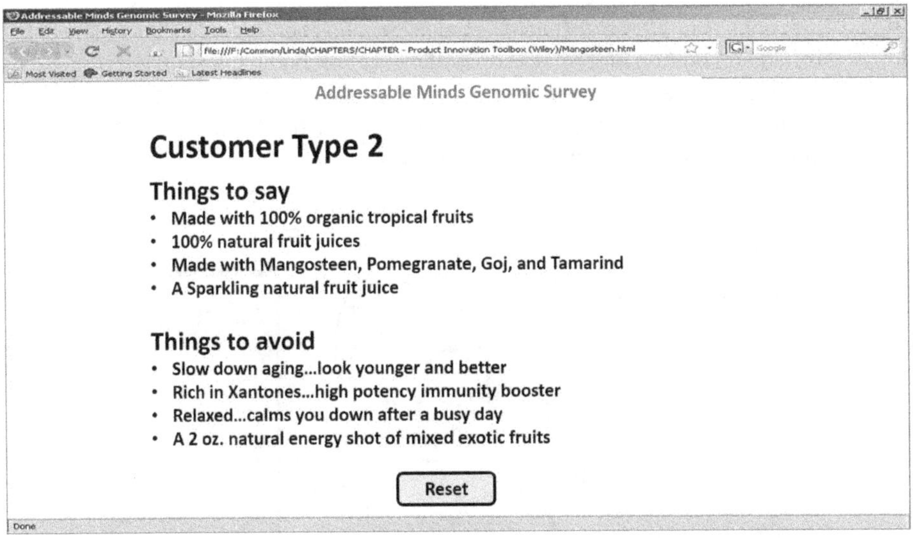

Figure 8.2.5 Example of output (what to say to the segment, what to avoid). Source: IT! Ventures LLC, a limited partnership between the Understanding and Insight Group and Moskowitz Jacobs Inc.

8

8.2.16.4 BimiLeap – DIY Mind Genomics™ with low cost, rapid turnaround, preformatted reports

By 2014 it was becoming increasingly clear that Mind Genomics studies of the traditional type, 36 elements, were taking simply too long to execute. Often, the first efforts themselves were perceived as daunting; the task of coming up with elements, the raw material that would be presented the vignettes. The reality was emerging that people simply did not have time to think through the problem. The pace of face decision-making had caught up with them. The culture of meeting after meeting to polish the element was simply not acceptable. The younger researchers realized the change, those who were growing up in the digital age. The older researchers did not. The former, the younger researchers, continued to ask for a revision so that they could use it.

The solution to the problem was to make the Mind Genomics process into a simpler, DIY, do it yourself system, first on the Android as an app, and then available by browser on smartphones, tablets, and laptops, as well as regular computers. The DIY and the new locations for set-up meant shortening the previously long process. Rather than a 4×9 (4 categories, 9 elements per category, 60 vignettes) or the popular and more recent 6×6 (6 categories, 6 elements per category, 48 vignettes), the revised Mind Genomics study became the 4×4 (4 categories, 4 elements per category, 24 vignettes).

The new version of Mind Genomics, called BimiLeap (Big Mind Learn app) guided the researcher through the study, as can be seen by the sequence of screenshots shown in Figure 8.2.6. The sequence was set up as a Socratic tutorial, in which the researcher had to pick a topic for the study, generate four questions (the aforementioned categories), and then provide four answers for each question (the aforementioned elements), along with a simple introduction and a rating scale. The new version permitted only one rating scale, three self-profiling classification questions (gender, age, a third under the researcher's control), and finally an open-ended question.

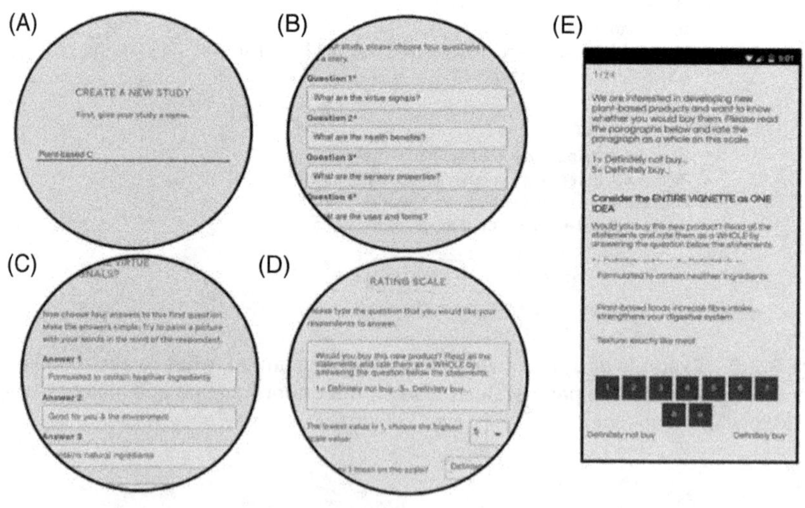

Figure 8.2.6 Sequence (from A to E) for Mind Genomics. Source: BimiLeap®.

The new version of Mind Genomics was accompanied by two more enhancements. The first was the measurement and analysis of response time, and the time between the presentation of the vignette and the respondent's rating. This would be analyzed to show how rapidly the messages were processed by the respondent and will be shown below with data. The second was the creation of a preformatted pair of reports, the first being an Excel file with the appropriate tables, the second being a presentation-ready PowerPoint that could be shown to the client.

These enhancements made the Mind Genomics process more popular. Researchers were slowly weaned away from the large, ponderous experiment designs comprising 36 elements, designs which required meeting after meeting to complete, and led to an increasingly insecure user who wondered whether the research had tested "all the relevant elements." Rather than becoming a tool for investigation, the large designs were subtly evolving into tools for confirmation, with many studies aborted because of the paralyzing reality that one had to select many answers, and that somehow, they had to be "right." The smaller 4×4 designs were designed to be iterative, fast, and noncomprehensive. With 16 elements, one could always tell the user that another iteration was just an hour away.

Beyond the Excel file was the PowerPoint report, formatted for user presentation, with so-called "boiler-plate" pages (text) explaining the method and how to read the data (tables), as well as dynamically replaced descriptive and the tabled data from the actual study. Figure 8.2.7 shows an example of some of the pages, giving a sense of what the user received.

During the period of development, around 2015–2017, it was becoming increasingly obvious that researchers did not want to pour through Excel files to pick up the story. It was necessary to distill the results into the key findings, yet not do the thinking for the respondent. The Excel file, which was generated, provides summarized data, first for total panel, then for gender, then for age, for subgroups defined by the researcher, and then for two and three mind-sets, respectively. Breaking out the data in this fashion provided the necessary format for the key findings to emerge.

In most of the cases, the key findings were shown by the one Excel tab on mind-sets (see Table 8.2.9). Over time, the reportage became more refined, both by format and by analysis. The negative coefficients were removed because they did not present much useful information. Then, the strong performing elements, with coefficients of 7.51 or higher were shown in shade. This way of presenting the data increased the comprehensibility of the data.

The newly developed BimiLeap was enhanced with other features as well. The first was the creation of a very quick, front to back system, in which the entire process, from set up to launch, would take no more than 20 minutes, and acquisition of respondents would take no more than a minute. Conventional conjoint measurement had turned into an extended process, with the set up requiring one to work with a specialty company, after which the respondents were acquired, and then the study run. The process might take up to a few days, or even longer. The "goal" of the BimiLeap was to shorten the time of experiment, from set up to analysis to less than 90 minutes. The aforementioned setup program helped, wherein the respondent typed in the questions in a certain format. What made a great difference however, was the integration of a panel selecting "API" or program, from Luc.id Inc., allowing the researcher to tailor the panel

Study Title	Plant-based C
Identification Number of the study:	06142020.Plant-
Date when the study was run:	(06/14/2020-06/14/2020)
Number of respondents:	47
Purpose of the study:	We are interested in developing new plant-based products for the Canadian consumer.
Keywords:	plant-based, sensory, sustainability
Study Info:	We are interested in developing new plant-based products and want to know whether you would buy them. Please read the paragraphs below and rate the paragraph as a whole on this scale. 1= Definitely not buy... 5= Definitely buy...
Preliminary question:	How do you feel about plant-based meat?
Possible answers:	1=Never tried it 2=Eat it very occasionally 3=Really like it 4=Changed my life for the better
Rating question:	Would you buy this new product? Read all the statements and rate them as a WHOLE by answering the question below the statements. 1= Definitely not buy...5= Definitely buy...

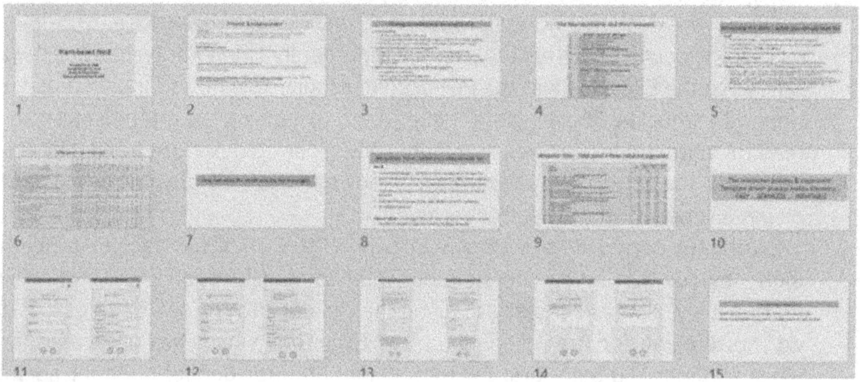

Figure 8.2.7 An example of output from BimiLeap® in Power Point format. Source: BimiLeap®.

Table 8.2.9 An example Excel table report on different Mind-set (MS) solutions (in this case 2 Mind-sets vs. 3 Mind-sets).

	Group (Binary Ratings)	Total	MS 1 of 2	MS 2 of 2	MS 1 of 3	MS 2 of 3	MS 3 of 3
	Base Size	51	25	26	16	21	14
	Additive Constant	54	46	63	42	70	43
A1	Formulated to contain healthier ingredients	1	8				10
A2	Good for you and the environment						12
A3	Contains natural ingredients		5		3		7
A4	Produced with sustainable ingredients						9
B1	Helps you to lose weight	3	1	6		8	

Table 8.2.9 (*Continued*)

	Group (Binary Ratings)	Total	MS 1 of 2	MS 2 of 2	MS 1 of 3	MS 2 of 3	MS 3 of 3
B2	Plant-based foods increase fiber intake... strengthens your digestive system	1		1		6	
B3	Plant-based foods are rich in vitamins	3	8		15		
B4	Plant-based foods reduce the risk of type-II diabetes and cardiovascular diseases	2	1	2	4	4	
C1	Appearance: exactly like meat		9		17		
C2	Aroma: exactly like meat		11		14		3
C3	Texture: exactly like meat	4	13		18		5
C4	Flavor: exactly like meat	2	14		21		2
D1	Use: pastry, bread and meat pie	4		8		4	14
D2	When: at a business lunch	8		17		9	20
D3	When: at home with the family	5		12		10	11
D4	Form: meat slices and patties	6		14		14	8

using general criteria such as gender, age, income, children, market, etc. The process, from start to finish, was made even more efficient by introducing credit card payments, an idea suggested by the author H Moskowitz 24 years before, in 1997.

8.2.17 The personal viewpoint identifier (PVI)

The information generated by the respondents through Mind Genomics provides valuable archival information about the way people respond to the stimuli of the outside world. Furthermore, one can observe the behavior of the individuals who are placed into mind-sets, and in doing so figure out whether their other behaviors provide some insight into the relevance of Mind Genomics to predict other behaviors.

In recent years, the notion of mixing Mind Genomics with epidemiological and sociological thinking has arisen, mainly from conversations both with marketers for applications, and among scientists regarding the nature of mind-sets and their covariation with other behavior, especially in health. Mind Genomics studies provide a unique way to understand a person's behavior from the bottom up, a level of granularity that is critical to understanding behavior because it deals in the specifics. Yet, because of cost, Mind Genomics studies are limited to a few hundred respondents, rather than to thousands of respondents. The ideal would be to run the same Mind Genomics experiment on thousands of people, an ideal that will not be readily achieved.

For Mind Genomics to be widely used, new people must be assigned to the proper mind-set in a rapid, inexpensive, viz., scalable fashion, open to everyone. This means that the assignment of new people to a mind-set cannot be left to the domain of those in possession of extensive, powerful computers and algorithms, but rather must be widely available. To this end, the authors and colleagues Dr. Attila Gere of Budapest, and programmer Robert Sherman of ShermanSoft at Las Niguelas, California, have developed a technology called the personal viewpoint identifier (PVI). The objective is to use the summary data from the Mind Genomics experiment, and through Monte Carlo simulation, identify the six elements which best and most robustly differentiate among two mind-sets or three mind-sets, in the presence of statistically induced error. The ingoing objective was to create an easy, automatically created tool that one could use to assign new people to one of the mind-sets, provide feedback to the person, provide an opportunity for the person to see an appropriate video (e.g. for sales), or even to be led to the proper landing page for e-commerce.

The tool begins with an easy to use template (www.PVI360.com). We will use the template in conjunction with the three mind-set solutions of Table 8.2.10. Furthermore, since the table is presented without the negative or 0 coefficients, will arbitrarily fill those missing coefficients with the value 0. Finally, we will create

Table 8.2.10 An example of a PVI project on Plant-based meat.

HEADER TEMPLATE					
ID	**Question Code**	**Question Text**	**Mind-set1**	**Mind-set2**	**Mind-set3**
1	xxx0				
2	xxx1	Mind-Set Name	MS1 Focus on wanting it to meat	MS2 Hamburger replacer for family	MS3 Health benefits and novel food forms and eating situations
3	xxx2	Mind-Set Feedback	Taste & Texture - exactly like meat	Meat patties for family at home	At business lunch, food for you and environment, use in pastry, bread, meat pies
4	xxx3	Mind-set Video			
5	xxx4	Mind-set Link			
6	xxx5	Additive Constant	42	70	43
7	xxx6	How do you feel about this idea	I want it	Not for me	
...	...	...	...	...	...

Table 8.2.10 (*Continued*)

SPECIALTY QUESTION TEMPLATE					
ID	**Question Text**	**Option A**	**Option B**	**Option C**	**Option D**
1	Do you Eat meat	Excitedly yes	Yes	No	Definitely no
2	Do you know about vegetable-based meat	Yes	Have heard but don't know much	Just starting to hear about it and am intrigued	Just starting to hear about it, but confused
3					
4					
...	...	...	...	...	...

QUESTION TEMPLATE					
ID	**Question Code**	**Question Text**	**Mind-set1**	**Mind-set2**	**Mind-set3**
1	E01	Flavor: exactly like meat	21	0	2
2	E02	Texture: exactly like meat	18	0	5
3	E03	Appearance: exactly like meat	17	0	0
4	E04	Plant-based foods are rich in vitamins	15	0	0
5	E05	Aroma: exactly like meat	14	0	3
6	E06	Form: meat slices & patties	0	14	8
7	E07	When: at home with the family	0	10	11
8	E08	When: at a business lunch	0	9	20
9	E09	Use: pastry, bread and meat pie	0	4	14
10	E10	Good for you and the environment	0	0	12

(*Continued*)

8

Table 8.2.10 (*Continued*)

QUESTION TEMPLATE					
ID	**Question Code**	**Question Text**	**Mind-set1**	**Mind-set2**	**Mind-set3**
11	E11	Formulated to contain healthier ingredients	0	0	10
12	E12	Produced with sustainable ingredients	0	0	9
13	E13	Contains natural ingredients	3	0	7
14	E14	Helps you to lose weight	0	8	0
15	E15	Plant-based foods increase fiber intake... strengthens your digestive system	0	6	0
16	E16	Plant-based foods reduce the risk of type-II diabetes and cardiovascular diseases	4	4	0

8

the PVI template as shown in Table 8.2.11. This PVI template will be inserted into the PVI program references above, and then PVI questionnaire will appear. We finish with an example of the feedback that a respondent received (Figure 8.2.8).

The research value of the PVI is knowledge development, to understand the mind-set membership of an individual for one or for up to 20 different topic areas, with the time needed to fill the PVI being approximately 15 seconds per

Table 8.2.11 An example of PVI template with Mind-sets for plant-based meat.

PLANT MEAT 1		
Mind-Set 1	**Mind-Set 2**	**Mind-Set 3**
• Focus on wanting it to meat • Taste and Texture – Exactly like meat	• Hamburger replacer for family • Meat patties for family at home	• Health benefits and novel food forms and eating situations • At business lunch, food for you and environment, use in pastry, bread, meat pies

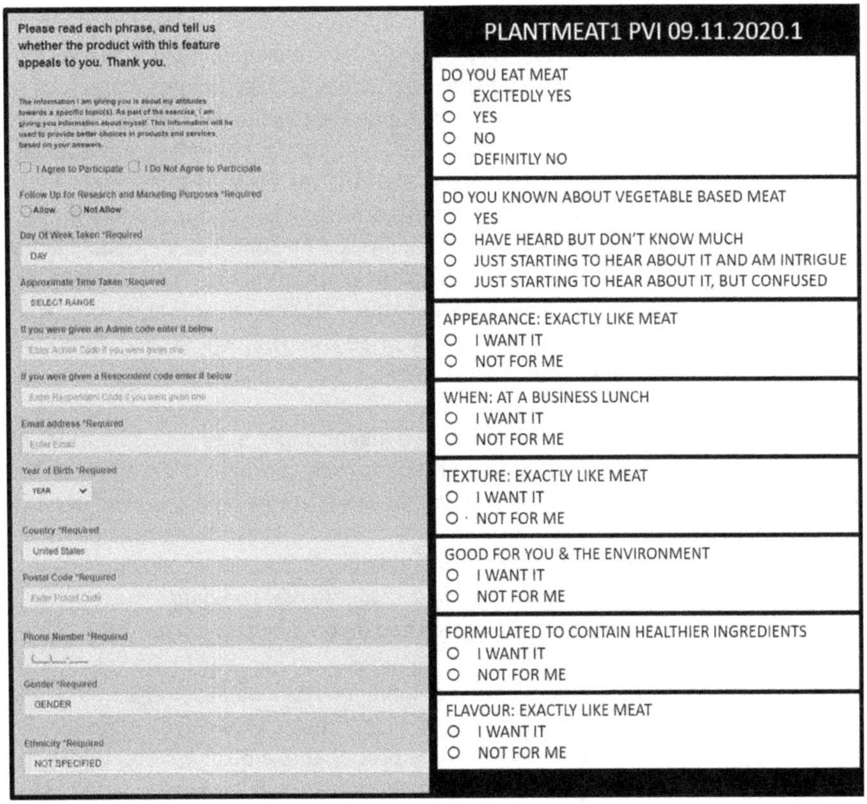

Figure 8.2.8 Example feedbacks that a respondent will get from the PVI program. Note: IdeaMap®.Net, Mind Genomics™, **and** BimiLeap® are registered trademarks of **Mind Genomics Associate, Inc.**

topic. Thus, to "sequence" the mind on 20 topics is an approximate five-minute, self-administered interview, with the data stored for future use, with the PVI acquiring the respondent's agreement (double opt in), with background information obtained about the respondent (but no name), and finally with additional information beyond mind-set membership acquirable. Such information can be used to understand the product needs of people of different mind-sets, and to help market products to these different mind-sets by knowing "what to say," "how to say it," and "to whom."

8.2.18 Four considerations dictating the future use of conjoint analysis

In the long run, any method of research with consumers will enjoy success when it can be done efficiently, is transparent, the cost is low, and most importantly, when the contribution is great. These four considerations apply to conjoint analysis as well:

(1) **Doing studies efficiently:** By efficiently, we mean that it is not onerous to run the study. Certainly, at the early stage of development in any research technique, implementation will be difficult. Easy-to-implement research methods are common. Yet, looking at their history will reveal that these now-easy methods had their beginnings in more tortuous methods, in experimental trials, and in false starts. Over time, as a research method gains traction and acceptance, ways are found by various users to make the process quick and easy. Conjoint analysis is one of those methods that have been simplified. Beginning with arcane methods in the 1960s, conjoint analysis has been simplified into a virtual cut/and/paste system (Moskowitz et al., 2001). Their system is, not of course, the only system, but it does represent a serious attempt to make conjoint analysis efficient and thus, attractive for general use.

(2) **Transparency:** By transparency we mean simple, easy-to-use methods; the opposite of normal business practice which obscures the method to "protect trade secrets." For conjoint analysis, transparency comes about by using well-defined, easy to replicate methods. One of these methods is ordinary least-squares regression as the analytic tool, which deconstructs mixtures to the contributions of their components. The happy outcome is the ease with which the utility values can be obtained, and their value communicated to management.

(3) **Low cost:** For many years, conjoint analysis was promoted as a high-visibility, high-effort approach and, thus, expensive method. This approach lent the method and its proponents a business and intellectual cachet, one hard to resist in a competitive world. Beginning with customized experimental designs, researchers made the method increasingly complicated, often by using hard-to-understand analyses. We believe, however, that creating an off-the-shelf version of conjoint analysis, morphing it into a web-based tool, and then expanding the use where possible is a better direction for the future, despite the fact that as a result conjoint analysis will become more ordinary, more pedestrian and less price-prohibitive.

(4) **Magnifying the contribution:** Read the books on statistical research methods and look at the way conjoint analysis is framed. Conjoint analysis is typically presented as a solution to a specific problem. As such, for many years, it developed its own literature, typically combining method and data. The papers that reported using conjoint analysis were as much expositions of methods to solve a particular problem as they were actual data of relevance to the science. The method was so difficult, time-consuming and complex that one had to spend as much time on the method as on the substantive results. With the development of web tools, however, methodological issues receded into the background. Conjoint analysis has now come to the fore as a tool to create a body of knowledge, and not simply as a state-of-the-art technology for one-off solutions to rare problems. Today, we see conjoint analysis joining the mainstream, as a business and science workhorse, rather than merely as a showcase capability, shown, but rarely used.

8.2.19 Conclusion

Mind Genomics has emerged over a period of 60 years, from the early 1960s, with the author H Moskowitz actively pursuing the development since 1981, a period of 40 years. If one were to summarize this 40-year experience and the nature of the research tools which have emerged, the lesson would be that it is the evolution of an ecological system which has allowed Mind Genomics to grow. What was at first an exploratory effort in response to the polite request of a corporate leader has emerged to become a new branch of psychology and consumer research, a branch having both science and business roots and reach. We are only at the beginning of Mind Genomics as an innovation machine. The future is bright, awaiting us, with a now easy-to-use, affordable, and remarkably rapid technology to help in the creation of new knowledge, new products, and new services.

Discussion Questions

- Where can you use conjoint analysis in your development process to replace and/or augment your current tools?
- How can you efficiently use the conjoint analysis results to identify drivers of acceptance and drivers of rejection?
- How can the additional tools of mind-set segmentation and consumer timed response measurement help guide the innovation team?

References

Beckley, J.H. and Moskowitz, H.R. (2002) "Databasing the consumer mind: the crave it!, drink it!, buy it!, protect it! and the healthy you! databases". *Paper Presented at the Annual Meeting of the Institute of Food Technologists*, Anaheim, California, June 2002.

Box, G.E.P., Hunter, W.G. and Hunter, J.S. (1978) *Statistics for Experimenters*. New York: Wiley.

Galanter, E., Moskowitz, H. and Silcher, M. (2011) *People, Products & Prices; Sequencing the Economic Genome of the Customer's Mind*. Sharjah, UAE: Bentham Science Publishers Ltd.

Gofman, A. (2006) "Emergent scenarios, synergies and suppressions uncovered within conjoint analysis". *Journal of Sensory Studies*, 21(4), 373–414.

Gofman, A. and Moskowitz, H.R. (2010) "Application of isomorphic permuted experimental designs in conjoint analysis". *Journal of Sensory Studies*, 25(1), 127–145. https://doi.org/10.1111/j.1745-459X.2009.00258.x.

Montgomery, D.C. (1991) *Design and Analysis of Experiments*. New York: Wiley.

Moskowitz, H.R. (1995) *Consumer Evaluation of Personal Care Products*. New York: Marcel Dekker.

Moskowitz, H.R. (2009) "Sequencing the genome of the CUSTOMER mind: application to food and drink". *Food Technology and Innovation Forum*. May, Chicago, IL.

Moskowitz, H.R. and Gofman, A. (2004) System and Method for Performing Conjoint Analysis. Provisional Patent Application, 60/538,787, filed 23 January 2004.

Moskowitz, H. and Gofman, A. (2007) *Selling Blue Elephants*. Upper Saddle River, NJ: Wharton School Publishing. (Translated into numerous foreign editions.).

Moskowitz, H.R., Gofman, A., Katz, R., Itty, B., Manchaiah, M. and Ma, Z. (2001) "Rapid, inexpensive, actionable concept generation and optimization - the use and promise of self-authoring conjoint analysis for the foodservice industry". *Foodservice Technology*, 1, 149-168.

Moskowitz, H.R., German, B. and Saguy, I.S. (2005) "Unveiling health attitudes and creating good-for-you foods: the genomics metaphor and consumer innovative web-based technologies". *CRC Critical Reviews in Nutrition and Food Science*, 45(3), 265-291.

Moskowitz, H.R., Gofman, A., Beckley, J.H. and Ashman, H. (2006) "Founding a new science: mind genomics". *Journal of Sensory Studies*, 21(3), 266-307.

Peryam D.R. and Pilgrim F.J. (1957) "Hedonic scale method of measuring food preference." *Food Technology*, 11, 9-14.

SYSTAT (2008) SYSTAT for Windows, Version 11. The system for statistics. User Manual. Evanston, IL: SYSTAT Corporation, Division of SPSS.

Wells, W.D. (1975) "Psychographics. A critical review". *Journal of Marketing Research*, 12, 196-213.

9

. . .There was nothing virtual about it. It was real.

From recent NPR Story on COVID-19 experiences
https://www.npr.org/2020/05/16/856819558/for-some-zoom-bar-and-bat-mitzvahs-highlight-most-meaningful-and-memorable-momen

What this is about

This chapter introduces research tools that do not rely on word expression. Starting with emotions, moods and motives in Chapter 9.1, we review key emotion theories and a sample of tools that try to identify emotions and their role in human behavior. Neuroscience approaches have evolved and are discussed in Chapter 9.2. Virtual reality and augmentation (VR and VA) are discussed in Chapter 9.3 where basic concepts and setup protocol are described. These tools provide additional means for Consumer Explorers (CEs) to identify, verify, and understand people's thinking from perspectives less dependent of word-based methods.

Product Innovation Toolbox: A Field Guide to Consumer Understanding and Research, Second Edition. Edited by Kannapon Lopetcharat, Dulce Paredes and Jacqueline H. Beckley.

Chapter 9

New Tools Beyond Conventional Qualitative and Quantitative Meanings

9.1

Emotions, Moods, and Emotives

Kannapon Lopetcharat and Dulce Paredes

Key learnings

- ✓ Common definitions for Affect, Mood, Emotion, Attitude, and Emotive
- ✓ Theories on emotion
- ✓ Appraisal theory and Critical Incident Technique
- ✓ Emotion regulation and coping
- ✓ Social media and emotion

9.1.1 Introduction

Emotional claims in new products and services have been growing steadily. According to Innova Market Insights database 2019–2020, the annual growth of emotional claims globally for food and beverages and for personal care packaged goods are 12% and 10%, respectively. The fastest growing claims relate to replenishment platforms such as reduce fatigue, stress, tiredness, refuel, and recharge. Product and services touting emotional connections and elevating moods are projected to grow as people prioritize holistic wellness in an increasingly stressful world. So how are consumer explorers (CEs) going to navigate this growing arena of capturing emotional experiences?

Product Innovation Toolbox: A Field Guide to Consumer Understanding and Research, Second Edition. Edited by Kannapon Lopetcharat, Dulce Paredes and Jacqueline H. Beckley.

Different consumption experiences elicit different emotional reactions and a person can have many different emotional responses for an experience. These facts are well accepted; however, it is still unclear "how" the experiences influence the nature of emotional responses and "which" emotional responses will be elicited from an experience. In addition, different emotional responses can be found from different people who encounter the same experience and that the same person can produce different emotional responses to the same experience over time. Furthermore, the influences of rational or cognitive emotional responses and the interactions between the two processes on a person's decision-making are not well understood. This chapter will introduce theories about emotion and several current methodologies to measure emotions or the idea of putting feelings into words.

For some of our CEs, here is a short background behind the "why" of understanding emotion in the context of a product innovation toolbox. The full embrace of integrating emotion understanding and therefore measurement in product research is relatively new. While emotion has been a core principle in advertising for a long time (and gathering strength during the 1950s with the advent of more social collective activities like television), discussion of emotion in science was not well accepted. Even though human psychology had to deal foundationally with human emotions, there was a sense that this would make science less scientific (Pert, 1997). The turning point that started the revival of emotion research can be viewed with the publication by Pine and Gilmore of The Experience Economy in 1999. Even though the authors do not overtly spell out emotion, they allude to the development of experiences that are individual and can be created, thus laying the groundwork for . . ." how do I do that?" This notation of experience that emerged from the transition of a supply to a demand economy which was building in the 1990s as "mass customization" became possible (today consider the array of apps one has on their smartphone which allows individual customization). Now we had more supply than we needed and thus had to figure out how to elicit demand. When one combines all of the global manufacturing, just-in-time concepts occurring along with house of quality innovations (Griffin and Hauser, 1993), the need to create emotive designs of things rather than just functional design leads to this "discovery" of the value of understanding human emotions and how they can create demand. Paralleling this shift from supply to demand side economics, we also had science, neuroscience, discovering that there were "chemicals" that we connected the brain, glands, and immune system probably representing the biochemical substrate of emotion, thereby making it relevant to the study of perception and product and service demand (Pert et al., 1985).

So the transition point from not considering emotion within the context of product innovation to where it has grown today is around 1999 and 2001. Popular textbooks such as Lawless & Heymann's *Sensory Evaluation of Food* (1998) and the 13th edition of Philip Kotler's *Marketing Management* (2009) do not speak to emotion as significant in either sensory evaluation or market research. The field is changing so much, at this writing, that Herbert Meiselman, a well-respected psychophysicist and editor of *Emotion Measurement*, is in his second edition in 2021.

9.1.2 Understanding differences between affect, attitude, mood, emotion and emotive

Fundamental issues plaguing consumer emotion research are the definitions and differentiation of emotion, mood, attitude, and affect. In several literatures, these terms (especially emotion, mood, and affect) are used interchangeably and can cause confusions without lengthy explanations of corresponding contexts from which the definitions are determined and decided. After several decades, the following definitions of and distinctions between affect, emotion, and mood are generally accepted as the following:

- **Affect** is a blanket term that encompasses emotion, mood and attitude.
- **Emotion** responds to a specific target in a particular, personally relevant situation with affect as an integral part of the experience. Emotion is a mental state experienced in relation to situations or a specific target such as a person, object, or event that has implications for an individual's goals or well-being. Therefore, emotion is "about something" and "specific," and it is not a generalized feeling or state of mind (Johnson and Stewart, 2005). Emotion can be related with intense autonomic arousal (e.g. "fight or flight" that are the decisions from the emotions of "anger or fear," respectively.
- **Mood** is merely an incidental or part of a more generalized response than emotion. Mood state is generally dissociated from any particular object or event and mood is less intense and lasts longer than emotion. Moods are rarely associated with similarly intense physiological arousal found in emotional reactions (few exceptions such as severe depressive moods or extreme euphoric moods can be found but only in extreme situations).
- **Attitude** differs from emotion because of its "evaluative nature" and its inclusion of cognitive, behavioral, and general affective components. Attitudes constitute one's beliefs' knowledge structure and the beliefs inform and shape emotional reactions. Therefore, attitudes can be antecedence of emotion. On the other hand, emotion can also influence attitudes by acting as an input for evaluative judgment formation (Batra and Ray, 1986).
- **Emotive or psychological motive** is an emotionally aroused state that drives people to act to achieve their goals (McClelland, 1965). The goals are based on basic need states (i.e. biological, social, and personal needs). Biological needs (e.g. hunger, thirst) and social needs (e.g. mate acquisition, disease avoidance, status, and kin care) are important for people to survive and thrive (Scott and Cohen, 2019; Neel et al., 2016). Social motives identified to drive people's behavior are things such as achievement, aggressiveness, power, affiliation (Gregariousness), acquisitiveness, and curiosity (Bardo et al., 1996; McClelland, 1965). Understanding emotions, underlying motives (emotives) along with personality (McCrae and John, 1992), live-stage, context, and situation will help CEs discern insights at fundamental levels especially decision-making and behavior.

9.1.3 Review of emotion theories

It is helpful to understand that there are a number of different theories around emotion, the number of emotions that are measurable, and how to understand emotions relative to the human subject. Six key theories are highlighted:

(1) Dimensional Theories (DT)
(2) Basic Emotions Approaches (BEA)
(3) Attribution Theory (AT)
(4) Autonomic vs. Controlled Nature of Emotion
(5) Appraisal Theory
(6) Emotion Regulation

Depending on how a researcher views the theories, they will then apply these theories to the creation of their approach to measure emotion.

9.1.3.1 Dimensional Theories (DT)

This is a well-known class of theories that has been applied to the emotion study of consumer behavior. These theories use simple representations of affective responses (a set of common dimensions of affect that can be used to distinguish specific emotions from one another). PAD (Russell and Mehrabian, 1977) and Circumplex models (Watson and Tellegen, 1985) are the most popular ones. The most basic DT approach simplifies emotions along two basic dimensions (affective valence and affective arousal).

- **Affective valence** (a.k.a. **valence**) classifying emotions into three groups: positive, negative, and neutral; meanwhile,
- **Affective arousal** (a.k.a. **arousal** or **intensity**) deals with the intensity of emotional responses (Johnson and Stewart, 2005).

The use of only two major dimensions (valence and intensity) can result in misclassification of very different emotions to have the same characteristics and miss the nuances of emotions that are a part of common experiences.

The Circumplex approaches to emotion upgrade the two-dimensional approach by defining four additional emotions from the interaction between the basic two orthogonal dimensions (valence and arousal). Consequently, the Circumplex model yields eight directions of emotions in total (Figure 9.1.1).

9

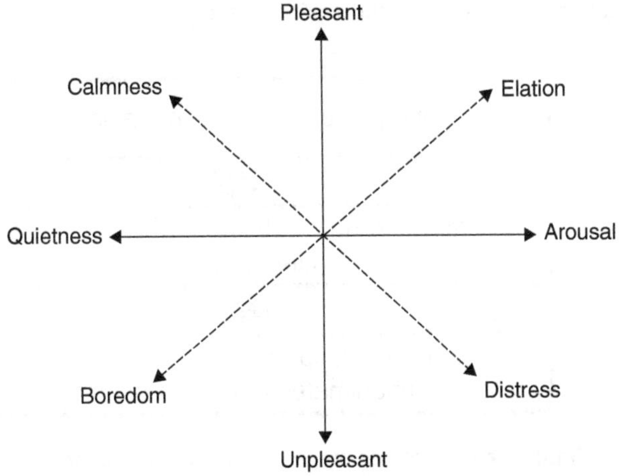

Figure 9.1.1 Circumplex Model of emotion with two main axes: valence (y-axis) and arousal (x-axis).

Circumplex models propose a variety of dimensions of affective response based on the relative similarity of emotions and their applicability of particular target setting or object.

Dimensional approaches (including the Circumplex model) are useful to identify and simplify emotional descriptions; however, the approaches DO NOT offer any insights to the "cause and consequences" of emotion. Moreover, dimensional approaches sometimes fail to capture the full range of emotions that consumers experience (Lazarus, 1991).

9.1.3.2 Basic Emotions Approaches (BEA)

These approaches have been developed based on the belief that there is a finite set of innate discrete basic emotions (such as joy, anger, sadness, and fear) that transcend cultural barriers (Plutchik, 1982). Consequently, other emotional responses are the results of the particular pattern of responses across these various basic emotions. BEA moves away from simple dimensional approaches to describing and differentiating subtle emotional states. BEA employs two major principles:

(1) There is a finite set of basic emotions (both verbal and nonverbal) and
(2) The basic emotions are cross-cultural. This means that these basic emotions are expressed and labeled similarly across cultures.

BEA practitioners usually begin their research by the following three steps (Figure 9.1.2).

Step 1: Identify a reference set of basic emotions: This usually refers to the work of Paul Ekman and friends on universal facial expression of emotion (Ekman et al., 1969) for nonverbal emotions and Baron-Cohen's English Emotion Vocabulary (EEV) contained in a software called Mind Reading (Baron-Cohen et al., 2010). For food (overall), Desmet and Schifferstein identified 22 emotion terms (11 positive and 11 negatives) (Desmet and Schifferstein, 2008); for food (odor), Porcherot and others developed SenseMove™ scale including 36 emotion terms in six dimensions (Porcherot

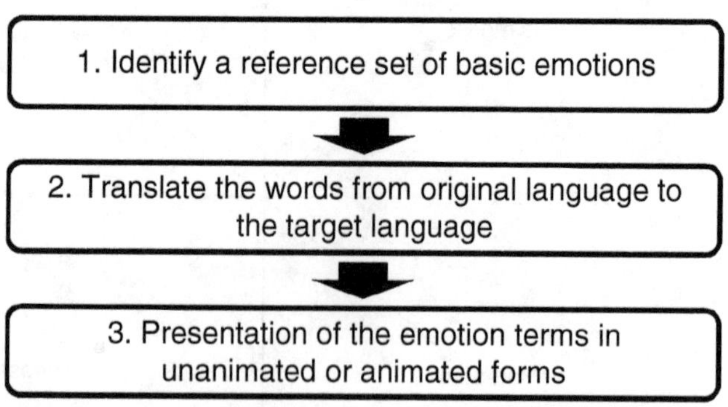

Figure 9.1.2 Three common steps for applying Basic Emotion Theory in emotion research.

et al., 2012); for product testing, King and Mieselman created EsSence Profile™ including 39 standard emotion terms (King and Meiselman, 2010).

Step 2: Translate the words from original language to the target language: This usually involves six-step standard translation practices that are the following:

- **Step 2.1 Translation** using two or three professional translators.
- **Step 2.2 evaluating the first translated words** by another independent team of professional translators and evaluate the consistency between the first set of professional translators. The word with high consistency will be selected and the others will be further discussed, changed, and included or discarded based on consensus from the group.
- **Step 2.3 Back-translation** of the selected terms by different professionals (at least 2) to check for the meaning accuracy. The consistency between the evaluators will be the indicator of the terms.
- **Step 2.4 Evaluating the comprehension of the terms** in the target group: this usually involves a survey using the terms selected in relevant situations. If children are involved, parents will be the evaluators of children's comprehension level.
- **Step 2.5 classifying the terms into groups** this usually involves sorting tasks and a multidimensional scaling analysis.
- **Step 2.6 classify the terms into affective valence groups** (positive, negative, or neutral) by using pleasantness rating on 9-point scale (1-3 = negative, 4-5 = neutral and 6-9 = positive). The presentation of the list can be one of regular formats (such as a question with choices to choose in a Check All That Apply (CATA) task or a list of questions with an intensity rating scale) or in a special arrangement format. For example, Geneva Emotions Wheel developed by Scherer (Scherer, 2005; Scherer et al., 2013; Sacharin et al., 2012) was recommended to be presented in Figure 9.1.3. This form includes both identification and intensity rating tasks.

9

Step 3: Presentation of the emotion terms: With computer, mobile technology and Internet advancement, facial expression and words/terms can be combined and presented with images in unanimated or animated forms. The following steps guideline the steps for developing such scales:

- **Step 3.1 Use actor/actresses to "act" and "express" the feeling of an emotion term** (one can use this as an output as well but the race or gender of the actor/actress can post problem in certain target population).
- **Step 3.2 Render the "live expression" to "cartoon version" by cartoonist** (this was to reduce the bias that comes with the use of actor/actress)
- **Step 3.3 Validate the cartoon (static or animated with/without sound) by matching tasks** (asking consumers to match the term with the animation). Note: one can use this result as output as well.
- **Step 3.4 Select the intensity scale** usually 3-point intensity scale (do not feel, feel lightly, and feel intensely) is used; however, for more information, 5-point intensity scale can be used (1 = do not feel the emotion and

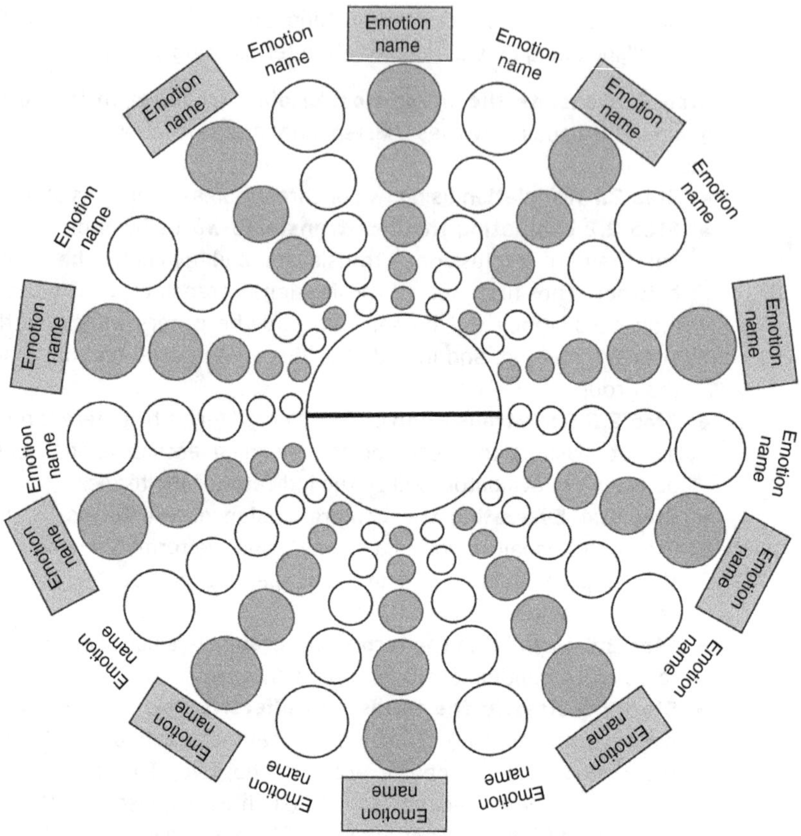

Figure 9.1.3 An illustration of a special arrangement for Geneva Emotions Wheel questionnaire. Source: Adapted from Scherer et al. (2005).

5 = feel the emotion intensely). There are many systems developed along these approaches such as PrEmo™ (www.tustudiolab.nl/desmet.premo (14 basic emotions: 7 positives and 7 negatives)) and Emoti*Scape™ by Ipsos-ASI.

BEAs are used to overcome the limitations of the DAs (missing many emotions) that are too parsimonious. However, the reliance on labels (either words or facial expression) of subjective emotional experiences is especially problematic because of the following reasons:

- Many emotions can possibly be expressed by the same facial expression.
- In some cultures, emotional reactions are not readily labeled.
- The requirement that basic emotions are cross-cultural. Hence, certain emotions that are readily labeled in one culture but not in other cultures will be dropped from the list of emotions. That will defeat the purpose of capturing the wide range of emotions in a particular target that, for sure, will be embedded within a culture.

However, BEAs and DTs do NOT provide "the process" and "consequences" of emotions.

9.1.3.3 Attribution Theory (AT)

To predict differentiated emotional responses elicited by an event or experience, AT has been used often with a hypothesis that, **within a certain event or experience, people are concerned about the cause(s) of an event.** This perceived cause influences emotional responses of an individual to a certain event or experience. For example, if a consumer attributes a product failure to the manufacturer, the consumer will experience "anger." However, if the consumer attributes the failure to be the consumer's own mistake, then the consumer will experience "regret or guilt."

Therefore, AT is established on the premise that emotional responses are influenced by the perception of causal factors of the actual experience. AT identifies three distinct dimensions of causal attributions that are the following:

(1) the ***locus*** of the cause (internal vs. external to the individual).
(2) the ***stability*** of the cause (likely vs. unlikely to recur).
(3) the ***controllability*** of the cause of the outcome of the relevant situation (controllable or not).

In summary, the AT provides the process of identifying and coping with causal factors and outcomes; meanwhile, the DAs and BEAs are about identifying which emotions arise during an experience within a situation.

9.1.3.4 Theories of the automatic vs. controlled nature of emotion

This class of theories believe that emotions can be influenced by either cognitive mediation (Johnson and Stewart, 2005) or automatic mechanisms (Bargh 1982; Clore and Ketelaar, 1997; Herr and Page, 2004). The automatic emotional responses are the ones that are difficult to measure and disrupt or change through cognitive manipulations compared to those with cognitive mediation (Johnson and Stewart, 2005).

9

9.1.3.5 Appraisal Theories (ApsTs)

Appraise is a process of evaluating available information that occurs in consumers' minds where the implications of the situation for the interests and goals of the individuals are evaluated (a.k.a. experience(s)). Consequently, appraisals are the conclusions of the information processing that define the tenor of the emotions experienced (Lazarus, 1991). Within the framework of the Appraisal Theories (ApsTs), the processing or appraising of personally relevant information gave rise to emotion(s) (Johnson and Stewart, 2005). The ApsTs specify the dimensionalities of "appraisal process" that give rise and, further, discriminate different emotional responses based on individual's idiosyncrasy in one's unique construction of the situation (Johnson and Stewart, 2005).

Appraising process happens consciously, unconsciously, or automatically. Also "situations" that are appraised can be either "real" or "imagined" (Johnson and Stewart, 2005). Therefore, within the framework of ApsTs, emotions are often "functional." Emotions have "functions" to motivate actions that serve the "goals of the individual," and other types of coping responses (Frijda et al., 1989). Appraisal Theory suggests a three-step process of emotion processing (Figure 9.1.4):

(1) the ***antecedents*** of the appraisal process
(2) the ***process of appraising*** personally relevant information, and
(3) the ***consequences*** of appraisal and emotions

9.1.3.5.1 Step 1: Antecedents

The ApsTs state that the situation is the cause of an emotion. *Situation is external, exists outside of the individual, and provides both a context for emotion and a guide to the meaning and consequences of an emotional response.* Based on their own unique experiences, expectations and goals, individuals enter every situation with their own idiosyncratic perspectives. Everyone has goals and expectations regardless of how refined or how vague the goals and expectations are. *The goals of a particular individual in a specific situation* are *the MOST IMPORTANT* determinants of emotional reactions.

Individual's knowledge including (1) *expectations* and *beliefs* about the situation and (2) relevant goals in that situation and the perception of personally relevant information are *antecedent of the appraisal process*. The perception of personal relevance depends on the individual's own goals and beliefs. Thus, the knowledge that an individual brings to the situation is an important antecedent of emotional processing. Knowledge about goals includes information regarding the desired outcome and about coping with the situation (avoiding an aversive situation or seeking an attractive situation). Attention to personally relevant information may be consciously controlled or automatically directed without consciousness.

Goals may also focus on avoiding a task, particular social relationships, or experiential outcomes. Goals indirectly influence the appraisal process and emotional response by focusing attention, but they do not directly enter into the process of appraising.

9.1.3.5.2 Step 2 Appraisal

Appraisal process happens when an individual gives meaning to a situation and its associated elements. Appraising the implications of the situation has six dimensions based on specific information processing on personally relevant information (Ruth et al., 2002) (Figure 9.1.5).

(1) **Expected goal congruency is the direction of goal congruence (a.k.a. Motive consistency):** This is an assessment of whether achievement of personally relevant goals is *facilitated or hindered in the current situation* and it is defined by whether the situation is perceived to move the individual closer to, or away from, desired goals. *The direction of goal congruence is "bipolar" in nature, but the emotional reactions are not necessarily bipolar.* When an individual appraises multiple goals in a situation, many emotional responses can arise.

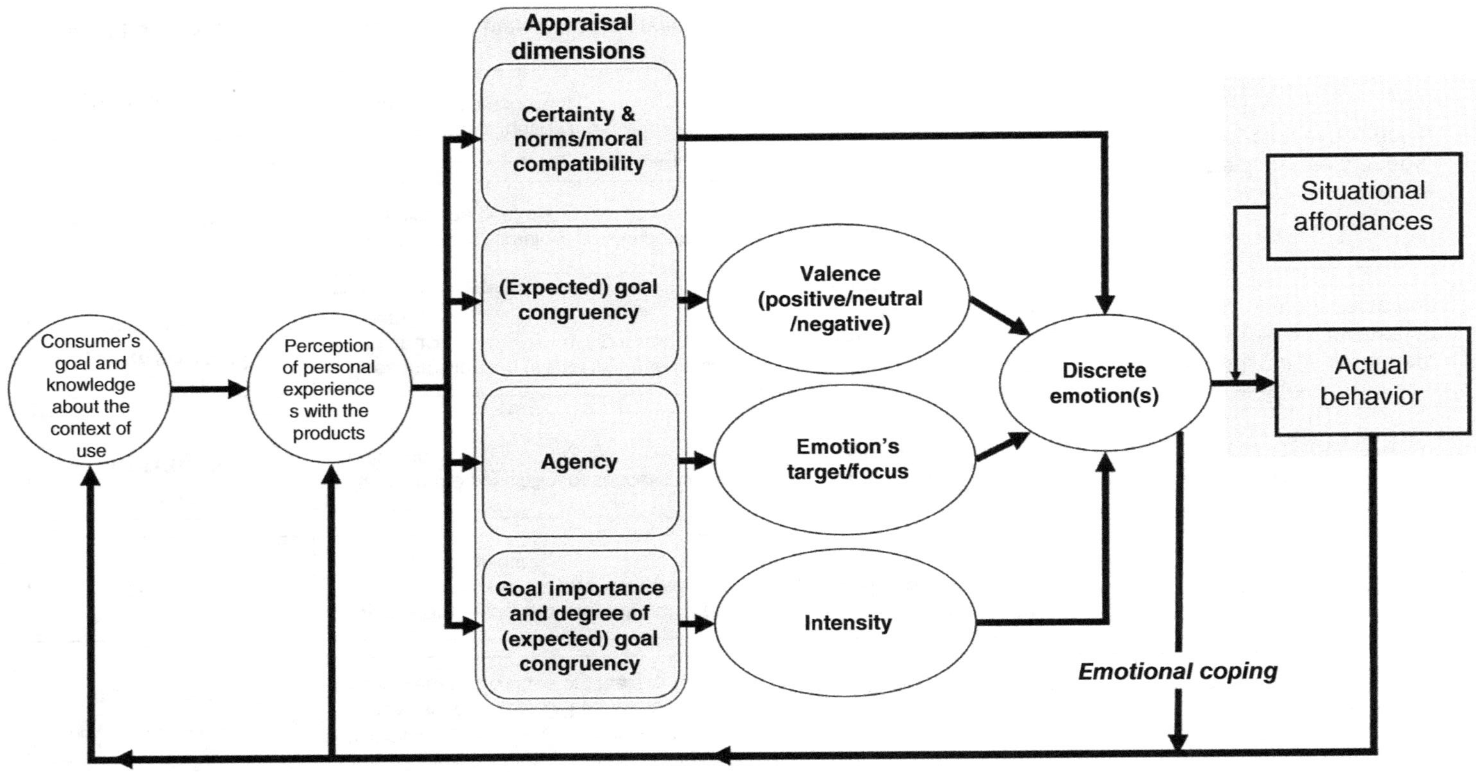

Figure 9.1.4 Appraisal model of emotion. Source: Adapted from Johnson and Stewart (2005).

Appraisal dimension	Meaning of appraisal dimension	Impacts on emotions	Emotion examples
(Expected) goal congruency	Within the situation, will the goal be successful or failure OR things are consistent with values or ideals or not	Valence (positive/neutral/negative)	Happy, sad, disappointed, pleasant, surprised
Agency	What is/are responsible for the outcome or in the control of the situation: self, other and/or object?	Who responsible for the outcome: seft, others or object (no person involve)?	Proud, angry, pitying, ashamed
Certainty	Gauging the possibility of successful outcome	Outcome-related *vs.* anticipatory emotions	Hopeful, joyous, anxious, sad
Norms/moral compatibility	Does the situation fit with my moral values and/or acceptable by my significant others?	Morality/Norms related emotion *vs.* emotions that do not reflect these concerns	Proud, guilty, ashamed, embarrassed, evalutaion, contemptuous, admiring
Goal importance	Within current situation, how important is achieving the goal(s)?	Emotion intensity	Happy, joyous, anxious, afraid, irritated, angry, hopeful, expectant
Degree of (expected) goal congruency	How likely the situation meet/likely to meet the expectations?		

Figure 9.1.5 Six appraisal dimensions and their impacts on emotions. Source: Adapted from Johnson and Stewart (2005).

(2) **Agency:** This is the assessment of the various entities (people, objects, products, brands, etc.) involved in a situation and the role each has played or might play in the outcome of the situation. *The agency dimension differentiates between emotions that are directed at the self, at other people, or at objects*. The appraisal of agency (a.k.a. appraisal of responsibility) involves a process that determines the focus or target of an emotion. Emotional reactions will vary depending on the appraisal of the target's role in the situation.

For example, a product was purchased to provide a certain benefit but failed to do so. If the consumers think that the product "fails to perform," then the consumers may feel "angry." However, if the consumers thought that it was their fault, then they may feel "guilt" and will not attribute "anger" to the product/brand. If there is no culprit identified for a situation, this phenomenon is called "circumstance-caused agency appraisals or indeterminate agency." This may result in emotional responses that focus on the "situation" itself (not person or product) (e.g. joy, relief, hope, and anxiety).

(3) **Certainty:** This is an assessment of probability (or extent) to which the situation will have a known outcome. Many emotions already demonstrate the degree of certainty when the words are used. For example, joy, anger, and relief demonstrate high probability of happening (goal is achieved or not as expected); and hope, anxiety, and dread demonstrate low probability of happening.

(4) **Certainty appraisal** is informed by goal achievement and it has a temporal dimension. Emotions that arise from this process serve a motivational function for individuals to continue or stop pursuing the goal(s). The critical issue is about the conviction of confidence that the outcome will occur in a given situation. Anticipatory emotions arise when appraisal leads to some degree of uncertainty and the valences of these anticipatory emotions are (1) positive (e.g. hope) or (2) negative (e.g. dread, anxiety) depending on the situation and the goals. The expectation (a.k.a. expected goal congruence) impacts type and strength of emotions that an individual will experience in a situation.

(5) **Normative/moral compatibility:** This appraisal involves an assessment of the situation in terms of what is "deemed to be normal and right" by the individual and within the specific social context. Violation of social norms or moral ideals can cue emotions like shame. This appraisal may be dominant in different cultures and, with the advent of social media, an individual can have different "cultures, norms and moral codes" beyond physical culture.

(6) **Goal importance:** The appraised importance of a goal increases the intensity of emotion because of the value or desirability or the severity of attaining or failing to achieve the goal. In conjunction with the degree of goal congruence (see below), goal importance influences "emotional intensity." If the actual outcome "exceeds" what has been "expected," then emotional intensity will be more "intense" than that of "as expected outcome." This

appraisal differentiates emotional intensity and enables prediction of discrete emotions that are intensity dependent such as happy, joy, and delight.

(7) **The degree of goal congruence:** This step is the appraisal of the extent to which the situation meets (or likely to meet) the desired state (at the moment of appraisal) or expectation.

At least ***40 basic discrete emotions*** (not including further classification for intensity level) were reported by Johnson and Stewart (2005) based on these appraisal steps. It is prudent to keep in mind that emotion responses can be verbal, nonverbal, and inexpressible (due to lack of words and these are the difficult ones to capture and measure).

9.1.3.5.3 Step 3 Consequences of emotions

After finishing the appraisal process (Figure 9.1.4), discrete emotional reactions arise and then are experienced by the individual (Figure 9.1.4). These emotions, in turn, may provide feedback to the individual (Figure 9.1.4) as additional information about the situation and behaviors (contemplated or performed) in response to the emotion may also provide feedback to the individual about goals and about the situation. Emotional reactions and the judgments will become part of the individual's knowledge (a.k.a. experience) about his/her goals in the context of the situation that influence future emotional processing.

The consequences of emotional experiences are defined by the process(es) of "**coping**" with the implications of appraisal of the individual's goals. Behavioral tendencies (a.k.a. action tendencies) associated with discrete emotions are generally adaptive responses to the situation and turn into actual behavior only if the conditions present in the environment allow the individual to ***take action to advance goal pursuit*** (a.k.a. <u>**situation affordances**</u>). If the conditions are not favorable, the individual must incorporate the information into knowledge structure and inspire further emotional processing such as *reinterpreting information in the context of the individual's other goals*, or *revising goals* to be more consistent with the affordances in the current situation. This leads to **Emotion Regulation.**

9

9.1.3.6 Emotion Regulation

Emotion regulation aims at coping with the implications of a given situation for the individuals and their goals and well-being. **Coping** is the means that an individual identifies and assesses the adaptive potential and significance of various actions or strategies. There are two types of coping based on its focus: (1) changing emotional response to the situation (Emotion-focused coping) or (2) taking action within the situation (Problem-focused coping).

(1) **Emotion-focused coping** involves behaviors or strategies aimed at moderating the emotional reactions because, after appraising, the situation affordance is low. Therefore, individuals "alter" their perspectives regarding the situation such as shifting the focus or appraisal process to a different or revised goal. In short, they change their minds.

(2) **Problem-focused coping (a.k.a. behavioral coping)** is associated with identifying and implementing behaviors and strategies that solve the problem or otherwise advance the goals of the individual (e.g. trying to solve problems, continuing the process to repeat positive experiences). Positive emotional experiences motivate individuals to pursue successful goals and broaden the range of potential goals that are considered.

Coping is relevant to consumer behavior because consumers must cope with their emotional reactions to consumption experiences and because coping with the results of other goals may "motivate goals related consumption behavior." It is important to create a product/advertising/experience that appeals to the types of emotions that might motivate consumers to use their products or services for emotional or behavioral coping. In addition, knowing which type of copings and emotional responses are the results from product consumption and experience is critical for product innovation.

The above six theories regarding emotions (Dimensional Theory, Basic Emotions Approach, AT, Automatic vs. Controlled Nature of Emotions, Appraisal theory and Emotion Regulation) have been used in developing tools to identify, measure, and evaluate emotions for individual within a situation (including product experiences) and, ultimately, to attempt to understand the sources of the emotions to predict the individual's behavior through emotion regulation. This has led to the creation of **methodologies on measuring emotion.**

9.1.4 Popular methodologies for the measurement of emotions

Many methodologies for measuring emotional responses have been developed based on the above theories and practicality in the fieldwork. Methods adapted from Dimensional/Basic Emotion philosophies are popular among product and market research because the theories focus on identifying emotion within a situation; meanwhile, Appraisal Theory is popular in clinical settings due to its attempt to identify both emotions and their corresponding causes. Emotion regulation theories are rising in interest in the clinical measurement area.

9.1.4.1 Methodologies that are influenced by Dimensional/Basic Theories

In practice, the instruments developed based on these theories need to be developed and validated for each target market before real practical use. The tools provide the reading on emotional state at a particular moment of consumption but do not take into account goals and perceptions of the situation in the study. Relying on labels available in the common vocabulary or facial expressions (image and animation included) for emotional experience limits the range of the emotions studied to those that have a clear definition in the native language or that researchers "have hypothesized" to be comprehensive for describing the emotional responses in a situation. Even though methodologies developed based on Dimensional and Basic Emotion approaches have their limitations mentioned above, the instruments are still very popular due to their ease of administration and relatively simple statistical analyses. The examples of these measurements are PreMo, EsSense Profile, Emoti*Scape, PANAS scale, Baron-Cohen English emotion Vocabulary (Mind Reading™ system), Geneva Emotion wheel, Noldus FaceReader™, Emotionomics™, etc.

Unlike PreMo, Emoti*Scape,PANAS scale or other standardized images or verbages developed and validated to capture emotions in many countries (PrEmo, 2000; PrEmo, 2009). Since the 1990s, unvalidated versions of words, phrases, and images (a.k.a. emojis, emoticons and, eventually, stickers) have been commercially used for the general public to express their emotions and attitudes toward certain events, products, or services in social media platforms. However, these stickers, emojis and emoticons were designed according to the designers' own understanding of emotion. Jaeger et al. (2020) presented findings recommending emojis to study human emotions. They found that their emoji questionnaire avoided many barriers found in word-only questionnaires (ecological validify, familiarity, and cross-cultural validity). The caution found was that regular emoji questionnaires can be easily misinterpreted, especially those for nuance emotions and arousal emotion and inappropriate for different cohorts (Jaeger et al., 2020).

9.1.4.2 Methodologies that are influenced by physiological measures of emotional responses

These measures such as Brain wave, EEG wave, or Pupillometer are very sensitive and objective systems to measure "what are believed to associated with emotional responses"; however, to use these systems, the contexts of situations are not "real-life." Therefore, it limits the generalization of the results to actual product experience in real situations. Like those instruments developed based on Dimensional and Basic emotion approaches, these physiological instruments provide only measures of emotional state at a moment and do not provide any insight into the causes or consequences of emotional responses. See Chapter 9.2 for more details about this set of tools and how to apply them successfully.

9.1.4.3 Methodologies that are influenced by Appraisal theories

Appraisal theories provide guidance for the development of ***multi-item measures*** of differentiated emotional reactions based on measuring "relevant appraisals" (at least 40 emotional responses) in addition to measuring emotional state only. ***Critical Incident method (CI)*** is the most used method to apply Appraisal theories of emotion (Flanagan, The Critical Incident Technique, 1954) (Butterfield et al., 2009). This method prompts participants to "recall" and "describe" a situation in which they experienced the emotional state of interest. The descriptions are then coded and analyzed to assess the appraisals that are relevant in situations where the emotion occurs as well as for emotion regulation attempts and other emotions that were experienced in the same situations. The CI method can be used in conjunction with more controlled measures of the phenomena of interest to provide insight into the occurrences of emotions in consumption situations. Applying Appraisal theories in research has some issues that are

(a) appraisals can occur unconsciously or automatically; hence, measuring appraisal may elicit reconstruction of a reasonable account of an emotional situation, rather than measuring the actual processing that gave rise to the emotion.
(b) asking questions or rating scales may coerce participants to create emotional reaction and leasing them to report emotions to please the experimenter. These emotional reactions can be newly created or different emotions from what were actually experienced.

These issues may be overcome by using several research methods and triangulating the phenomena to investigate consumers' "internal appraisal and emotion processes." Table 9.1.1 compares eight types of popular emotion measurement methodologies currently used.

9.1.5 Impact of social media on emotion research

Social media such as Twitter®, Facebook®, Line®, WhatsApp®, KakaoTalk®, or WeChat® are platforms for people to express themselves. With millions of posts and dialogues between people, social media platforms produce extremely large amounts of natural text and data across the world. Meiselman (2015) pointed out the advantages of using social media data in emotion research were (1) large amounts of data that conventional methods cannot produce, (2) easier ways of conducting cross-cultural studies, (3) relatively inexpensive cost, and (4) free-from-bias data compared to data obtained from questionnaires.

There are several limitations and cautions for using social media as a data source for emotion research. These limitations are detrimental for emotion research which deals with cross-culture, gender identity, or global projects. The limitations and cautions are:

(1) regional specific meanings of words and jargons in the same language within a supposedly one target: this issue raises complexities in a study especially cross-cultural study where one uses only geographical factors to define culture when there are many subcultures within a supposedly homogeneous target. A word or a jargon can mean different things in multiculture society and, without prior knowledge, it will cause misinterpretation especially when using AI.
(2) differences in the laws in different countries: sentimental analysis counts positive and negative words from social media content. However, different countries have different laws regarding freedom of expression. For example, Thailand has a computer and communication law specifically designed to prevent and censor posting slandering comments toward someone or brands publicly. Therefore, having more positive than negative word counts for a brand in Thailand does not mean the majority of people like the brand.
(3) short-time nature of emotion: emotion is short-lived by nature; hence, an emotion trend is not stable and changes quickly. Golder and Macy (2011) reported that, among Twitter users, positive and negative effects changed

Table 9.1.1 Comparison between eight types of emotion measurement methodologies.

Scope and principle of measurement	Potential uses	Advantages	Limitations
1. Balanced Verbal Emotion scales (e.g. Profile of Mood States (POMS), the MAACL questionnaire, Positive and Negative Affect Schedule (PANAS)			
References: Watson et al. (1988)			
• Based on Dimensional and Basic emotion Approaches • Verbal emotion scales borrowed from standardized questionnaires of mood	• Identify and measure feelings (affects, moods, or emotions) using "suppose-to-be-common" adjectives • Use as measuring tools in an experiment to estimate emotional state or classify people based on their patterns of responses	• The administration of these scales can be done with minimum technological efforts (paper and pencil will be just fine)	• The balanced number of positive and negative emotions may not be suitable for product or food testing situations due to the Hedonic Asymmetry phenomenon (Desmet and Schifferstein, 2008; Schifferstein and Desmet, 2010) • Difficulties in translating to other languages
2. Imbalanced Verbal Emotion scales (e.g. EsSense Profile™, Differential Emotions Scale [DES])			
References: King and Meiselmann (2010); Galanakis et al. (2016)			
• Based on the belief that consumers can express their food-consumption experience through matching their feeling to the list of standardized emotion terms	• Specially developed to identify and measure intensity of emotions during actual product (especially food) consumption	• Tailored to fit with food consumption with more positive terms than negative terms • Good for exploring elicited emotions from food consumption	• Not suitable for any food testing under "realistic situation" • Lacks many terms that reflect the goal of "overall experience" or "branding identify" • Not include more "refined or sophisticated words" as it was limited to literacy level of consumers • NOT appropriate for investigating targeted emotion terms • Direct translation may be inappropriate as "official translation" may not actually reflect "actual feeling" in many languages

<table>
<tr><td colspan="4">3. Self-Assessment Mannequin (SAM) scale (see example of use in Chapter 9.2)</td></tr>
<tr><td colspan="4">References: Bradley and Lang (1994)</td></tr>
<tr>
<td>• Based on Basic emotion and Dimensional Approaches</td>
<td>• Use as measuring tools in an experiment as SAM was developed to measure intensities of three basic dimensions of emotions
• Cross-cultural study</td>
<td>• Easy to use: can be used with and with illiterate subjects (such as children)
• Enable cross-cultural comparison (Caicedo and Desmet, 2009)
• Has special version for food craving (Muzoz et al., 2010)</td>
<td>• SAM does not measure and differentiate specific emotions but rather measures the intensity of three key basic dimensions (Pleasure, Arousal, and Dominance)</td>
</tr>
<tr><td colspan="4">4. Facial Emotion Coding System and Facial Expression Scales</td></tr>
<tr><td colspan="4">References: Ekman et al. (1969)</td></tr>
<tr>
<td>• Based on Basic emotion and Dimensional Approaches
• Relies on predetermined set of rules based on Ekman's principle of facial expression stating that facial emotion expressions are "innate" in human</td>
<td>• Use in ethnography and observation type studies
• There are two different formats:
(1) Actual real person face coding system (Noldus FaceReader (2007); Emotionomics (2007)) and
(2) Facial scaling approach (Mind Reading system (Baron-Cohen (2003)); PrEmo™ (2000; 2009) & Emoti*Scape™ by Ipsos-ASI</td>
<td>• Suitable for live situations and online testing using digital image recording
• Responses can be instantaneous and not from recall
• Easy, fun, and interactive
• Can be treated as emotion measurements and statistically analyzed</td>
<td>• Requires computer interface to capture the emotional responses
• Not suitable for refined emotions such as indulgent, refreshing, etc.</td>
</tr>
</table>

(Continued)

Table 9.1.1 (*Continued*)

Scope and principle of measurement	Potential uses	Advantages	Limitations
5. OLAF, IAPS, and product cue reactivity paradigm systems			
References: Miccoli et al. (2014)			
• Based on Basic emotion and Dimensional approaches • Based on the affective picture viewing paradigm and product cue reactivity paradigm	• For studying affective reactions to food • Identify the intrinsic motivational significance of different types of food (represented by the images in the set • Consumer segmentation based on their emotional responses toward the food pictures	• Specifically developed for food • Standardized nature of the images that can be used as either stimuli in an experiment or measuring tools to capture emotional responses • Can be tailored for cross-cultural studies	• Need standardized computer screen and interface system for picture presentation • Need to control printing quality, if printed images will be used (this is not an easy task)
6. Critical Incident Technique (CIT)			
Reference: Flanagan (1954); Gremler (2004)			
• Based on Appraisal Theories • Discipline origin is from industrial and organizational psychology	• Research on sparingly documented topics • Used as an exploratory method to increase knowledge about a little-known phenomenon • Used as a tool for a thorough understanding of a phenomenon • Used as tool to develop conceptual structures or hypotheses) to be used and tested in subsequent research	• Semistructured interview with flexible procedure that can be modified • Can be done in 1-on-1 or group interview setting • Data collected are from the respondent's perspectives and in his or her own words • Engaging for participants at a personal level • Allowing free range of responses within an overall research framework • Reflects the normal way that consumers think and does not force them into any given framework	• Quantify the result for statistical analysis is quite cumbersome • Lengthy interview time • Stories reported in incidents can be misinterpreted or misunderstood • Ambiguity associated with category labels and coding rules within a particular study • Relying on events being remembered; hence, description may lead the respondent to reinterpret the incident

7. Emotion Prototype Method (see example of use in Chapter 6.2 and Chapter 7.1)			
References: Flanagan (1954); Shaver et al. (1987)			
• Based on Basic emotion and Dimensional Approaches in conjunction with Hierarchical Organization of Emotion	• Designed to manifest the "experiences/episodes" that constitute the feeling of a fuzzy emotion • The technique itself can be applied for words that do not have concrete meaning for a certain target consumer.	• Fairly simple • With some preparation, it can be conducted using paper-and-pencil or adapted to blogging platform easily	• Everyone must know the target emotion • Target emotion should have fuzzy boundaries • Difficult to put a set number of fundamental issues concerning the target emotion such as the definition of the emotion. • Generates a lot of texts and contexts and quantifying results is quite cumbersome
8. Emotion Appraisal, Regulation and Coping protocols			
References: Kang and Cappella (2008); Gratch and Marsella (2004); Scherer (1993); Garnefski et al. (2001); Gross and John (2003); Galanakis et al. (2016); Ruth et al. (2002); Kang and Cappella (2008)			
• Based on Appraisal theory of emotion	• Designed to identify emotions, investigate the causes of the emotions and identify coping decision and behavior for the emotions • Can be used for product design after certain concepts, product features and target emotions have been crystallized to a certain level through other methods	• Offer the understanding of the target emotion and potential behavior that the target consumers will emotionally and/or physically take	• Lengthy probing (can be done in the form of long questionnaires, blogging or interview) • Preparation must be carefully done especially the phasing of the statements to fit with the context, cover the ranges of the target experiences and appropriate for the target consumers to use (literally and socially)

within 24 hours. In addition, the changes' patterns were different from country to country. Therefore, the use of snapshot results may not be that valuable for projects that need a long time frame to finish such as new product development.

(4) interdependent nature of emotions to contexts: COVID-19 pandemic is a very good example for this issue. Fear was the predominant emotion during the very first few days of the pandemic. Social media were flooded with fear-type messages; however, now the pandemic has passed its first year, the fear is no longer the dominant emotion but the mood of the society. With the announcement of vaccines, hope turned to be the emotion amid the fear mood. Interpreting emotion without proper explanation about its accompanying context will do more harm than good for the stakeholder.

In conjunction with the advancement in text analysis and computing power through artificial intelligence (AI) technologies, sentiment analysis on social media messages will advance and be more popular in the coming years with the caveat that inherent bias of algorithms would be better understood and managed. However, caution should be heeded as the limitations of using social media as a data source for emotion research can lead to misinterpretations.

9.1.6 Conclusion and recommendations

Emotion research has been focusing on identifying which emotion responses are elicited during an episode within a situation. Dimensional and Basic emotion approaches are the fundamental framework of current emotion measurements such as SAM scale, PANAs scale. The Circumplex-type theories of emotions are the most commonly followed and applied. Emotion scales developed for product design and food consumption are PreMo and EsSense Profile that contain more positive emotion labels appropriate for product consumption than other standard scales due to the Hedonic Asymmetry phenomenon.

Facial coding and Pictorial scales are also based on the Circumplex and Basic emotion framework. While the scales based on these basic dimensions of emotion responses are very simple to use (with some requiring computer-aid technology), the simplicity of the model may neglect nuanced emotions that are important. Since these emotions are usually considered "fuzzy" and mixed together, the actual feeling can be lost. Additionally, certain emotions cannot be expressed by word or nonverbal cues and many nonverbal cues have multiple emotional meanings. Workarounds such as triangulation and emotional prototyping may be recommended, yet, it is clear that we cannot be CERTAIN, how well emotions are being measured or connected to actions.

From the clinical side, the latest and the most popular theory and approaches in emotion research is Appraisal Theory. This theory concerns the appraisal that gives rise to emotion responses and, consequently, emo-tion coping and emotion regulation that greatly constitute behavioral and attitudinal responses. Applying Appraisal Theory in consumer-product research will require researchers to formulate hypotheses and decision models about consumption situations and decision rules thoroughly before conducting any research. Expected outcomes of Appraisal Theory based research are antecedents of the situation, important appraisal steps, emotion responses, coping and emotion regulation process and, ultimately, behavior prediction.

In her book, *How Emotions are Made*, Lisa Feldman-Barrett (2017) introduced the theory of constructed emotion and differentiated it from classical views such as the Basic Emotion and Appraisal Theories. Emotions are not reactions to the world but a person's construction of the world from experiences and perceptions without the need for consistent biological fingerprints or universal labels. The brain is continuously predicting, fine tuning concepts, and learning from internal and external sensory inputs so it understands what they mean and what to do with them. Emotions come from this rich set of concepts built in social reality. This paradigm shift proposed by Feldman-Barrett in studying the science of emotion will continue to evolve with neuroscience learnings about the structure of the brain.

As for CEs, this brings us back to the importance of understanding the individual experience to a stimulus within a context and the challenge of aggregating multiple experiences to identify meaningful patterns for generating insights.

Discussion Questions

- How might you design a study to understand the root cause of a certain emotion after actually trying a product?
- What should the concerns of a CE be around the trend to apply emotion metrics to product attribute evaluations?
- What are limitations of using emotion scales in multinational studies?

References

Bardo, M.T., Donohew, R.L. and Harrington, N.G. (1996, May) "Psychobiology of novelty seeking and drug seeking behavior". *Behavioural Brain Research*, 77(1-2), 23-43. https://doi.org/10.1016/0166-4328(95)00203-0.

Bargh, J.A. (1982) "Attention and automaticity in the processing of self-relevant information. *Journal of Personality and Social Psychology*, 43(3), 425-36.

Baron-Cohen, S. (2003) *Mind Reading: The Interactive Guide to Emotions*. London, UK: Jessica Kingsley.

Baron-Cohen, S., Golan, O., Wheelwright, S., Granader, Y. and Hill, J. (2010) "Emotion word comprehension from 4 to 16 years old: a developmental survey". *Frontiers in Evolutionary Neuroscience*, 2(109).

Batra, R. and Ray, M.L. (1986) "Affective responses mediating acceptance of advertising". *Journal of Consumer Research*, 13(20), 234-48.

Bradley, M. and Lang, P.J. (1994) "Measuring emotion: the self-assessment manikin and the semantic differential". *Journal of Behavior Therapy and Experimental Psychiatry*, 25(1), 45-59. https://doi.org/10.1016/0005-7916(94)90063-9.

Butterfield, L., Borgen, W., Maglio, A. and Amundson, N. (2009) "Using the enhanced critical incident technique in counselling psychology research. *Canadian Journal of Counselling*, 43(4), 265-283.

Caicedo, D.G. and Desmet, P. M. (2009) *Designing the New PrEmo: An Empirical Research on how to Improve the Emotion Measuring Tool*. Delft University of Technology (TUDelft).

Clore, G. and Ketelaar, T. (1997) "Minding our emotions: on the role of automatic, unconscious affect". In J. Robert, S. Wyer (eds). *The Automaticity of Everyday Life: Advances in Social Cognition* (Vol. 10). Mahwah, New Jersey: Lawrence Erlbaum Associates.

Desmet, P. and Schifferstein, H. (2008) "Positive and negative emotions associated with food experience". *Appetite*, 50(2-3), 290-301.

Ekman, P., Sorenson, E.R. and Friesen, W.V. (1969) "Pan-cultural elements in facial displays of emotions". *Science*, 164(3875), 86-88.

Feldman-Barrett, L. (2017) "*How Emotions are Made, The Secret Life of the Brain*. Boston: Houghton Mifflin Harcourt.

Flanagan, J.C. (1954) "The critical incident technique". *Psychological Bulletin*, 51(4), 327-358.

Frijda, N.H., Kuipers, P. and ter Schure, E. (1989) "Relations among emotion, appraisal, and emotional action readiness". *Journal of Personality & Social Psychology* 57(2), 212-228.

Galanakis, M., Stalikas, A., Pezirkianidis, C. and Karakasidou, I. (2016) "Reliability and validity of the modified differential emotions scale (mDES) in greek sample". *Psychology*, 7, 101-113.

Garnefski, N., Kraaij, V. and Spinhoven, P. (2001) "Negative life events, cognitive emotion regulation and depression. *Personality and Individual Differences*, 30(8), 1311-1327.

Golder, S.A. and Macy, M.W. (2011) "Diurnal and seasonal mood vary with work, sleep, and daylength across diverse cultures". *Science*, 333(6051), 1878-1881.

Gratch, J. and Marsella, S. (2004) Evaluating a General Model of Emotion Appraisal and Coping. *AAAI Spring Symposium on Architectures for Modeling Emotion: Cross-disciplinary Foundations*. Palo Alto, CA.

Gremler, D.D. (2004) "The critical incident technique in service research". *Journal of Service Research*, 7(1), 65-89.

Griffin, A. and Hauser, J. (1993) "The voice of the customer". *Marketing Science*, 12(1), 1-27.

Gross, J.J. and John, O.P. (2003) "Individual differences in two emotion regulation processes: implications for affect, relationships, and well-being. *Journal of Personality and Social Psychology*, 85(2), 348-362.

Herr, P. and Page, C. (2004) "Asymmetric association of liking and disliking judgments: so, what's not to like?" *Journal of Consumer Research*, 30(4), 588-601.

Jaeger, S.R., Vidal, L. and Ares, G. (2020) "Should emoji replace emotion words in questionnaire-based food-related consumer research?. *Food Quality and Preference* 92, 104121. https://doi.org/10.1016/j.foodqual.2020.104121.

Johnson, A.R. and Stewart, D.W. (2005) "A reappraisal of the role of emotion: traditional and contemporary approaches". In N.K. Malhotra (ed). *Reviews of Marketing Research (Volume 1)*. NY: M.E. Sharpe.

Kang, Y. and Cappella, J.N. (2008) "Emotional reactions to and perceived effectiveness of media messages: appraisal and message sensation value". *Western Journal of Communication*, 7(1), 40-61.

King, S. and Meiselman, H. (2010) "Development of a method to measure consumer emotions associated with foods". *Food Quality and Preference*, 21(2), 168-177.

Kotler, P and Keller, K. (2009 *Marketing Management*, (13th edition). Upper Saddle River, NJ: Pearson/Prentice Hall.

Lawless, H.T. and Heymann, H. (1998) *Sensory Evaluation of Food: Principles and Practices*. NY: Springer Science & Business Media.

Lazarus, R.S. (1991) "Cognition and motivation in emotion". *American Psychologist*, 46(4), 352-367.

McClelland, D.C. (1965) "Toward a theory of motive acquisition". *American Psychologist*, 20(5), 321-333. https://doi.org/10.1037/h0022225

McCrae R.R. and John O.P. (Jun. 1992) "An introduction to the five-factor model and its applications". *Journal of Personality*, 60(2), 175-215. https://doi.org/10.1111/j.1467-6494.1992.tb00970.x. PMID: 1635039.

Meiselman, H.L. (2015) "A review of the current state of emotion research in product development", *Food Research International*, 76(Part 2), 192-199.

Meiselman, H.L. (ed) (2021) *Emotion Measurement*, (2nd edition). Sawston, UK: Woodhead Publishing.

Miccoli, L., Delgado, R., Rodriguez-Ruiz, S., Guerra, P., Garcia-Marmol, E. and Fernandez-Santaella, M. (2014) "Meet OLAF, a good friend of the IAPS! The open library of affective foods: a tool to investigate the emotional impact of food in adolescents". *PLoS One*, 9(12), e114515. https://doi.org/10.1371/journal.pone.0114515.

Muzoz, M.A., Viedma-del-Jesus, M.I., Fernandez-Santaella, M.C., Peralta-Ramirez, M.I. and Cepeda-Benito, A. (2010) "Assessment of tobacco craving by means of the effective image visualization paradigm". *Motivation and Emotion*, 34(1), 93-103.

Neel, R., Kenrick, D.T., White, A.E. and Neuberg, S.L. (2016) "Individual differences in fundamental social motives". *Journal of Personality and Social Psychology*, 110(6), 887-907. https://doi.org/10.1037/pspp0000068

Noldus FaceReader (2007), www.noldus.com and Emotionomics (2007), www.sensorylogic.com were developed based on this approach.

Pert, C. (1997) *Molecules of Emotion their Receptors: A Psychosomatic Network*. NY, NY: Touchstone Book/Simon & Schuster.

Pert, C., Ruff, M., Weber, R. and Herkenham, M. (1985, August) "Neuropeptides and their receptors: a psychosomatic network". *Journal of Immunology*, 135(2 Suppl), 820s-.

Pine, J and J. Gilmore (1999) *The Experience Economy*. Boston, MA: Harvard Business School Press.

Plutchik, R. (1982) "A psychoevolutionary theory of emotions". *Social Science Information/sur les sciences sociales*, 21(4-5), 529-553. https://doi.org/10.1177/053901882021004003.

Porcherot, C., Delplanque, S., Planchais, A., Gaudreau, N., Accolla, C. and Cayeux, I.(2012) "Influence of food odorant names on the verbal measurement of emotions". *Food Quality and Preference*, 23(2), 125-133.

Russell, J.A. and Mehrabian, A. (1977) "Evidence for a three-factor theory of emotions". *Journal of Research in Personality*, 11(3), 273-294. https://doi.org/10.1016/0092-6566(77)90037-X.

Ruth, J.A., Brunel, F.F. and Otnes, C.C. (2002) "Linking thoughts to feelings: investigating cognitive appraisals and consumption emotions in a mixed-emotions context". *Journal of the Academy of Marketing Science*, 30(1), 44-58.

Sacharin, V., Schlegel K. and Scherer K.R. (2012) *Geneva Emotion Wheel Rating Study*. Unpublished report.

Scherer, K.R. (1993) "Studying the emotion-antecedent appraisal process: an expert system approach". *Cognition and Emotion*, 7(3-4), 325-355.

Scherer, K.R. (2005) "What are emotions? And how can they be measured?. *Social Science Information*, 44(4), 693-727.

Scherer, K.R., Shuman, V., Fontaine, J.R.J and Soriano, C. (2013). "The GRID meets the wheel: assessing emotional feeling via self-report". In J.R.J. Fontaine, K.R. Scherer and C. Soriano (eds). *Components of Emotional Meaning: A Sourcebook*. Oxford: Oxford University Press. pp. 281-298.

Schifferstein, H.N. and Desmet, P.M. (2010) "Hedonic asymmetry in emotional responses to consumer products". *Food Quality and Preference*, 21(8), 1100-1104. https://doi.org/10.1016/j.foodqual.2010.07.004.

Scott, M.J. and Cohen, A.B. (2019) *Surviving and Thriving: Fundamental Social Motives Provide Purpose in Life*. https://www.researchgate.net/publication/336242830_Surviving_and_Thriving_Fundamental_Motives_Provide_Purpose_in_Life, accessed 28 March 2021.

Shaver, P., Schwartz, J., Kirson, D. and O'Connor, G. (1987) "Emotion knowledge: further exploration of a prototype approach". *Journal of Personality and Social Psychology*, 52(6), 1061-1086.

Watson, D. and Tellegen, A. (1985) "Toward a consensual structure of mood". *Psychological Bulletin*, 98(2), 219-235. https://doi.org/10.1037/0033-2909.98.2.219.

Watson, D., Clark, L.A. and Tellegen, A. (1988) "Development and validation of brief measures of positive and negative affect: the PANAS scales. *Journal of Personality and Social Psychology*, 54(6), 1063-1070.

9.2

Applied Consumer Neuroscience and Behavioral Approaches for Innovation, Product Development, and Communications

Michelle Niedziela and Kathryn Ambroze

Key learnings

✓ Behavioral science in innovation
✓ Neuroscience in innovation
✓ Applying Novel Methods to Innovation: Choosing the Right Tool

9

It's often said that the definition of insanity is doing the same thing over and over and expecting a different result. Yet, when exploring white space, or innovation opportunities, researchers tend to rely on the same tools that were unable to provide truly innovative insights in previous research and explorations. According to Harvard Business School professor Clayton Christensen, over 30 000 new products are introduced every year, and 80% fail (2016). Christensen, the author of the theory of disruptive innovation, is widely regarded as one of the world's top experts on innovation and growth. He has suggested that companies often fail in introducing new products because they study the wrong product and consumer data, leading them to unwittingly design innovation processes that "churn out mediocrity."

These unfavorable statistics may be because different types of innovation require unique strategic approaches. To be truly innovative, product developers, and brand managers may need to be open to new approaches for better understanding of the consumer experience with brands and products. One potential opportunity to improve innovation may be in looking to the brain's role in human

Product Innovation Toolbox: A Field Guide to Consumer Understanding and Research, Second Edition. Edited by Kannapon Lopetcharat, Dulce Paredes and Jacqueline H. Beckley.

behavior and preference. Emotional and perceptual congruence in the consumer experience is key to consumer satisfaction, and therefore crucial for successful products. Unfortunately, traditional survey tools are often unable to effectively capture these aspects of consumer reactions (Köster and Mojet, 2015). Therefore, the use of traditional tools in conjunction with additional emotional and perceptual techniques from psychology and neuroscience may help to gain a comprehensive understanding of the consumer experience. The ability to gain insight into the emotional product journey can provide useful information of the consumer perceptive-hedonic experience.

This chapter begins by reviewing the importance of emotions and expectations during the consumer experience of products and brands, as well as how these reactions drive consumer behavior and habits. We will discuss neuroscientific, psychological, and behavioral approaches with a review of the methodologies in applied consumer neuroscience and behavioral science for innovation research paired with demonstrative case studies. Finally, we will outline a research design pathway to develop consumer technical models for driving truly innovative products, new product pipelines, and valuable communications.

9.2.1 A behavioral approach: behavioral and consumer neuroscience science

Neuroscientific and behavioral research approaches have become increasingly important in understanding how we, as humans, respond emotionally and physically to the world around us. One of the intriguing reasons to include neuroemotional techniques is that research widely recognized measuring likeability alone fails to predict product performance in the market (Thomson, 2007; King et al., 2010; Thomson and Crocker, 2015). However, measuring self-reported experiences also often failed to predict behaviors, such as purchase intent (Thomson, 2007; King et al., 2010; Thomson et al., 2010). Adding behavioral science and applied neuroscience proposes an interdisciplinary dimension to consumer market research where, traditionally, researchers have relied on self-reported reactions, leading to better research designs, innovation insights, and ultimately better product and communication launches.

While this expanded understanding has added valuable information for developing innovative products and communications, it is important to recognize that these methodologies are not a replacement of traditional testing. Nor should these methodologies be done independently of other forms of consumer research, but rather used in conjunctions with and in addition to traditional insights. These additions broaden the possibilities for understanding underlying consumer motivations, with each potential tool providing a different sort of insight. Using the right tool for the appropriate research question is crucial. In this section, we will review several methods within behavioral sciences, neurosciences, and psychology to provide a foundational framework for deciding which tools (if any) should be integrated into a research proposal. Through critical thought and awareness of the benefits and limitations of each measure, the process of selecting parameters within the research design can be a constructive and reflective component to the innovation research process.

9.2.1.1 Behavioral sciences in innovation: driving behavior and decision-making

When discussing behavioral approaches to research, it seems appropriate to start with behavioral economic theory. Behavioral economics is the intersection of economics and psychology. Traditional economics subscribes to the notion that markets are made up of rational actors, individuals with concrete preferences and stable decisions when comparing the among the available options. These actors have an inherent knowledge and capacity for maximizing their happiness and make their choices accordingly. Yet, this is unrealistic in the natural world. Individuals are sometimes irrational and deviate from the normative, expected behavior. Understanding these patterns and trends of human idiosyncrasies helps marketers make better predictions about how consumers will behave in real-world conditions (Binmore and Shaked, 2010). Marketers have been using the theories of behavioral economics to understand and exploit the irrationality of consumers long before the golden age of advertising (think simple marketing tricks such as "buy now, pay later," inflated value pricing, or decoy alternatives) and have been able to effectively direct or, in some cases, manipulate consumers to buy products or services they may not need or otherwise or intended to buy. These tactics, however, are more than just marketing tricks. Broad evidence suggests the behavioral science movement is beginning to flourish, with more and more educated professionals taking up consumer experience roles in behavioral economics (Samson, 2019).

To summarize the field of behavioral economics and behavioral science into one word, it would be choice. Choice is the driving interest in these conjoined fields. Theorists, marketers, and researchers alike are all intrigued by the forces which shape decisions. Behavioral economics uses controlled interventions to better understand the way people react the way they do. Through continuous experimentation, research in behavioral economics chips away at its overall goal to make descriptive models of human behavior. Further, principles from behavioral economics can uncover innovation opportunities in applying these models to the market.

9.2.1.1.1 Nudging behaviors and decisions

Manipulating the environment can impact choice architecture, and ultimately behavior. The evaluation of preferences, such as risk, time, and social comparisons, drive human responses in different situations. Researchers and product developers can apply these preference considerations into their designs. For example, one of the most common tactics utilized in behavioral science is nudging. Thaler and Sunstein (2008) refer to nudging as "any aspect of the choice architecture that alters people's behaviors in a predictable way without forbidding any options or significantly changing their economic incentives." These interventions are nonintrusive and are frequently small alterations that make large impacts on behavior. Nudges can be created in the way information is presented, such as suggested recommendations or highlighting desired features about a product.

The big secret behind the success of nudges is quite simple - just make things easy. In making concepts or products more intuitive, consumers can decide something quickly without exuding too much effort. A mental shortcut, or a heuristic, is a fundamental component of nudging (Kahneman, 2013). Simplifying the judgment process to streamline information can lead to bias, which is not necessarily a bad thing when it comes to products and communications. Quick reactions to any type of change are essential for individuals to function in life. As humans gain new evidence through experiences, associations are strengthened and weakened, resulting in an evolution of heuristics and biases. Nudging is intended to assist in the evaluation of something quickly by alleviating some of the cognitive tolls of uncertainty (Sunstein, 2014; Samson, 2019). Promoting a straightforward experience is preferred by consumers. No one wants to struggle to open a package or navigate a website. An innovative idea or design is something effective at meeting consumer demand, but a product or message is only as effective as the consumer's ability to utilize it. Don't make your consumer think; steering consumers toward a product or a concept, nudging reduces the barriers in the consumer journey.

Heuristic strategies give consumers, as well as marketers, an opportunity to work smarter, not harder. By streamlining trivial tasks, more time can be spent enjoying (or developing) a product or communication. Although nudge interventions are extremely widespread, some of the most frequently implemented strategies for innovation are listed in Table 9.2.1.

Each heuristic acts as a tool emphasizing specific components of a product, package, service, or communication. However, regardless of the nudge or heuristic employed, the consumer's freedom is never jeopardized (Sunstein, 2014). The design of a nudge intervention simply encourages a desired action but never forces a decision. The consumer must decide independently what decision to make.

9.2.1.1.2 Driving habits

Our lives are full of small habits that make up the behaviors of the day and products and brands play a large role in this. Whether it's using the same product each time to initiate a behavior (such as having a favorite shampoo) or having behaviors influenced by a product or a service (such as television show recommendations from Netflix), products are integral to daily living. Product developers then benefit by understanding the ways people incorporate products and brands into their habitual routines and habit loops. Breaking down the formation of habits can not only help understand how daily activities play into lifestyles but also how modifications to the habit loop can influence behavior.

Cues, routines, and rewards are fundamental components of the habitual behavioral and decision-making process (See Figure 9.2.1). Learned associations guide habits by linking certain triggers to specific outcomes (Wood et al., 2005). The steps taken after the initial cue to reach the outcome is the routine. The sensation of the reward further motivates individuals to pursue the same actions again. The gratification of the dopamine release from rewarded behaviors encourages repeated actions which eventually evolves into habit (Berkman, 2018).

Important to marketers and product developers, the routines that occur between the cue and the reward can be manipulated within the habit loop. Multiple responses to a trigger can provide similar (or more optimal) outcomes. If a brand

Table 9.2.1 Examples of heuristic interventions and how they are applicable to the innovation process.

Heuristic Intervention	How it works	Innovation benefit	Example
Default	Presetting options, which take effect if unchanged by the decision-maker	Time and energy are saved in preselecting options, while still allowing the decision maker to adjust preferences as necessary Suggests the consumers unmet need by preselecting the new option and increases product or concept onboarding	If you indicate preferring a cruise, vs. a camping vacation, the displayed options can adjust in real time to suggest the best trip to book
Framing	Emphasizes specific information to highlight positive or negative aspects of a decision	Highlighting benefits as oppose to the less desirable traits to show how it differentiates among competitors	"90% fat free" vs. "10% fat"
Social Norms	Creates an awareness of preferences and judgments of similar demographics	Showcases likeminded consumers (who may have similar unmet needs) implementation of a new product or concept into their lifestyles	Comparing the consumer to buyers with similar interests (i.e. "People who bought X also bought. . .")
Warnings	Informs the consumer of an upcoming change or an alteration to a service or product	Disrupts the consumer habit by reinforcing the new behavior through a visual, educational alert	Having a "Healthy, new recipe!" stamp on the packaged good Push notifications for a streaming app when new music is released
Implementation	Emphasizes future plans to encourage an action	Projections into the future help set the intentions of product adoption	"When is the next time you want to prepare a healthy dinner, but are in a rush?"
Simplification	Removes behavioral burdens or confusion	Showcases the benefits of the new concept without any hassle	Packaging with symbols to indicate how to condense the box after use for recycling
Anchoring	Uses the first exposure to a number to create a reference point, influencing subsequent judgment	Helps to appreciate the value proposition of a new concept or product	Showing the full value for an item crossed out, as well as the reduced price

Adapted from Sunstein (2014), Samson (2019).

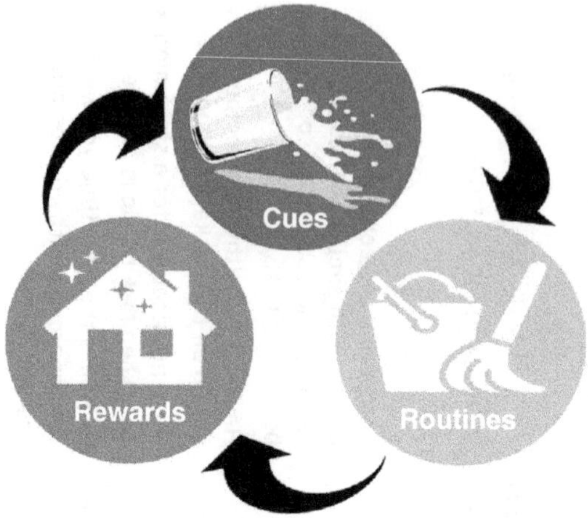

Figure 9.2.1 Habits play a major role in personal lives and create a loop of behavior and decision-making consisting of cues, routines, and rewards. The cue, or the mess, initiates the behavior to use a cleaning product ultimately results in a reward, a cleaner space. Source: Adapted from HCD Research.

outperforms its competition, the consumer can use the same cue to get an equal (or better) reward. Brands must entice the consumer to stray from the familiar for the opportunity to expand to their products by detecting the cue. In short, diagnosing the habit loop provides the opportunity to change or reinforce it and presents marketers and developers with key innovation opportunities.

Habits can be a window into understanding consumer lifestyles and decisions. Brand loyalty is a great example as to how the patterns are enforced when a product not only meets the needs of the consumer but also contributes to a life experience. Depending on how the experience is interpreted by the consumer, it encourages the consumer to continue (or discontinue) the relationship with a brand. The product experience is composed of so many elements that can impact the way a product or a brand is perceived. Using cues, routines and rewards can encourage positive habits and decision-making. Furthermore, detecting cues in the habit loop can aid in achieving higher-order benefits in emotional and physical well-being through indirect suggestions (aka behavioral nudges). Emphasizing certain perceptions through communications or packaging can build up a consumer's reason to believe in the product and encourage its adoption into regular use.

9.2.1.2 Neuroscience in innovation: understanding motivations

Published in 2013, Kahneman's book *Thinking, Fast and Slow* has become dogma to neuromarketers and applied consumer neuroscientists, dividing consumer decision-making into two processes: System 1 and System 2. System 1 being the fast and emotional (largely defined as the unconscious part of decision-making), reactive decision-making process and System 2 being the slow and deliberate, purposeful (and conscious, cognitive) decision-making process. According to the theories, System 1 is the emotional decision-maker that marketing should target for stronger appeal and to ultimately drive consumer purchase.

Individuals navigate through the day using a combination of System 1 and 2 to help make decisions, from what to wear to how to drive. As individuals partake in their daily lives, both conscious and nonconscious sensations occur. While interacting with the environment, information is taken in nonconsciously through the five senses - taste, smell, sight, sound, and touch. In finding meaning and importance to these inputs, individuals become consciously aware of them, deciding how to react both behaviorally and emotionally. For example, when a cell phone rings, the hair cells in the inner ear react to sound vibrations and send neural signals to the brain (Lumpkin et al., 2010). This happens without the person being consciously aware of the sound quite yet (nonconscious). But as the brain receives the signal, it classifies its meaning and value and then deliberates on that information to finally decide, consciously, how to react. The utility of understanding the nonconscious actions (the sound of the cell phone) of this process for marketers and product developers is that through better understanding the nonconscious response, researchers may be able to influence the conscious effort and behaviors (answering the phone call or not). For example, changing the tone, pattern, or length of the ring tone may influence the response or changing the color of a package may influence perceptions of a product.

Neuroscience is, of course, not limited to designing innovative products and communications, but is rather primarily focused on understanding how the bodies respond emotionally and physically to the environment and experiences. Every thought is created by the brain. And yet surprisingly, exactly how the human brain functions is still a mystery. As neuroscientists continue to question and explore the mechanisms of the brain, the more surprises they uncover and the more they find they must appreciate the brain's complexity. This complex biological organ possesses immense computational capability, with 100 billion neurons communicating with 100 trillion connections all within only about 3 lbs., while also constructing our sensory experiences, regulating our thoughts and emotions, and controlling our behaviors (Parent and Carpenter, 1995). Knowing this, it is safe to say neuroscience is a very broad discipline consisting of a combination of all the sciences, from psychology to biochemistry, and pharmacology to physics, all dealing with the structure or function of the nervous system and brain. Each discipline uses different schools of thought, all trying to answer the same question: How do people see, interpret, and behave in the world around them?

9

The addition of neuroscientific methodologies to traditional consumer research has become more and more frequent in the last 10 years, as seen in the increasing number of both scientific papers and approaches developed in private companies. A push has been made to use neuroscience approaches more and more often, which has led to some interesting applications as well as some concerning misuses. The recent popularity has pushed the field to explore more accessible and consumer grade neurotechnologies and found new applications for academic and nonacademic research questions. However, studies are still needed to investigate more deeply the strict links among experience characteristics, hedonics, and emotional responses as well as reliable biomarkers of emotional reactivity.

The more popular neuroscientific methodologies (see overview in Table 9.2.2) include a range of technologies from biometrics or autonomic nervous system measures (heart rate [HR] variability, galvanic skin response [GSR], facial

Table 9.2.2 An overview of tools utilized to measure emotions discussed in this chapter with key studies highlighted.

	Measurement tool	Dimensions of emotion	References
Peripheral nervous system	Galvanic skin response (GSR)	• Physiological arousal	Boucsein, 2012
		• Engagement	Dawson et al., 2007
		• Intensity	Gouizi et al., 2011
		• Excitement	
	Heartrate/Heartrate variability (HR/HRV)	• Motivation	de Geus et al., 2019
		• Emotional arousal	Poels and Dewitte, 2006
		• Attention	Quintana and Heathers, 2014
		• Physical activity	Vescio et al., 2018
		• Mental activity	Williams et al., 2015
	Facial electromyography (fEMG)	• Positive/negative emotional valence	Larsen et al., 2003 Magnée et al., 2007
	Eye tracking	• Visual attention	Rayner, 2009
Central nervous system	Electroencephalogram (EEG)	• Motivation	Beres, 2017
		• Emotional valence	Biasiucci et al., 2019
		• Affective states	Britton et al., 2016
		• Cognitive processes	Liu et al., 2011
		• Long-term recall	Macy, 2015
		• Attention	Nunez and Srinivasan, 2006
	Functional near-infrared spectroscopy (fNIRS)	• Affective states	Balconi et al., 2015
		• Cognitive processes	Gruber et al., 2020
			Jackson and Kennedy, 2013
			Villringer et al., 1993
	Functional magnetic resonance imaging (fMRI)	• Affective states	Gross and Thompson, 2007
		• Cognitive processes	Singleton, 2009

electromyography [fEMG]) to brain imaging (functional magnetic resonance imaging [fMRI]). However, the usefulness and validity of some technologies (fMRI or electroencephalography [EEG]) for consumer research is a hot topic for debate in the field and can often depend on cost and variability in the quality of the technology (Niedziela and Ambroze, 2020).

In a way, industrial research (consumer or market research) has pushed the popularity of applying neuroscientific approaches to understanding emotional and perceptual experiences. The idea behind using applied consumer neuroscience (also referred to as neuromarketing, behavioral science, or System 1- and System 2 thinking) to understand consumer preferences, attitudes, and behaviors is to delve into the nonconscious or noncognitive causes of these factors and that the combined effect of these factors on behavior is mediated by emotional responses. And while academia has always been interested in these ideas, industry has been quicker to push the limits of the neuroscientific methodologies to be cheaper, better, and faster. This has resulted in some leaps of faith, extrapolating data and interpretations further than academia may have intended (as will be discussed later in this chapter). Commercial adoption has also acted as the catalyst for more applied methodologies and technologies to be developed.

9.2.1.3 Psychological sciences in innovation

Measuring emotions and perceptions is one of the most complex, but important, considerations when looking at influences of decision making. Behavioral economics has demonstrated that consumers do not always prioritize saving the maximum amount of money but do try to maximize utility. Learning consumer affective states allows the designers to have a better grasp of a product's utility. This research gives insight into what the consumer believes is worth their money and time. To make a successful innovation, consumer's psychological state and emotional status is critical for determining if consumers enjoy a product experience enough to make the same repeated trade-off (a.k.a. the price). Emotional and perceptual information about consumers should not be neglected since it reveals the value of the stimulus to the consumer in various conditions. Further, this research can also suggest how the value may change based on the modifications of the product or the circumstances in which they are needed.

Every consumer wants to maximize their quality of life. It is the job of the research and development team to investigate what is the best way to analyze thoughts or feelings driving consumer intent. Through understanding these insights into consumer psychology and adapting to meet consumer expectations, current and potential consumers may think more favorably about what is being offered.

9.2.1.3.1 Implicit associations and perceptions

In 1998, Greenwald, McGhee and Schwartz released a paper in the Journal of Personality and Social Psychology introducing the Implicit Association Test (IAT). IAT is a computer-based psychological testing methodology developed and validated in academic research to uncover mental biases and associations between concepts in memory. The goal of IAT is to discover the automatic associations of target and attribute categories (Greenwald et al., 1998). While the traditional IAT is standard in academia, implicit response can be analyzed in many ways. Each method has its benefits and disadvantages, depending on what type of research in question. Two common methods include:

- **GNAT (go-no-go brief IAT):** This method allows a single category to be analyzed rather than a comparison. The participants are instructed to press the spacebar if the descriptor and the concept should be categorized together. The amount of time it takes for a participant to click the button reveals the strength of the association.

- **Affective misattribution procedure (AMP):** This nonreaction time technique relies on participant ratings to determine the implicit measurement. When shown a neutral stimulus, participants are asked to determine which are pleasant or unpleasant. If a neutral stimulus is positively rated, it suggests the preceding concept primed the participant, and therefore, has a positive valence (Goodall, 2011).

Implicit association has a major benefit over more traditional methods in its nature of being so specific. It can assess higher-order emotions or associations more effectively through its focus on each individual attribute being tested. Check all that apply (CATA) is a popular method that is more traditionally utilized to address a lot of descriptors quickly (see Chapter 7.2 for leveraging CATA methodology in decision making). While using CATA, participants simply check all the emotions or attribute words that relate to a concept. Intensity levels do not exist for any of the items - it is either yes or no, making it impossible to differentiate between attributes. By giving a specific value to the strength of the association, implicit measures can more accurately scale the items according to the responses.

Experiences are a large part of what shape associations. Semantic memories are the links between concepts that are kept in the brain (Calvert, 2015). Remember, these connections develop in the brain and change as individuals become more familiar with the concept. Familiarity affects the biases that humans intuitively have, which is why consumer behavioral research is constantly trying to build a strong connection, especially in comparison to a competitor. For example, Gattol et al. (2011) used implicit association to compare three different car competitors: Audi, Ford, and BMW. In this study, six attributes were examined and revealed trends about each comparison, giving insight into brand associations and providing a better sense of how consumers perceive brands.

The IAT methodology can be applied to various brands, logos, packaging, fragrance, taste, etc. Often, these mental biases and associations cannot be articulated since it is inaccessible to conscious awareness. Further, implicit association is extremely advantageous for companies within these markets because it is very challenging to find comparable descriptors for intangible products such as smell. Cross-modal perceptions involve multiple sensory modalities that influence observations. Additionally, implicit association is an easy task to implement, creates large effect sizes, and has strong reliability (Greenwald and Nosek, 2001). Implicit association allows companies to understand the intuitive perception of a brand and consumer behavior based on associations. These insights provide an opportunity to reflect on how a company can best position itself to reinforce or change the target associations, acting as a gut check to brand identity and future growth or a way to assess white-space, or innovation opportunities missing from current marketing or product perceptions.

9.2.1.3.2 Self-Assessment of Emotions (SAM)

Emotional batteries (which are lists of emotional words - happy, sad, irritated, etc. - for the consumer to choose from) are frequently used to measure felt emotions. However, these words may not fully capture what the participants are experiencing and often do not account for variations in intensity of their emotions. Further, explicit lists of words may be influential to those experiences, biasing consumer responses.

The self-assessment manikin (SAM), is a validated measure of psychological, emotional responses to stimuli such as images, videos, colors, sounds, and words. This nonverbal, pictorial measure has proven to be a direct and quick way to gauge consumer emotional reactions without introducing bias while also accounting for intensity (Morris et al., 2002). Having the freedom and flexibility to explore so many mediums make SAM an ideal methodology for assessing noncognitive emotional responses. SAM applies both valence (positive to negative emotion) and arousal (mild to excitement) to common dimensions that describe human reactions. Developed by Bradley and Lang (1994), SAM utilizes graphic scales based on the Pleasure-Arousal-Dominance (PAD) theory of affective response. According to the PAD theory, all emotional states can be summarized through a combination of the three basic emotions within the PAD acronym (Morris et al., 2002). Multidimensional mood maps, as seen in Figure 9.2.2, are used to graphically display multiple descriptors of the three dimensions employed by PAD. This visual map provides insight into how the

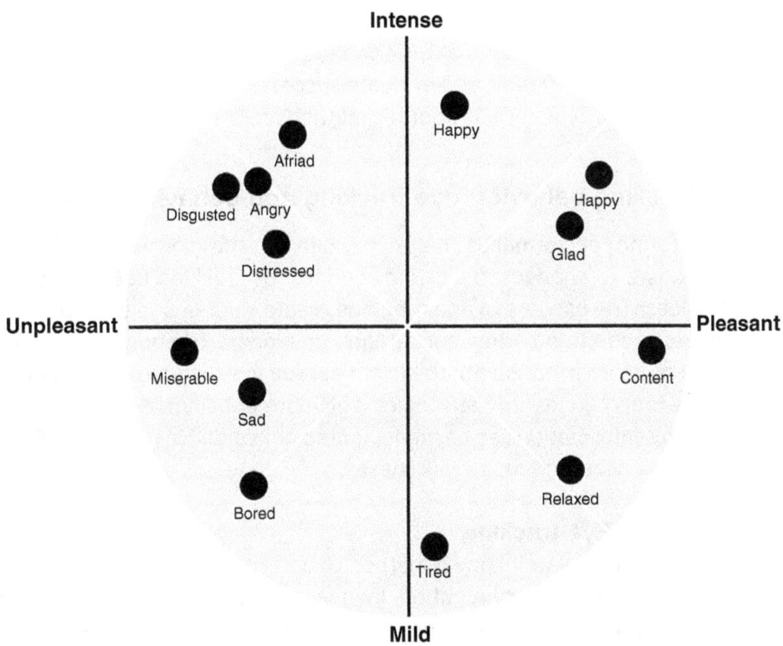

Figure 9.2.2 Above is an emotional mood map based on PAD. For example, when a participant feels aroused and pleasant, it could be described as "glad," "happy," or "astonished." A third (not pictured) dimension of dominance helps to further describe and differentiate emotional states. For example, feeling aroused and unpleasant could be described as feeling fear or anger. The third dimension of dominance (or motivation) reveals that the participant feels more dominant or motivated to act when angry, while less in control or motivated when afraid. Source: Adapted from HCD Research.

9

three dimensions can be used to represent the entire spectrum of emotional experience.

When using SAM, three rows with five different images illustrate the levels of an emotion (Bradley and Lang, 1994). Underneath the pictures are a point scale (typically nine points) for participants to respond where his or her emotion would best align. For example, pleasure in the SAM model ranges from a smiling face to a frowning face. The images for arousal go from a picture of a body being very awake to asleep. Dominance grows from a small to a large figure to represent control. The SAM images have been proven to be scientifically valid in conveying various types of emotional behavior.

The simplicity of SAM makes it very easy for participants to use - participants typically complete ratings in less than 15 seconds (Morris et al., 2002). Working with such a quick time frame allows for numerous stimuli to be tested and less participant fatigue. The fast pace of responses keeps the participant attentive, which is another major advantage. This proves to be a strong, noncognitive approach suitable for reporting on intuitive emotions.

Additionally, SAM makes research more inclusive for non-English speakers, children, and participants with a language disorder or clinical syndromes. Excluding words also allows SAM to be applicable regardless of cultural differences. The widespread use of SAM is valuable when trying to understand a global or diverse market for introducing a product or communication. Additionally, applications of SAM can provide insights into a product's positioning, a creative strategy, and evaluate success of brand harmony across multiple markets (see Chapter 9.1 for other methods for emotion research).

9.2.1.4 Tracking behavior: eye tracking and behavioral coding

9

Monitoring performance helps to identify the consumers' behaviors and responses to changes in existing routines. By tracking behavior, the designers can learn the barriers to adoption and create solutions to overcome these roadblocks. Addressing what encourages or hinders traction gives the designers even more information about how to approach satisfying unmet needs. Further, these tools can identify strategies which are enticing or distasteful to consumers. This information can be implemented or avoided in future designs to sustain use and encourage repeat purchases.

9.2.1.4.1 Eye tracking

Eye tracking is a well-known method to determine an objective measure of visual attention. Visual perception, language processing, reading, marketing, and cognitive development are just a few fields that take advantage of employing eye tracking within their research. Common eye tracking metrics include gaze points, fixations and saccades. Gaze points are simply one raw image sample captured by the eye tracker. When multiple gaze points are close in time and location, it creates a fixation (Rayner, 2009). Fixations are brief pauses from scanning a scene to gather a more comprehensive idea of what is being viewed. The eye remains relatively still, spending time in one location to understand the visual world. Most fixations last between 50 and 600 ms, but the amount of time spent on a specific fixation is dependent on both the task and stimulus

(Matos, 2015). Fixation points are likely to involve meaningful information that is processed; however, that is not necessarily true for every fixation. It is possible that fixations occur while shifting attention. Yet, determining the pattern of visual attention or the inhibition of the future search in visual space can be accomplished with fixations, proving itself to be a meaningful metric.

Eye tracking data allows researchers to dive deeply into evaluating the stimuli. One common data visualization is a heatmap. The colorful output shows the distribution of gaze points and fixations, displaying the areas of highest visual attention highlighted in certain colors. Red implies that there is a high number of gaze points, followed by diminishing visual attention in yellow and green. A lack of color suggests little to no visual attention. For an example of eye tracking heatmaps, please see the Love at First Sight case study# 3 in this chapter. Other common metrics in eye tracking include fixation sequence, time to first fixation (TTFF), and areas of interest (AOI). A fixation sequence notes where and how long a participant focuses on a specific area. Fixation sequences are very helpful in determining salient elements, and time to first fixation is the amount of time it takes for a participant to look at a specific area of interest. An AOI is a subregion that can be subjective based on what the researchers are trying to uncover (Foster, 2019). These areas typically are of high importance, thus providing information on a specific region in comparison to the overall stimulus. For example, if an AOI is created on a label, the software can track various metrics about that specific area such as TTFF, time spent on an AOI or revisits to a given area.

Eye tracking has proven itself to be an extremely useful experimental technique to gain insight into covert human reactions. While eye tracking can produce a substantial amount of data about human perception, it is important to know how to analyze it to best explain a study. Using it in conjunction with other biometric features, such as skin conductance or facial EMG, or traditional surveys can help create a comprehensive understanding about where and how the participants react. By understanding what is grabbing a participant's focus, as well as what is distracting, can help designers implement certain features (or nudges) to not only attract the consumer to the product but also to create visuals to encourage sustained engagement with it.

9.2.1.4.2 Behavior coding

Behavioral coding is a simple, noninvasive, observational research method used to quantify a participant's actions within an environment or with a particular stimulus. This tool is used in research to identify and observe notable behaviors that then are recorded (a.k.a. coded) to track trends and patterns in behavior. Behavioral coding can provide an indication of the participant's experience or demonstrate how circumstances dictate certain responses (Chorney et al., 2014). Behavioral coding can be useful when trying to understand the user experience without disrupting it.

To ensure good and actionable data, the research objectives should be defined to determine what behaviors are of most interest. Certain behaviors, or lack thereof, are categorized based on predetermined sets of rules. Similarly, multiple behaviors can occur simultaneously, and the coder must be prepared to record all information as well as provide accurate start and stop times for each

Table 9.2.3 Common behavioral coding strategies defined with an example.

Behavioral code	What it examines	Example
Specific interactions	The amount or duration of time a behavior occurs	How frequently a participant's mouse hovers on a logo
Binary choice	The result of a either a yes or no decision	If the participant uses a phone case during the session
Group comparisons	The nature of systematic behaviors	The percentage of male to female coworkers on financial projects

behavior. This is often accomplished by using coding software and multiple coders. The duration of the observation is intentional (for example, from entering a room until leaving or from the start of a meal to the end), often based on a stimulus, clinical restrictions, past research, or a combination of these options (Chorney et al., 2014). Observations can be conducted via an audio/video recording or live. Regardless of which form of data collection is chosen, research with behavioral coding should include a review period to ensure that the information is correctly gathered. By analyzing and reporting the participant's responses, a visual output of sequential behaviors can be developed to identify patterns and motivations.

When specific labels represent behaviors, it is known as a behavioral code. Coding has multiple types of dimensions, ranging from a general context to specific behaviors of interest. They can be physical codes, such as squinting, or social codes, such as sensitivity to a topic (Chorney et al., 2014). The codes are dependent on context; therefore, the best code to capture the target behavior is subjective. Table 9.2.3 provides an idea of some of the most frequently utilized codes.

When a participant is being observed for behavioral coding, it is crucial that the researcher does not intervene or influence the behaviors in any way to ensure the validity of the results. The information gathered in behavioral coding gives insight not only into the final reaction but also a timeline of the experience in its entirety. Behavioral frequencies, changes in behavior over time, or changes in behavior when experiencing different exposures can all be used to see how behaviors may relate to one another sequentially.

Behavioral coding is ideal for market and consumer research. With behavioral coding, researchers can quantify behavior without interrupting the experience to gain a strong understanding of the interactions taking place. Observing the consumer interacting with a product or concept can demonstrate pitfalls or advantages of a product experience, thus providing a springboard of ideas and modifications to continue to improve the consumer journey (read Chapter 7.1 for using network analysis on behavioral coding data).

9.2.1.5 Bringing it all together: integrating methodologies

Qualitative and quantitative market research methodologies, such as interviews and surveys, are the backbone of strategic business strategy. Getting the voice of the consumer is one of the key factors used in maintaining competitiveness against rival brands. Research does not have a single approach that works for every question. By combining methods that integrate well, a global perspective of the consumer's reaction is obtained. Objectively discovering trends in human behavior gives us the ability to reliably predict if something will reoccur. Having a window into an unfiltered, unbiased perspective gives so much value to understanding human behavior.

Undoubtedly, it is still important to ask people what they think and how they feel and not rely solely on physiological or neuroscientific measures. At the very least, this is important for construct validity, but it is also important to consider that any one measure, whether from neuroscience or self-report, is not capable of assessing all aspects of any emotional experience. To get a more complete representation of the experience, it is necessary to measure from multiple angles. Conscious and nonconscious measures aim at providing very different perspectives and answers (or should if done correctly). The approaches should not be at odds with one another, but rather they should be providing added and synergistic information to help companies make better business decisions. If applied neuroscience measures simply repeated results from explicit testing, then it would not be worth doing. The comprehensive results should help tell a more complete story of the experience. Understanding emotion through using a combination of qualitative and quantitative research with applied neuroscience provides the possibility to gain a deeper insight of the drivers of behavior and emotional response.

Research does not have a single approach that works for every question. By combining methods that integrate well, a global perspective of the consumer's reaction is obtained. Objectively discovering trends in human behavior gives us the ability to reliably predict if something will reoccur. Having a window into an unfiltered, unbiased perspective gives so much value to understanding human behavior.

9.2.2 Applying novel methods to innovation: choosing the right tool

With the addition of behavioral and neuro sciences, innovators have quite a few options to choose from the toolbox. Psychological and neuroscience measures can range from fMRI to psychological questionnaires. Ultimately, no one tool will cover all research possibilities. Practitioners must be willing to accept that certain techniques may be better at collecting specific types of information over other tools. It is perfectly reasonable to use any one of these measures if the practitioner is clear on all the potential limitations *and* is sure to use the tools properly.

The process of vetting which tools to use is complicated. While it is important to make sure the tools in question are validated through scientific research and peer reviews, it is even more critical to make sure it is useful for the specific research. Investigate existing evidence about the tool's claims and contexts where the tool is most beneficial. Then, reflect on the research question and make sure each tool is adding information of value. The research should be customized to the question, not the other way around. The research program should cater to the whole consumer experience, with considerations toward context. The stimuli being tested is an opportunity to communicate with the consumer and evaluate how they respond. Impressions of the manipulated environment will expose how consumers interpret the changes but may also provide feedback useful for future innovations. By analyzing the mechanisms behind behavior through the proper tools, innovators can find actionable benefits from the research to expand on the next (small improvement or) big idea. The following case studies demonstrate the various ways that some of these tools can be applied to impact innovation and help give consumers a reason to believe in the product, packaging, concept, or communication.

9.2.3 Case studies using behavioral science and applied consumer neuroscience

Case Study #1 Exceeding or meeting expectations – using implicit testing

Established products have a certain profile, encompassing a specific look, sound, feel, smell, and taste. A distinctive product design helps to differentiate from competitors, thus adding additional value to consumers. These features add brand personality in addition to communicating and reinforcing messages about the product experience. Products and messaging must stay relevant and aligned to keep consumer interest and when making product changes there can be a real risk to consumer satisfaction if expectations are not met. To keep up with new technology, shifts in regulation, or product trends, companies reposition and reformulate often. Alterations to a product can lead to a mismatch between expectations and product experience, influencing product acceptability, and ultimately liking of the product (Cardello, 2007). By defining consumer expectations and perceptions of the product or brand category, this research ensures the renovated stimulus does not alienate the consumer, but rather meets their needs.

The problem

Leveraging the expectations of the consumer can be challenging, especially when a component of a product is being changed. Although sometimes unavoidable, these changes force a company to pivot and revisit the emotional reactions consumers have toward a new concept. Companies strive to adjust in such a way that preserves or enhances the unique associations and qualities of the product or the brand in question. Measuring the fundamental characteristics of a concept's identity via implicit testing shares not only areas of unmet need (need-gaps) but also validates where the expectation and performance is congruent.

Consumers also may struggle to articulate innate features of a product being reviewed. Established positive associations must be sustained or exceeded when shifting product design, or else undesirable associations may replace existing beliefs and brand reputation may be damaged. Learning the perspective of the consumer gives insight into the profile which must be maintained to successfully

Case Study #1 (*Continued*)

adjust to the new circumstances. The implicit approach provides a clear example of redesigning a product to continue to fit the brand personality.

The solution

Implicit (IAT) measurement of consumer perceptions combined with traditional surveys aids the development and innovation of the product to ensure the signature profile is satisfied.

The case study

Administering an implicit association test provides an unbiased evaluation regarding how well the product is meeting consumer expectations. Detecting the strength (speed of response) and certainty (percent of consumers who agree) of consumer's automatic associations between determined descriptors and the product helps reveal differentiating factors that need to be upheld or improved. When a natural disaster forced a premium candy bar company to source their cocoa beans from its original site to a new location, the implicit test paired with self-reported questionnaire proved to be valuable tools in differentiating the nuance modifications in the taste of the candy bar. Additionally, the responses share insights into perceptions, expectations, and associations the consumer has regarding the product experience of the candy bar.

The results revealed the mental associations across descriptors for the prototypes: Prototype B had the most similar associations to the original candy bar. Descriptors ranking in the 80th percentile or higher in a perceptual area are noted as being highly associated (See Figure 9.2.3; Figure 9.2.4).

Current Product Associations	High	Medium	Low
Original	Luxurious taste Delicious Indulgent	Sweet Satisfying	Healthy Artificial
Prototype A	Artificial Sweet	Delicious Satisfying Luxurious taste	Healthy Indulgent
Prototype B	Luxurious taste Sweet Indulgent	Artificial Satisfying	Healthy Delicious

Figure 9.2.3 The implicit scores for Original, Prototype A, and Prototype B broken down into high, medium, and low associations. Source: Adapted from HCD Research (2021).

9

Case Study #1 (*Continued*)

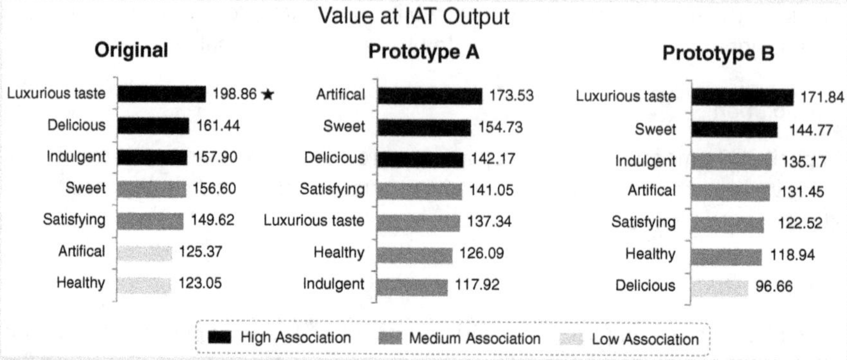

Figure 9.2.4 The values of the implicit test for Original, Prototype A, and Prototype B broken down into high, medium, and low associations in descending order. Source: Adapted from HCD Research (2021).

The Original and Prototype B samples shared high associations of "Luxurious taste" and "Indulgent." Prototype A was the only sample evaluated to have a low, rather than a high, association with "Indulgent." Prototype A also had the least number of high associations (1) compared to the Original and Prototype B (2). The Original was differentiated from the prototypes by being the only sample highly associated with "Delicious." Unlike the Original, both prototypes were highly associated with "Sweet." The information from this output provided an understanding of the consumer landscape regarding where the candy bars were lacking and succeeding.

The timed reaction scores allowed for statistical differences to be interpreted, exposing what drives interest as well as ways to leverage it to improve aspects of the prototype to better suit expectations (See Figures 9.2.5 and 9.2.6).

Self-reported surveys provide context into the drive for liking, revealing that consumers rated the sample of the original significantly more delicious than the sample from Prototype A. This information combined with the implicit testing also shares the brand cohesion, developing an understanding of the extent to which the new cocoa can be utilized.

Implicit testing and self-reported measures present the associations and perceptions which are either successfully or unsuccessfully communicated to the consumer. When reviewing the dataset in its entirety, the results suggest Prototype B is a better fit because it meets the signature expectations and matching brand personality. The circumstances surrounding the necessary innovation provided an opportunity to reflect on the product's signature taste, as well as ways to improve the new product to maximize the opportunity for a seamless transition with the new cocoa bean distributor.

Case Study #1 *(Continued)*

Mean Statistical Result Compared to the Original	Prototype A	Prototype B
Sample Liking *(1: Dislike Extremely; 9: Like Extremely)*	=	=
Sample Quality *(1: Very Poor Quality; 7: Very Premium Quality)*	=	=
Sweetness *(1: Not at all Sweet, 7: Extremely Sweet)*	=	=
Deliciousness *(1: Not at all Delicious, 7: Extremely Delicious)*	<	=
Cocoa Flavor *(1: Too Much Cocoa Flavor, 7: Not Enough Cocoa Flavor)*	=	=
Future Interest *(1: Not at all Interested: 7: Extremely Interested)*	=	=

*Note "<" symbol represents a significantly lower result compared to the Original, whereas "=" represents no significant difference.

Figure 9.2.5 The statistical results comparing the Original sample to Prototype A and Prototype B. Source: Adapted from HCD Research (2021).

	Original	Prototype A	Prototype B
Sample Liking *(1: Dislike Extremely; 9: Like Extremely)*	8.7	8.4	8.6
Sample Quality *(1: Very Poor Quality; 7: Very Premium Quality)*	6.1	5.7	6.1
Sweetness *(1: Not at all Sweet, 7: Extremely Sweet)*	5.8	6.0	5.7
Deliciousness *(1: Not at all Delicious, 7: Extremely Delicious)*	6.4	5.4	6.5
*Cocoa Flavor *(1: Too Much Cocoa Flavor, 7: Not Enough Cocoa Flavor)*	5.1	5.1	5.1
Future Interest *(1: Not at all Interested; 7: Extremely Interested)*	6.1	6.0	6.3

*Note statistical testing is shown based on which sample is higher; however, the ideal mean for this question is 4.

Figure 9.2.6 The mean self-reported survey results summary for Original, Prototype A, and Prototype B. Source: Adapted from HCD Research (2021).

Case Study #2 Habit loops of a dirty diaper – combining implicit and explicit methods

Habits beget lifestyle; therefore, developing a product intended to fit within an established routine and encourages consumer loyalty and repurchases. The cues, routines, and rewards guide consumers in making decisions. By exploring these repeated performances, product and development teams can learn how to best encourage consumers to engage with the product. Learning both the consumer language and associations can help understand the impressions formed in the brain. Furthermore, investigating human behaviors surrounding a product experience gives insight into how the product can be more seamlessly integrated into established routines. Measuring explicit and implicit responses allows companies to innovate around the whole experience, in addition to identifying and addressing any penetration barriers for product adoption. Incorporating the preferences and perceptions of the consumer into the innovation process will help develop designs lacking in the consumer space which meets consumer ideals.

The problem

The design and optimization process of products involve a deep dive into understanding the consumer experience. Traditional survey measures such as Check All That Apply, open end responses, and self-reported scales have typically been used to try to assess these experiences. However, it is challenging to fully grasp the unmet needs of the consumer, especially when they may struggle to express or conceptualize what they want, as well as how it would best integrate into their lifestyles. Measuring explicit preferences with implicit perceptions can help to provide context about the improvements and innovations necessary to ensure an effortless product adoption.

The solution

Using the synergistic combination of MaxDiff and implicit testing creates a multiphase research design ensuring the product meets action standards of company ideals, as well as the needs of the consumer. Furthermore, by learning the habit loops of target consumers, innovation strategies can cater to the motives of buying behavior.

The case study

Targeting the consumer to understand expectations is crucial. Qualitative exploration, past research, or social listening can provide an understanding of consumer needs, perceptions, and expectations. Using the information gleaned from qualitative research helps pair the consumer lexicon with research targeting important priorities and perceptions. The awareness of unmet needs in the market allows companies to modify the product accordingly. For example, a baby

Case Study #2 *(Continued)*

diaper company is looking to understand not only what inconveniences to avoid but also the ideals sought after by parents. Using MaxDiff, a forced choice exercise to determine what descriptor is most and least important, the company can prioritize the needs involved in a diaper. The trade-offs embedded in the survey design provide context into consumer habits, exposing the expectations of product performance. The key descriptors developed from the MaxDiff are then utilized in an implicit test to expose perceptions and reveal how well the different brands meet the expectations of the consumer. The combination of MaxDiff and implicit share how the current competitive landscape satisfies or neglects drivers of purchase.

As seen in Figure 9.2.7, the top ranked features (left) show the top priorities involved in product experience, with the implicit responses (right) providing the perceptions of different companies. Company C is not highly associated with any of the top-ranked attributes. Natural, affordable, safe, premium, reliable, and soft got strong responses (>60%) for Company C, suggesting an opportunity to strengthen those associations. The features are categorized into cues, routines, and rewards to help in identifying exactly where the feature is within the habit loop to best design and market the new product to fit unmet needs.

Perceptions of Company C match reasons for choosing Company C on the following descriptors: overnight, block odor, easy use, and (somewhat) premium. Learning where the uncertainty may lie in these perceptions helps to identify attainable opportunities for nudges to drive behavioral and habitual

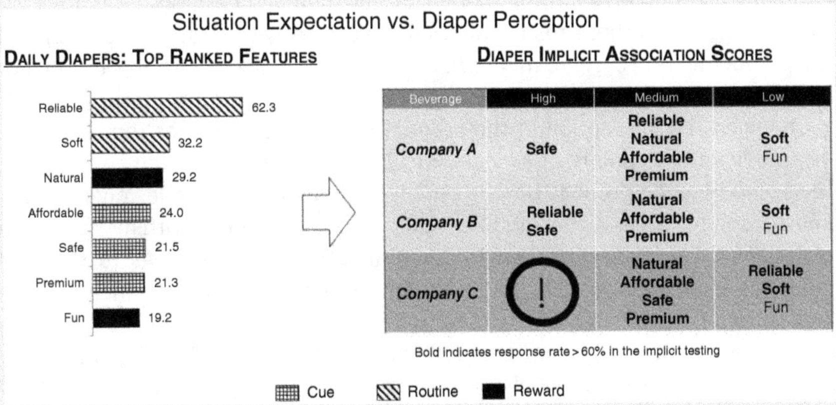

Figure 9.2.7 The MaxDiff results (left) show the top-ranked priorities consumers have toward diaper purchases. The cues, routines, and rewards of the product experience are depicted in the colors associated with the ranked feature. The implicit scores (right) for Company A, Company B, and Company C are broken down into high, medium, and low associations. Source: Adapted from HCD Research (2021).

Case Study #2 (Continued)

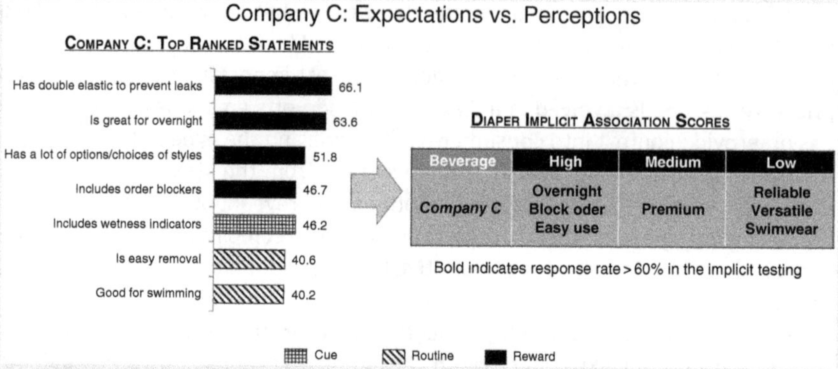

Figure 9.2.8 The MaxDiff results (left) show the top-ranked priorities consumers have specifically towards Company C. The cues, routines, and rewards of the product experience are depicted in the colors associated with the ranked feature. The implicit scores (right) for Company C are broken down into high, medium, and low associations. Source: Adapted from HCD Research (2021).

change. For example, as seen in Figure 9.2.8, Company C is not currently seen as reliable, versatile, or swimwear, yet these perceptions did have a high response (>60%). Consumers' responses around these product category descriptors suggest an opportunity to strengthen those associations. Using the information from the MaxDiff and implicit testing helps to establish how Company C is perceived in the market by consumers. Learning the motives of the purchase from these results allowed the company to strategize throughout the product testing cycle to minimize the risk of overlooking the new product or disappointment in its ability.

Changing the perception of the product is multifaceted since the product is layered in various points of association. To keep consistency and congruency, reviewing how the product flows through the consumer lifestyles give context and reasoning behind the modifications or developments within the design experience. The behavioral, environmental, and product nudges further the innovation opportunities to be used for evaluating product concepts (see Figure 9.2.9).

Considering the lifestyle and unmet needs of the consumer, the diaper industry would benefit by having a swimwear product that is hypoallergenic, safe, and effective. Common opportunities for Company C included enhancing the descriptors: reliable, premium, affordable, safe, natural, and soft. Furthermore, natural and reliable rewards stand out as key innovation opportunities to satisfy the unmet need in the competitive marketspace. Company C has a clear product prospect with the potential to distinguish themselves from competitors, while also increasing consumer satisfaction, connection, and brand equity.

Case Study #2 (*Continued*)

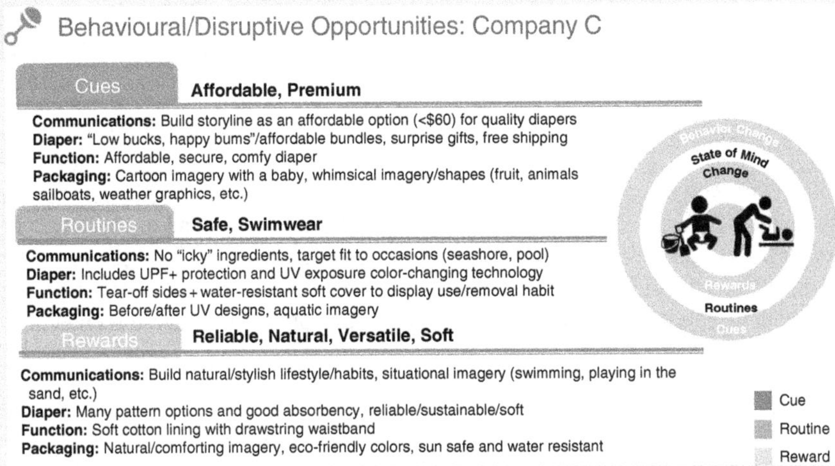

Figure 9.2.9 The cues, routines, and rewards of the product experience are depicted above with the corresponding colors. Combining the information provided from consumers' priorities and perceptions allowed for the design of behavioral and disruptive opportunities above. The breakdown includes communication, function, and packaging strategies, as well as the diaper purpose to give consumers a reason to believe in the product. Source: Adapted from HCD Research (2021).

Case Study #3 Love at first sight – eye tracking for innovation

A physical product has many notable factors directing consumer attention and driving purchase. Product design and presentation through packaging communicates brand identity through conceptual and sensory information (Rettie and Brewer, 2000). The expectations created by the sensory and branding aspects can impact the valuation of quality, price, and ultimately consumer satisfaction over time (Silayoi and Speece, 2007). Expectations can elicit specific emotions, such as pleasant surprise (when the actual experience is better than the expected one) or disappointment (when the actual experience is worse than the expected one). Further, the product experience may elicit emotions unsubstantiated by the brand or package messaging, thus altering the product performance. For example, if packaging includes recyclable or sustainable messaging, yet as a product provokes an association of single-use or wastefulness, it may result in confusion or disappointment. The package provides a major opportunity to communicate with consumers. However, for the product experience to even occur, it must differentiate itself among its competitors at the store. Thoughtful product and packaging design is effective at both increasing findability and conveying brand identity

9

Case Study #3 (Continued)

when the emotions elicited are congruent with the intended experience (Krishna, 2012). Through an enticing package, brands and companies can establish truly disruptive innovation, taking the product from being on the shelf to in the cart.

The problem

Interfering with the consumer experience at the shelf by asking questions or having consumers take surveys can be disruptive and distracting, making it difficult if not impossible to assess the naturalistic consumer experience at shelf. Passive measures, such as those from behavioral science, combined with a postexposure survey can help researchers analyze the experience from start to finish uninterrupted.

The solution

A combination of passive methodologies, such as eye tracking and behavioral coding, with a postexposure survey can be used to unobtrusively dive into the consumer experience.

The case study

To gain insight into the findability and effectiveness (stand-out) of bottle displays, the consumer experience was analyzed with behavioral eye-tracking measures. Evaluating how a shelf display is integrated into the aisle journey, behavioral research provides information about packaging impacts on the shopper experience and influence on shopping behavior. Eye-tracking captures where visual attention is drawn initially as well as the following sequence of fixations. Behavioral eye gaze coding quantifies the amount of time spent looking at certain items and reviews how viewers search for information, giving a timeline of the experience and sharing covert consumer behavior.

After being set-up with the eye-tracking glasses, shoppers were instructed to shop for a high-end beverage and choose one to purchase (see Figure 9.2.10).

Once this task was completed, shoppers were asked to browse the shelf like they normally would and then asked which three brands stood out (see Figure 9.2.11). Following this as seen in Figure 9.2.12, shoppers were asked to identify the Bottle X being evaluated for this study. Finally, a postexposure survey was administered asking questions about the overall shopping experience: what stood out, likes, dislikes, uniqueness, and ease of finding Bottle X. The two conditions in the experiment were for "New" and "Current" designs. The New condition included the prototype design being considered and the Current condition was the bottle design currently used in-market.

When reviewing and comparing datasets in entirety, overall results from this research indicated Bottle X stood-out because it seemed unique; however, the three main competitors also stood out due to brand familiarity. Bottle X was rarely seen early-on in the shopping experience and struggled initially capturing shoppers' gaze. When it was seen, Bottle X generally retained visual attention. This information suggested that the perceptions of Brand X were driven by color and shape. The bottle design caught shoppers' attention regardless of the condition but lacked stand-out when compared to competitors.

Case Study #3 (Continued)

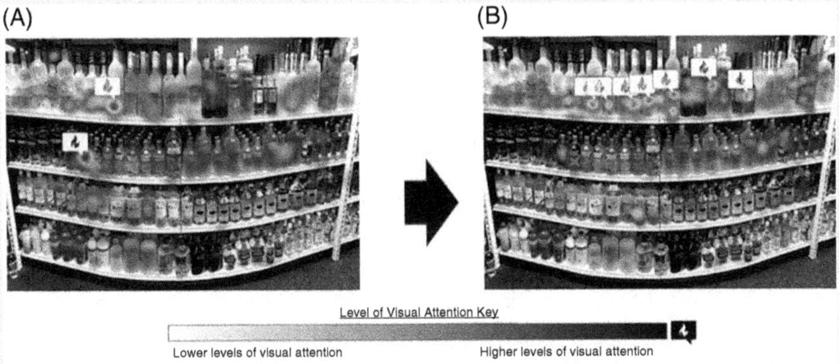

Figure 9.2.10 Shoppers could naturally shop as they would for high-end beverages. This heat map is representative of what captured the most visual attention during the first 5 seconds (A) and 20 seconds (B) of their shopping experience for the Current condition. The deeper the color (red) indicates more attention. Source: Adapted from HCD Research (2021).

Stand-Out Interview Results

Current 750	Bottle X	Competitor A	Competitor B	Competitor C	New 750	Bottle X	Competitor A	Competitor B	Competitor C
1st Mention	**3%**	**20%**	**10%**	**10%**	**1st Mention**	**14%**	**31%**	**21%**	**7%**
2nd Mention	13%	33%	3%	7%	2nd Mention	10%	21%	3%	24%
3rd Mention	3%	3%	10%	10%	3rd Mention	3%	17%	14%	0%
All Mention	**20%**	**57%**	**23%**	**27%**	**All Mention**	**28%**	**69%**	**38%**	**31%**

Figure 9.2.11 This is an example of the interview results recorded after shoppers naturally shopped the aisle, shoppers were asked what stood out the most, second most, and third most. Overall, Brand A stood out most amongst all four brands; however, the New Bottle X stood out slightly more than the Current Bottle X. Source: Adapted from HCD Research (2021).

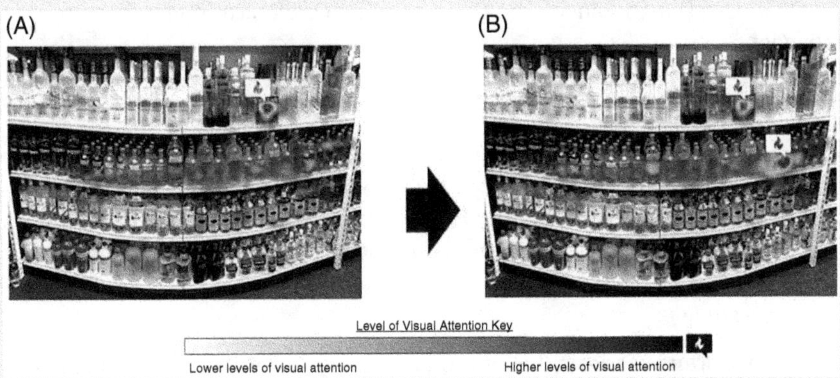

Figure 9.2.12 Shoppers were specifically tasked to find Bottle X. This heat map is representative of what captured the most visual attention during the first 5 seconds (A) and 20 seconds (B) of their shopping experience for the Current condition. The deeper the color (red) indicates more attention. Source: Adapted from HCD Research (2021).

Case Study #4 Sensing the change – physiology to prove innovation claims

It is crucial during innovation to reflect on the intended takeaway experience and revisit that goal frequently throughout development. Each component of the product plays a role in the overall experience; therefore, any changes involve a risk of imbalance. While innovating, the adjustments must be reviewed to determine if the alterations hinder or enhance the overall message.

To ensure the product is meeting the intended claims, the message must be backed and proven by consumer research. Claim substantiation research must be carefully crafted to ensure findings are not overstated. If claims are questionable, groups (or competitors) can challenge the claim and request the research supporting it. Researchers must provide objective data which validates the claims. Depending on the claim being made, certain tools are preferable for providing more insights. For example, if the claim suggests an external consumer response (i.e. "visibly less wrinkles"), the use of certain tools is needed (skin images characterizing wrinkle depth). Utilizing the right tools to gain an objective, honest measure provides feedback about not only the claim but also ensuring the new product is correctly targeting the innovation goals.

The problem

Assessing consumer experience of products is often done through surveys, interviews, and open-ended questions after consumers have experienced the product. While the qualitative measure is certainly important to assess, it misses the passive, undisrupted product experience that is full of reactions evoked from specific stimuli, such as feeling the consistency or smelling the fragrance. Gaining information from the unobtrusive moment-by-moment product experience provides a level of detail impossible with surveys. In addition, surveys, interviews, and open-ended responses are also inherently biased to the perspective of the consumer. When trying to comprehend the full effect of the innovation, the consumer may struggle to explicitly acknowledge the minute product changes. Furthermore, qualitative research cannot show the psychophysiological effect in innovation changes.

The solution

Measurements within psychophysiology can be used to dive into the consumer experience, while providing unbiased and involuntary responses. These measures can be used as proof of concept or for claims substantiation, especially for product features that may be difficult for consumers to articulate clearly.

The case study

A skincare brand carries a line of products using Essential Oil Mix A (EOMA) in seeking a sensate composition to provide a fresh, cool feeling. The company would like to promote the line with predetermined claims about the aroma

Case Study #4 (*Continued*)

"sharpening focus and improving concentration." The research requires clarification and substantiation on the effect of olfactory stimulation by EOMA compared to a control. The research design includes the following physiological measures to approach the objective: HR, heart rate variability (HRV), and GSR.

The HR response is used as a measure of focused attention. As HR decreases, focused attention increases (Lang et al., 1999). The HR significantly decreases from baseline; therefore, implying significant increases in cognitive attention occur during the first half of the exposure to the EOMA essential oil mix (see Figure 9.2.13). The means also demonstrates EOMA exposure increasing focused attention as the duration of the experience continues. Furthermore, the control measure had no significant changes from baseline.

When evaluating the output of this psychophysiological measure, arousal increases as GSR increases (Braithwaite et al., 2013). As seen in the data visual on Figure 9.2.14, the first half of the exposure to EOMA essential oil mix resulted in a significant increase in arousal. Furthermore, significant increases also were seen in arousal compared to control during the first half of the exposure to the EOMA. Unlike the EOMA, no significant changes from baseline were noted for the control group. A significant difference from the control was also determined from the 15 seconds arousal mean.

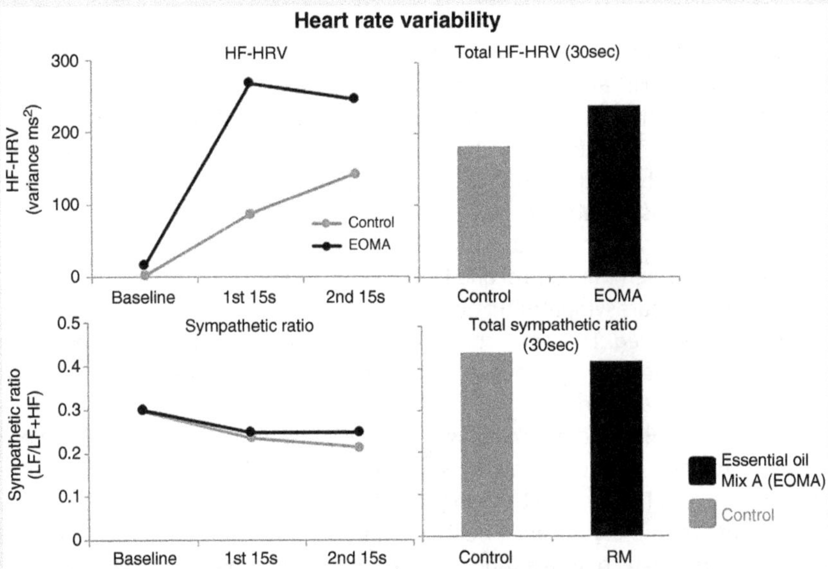

Figure 9.2.13 The heart rate response measures the changes in heart rate from baseline measurements over 30 seconds (top) for both the control and EOMA. Additionally, the heart rate means of 30 seconds (bottom, left) and 15 seconds (bottom, right) were graphed. Source: Adapted from HCD Research (2021).

9

Case Study #4 (*Continued*)

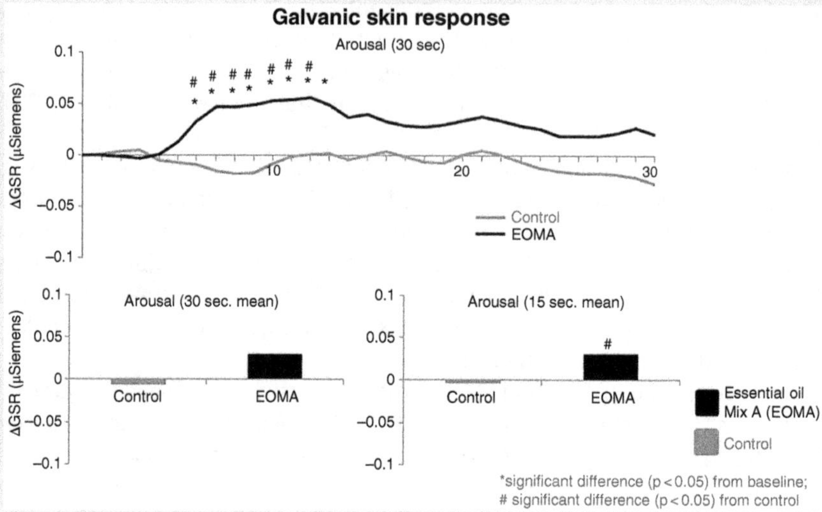

Figure 9.2.14 Galvanic skin response (GSR) measures skin conductance from a baseline measurement over 30 seconds (top) for both the control and EOMA. Additionally, the GSR means of 30 seconds (bottom, left) and 15 seconds (bottom, right) were graphed for both control and EOMA. Source: Adapted from HCD Research (2021).

HRV is the change in time between heart beats. The fluctuations in the beat-to-beat intervals of HR are an indicator for cognitive performance. An increase in HRV (or high-frequency HRV) is associated with the parasympathetic nervous system activities. Therefore, high HRV is an indicator for an increase in cognition and relaxation, while a decline in HRV (or low-frequency HRV) reflects both sympathetic and parasympathetic nervous activities (Appelhans and Luecken, 2006). From the information provided in the data shown on Figure 9.2.15, no differences were noted between the control and EOMA sympathetic ratios. The lack of differences indicates parasympathetic and not sympathetic activity were driving the differences observed. This confirms and supports the findings from HR and GSR, suggesting an increase in focused attention and arousal as opposed to some other physiological change.

While no significant differences were found between the control and EOMA exposure for HRV measures, the HF-HRV was higher during EOMA exposure than control (not significant), most notably during the first 15 seconds corresponding to (and confirming) significant effects seen in HR and GSR measures. This indicates an increase in cognitive function during this time point.

Considering the overall findings of this research, effects of the EOMA fragrance were provided to the skincare brand. In conclusion, the EOMA fragrance increased focused attention, which was determined by the significant decrease in HR and significant increase in arousal via GSR. The parasympathetic response

Case Study #4 *(Continued)*

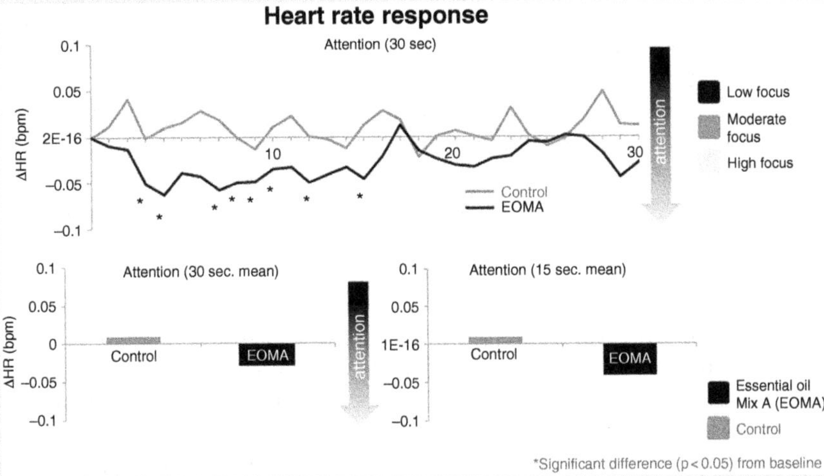

Figure 9.2.15 Heart rate variability (HRV) measures cognitive performance of respondents while exposed to the control and EOMA. The HRV (top, left) and sympathetic ratio (bottom, left) is segmented into baseline measurement, followed by two 15 seconds exposures. The overall High Frequency (HF)-HRV (top, right) for 30 seconds and the total sympathetic ration (bottom, right) for 30 seconds were graphed to compare the control and EOMA. Source: Adapted from HCD Research (2021).

determined by the HF-HRV activity further confirmed the response to the fragrance. With this research, the claim regarding sharpened focus and promoted concentration is validated. The specific use of EOMA as a technological innovation caused the desired effect, further promoting the intended product experience for the consumer. Using the sensate composition in EOMA fragrance, it imparted desirable characteristics the brand was trying to achieve. The addition of the EOMA fragrance elevates the product experience to encompass the brand messaging fully and harmoniously.

9.2.4 Conclusions: conceptual framework for behavior-led Innovation

The avenues for application of neuroscience and behavioral science in business are countless. By starting with the research question, reflecting on insights needed about the product, communication, or consumer, and then designing a proper study to answer the research question will lead to useful and actionable to specific business needs.

Product consumer research examines how consumers perceive products holistically, as well as by attributes. Teasing apart product attribute effects can effectively help with product innovation. Understanding the impact of taste, smell, sight, sound, and feel on overall product perception can help in product design. Consumer emotional, psychological, behavioral, and physiological responses can help guide innovators to design better and more successful products.

This chapter reviewed the ways in which novel methodologies from psychology and neuroscience can help build better and more innovative products. But it has also exposed that applied consumer neuroscience research isn't all about emotional impact. For example, how will this flavor innovation affect brand perception? Is a product attribute change congruent to packaging design? Does the feel of this lotion on the skin give benefits that signal moisturization? Does this smell connote clean? How does the new fragranced product perform against a competitor's fragranced product? And perhaps most importantly, do these innovations fit with brand perceptions?

Understanding consumer needs for products can help build better products (a top-down as opposed to a bottom-up approach to research) and lay the basis for building consumer technical models (CTMs) for innovation programs. By starting with understanding the consumer through a combination of qualitative and quantitative research with applied consumer neuroscience, then a real story can be built into the drivers of behavior and liking of consumer products and ultimately consumer belief and in a product.

Referencing back to the discussion of the importance of understanding consumer habits and the habit loop (Section 9.2.1.1.2), building products that meet needs, drive perceptions, and provide meaningful rewards for behavior become successful products, integrated into consumer lifestyles, and develop consumer loyalty. Building a successful CTM (Figure 9.2.16), or research strategy, means starting with understanding consumers' base needs and expectations in a product category, through a combination of traditional and nontraditional methods (such as MaxDiff and Implicit Associations; see Case Study #2 in Section 9.2.3). Then product and brand need gaps can be identified from that data to guide and reveal innovation opportunities and white space. Prototypes can be developed and then tested to ensure these perceptions are being driven, giving consumers

9

Figure 9.2.16 The consumer technical model (CTM) is a market research strategy used to build more in-depth consumer insight for product innovation to ensure that the product, concept, communication, or package can achieve key benefits to fill the identified need-gaps. The figure is a general representation of the process of developing the CTM. Source: Adapted from HCD Research (2021).

a reason to believe the improvements/innovations. This process provides a basis or a chassis for long-term innovation pipelines and research programs.

Below are two different approaches useful for exploring innovation opportunities:

(1) **Top-down approach:** this is focused on the consumer's needs/routines/behaviors. Better understanding the consumer's experience with a product can help to identify the need-gaps and innovation opportunities by creating consumer technical models, a consumer-centric approach. This approach is typically larger-scale research and involves three phases (exploration - exploring the consumers' needs/behaviors; prioritization - uncovering what needs and behaviors are most important and missing for the product; validation - creating proof of principle).

(2) **Bottom-up approach:** This is focused on the product's performance. By understanding how the different elements of the product affect the consumer, clients can use the findings to make product improvements. This can involve an exploration of the sensory elements (fragrance, colors, sounds, etc.) or the functional elements of the product (product ease of use, etc.).

9.2.5 Future of neuroscience

Using a combination of different traditional, neuroscientific, and behavioral science research approaches will assist in identifying innovation opportunities. The marriage of these tools can provide actionable benefits by considering the holistic consumer experience. It is important to convey a message that resonates with the consumer, and this can be accomplished by adjusting innovation goals based on feedback and findings. Through the application of a cohesive and comprehensive CTM, specific research programs can build on the understanding of consumers across multiple projects, thus keeping consumer language, habits, and lifestyles at the forefront of product and communication development. Additionally, the knowledge gained in the exploration and prioritization of need-gaps, certainty of interest, and developmental nudges can be useful in multiple different domains, creating long-lasting use of the CTM. The foundational findings of consumer behavior can be useful for various projects to develop positive pivots for future campaigns, while also encouraging smaller-scaled assessments to keep a pulse on relevant, emerging trends. By developing an effective research design, with appropriate measures, the information gleaned from these efforts will result in consumer-focused innovation, ultimately creating higher satisfaction and stronger relationships with the consumer.

Discussion Questions

- Which steps in NPD can neuroscience measurements be leveraged for innovation?
- How can you correlate neuroscience tools' results with consumers decision-making?

References

Appelhans, B.M. and Luecken, L.J. (2006) "Heart rate variability as an index of regulated emotional responding". *Review of General Psychology*, 10(3), 229-240.

Balconi, M., Grippa, E. and Vanutelli, M. (2015) "Resting lateralized activity predicts the cortical response and appraisal of emotions: an fNIRS study". *Social Cognitive and Affective Neuroscience*, 10(12), 1607-1614.

Beres, A.M. (2017) "Time is of the essence: a review of electroencephalography (EEG) and event-related brain potentials (ERPs) in language research". *Applied Psychophysiology and Biofeedback*, 42(4), 247-255.

Berkman, E. (2018) "The neuroscience of goals and behaviour change". *Consulting Psychology Journal*, 70(1), 28-44.

Biasiucci, A., Franceschiello, B. and Murray, M.M. (2019) "Electroencephalography". *Current Biology*, 29(3), R80-R85.

Binmore, K. and Shaked, A. (2010) "Experimental economics: where next?". *Journal of Economic Behavior & Organization*, 73(1), 87-100.

Boucsein, W. (2012) *Electrodermal Activity*. New York, NY: Springer Science & Business Media. https://doi.org/10.1007/978-1-4614-1126-0.

Bradley, M.M. and Lang, P.J. (1994) "Measuring emotion: the self-assessment manikin and the semantic differential". *Journal of Behavior Therapy and Experimental Psychiatry*, 25(1), 49-59.

Braithwaite, J.J., Watson, D.G., Jones, R. and Rowe, M. (2013) "A guide for analysing electrodermal activity (EDA) & skin conductance responses (SCRs) for psychological experiments". *Psychophysiology*, 49(1), 1017-1034.

Britton, J.W., Frey, L.C., Hopp, J.L., Korb, P., Koubeissi, M.Z., Lievens, W.E., et al. (2016) *Electroencephalography (EEG): An Introductory Text and Atlas of Normal and Abnormal Findings in Adults, Children, and Infants*. Chicago: American Epilepsy Society.

Calvert, G. (2015) Everything you Need to Know About Implicit Reaction Times (IRTs). *Gemma's Blog* (2015).

Cardello, A.V. (2007) "Measuring consumer expectations to improve food product development". In *Consumer-Led Food Product Development*. Cambridge, UK: Woodhead Publishing. pp. 223-261.

Chorney, J.M., McMurtry, C.M., Chambers, C.T. and Bakeman, R. (2014) "Developing and modifying behavioral coding schemes in pediatric psychology: a practical guide". *Journal of Pediatric Psychology*, 40(1), 154-164.

Christensen, C.M., Hall, T., Dillon, K. and Duncan, D.S. (2016) *Competing Against Luck: The Story of Innovation and Customer Choice*. New York: Harper Business.

Dawson, M.E., Schell, A.M. and Filion, D. L. (2007) "Chapter 7: The electrodermal system". In J.T. Cacioppo, L.G. Tassinary, G.G. Berntson (eds). *Handbook of Psychophysiology* (3rd edition), 2. Cambridge, UK: Cambridge University Press, 159-181.

Foster, G. (2019) *Biophysical Measurement in Experimental Social Science Research*. London, UK: Academic Press.

Gattol, V., Sääksjärvi, M. and Carbon, C.C. (2011) "Extending the implicit association test (IAT): assessing consumer attitudes based on multi-dimensional implicit associations". *PLoS One*, 6(1), e15849.

de Geus, E.J., Gianaros, P.J., Brindle, R.C., Jennings, J.R. and Berntson, G.G. (2019) "Should heart rate variability be "corrected" for heart rate? Biological, quantitative, and interpretive considerations". *Psychophysiology*, 56(2), e13287.

Goodall, C.E. (2011) "An overview of implicit measures of attitudes: methods, mechanisms, strengths, and limitations". *Communication Methods and Measures*, 5(3), 203-222.

Gouizi, K., Bereksi Reguig, F. and Maaoui, C. (2011) "Emotion recognition from physiological signals". *Journal of Medical Engineering & Technology*, 35(6-7), 300-307.

Greenwald, A. and Nosek, B. (2001) "Health of the implicit association test at age 3". *Zeitschrift für Experimentelle Psychologie*, 48(2), 85-93.

Greenwald, A.G., McGhee, D.E. and Schwartz, J.L. (1998) "Measuring individual differences in implicit cognition: the implicit association test". *Journal of Personality and Social Psychology*, 74(6), 1464.

Gross, J.J. and Thompson, R.A. (2007) "Emotion regulation: conceptual foundations". In J.J. Gross (ed). *Handbook of Emotion Regulation*, Vol. 3. New York: Guilford Press, pp. 3-24

Gruber, T., Debracque, C., Ceravolo, L., Igloi, K., Bosch, B.M., Frühholz, S. and Grandjean, D. (2020) "Human discrimination and categorization of emotions in voices: a functional Near-Infrared Spectroscopy (fNIRS) study". *Frontiers in Neuroscience*, 14 (Article 570), 1-14. https://doi.org/10.3389/fnins.2020.00570.

Jackson, P.A. and Kennedy, D.O. (2013) "The application of near infrared spectroscopy in nutritional intervention studies". *Frontiers in Human Neuroscience*, 7 (Article 473), 473-473.

Kahneman, D. (2013) *Thinking, Fast and Slow*. London, UK: Penguin Books.

King, S.C., Meiselman, H.L. and Carr, B.T. (2010) "Measuring emotions associated with foods in consumer testing". *Food Quality and Preference*, 21(8), 1114-1116.

Köster, E.P. and Mojet, J. (2015) "From mood to food and from food to mood: a psychological perspective on the measurement of food-related emotions in consumer research" *Food Research International*, 76, 180-191. https://doi.org/10.1016/j.foodres.2015.04.006.

Krishna, A. (2012) "An integrative review of sensory marketing: engaging the senses to affect perception, judgment and behavior". *Journal of Consumer Psychology*, 22(3), 332-351.

Lang, A., Bolls, P., Potter, R.F. and Kawahara, K. (1999) "The effects of production pacing and arousing content on the information processing of television messages. *Journal of Broadcasting & Electronic Media*, 43(4), 451-475

Larsen, J., Norris, C. and Cacioppo, J. (2003) "Effects of positive and negative affect on electromyographic activity over *zygomaticus major* and *corrugator supercilii*". *Psychophysiology*, 40(5), 776-785.

Liu, Y., Sourina, O. and Nguyen, M.K. (2011) "Real-time EEG-based emotion recognition and its applications. In *Transactions on Computational Science XII*. Berlin, Heidelberg: Springer. pp. 256-277.

Lumpkin, E.A., Marshall, K.L. and Nelson, A.M. (2010) "The cell biology of touch". *Journal of Cell Biology*, 191(2), 237-248.

Macy, A. (2015, July 15). Electroencephalography (EEG) Pt. 1. Retrieved February 22, 2019, from https://blog.biopac.com/electroencephalography-eeg/

Magnée, M., De Gelder, B., Van Engeland, H. and Kemner, C. (2007) "Facial electromyographic responses to emotional information from faces and voices in individuals with pervasive developmental disorder". *Journal of Child Psychology and Psychiatry*, 48(11), 1122-1130.

Matos, R. (2015) Types of eye movements. August 06 Retrieved from https://www.tobiipro.com/learn-and-support/learn/eye-tracking-essentials/types-of-eye-movements/

Morris, J.D., Woo, C., Geason, J.A. and Kim, J. (2002) "The power of affect: predicting intention". *Journal of Advertising Research*, 42(3), 7-17.

Niedziela, M.M. and Ambroze, K. (2020) "The future of consumer neuroscience in food research". *Food Quality and Preference* 92, 104124.

Nunez, P.L. and Srinivasan, R. (2006) *Electric Fields of the Brain: The Neurophysics of EEG*. USA: Oxford University Press.

Parent, A., Carpenter, M.B. (1995) Chapter 1. In A. Parent (ed). Carpenter's Human Neuroanatomy (9th edition). Baltimore, MD: Williams & Wilkins. ISBN 978-0-683-06752-1.

Poels, K. and Dewitte, S. (2006) "How to capture the heart? Reviewing 20 years of emotion measurement in advertising". *Journal of Advertising Research*, 46(1), 18–37.

Quintana, D. and Heathers, J. (2014) "Considerations in the assessment of heart rate variability in biobehavioral research". *Frontiers in Psychology*, 5 (Article 805), 1–10. https://doi.org/10.3389/fpsyg.2014.00805.

Rayner, K. (2009) "Eye movements and attention in reading, scene perception, and visual search". *The Quarterly Journal of Experimental Psychology*, 62(8), 1457–1506.

Rettie, R. and Brewer, C. (2000) "The verbal and visual components of package design". *Journal of Product and Brand Management*, 9(1), 56–70. ISSN (online), 1061-0421. http://dx.doi.org/10.1108/10610420010316339.

Samson, A. (ed) (2019) *The Behavioral Economics Guide 2019 (With an Introduction by Uri Gneezy)*. Retrieved from https://www.behavioraleconomics.com

Silayoi, P. and Speece, M. (2007) "The importance of packaging attributes: a conjoint analysis approach". *European Journal of Marketing* 41(11/12): 1495–1517, Available at SSRN: https://ssrn.com/abstract=2576094.

Singleton M.J. (2009) "Functional magnetic resonance imaging". *The Yale Journal of Biology and Medicine*, 82(4), 233.

Sunstein, C. (2014). "Nudging: a very short guide". *Journal of Consumer Policy*, 37(4), 583-588.

Thaler, R. and Sunstein, C. (2008) *Nudge: Improving Decisions About Health, Wealth, and Happiness*. New Haven, CT: Yale University Press.

Thomson, D. (2007). "SensoEmotional optimisation of food products and brands. In H. MacFie (ed). *Consumer-Led Food Product Development*. Cambridge, UK: Woodhead Publishing. pp. 281–303.

Thomson, D. and Crocker, C. (2015) "Application of conceptual profiling in brand, packaging and product development". *Food Quality and Preference*, Vol 40 (Part B), 343–353.

Thomson, D.M., Crocker, C. and Marketo, C.G. (2010) "Linking sensory characteristics to emotions: an example using dark chocolate". *Food Quality and Preference*, 21(8), 1117–1125.

Vescio, B., Salsone, M., Gambardella, A. and Quattrone, A. (2018) Comparison between electrocardiographic and earlobe pulse photoplethysmographic detection for evaluating heart rate variability in healthy subjects in short-and long-term recordings. *Sensors*, 18(3), 844.

Villringer, A., Planck, J., Hock, C., Schleinkofer, L. and Dirnagl, U. (1993) "Near infrared spectroscopy (NIRS): a new tool to study hemodynamic changes during activation of brain function in human adults". *Neuroscience Letters*, 154(1-2), 101–104.

Williams, D.P., Cash, C., Rankin, C., Bernardi, A., Koenig, J. and Thayer, J.F. (2015) "Resting heart rate variability predicts self-reported difficulties in emotion regulation: a focus on different facets of emotion regulation". *Frontiers in Psychology*, 6 (Article 261), 1-8.

Wood, W., Tam, L. and Witt, M. G. (2005) "Changing circumstances, disrupting habits". *Journal of Personality and Social Psychology*, 88(6), 918.

9.3

Review of Applications of VR Tools, New Opportunities, and Limitations

Alina Stelick

Key learnings

- ✓ Understanding the intricacies of context influence on consumer perception
- ✓ Ecological validity
- ✓ Build your own DIY multisensory media environment

9.3.1 Importance of context in consumer product research

In developing a new product, most of the initial product testing is done in the laboratory setting, whether by the lead product developer on the bench or by the sensory team in the sensory booths where everything, from what is being tested to who is testing it, how and where the test is conducted, is tightly controlled. This is by design. Typically, the focus of testing at that preliminary stage is on the physical product attributes. Once the rough prototype of the product exists, the product developer's focus shifts. The researcher wants to understand the product-user interface and evaluate whether the product functions as intended when the user chooses to use it. Thus, at this juncture, the research context becomes important.

To use food and beverage as an example, there is a significant body of research that indicates that what we choose to eat, how much of it and how pleasant or tasty it seems to us changes depending on the eating context: the physical, temporal, and/or social environment we are in at the time.

The physical place where we are eating or drinking makes a difference on how food or beverage tastes. Olabi et al. (2002) reviewed sensory studies done in space (*Yes! Scientists do studies in space!*) and reported decreased odor and taste sensitivity between subjects who were in orbit compared to those on the surface of the planet. That and the relatively monotonous diet was attributed to

Product Innovation Toolbox: A Field Guide to Consumer Understanding and Research, Second Edition. Edited by Kannapon Lopetcharat, Dulce Paredes and Jacqueline H. Beckley.

increased weight loss during space missions. To remedy that, scientists suggested introducing spicier foods or foods with stronger flavors for astronauts during their missions. Here on Earth, umami foods were shown to taste more umami and sweet foods were shown to taste less sweet when tasted in an environment that simulated airplane noise compared to the control quiet condition (Yan and Dando, 2015). This means that, when flying across the Atlantic at 30 000 feet, tomato juice may seem tastier to you compared to an apple juice. Alcohol beverages were shown to taste sweeter and less bitter in noisier environments, such as a dance club, compared to quiet ones (Gueguen et al., 2008; Stafford et al., 2012; Gueguen et al., 2013). Therefore, you might consume more alcohol than you intended when it is presented in a highly sensory-stimulating atmosphere. It can be more difficult for you to judge the strength of the drink in your hand. A place can also influence the type of product we crave to consume there. García-Segovia et al. (2015) observed differences in consumer responses to the same food plated in a gourmet restaurant compared to a plastic tray in the sensory booth. Additionally, through repeat conditioning (*think Pavlov's dog and bell experiment*), people can learn to crave a particular product in a particular environment. Robinson and Berridge (2003) and Townshend and Duka (2001) showed that a simple image of a pub or a beer glass, even in the absence of an alcohol, can make you crave a beer.

We often are eating "on the go" rather than waiting to settle in one spot at our dinner table at home. When "on the go," we are more likely to be distracted (e.g. driving a car, talking on the phone). When distracted, one doesn't remember as vividly what was eaten or how it tasted. Studies have shown that being present in the moment and taking time to focus on the sensory properties of food being consumed made it taste more intense (more bitter, salty, or fatty), more pleasant and stand out in our memory (Higgs and Donohoe, 2011; Cornelis et al., 2017). Thus, making this mindful eating a good strategy for weight loss and management (Robinson et al., 2013). And for the "on-the-go" snack developer, this might mean that a more flavorful sensory profile is needed to make the snack stand out and memorable.

It is not just where but when we are eating that makes a difference. Time of day also influences what we choose to consume. Caffeine in coffee negatively impacts our sleep quality and quantity (Shilo et al., 2002). So many of us drink a cup of coffee every morning but will switch to decaf or tea in the afternoon. This also means that when developing a new coffee-based product, the product developer should convene a taster panel earlier in the day rather than after lunch.

The time of the year also has an effect on what and how much we choose to eat. Multistudy review by Stelmach-Mardas et al. (2016) showed that our energy intake increases in after-harvest months. It turns out that humans eat more fruits, vegetables, eggs, meat, and cereal and consume more alcohol in the post-harvest season than any other time of year. Even in today's global economy, some fruits and vegetables may not be available outside their harvest season, and there is definitely a difference in how they taste. Remember the sweet and juicy glorious tomatoes from your local farmer's market in July and the puny watered-down red objects sold as tomatoes in the middle of January? There is a difference.

Social environment also influences our behavioral choices and sensory perceptions. We consume less calories, less sodium, less water, and less alcohol when we are eating alone than when we are in the company of others (de Castro and de Castro, 1989; de Castro and Brewer, 1992; Redd and de Castro, 1992). Some attributed this to peer pressure, the desire to match what others are consuming (Kumar et al., 2016). Others found a more nuanced response: people eat more calories with friends than with strangers, potentially due to added stress of managing their image (Herman et al., 2003). If we are worried about how we are perceived by others, we have less bandwidth focusing on the sensory properties of the food or drink.

Understanding the intricacies of context influence on consumer perception could help product developers design a better product that would work in or across multiple specific situations of product use, whether they were designing food for astronauts, coffee beverage, a new lipstick, or a vacuum cleaner.

Two similar products can theoretically work well in two different situations, but, depending on those exact contexts, one may be viewed as more suitable in particular circumstances than the other. Why is that? It is because the expectations for product performance and, therefore, the optimal settings of the product attributes or features change. To ensure that products perform as intended across multiple environments, researchers strive to test them in those same environments and circumstances of use.

9.3.2 Means of creating context

Before we dive into how context can be created, let us first discuss the concept of ecological validity (Koster, 2003). Ecological validity in product testing is the idea that consumer testing should be done in the "natural" product consumption environment rather than in the laboratory setting that removes all the environmental or contextual cues, thus, making the findings less reflective of the real world. Multiple studies have found that consumer liking can vary depending on the setting or location of the test. Researchers have found that testing products in their natural usage environments (at home, in the cafeteria or restaurant) increased consumer product liking compared to when tested in the laboratory or sensory booths setting (Pound et al., 2000; Edwards et al., 2003; Meiselman, 2003; Meiselman, 2006; King et al., 2007; Boutrolle et al., 2007; Meiselman, 2013; García-Segovia et al., 2015). When it is not possible to produce the real usage conditions for product testing, researchers must get creative in the way context is simulated.

The idea that context can be created by using different sensory modalities is not new. The entertainment industry has long been creating Sensorama type rides to thrill the holidaymakers in their theme parks (https://en.wikipedia.org/wiki/Sensorama). By sequestering a person in a small booth and stimulating several senses simultaneously at specific times, for example, using static or moving images (visual), stereo sound (auditory), simulating skin receptors to convey the shape, weight, texture of the object (haptic), diffusing background scents (olfactory), the operators of the ride could make their customer feel fully transported into a different "reality." The more seamless the simulation, the greater sense of "presence" in that environment. Thus, the feeling of "realness" of the situation is brought forth by the quality of the simulated context. Whitmer and Singer (1998) developed a questionnaire to measure the "realness" or presence in the simulated environment where they focused on four types of factors: subject's control of/within the environment, sensory richness of the environment, awareness of actual hardware/equipment (e.g. virtual reality (VR) headset or edge of the computer monitor), and realism of the scene or subject's own body orientation within that environment. The researchers believed that the true measurement of presence is a combination of objective (environment specific, e.g. pixel resolution of the images or quality of aromas being diffused in the simulated space) and subjective (internal to the participant willingness to image or believe that they are present in the environment). Depending on the resources the product developer can use, given currently available technologies, one can get pretty close to creating really "real" context. When developing your own context for product testing, consider using the elements of the Whitmer Singer Questionnaire to confirm the degree of context realness or believability to the study participant (Whitmer and Singer, 1998).

There are multiple tools that can be employed for creating a truly authentic-feeling albeit artificial environment for product testing when it is not possible to duplicate the actual physical location or situation in a product test, depending on the type of the context that needs to be created, the researcher's technical skill level and resources at hand.

Table 9.3.1 gives a broad overview of tools one can employ to increase the "realness" of the context to life in a laboratory setting. Piqueras-Fiszman and Jaeger (2014) demonstrated that asking a panelist to imagine a particular scenario of food consumption changed panelist's emotional response and perception of food compared to the control condition (when no scenario was given). Including elements of physical environment that could evoke a particular usage occasion or staging physical location can bring the scenario to life even more so. In the past several years, new technology developments made it possible for researchers to start adapting relatively low-cost mass-market technologies for product sensory testing needs, specifically in the context creation. For example, Bangcuyo et al. (2015), in addition to physically staging their product testing space with table cloths and coffee mugs, also projected a video of a "real" coffee house on the wall of nine 48-in. high definition LCD monitors, while the panelist made their test coffee evaluations. Others have created a 3-D set up called a CAVE, where three walls and the floor can display stereo imaging (video and audio) to simulate the desired environment (Bigne et al., 2016). In the CAVE study, participants can move around this room and observe the "store shelves" in this 3-D store simulation. An example of the CAVE set up not related to the Bigne et al. (2016) work can be seen on https://www.youtube.com/watch?v=M16mz8Pc_OM.

More recently, researchers used 360° videos projected in VR headsets to study perceptions of blue cheese (Stelick et al., 2018), nonalcoholic cocktail mixes (Zhou et al., 2020), alcoholic beverages (Picket and Dando, 2019), chocolate (Kong et al., 2020), wine tasting experiences (Carulli et al., 2018), and more. The advantage of this approach is its relatively lower cost and ease of production compared to the previous example, as well as the ability to film real-life

Table 9.3.1 Context creation strategies.

Type	Level		Example	Source
Duplication	Gathering data in an actual "real" physical location (not laboratory)		• Going to a coffee shop and asking patrons to complete a questionnaire about their cup of joe • Observing coffee shop patron behavior without intervening in their activities	Kanjanakorn and Lee, 2017
Evocation or use of memory & imagination	Panelist's own memory recall		• *Please remember the last time you . . .*	Hein et al., 2010
	Text description		• *Imagine that you are. . .*	Piqueras-Fiszman and Jaeger, 2014
	2-D images as physical photo, rendering or graphics		• Picture of the cafe on a computer screen	Van Herpen et al., 2016
Physical object or setting imitation	Physical element of environment in laboratory setting		• Beverage in the glass cup, glass bottle, aluminum can vs. plastic cup	Lefebvre and Orlowski, 2019
	Staging physical location with appropriate physical objects and decor		• Small round tables covered by white table cloth and dining room chairs, test product is served in a mug (not paper cup) by a researcher dressed as barista (apron, name tag, etc.)	Kim et al., 2016
Immersion	Static, panelist cannot interact with environment (typically vision, audition, olfaction)	3-D static images (CG or real environment)	• Presenting different meal type, location and situation	Ferrer et al., 2013
		Wall of HD LCD monitors for audio/video (real environment); Added specific aroma piping	• Placing a respondent in staged location with congruent video and audio (e.g. music) and aromas (e.g. coffee or cinnamon and vanilla spices in the air)	Bangeuyo et al., 2015
		CAVE set up (3X3 immersive reality room with stereoscopic imaging); CG or real environment; Video & audio	• Showing different environments, e.g. a store layout (aisles, shelving, product displays) where respondents can "freely" move in the environment or a natural scenery (e.g. mountains)	Bigne et al., 2016 Pennanen et al., 2020

(Continued)

Table 9.3.1 (*Continued*)

Type	Level		Example	Source
Immersion		360° video of real environment presented in Samsung Gear VR headsets	• Tasting the same blue cheese in three different VR environments and scaling responses	Stelick et al., 2018 Kong et al., 2020
	Dynamic, panelist can interact with environment (adding haptic/touch and taste)	Software-generated environment with ability to indirectly interact with the product (VR headsets, HTC Vive + hand held controller, two-beacon Vive Lighthouse system, Vizard 5, NvidiaGeforce GTX980 graphics card)	• Allowing the panelist to see their hand and body interacting with the test product (foam donut with chocolate aroma infused cotton ball) in the CG environment	Li and Bailenson, 2018
		Software-generated environment, virtual participant avatar being able to taste the virtual product	None, does not exist at this time	Active area of research

environments without any software-created renderings that can feel fake. Figure 9.3.1 shows an example of a 360° video, go to this short YouTube video of Cornell University Chimes Concert, or here's a screenshot of a university auditorium from Stelick et al. (2018) (Figure 9.3.2).

Production of 360° videos is also more novice friendly compared to the computer science knowledge-intensive set ups, such as one described in research by Li and Bailenson (2018). They used HTC Vive System with hand-held controller and two-beacon Vive lighthouse system to emit infrared light through LED positioned at opposite ends of the test space to track the movement of the participants in physical space and adjust the Vizard 5 software-programmed environment on a 3.4-GHz Intel i5 computer with an NVIDIAGeforce GTX 8 = 980

Figure 9.3.1 A screen shot from the Cornell University Chimes Concert taken in 360° video format (https://www.youtube.com/watch?v=ot-4zdr4Upk) (viewing in VR headset is preferred). Source: YouTube.

Figure 9.3.2 A screen shot of a university auditorium. Source: Stelick et al. (2018)/with permission of Institute of Food Technologists.

graphics card to their position and movements. This set up allowed researchers to create computer-generated environment where participants were able to pick up a computer-generated donut, an object that had a shape and weight similar to a real donut, with the controller. The panelists were able to see a simulated version of their hand moving and holding the virtual donut, essentially mimicking their movement in the real world, in the computer-generated environment. The researchers asked participants to either pretend to "eat" a virtual donut without having anything in their hand or hold objects weighing and shaped as a real-life donut with a chocolate aroma-infused cotton ball attached to it. They found that, even in this clearly artificial environment, the participants "ate" significantly less of this virtual donut, when not having a physical object that smelled like chocolate donut in their hand. So having multiple sensory cues, haptic in addition to audio-visual, seemed to increase the "realness" of the scenario to study participants. In the scenarios discussed above, the participant largely stayed in one spot in the virtual environment, able to turn around and interact with it in some way, but not able to go for a walk in it.

In the next few examples, the study participants were able to move in virtual spaces somewhat freely. A system similar to one of the previous examples (Unity software version 5.4.3 to program the environment along with ReWaVe (Real Walking in Virtual Environments) system, Intersense IS-1200 positioning and orientation tracking system at 180 Hz, which ran on an XMG U506 with NVIDIA 980M GPU, Oculus DK2 display with SMI eye-tracking and HTC Vive controller) was used by Siegrist et al. (2019) to track consumer food selection in a virtual supermarket. They found that participants choosing cereals behaved similarly in real life and in VR. In contrast, Harley et al. (2018) used a hybrid of high-tech and nondigital means to create VR sensory experiences in order to keep the cost of the VR production as low as possible. They also used Unity 3D software to create the environments and Oculus DK2 to display it, along with addition of strategically placed sound sources and physical objects (like a sand box to illustrate a beach) for haptics. They found this approach effective in creating the feeling of presence for the participants. Yet, another approach was described by Lonbart et al. (2019), where they studied consumer emotional reactions to misshapen fruits and vegetables in the grocery store. They used Oculus Rift DK2 HMD to create a virtual supermarket where respondents could move back and forth, picking up groceries with an Xbox controller.

Enhancing the virtual audio-visual landscape with olfactory cues can increase the feeling of presence, as illustrated by many studies. For example, Tuanquin et al. (2018) found that adding olfactory cues to the virtual environment made panelists crave the presented virtual foods more than in their absence, and allowing them to interact with virtual foods increased the virtual food cravings even more. Jones and Dawkins (2018) demonstrated that adding changes to ambient temperature and scent at certain points of the 360° immersive film viewed in the virtual environment increased the sense of presence for participants, whereas lack of those additional cues detracted from the realness of their experience. A lot of research has been done recently trying to perfect the olfactory cue delivery in the VR environments, from having air pumps attached to the headset that delivered the aroma close to the study participant's nose precisely when an object of interest appeared on the screen (Narumi et al., 2011), to temperature-responsive hydrogel to allow measured aroma release with tightly controlled temperature changes (Kim et al., 2009).

Thus far, we focused most of our discussion on VR- derived context tools where the panelist is transported into a digitally generated world that is static or interactive, albeit that interactivity is somewhat limited. There are, however, limitations to what is currently possible in VR testing sessions. Not every product naturally lends itself to being studied in VR environments. Zandstra and Lion (2019) discussed haptic-heavy objects that change their shapes easily, semisolids, or liquid foods would still be quite difficult to test in VR. Additionally, one must be aware of the certain side effects of being immersed in the VR systems, especially for prolonged periods of time. Israel et al. (2019) recently discussed the prevalence and dangers of cybersickness, a health condition that can happen after prolonged use of VR technologies, where the subject can start feeling motion (even when standing still), headaches, blurred vision, excessive salivation, eye strain, dizziness, and even vomiting. This side effect of technology use can undermine participant engagement with the VR system, therefore, taking all the effort of creating an enjoyable testing experience to null.

As we are moving along in the "realness" ladder, the next step in the context creation would be to create an environment, where an actual participant or their avatar is able to interact with their environment using all their senses, including the sense of taste. While the use of avatars is routine in many non-VR and VR video games and has been used in research before (Dean et al., 2009), the player cannot fully sense their VR environment via their avatar. Creation of virtual taste devices is an active area of research with some quite clever devices being introduced (we'll discuss them later on in this chapter). However, for now, the most participants can experience is being present in a "real-enough" environment and able to interact with it on a limited basis by, e.g. seeing their hand with a fork bring the real food to them while the plate and the background is pure fiction. Significant investment into new software and hardware development and programming is needed to create the vivid popular culture worlds of the Matrix franchise (1999, 2003), Surrogates (2009), Avatar (2009), or more recent Ready Player One (2018). At this time, it is still all fiction.

9.3.3 How to create a study using VR/AR tools

As with any research study, the number one question for the researcher should be "What is the objective of the study?" Understanding the objective and defining, based on the objective, the type of context, and the level of immersion needed to be created is the most important factor in the study success. Once that is defined, one can start thinking about the actual study parameters needed to answer the main objective:

(1) Test samples: number and type or form of samples, how they are going to be served
(2) Study participants: number of participants, sociodemographic requirements

(3) Type of context: passive vs. interactive with test product, etc.
   (a) Historical data or new research may be needed (e.g. ethnography, shop-along, video diary with daily or weekly check-ins over a period of time, a survey to understand the product usage context or consumption patterns)
(4) Resources and equipment needed
   (a) For creating the context: study designer's technical skill level, associated budget to source necessary equipment, and props purchases
   (b) For study administration: space available, staffing levels to support test fielding
(5) Questionnaire design: Number and types of scales, if any, needed to meet study objectives
(6) Data analysis plan: Largely dependent on study design and questionnaire type

To illustrate the flow of study design, let us work within the following scenario:

> Let's say that we are snack bar manufacturer "Healthy-Snacks-R-Us," developing a new product line. Our brand focuses on nutrition and fitting the needs of our general younger active lifestyle loving customer. To make ourselves different from our competitors, we'd like to link specific flavors with specific settings, for example, office lunchroom, relaxing day at the beach, hiking in the state park, etc. The product developer has several different flavor options with trendy floral notes. Our best selling flavor currently is chocolate vanilla. For starters we'd like to focus on one of the flavors: strawberry rose.

Given this scenario, first, we will start by defining the objective of the study: to determine consumer acceptance of the strawberry rose snack bar product in several different occasions or situations.

Based on the example, we have two products that need to be tested: a test, a new strawberry rose snack bar, and a control product, our current chocolate vanilla bestseller. This would allow us to compare the new product's performance to our bestseller. The study requires participants to taste the actual prototype we have already developed. Being a solid snack bar, as researchers, we have minimal concerns about how the product would be sampled in the virtual environment. We think that it would be easy for us to blind and present the test product to study participants in any virtual environment we decide to create, even if they won't be able to see the product or their hand interacting with the product in virtual space.

Our current market research created the profile of our target consumer: under 34 years old, GenZer or Millennial; student or young professional who leads an active lifestyle; likes to travel; enjoys flavors from far-away places; and is environmentally conscious. We have reviewed our historical data files and found that our consumers are very busy people that eat our product while alone, mostly "on-the-go," as a postworkout snack or a 3 pm pick-me-up in the office. Therefore, the virtual environment should reflect those spaces as realistically as possible. However, we'd like to limit the number of virtual environments we present to participants to two or three, that are most representative of the consumption contexts and most different from each other, to reduce the length of evaluation session and not tire the participant. Using 360° video of real environments presented in Samsung Gear VR headsets may be the best and most cost-effective option for us. Additionally, since our consumers are relatively young, we expect them to be somewhat more comfortable using the VR headset and equipment set up that we have in mind, than, for example, if we were to study older adults with health conditions such as arthritis. With regard to the number of participants needed for this kind of test, ideally, we'd like to aim for having at least 100 consumers to achieve the necessary statistical power.

Since the main objective is to determine consumer acceptance, all we need is one Overall Liking question to be asked while the subject is in the VR environment. For that, we plan to use traditional 9-pt Hedonic Scale (Jones et al., 1955; Peryam and Pilgrim, 1957). We plan to calculate the mean Overall Liking for each of the products and contexts they were presented in, and run traditional ANOVA followed by a statistical significance test (e.g. Tukey's) to determine whether differences in means were statistically significantly different at 95% Level of Confidence.

9.3.3.1 Scene creation considerations

In our example, 360° videos are used to place the study participant in the desired virtual environment.

To increase the sense of plausibility of the scenario, Korsgaard et al. (2019) recommended thinking about the following aspects when creating desired test virtual environment:

- Perceived personal safety of the panelist in the environment (Consider subject's potential phobias, e.g., claustrophobia, fear of heights)
- Duplication of the actual environment (Does it actually exist in real life?)
- Plausibility and practicality of food consumption in the VR environment (Can you really mountain climb and eat soup from the bowl in real life? Do you eat breakfast cereal in a 4 Michelin star restaurant?).

9

Before creating any virtual scenes, one should consider the real environment the scene is based upon, including the visual imagery, audio, and all other subtle cues that tell you where and when you are. For example, a garden in late June may smell like roses. Or when you are at the beach, there's sand on your feet. So when you position the panelist in a room for the first example, you would diffuse some rose aroma and make sure that the room temperature is at a comfortable 68–70 °F. Whereas, when you're simulating the beach, you may want to ask your panelist to take their shoes off and step into a box filled with sand. Using the additional sensory cues in addition to what is viewed and heard in the headset has been shown to increase the sense of presence in the virtual environment.

Another very important thing to consider before you start recording any videos, please make sure to seek the permission from the owner of the space you are recording at and all that are present in that space at the time of the recording. For example, if you're looking to record a video at the pub, ask the permission of the pub owner ahead of time and schedule the recording session in advance. Then ask permission to record the scene from each person who happens to be in the frame. Follow your local laws and be considerate of people's property and privacy. Ask everyone to sign a consent and video release form, which informs them of the purpose, risks, and benefits of your activities.

It is important to consider the social aspect of a particular context or scenario. In most of the recent research I reviewed, the subject is typically alone in the virtual space. However, it is possible to position the subject in a social situation (e.g. a cocktail reception or public transportation) as a passive observer where the participant is almost as if a ghost observing the space but invisible to others, the participant cannot see their own body and interact with others. This can feel very awkward for any human, as we are very social creatures, and subsequently would probably decrease the feeling of presence in that type of an environment. This type of a social scenario could also feel awkward for those humans "populating" the video. For example, in our pub example, patrons of the pub may stare at the camera while it is recording or make comments about it. As a result, the subject placed in that virtual environment later on would not feel comfortable there, as staring is typically not socially acceptable and can feel intimidating, thus, negating the generally warm and friendly pub atmosphere the researcher would want to convey to the study subject. Additionally, as a passive observer, the subject does not have the ability to directly interact with those in the virtual environment created through 360° videos, therefore, nothing should be spoken to the respondent or require them to respond. "Hey, can you pass that casserole dish?" type of scenario would be impossible to execute within current technology systems.

Each 360° video should be about 2–3 minutes long. This is just enough to get familiar with the setting to be "transported" there and not too long when the magic of being present there starts to fade. However, this is more due to practical considerations rather than scientifically backed findings, as no studies have been done thus far on the length of the video and its effect on the perception of realness of the experience. The headsets, earphones can also feel heavy and bulky, especially when they have to be used for extended periods of time.

A practice environment with study instructions so that the study participant can get comfortable with the hardware and the software. In our experience, about 5% of the subjects report experiencing dizziness while wearing the VR headset and as much as a quarter can have difficulties operating the equipment (getting to the right video, if testing with multiple videos; or understanding how to use the cursor within the video to answer the question, etc.).

To gauge individual comfort during the study, modeling after Zhou et al. (2020), one can simply ask the respondents:

- How immersed did you feel during the VR experience?
- Did you feel dizzy or lightheaded during the VR experience?

This follow-up survey can help researchers understand participants' experiences during the study to improve it for the future.

9.3.3.2 Questionnaire/scales

Based on most of the research reviewed, researchers typically do not ask their subjects any questions while they are immersed in the virtual environment. Participants may have answered a structured questionnaire before and/or after wearing the VR headset but not while they were in the VR. This may be by design. If a 360° video is used for immersion, the question and the associated scale would be artificially floating in thin air.

In Stelick et al. (2018), we had to superimpose the question onto the video and position it in the back of our panelist so that they had to first see the space in front of them, taste the sample, and only then were instructed to look back and answer the question. The panelists used a cursor controlled by a specific keyboard key and took a screenshot of their responses to answer the questions. This set up worked reasonably well for us. The drawback, besides its general strangeness, was that one could not ask very many questions about the product being tested. It could get overwhelming for participant, especially with more complex sensory scale types such as 9-pt Hedonic, general Linear Magnitude Estimation Scale (gLMS), line, Check all that apply (CATA) or rate-all-that-apply (RATA), time intensity (TI), etc. One must take careful consideration when converting panelist ratings on line or gLMS scales to numeric data. Additional conversion equations may need to be used (see example in Stelick et al., 2018). While we used 9-pt Hedonic and gLMS scales in our research as proof of principle, simpler, "yes/no," 3- or 5-point scales would have worked better from the panelist point of view.

Alternative set ups could be used. One can imagine asking panelists the questions recorded within the video soundtrack, panelists verbalizing the answers for researchers in the room to capture their answers. The obvious drawback in this scenario is that each participant would have a personal interviewer and that the soundtrack would not be, in any way, naturally connected with the virtual environment, again, decreasing the feeling of presence in virtual space.

9.3.3.3 Hardware/software needs

9

Hardware and software will depend on the type of the virtual environment you want to create. Using Stelick et al. (2018) example, Table 9.3.2 shows the lists of equipment and software used in the study.

Supplemental materials of the Stelick et al. (2018) paper contain step-by-step instructions on how to stitch, edit, save, and transfer the 360° video environments for further use. Saleme et al. (2019) also provided a great survey of devices and low-cost tutorial of how to build your own DIY multisensory media environment.

9.3.4 Looking ahead: what are the current technology limitations and what might be coming up next

Most of the material so far focused on VR. Now let us discuss other technologies used in similar space, starting with new developments in augmented reality (AR). Pennanen et al. (2020) defined VR as "realistic alternative environments" used

Table 9.3.2 Examples of hardware and software for creating a virtual environment.

	Hardware/software	Purpose
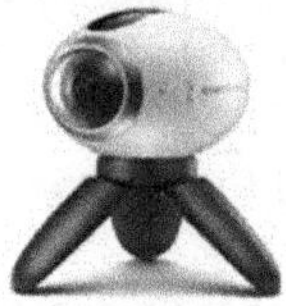	Samsung Gear VR 360° camera (Samsung Group, Seoul, SK)	Record 360° videos of the environment
	Samsung Gear 360 Action Director	Remove the joining line between "front" and "back" of the recording, aka stitch video
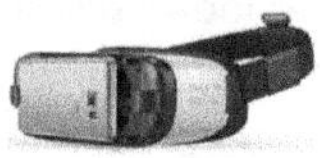	Adobe Premiere Pro CC 2017 (Adobe Systems, San Jose, CA, USA), Adobe Photoshop 2017, Microsoft Paint (Version 1607)	Cropping videos to equal length, overlaying audio instructions and visual scales at specific positions within video, exporting video to each VR headset/smartphone combo and on Google Drive
	VR Oculus Headset	Project the videos
	Samsung S6 smartphones	Project the videos Store panelists' answers (screenshots)
	Bluetooth keyboard with product placeholder, one key obscured and connected to particular phone/ headset combination	Serve as food tray to serve test product as a controller to record the answers to scaled questions.
	Dropbox (Dropbox Inc., San Francisco, CA, USA)	Automatically save answers (screenshots) in real time so that study personnel can validate each response

"in place of a real" one, whereas augmented reality allows to add information or "digital sensations to real environments, enabling an environment or any object to be augmented or modified." Augmented reality creates an illusion that a physical object and a virtual one exist in the same physical space (Ong and Nee, 2004). Crofton et al. (2019) recently reviewed a series of new developments in the virtual and augmented reality technologies specific to sensory science and consumer product testing. At the moment, AR has not yet been widely used in the food industry. Etherington (2016) wrote that Mercedes Benz has been using AR since 2013 to support accident and emergency services by placing quick response (QR) codes in specific locations on all new cars. That enabled the first responders to quickly view the car wiring and fuel system with an AR mobile application. Since it was meant to be interactive, used in real-time in real physical spaces, this "on-the-go" AR information could be displayed on smartphones or tablets simply using GPS, camera, hybrid vision, and accelerometer to orient the user and the object of interest in the space (Comport et al., 2006; Kipper and Rampolia, 2013). However, even more sophisticated versions now exist that use head mounted stereoscopic displays that are typically used for VR projection because they allow voice recognition, spatial mapping, and gesture detection (Dickson, 2018). However, unlike Google Glass, the headsets are still quite bulky and uncomfortable to wear for extended periods of time, and not to mention can be quite still quite pricey for an average user. Currently, there is lot of research going on in this area as developers see endless potential from displaying nutritional labels (Bayu et al., 2013; Csakvary, 2017) or food origin, to accurate estimation of food serving size (Rollo et al., 2017) to all kinds of other information about everyday products and objects. Fujimoto (2019) used image-processing software to images of food on their pixel level to maximize the deliciousness perception of the respondents tasting the real version of that food, hence, providing an enhanced sensory experience to the study participants. Imagine how consumer behavior is going to change once these types of technologies go mainstream. It is well known that information about food can and does change consumer behavior and purchase intent (Kwak et al., 2017; Pinto et al., 2020, and many more).

Another fast growth technology area in the field of sensory and consumer product research is biometric data tracking. Biometrics are thought of as means to measure implicit response to a product in contrast to having respondents explicitly scale the myriad of different sensory attributes as traditionally done. Scientists are looking to link involuntary physiological responses such as heart rate or skin temperature to exposure or experience with specific products, for example, beer (Beyts et al., 2017), chocolate (Torrico et al., 2018), and juice (Verastegui-Tena et al., 2019). Researchers are trying to capture facial expressions (de Wijk et al., 2012; He et al., 2016; Beyts et al., 2017), track eye movement (Siegrist et al., 2019), and even record brain activity via electroencephalography (EEG) (Viemose et al., 2013) in an effort to capture the unconscious responses to products (see Chapter 9.2 for more details on neuroscience tools and consumer research). Burn (2017) reported the development of the Mask, a foam-based insert fit with biometric technology that can detect users' facial expressions and translate them into emotion expression of their virtual avatars in real time. Balbin et al. (2019) created an augmented reality device to monitor customer facial expressions while eating to measure customer satisfaction in place of asking explicit surveys. The argument is being made that traditional, explicit response seeking, methods of research introduce too much bias and can mislead the researcher about consumers' true feelings about product prototype. However, an individual person's privacy should be carefully safeguarded when looking to develop or utilize any of these approaches.

A lot of research has been going on in the last decade or so around the idea of increasing the realness of the VR experience by improving the virtual aroma and taste delivery systems. Researchers see a lot of applications beyond VR for these types of systems or devices. Imagine a world where you can click on the perfume ad on Facebook and actually smell the aroma of that perfume sitting on your couch at home. Yoshimura and Sakashita (2006) developed a device that would release a real perfume in the subject's immediate space when an e-mail with a key program is read on their device. Viswadhara Meenakshi and Sowmya (2003) created Experiencing Smell through Electronic Reality (ESTER) device that has four to eight different scents stored on its hardware, ready to be released through an electronic media. While real taste and flavor requires simulating tastes buds on the tongue in the mouth or olfactory bulbs retronasally using chemical substances such as sucrose (sugar) for sweet taste or isoamyl acetate (banana aromatic) for banana flavor, it is much more difficult to create the same sensations virtually, in the absence of these substances. In 2005, Maynes-Aminzade introduced a TasteScreen that allows panelists to taste 20 different chemical flavorants sprayed on a computer display by licking the screen. There have been some success using low electric current to stimulate the sour taste receptors to create sourness perception in the mouth with a device called Digital Lollipop (Ranasinghe and Do, 2016). Karunanayaka et al. (2018) developed a Thermal Taste Machine that, they claimed, can produce 20 different taste and nontaste sensations without use of chemical substances, such as sweet, fatty/oily, warm, electric, and metallic when warming the tongue, and mint, pleasant, cold when cooling it. In 2020, Homei Miyashita released Norimaki Synthesizer, a handheld digital device that uses five gels and low electronic current to recreate the five basic tastes (Miyashita, 2020). Other researchers tried to include cross-modal correspondence to superimpose different colors on a real beverage to create different flavor sensations (Narumi et al., 2010). Ranasinghe et al. (2017) created a Vocktail device that combined the visual, olfactory, and taste sensations to create virtual cocktail experience by changing the color of the liquid in the cup, pumping certain aromas and simultaneously using electric current to produce sweet, sour, salty, bitter, and umami perceptions. The same team of researchers proposed utilizing everyday objects (chopsticks, cigarette, straw, spoon, or fork) equipped with electrodes to produce controlled low electrical pulses to imitate the food taste in VR settings (Ranasinghe and Do, 2016). More is yet to come in this area.

9.3.5 Summary

To summarize, Virtually Reality applications allowed for in-context testing in controlled product usage conditions, and delivering consistent user experience across the study participants. Researchers are able to expose subjects to multiple contexts within the same study session through engaging and dynamic experience. Hartmann and Siegrist (2019) support the notion that VR can be successfully used to study consumer behavior, individual product performance, and food sensory perceptions. VR still does not fully replace the actual reality, because, in the words of one "Ready Player One" character, "Reality is Real." Korsgaard et al. (2019) exposed mobility impaired older adults living alone to eating meals in different VR settings. While the participants successfully operated the hardware and performed all study tasks satisfactorily, in the end, only one participant said they would consider eating in virtual environments in the future because it was "unnatural." Thus, its drawbacks for the researcher, product developer, could be the following:

- Required computer science knowledge and confidence in handling complex software & hardware
- Cost of software/hardware and environment staging (e.g. upscale restaurant vs. public park setting)

- Current inability to have the panelist fully engage with the product due to its texture/haptics or state (semisolid, liquids, etc.)
- Current inability to engage with other panelists in a social situation

As a panelist in the VR setting, one might experience cybersickness or diminished sense of realness of the experience. Others may have difficulties handling hardware (due to weight of the headset) or test food since, in many cases, one cannot see their hands holding the test sample.

Discussion Questions

- What research questions can the VR/VA system help answer?
- Which context in real-life can VR/VA systems be used?

References

Balbin, J.R., Paglinawan, C.C., de Castro, M.J.A., Llamas, J.K.C., Medina, M.E.T., Pangilinan, J.J.O. and Valiente, F.L. (2019) Augmented Reality Aided Analysis of Customer Satisfaction Based on Taste-Induced Facial Expression Recognition Using Affdex Software Developer's Kit. *ICBET' 19: Proceedings of the 2019 9th International Conference on Biomedical Engineering and Technology*, Tokyo, Japan, 28-30 March 2019. pp. 204-209. https://doi.org/10.1145/3326172.3326221.

Bangcuyo, R.G., Smoth, K., Zumach, J.L., Pierce, A.M., Guttman, G.A. and Simons, C.T. (2015) "The use of immersive technologies to improve consumer testing: the role of ecological validity, context and engagement in evaluating coffee". *Food Quality and Preference*, 41, 84-95.

Bayu, M.Z., Arshad, H. and Ali, N.Z. (2013) "Nutritional information visualization using mobile augmented reality technology". *Procedia Technology*, 11, 396-402.

Beyts, C., Chaya, C., Dehrmann, F., James, S., Smart, K. and Hort, J. (2017) "A comparison of self-reported emotional and implicit responses to aromas in beer". *Food Quality and Preference*, 59, 68-80.

Bigne, E., Llinares, C. and Torrecilla, C. (2016) "Elapsed time on first buying triggers brand choices within a category: a virtual reality-based study". *Journal of Business Research*, 69, 1423-1427.

Boutrolle, I.J., Delarue, J., Arranz, D., Rogeaux, M. and Köster, E. (2007) "Central location test vs. home use test: contrasting results depending on product type". *Food Quality and Preference*, 18, 490-499.

Burn, J. (2017) *Neuro-sensitive foam shows your expression in VR before your face does.* https://www.forbes.com/sites/janetwburns/2017/04/12/this-neuro-sensitive-foam-insert-is-changing-the-face-of-vr/#2ddbc5963773, accessed 07 December 2020.

Carulli, M., Bordegoni, M., Ferrise, F., Gallace, A., Gustafsson, M. and Pfuhl, T. (2018) Simulating Multisensory Wine Tasting Experience. In D. Marjanovic, M. Storga, S. Skec, N. Bojcetic, N. Pavkovic (eds). *DS 92: Proceedings of the DESIGN 2018 15th International Design Conference*, Dubrovnik, Croatia, 21-24 May 2018. pp. 2177-2186. https://doi.org/10.21278/idc.2018.0485

Comport, A., Marchand, E., Pressigout, M. and Chaumette, F. (2006) "Real-time markerless tracking for augmented reality: the virtual visual serving framework". *IEEE Transactions on Visualization and Computer Graphics*, 12, 615-628.

Cornelis, M.C., Tordoff, M.G., El-Sohemy, A. and van Dam, R.M. (2017) "Recalled taste intensity, liking and habitual intake of commonly consumed foods". *Appetite*, 109, 182-189.

Crofton, E.C., Botinestean, C., Fenelon, M. and Gallagher, E. (2019) "Potential applications for virtual and augmented reality technologies in sensory science". *Innovative Food Science & Emerging Technologies*, 56 (August), 102178. https://doi.org/10.1016/j.ifset.2019.102178

Csakvary, B. (2017) *Promoting Healthier Food Choices with the Application of Augmented Reality* (Masters Thesis). Netherlands: Wageningen University.

Dean, E., Cook, S., Keating, M. and Murphy, J. (2009) "Does this avatar make me look fat? Obesity and interviewing in second life". *Journal of Virtual Worlds Research*, 2(2), 3D Virtual Worlds for Health and Healthcare, 4-11.

De Castro, J. M and Brewer, E.M. (1992) "The amount eaten in meals by humans is a power function of the number of people present". *Physiology & Behavior*, 51(1), 121-125.

De Castro, J.M and de Castro, E.S. (1989) "Spontaneous meal patterns in humans: influence of the presence of other people". *The American Journal of Clinical Nutrition*, 50(2), 237-247.

Dickson, B. (2018). *4 reasons AR Glasses Won't Replace Your Smartphone (yet)*. https://uk.pcmag.com/opinions/117014/4-reasons-ar-glasses-wont-replace-your-smartphone-yet, accessed 07 December 2020.

Edwards, J.S.A., Meiselman, J.L., Edwards, A. and Lesher, L. (2003) "The influence of eating location on the acceptability of identically prepared foods". *Food Quality and Preference*, 14, 647-652.

Etherington, D. (2016) *AR in Mercedes-Benz's rescue assist app gives first responders an inside look*. https://techcrunch.com/2016/07/27/ar-in-mercedes-benzs-rescue-assist-app-gives-first-responders-an-inside-look/, accessed 07 December 2020.

Ferrer, M.F.G., Gutiérrez Maldonado, J. and Pla, J. (2013) "Cue-elicited anxiety and craving for food using virtual reality scenarios". *Studies in Health Technology and Informatics*, 191, 105-109.

Fujimoto, Y. (2019) "Projection mapping for enhancing the perceived deliciousness of food". *IEEE Access*, 6, 59975-59985. https://doi.org/10.1109/ACCESS.2018.2875775

García-Segovia, P., Harrington, R.J. and Seo, H.S. (2015) "Influences of table setting and eating location on food acceptance and intake". *Food Quality and Preference*, 39, 1-7.

Gueguen, N., Jacob, C., Le Guellec, H., Morineau, T. and Lourel, M. (2008) "Sound level of environmental music and drinking behavior: a field experiment with beer drinkers". *Alcoholism: Clinical and Experimental Research*, 32, 1795-1798.

Gueguen, N., Le Guellec, H. and Jacob, C. (2013) "Sound level of background music and alcohol consumption: an empirical evaluation". *Perceptual and Motor Skills*, 99, 34-38.

Harley, D., Verni, A., Willis, M., Ng, A., Bozzo, L. and Mazalek, A. (2018) Sensory VR: Smelling, Touching, and Eating Virtual Reality. In *Proceedings of the Twelfth International Conference on Tangible, Embedded, and Embodied Interaction (TEI '18)*. New York, NY, USA: Association for Computing Machinery. pp. 386-397. https://doi.org/10.1145/3173225.3173241

Hartmann, C. and Siegrist, M. (2019) "16-Virtual reality and immersive approaches to contextual food testing". In H.L. Meiselman (ed). Context. Woodhead Publishing. pp. 323-338. https://doi.org/10/1016/B978-0-12-814495-4.00016-7.

He, W., Boesveldt, S., de Graaf, C and de Wijk, R.A. (2016) "The relation between continuous and discrete emotional responses to food odors with facial expressions and non-verbal reports". *Food Quality and Preference*, 48, 130-137.

Hein, K.A., Hamid, N., Jaeger S.R. and Delahunty, C.M. (2010) "Application of a written scenario to evoke a consumption context in a laboratory setting: effects on hedonic ratings". *Food Quality and Preference*, 21, 410-416.

Herman, C.P., Roth, D.A. and Polivy, J. (2003) "Effects of the presence of others on food intake: a normative interpretation". *Psychological Bulletin*, 129, 873-886.

Higgs, S. and Donohoe, J. (2011) "Focusing on food during lunch enhances lunch memory and decreases later snack intake". *Appetite*, 57, 202-206.

Israel, K., Zerres, C., Tscheulin, D., Buchweitz, L. and Korn, O. (2019) "Presenting your products in virtual reality: do not underestimate cybersickness". In F.H. Nah, K. Siau (eds). *HCI in Business, Government and Organizations. eCommerce and Consumer Behavior. HCII 2019. Lecture Notes in Computer Science*. Cham: Springer. pp. 11588. https://doi.org/10.1007/978-3-030-22335-9_14

Jones, S. and Dawkins, S. (2018) The Sensorama revisited: evaluating the application of multi-sensory input on the sense of presence in 360-degree immersive film in virtual reality. In T. Jung, M. Dieck (eds). *Augmented Reality and Virtual Reality*. Cham: Springer. https://doi.org/10.1007/978-3-319-64027-3_13

Jones, L.V, Peryam, D.R. and Thurstone, L.L. (1955) "Development of a scale for measuring soldiers' food preferences". *Journal of Food Science*, 20(5), 512-520. https://doi.org/10.1111/j.1365-2621.1955.tb16862.x

Kanjanakorn, A. and Lee, J. (2017) "Examining emotions and comparing the EsSense Profile → and the coffee drinking experience in coffee drinkers in the natural environment". *Food Quality and Preference*, 56, 69-79.

Karunanayaka, K., Johari, N., Hariri, S., Camelia, H., Bielawski, K.S. and Cheok, A.D. (2018, April) "New thermal taste actuation technology for future multisensory virtual reality and internet". *IEEE Transactions on Visualization and Computer Graphics*, 24(4), 1496-1505. https://doi.org/10.1109/TVCG.2018.2794073

Kim, S.-E., Lee, S.M. and Kim, K.-O. (2016) "Consumer acceptability of coffee as affected by situational conditions and involvement". *Food Quality and Preference*, 52, 124-132.

Kim, D.W., Nishimoto, K., Kunifuji, S., Cho, Y.H., Kawakami, Y. and Ando, H. (2009) *The 6th IEEE International Conference on Electronics, Circuits, and Systems, ICECS 2009*, Yasmine Hammamet, Tunesia, December 13-19, 2009.

King S.C., Meiselman, H.L., Hottenstein, A.W., Work, A.M. and Cronk, V. (2007) "The effects of contextual variables on food acceptability: a confirmatory study". *Food Quality and Preference*, 18, 58-65.

Kipper, G. and Rampolia, J. (2013) *Augmented Reality: An Emerging Technologies Guide to AR*. USA: Elsevier.

Kong, Y., Sharma, C., Kanala, M., Thakur, M., Li, L., Xu, D., Harrison, R. and Torrico, D.D. (2020) "Virtual reality and immersive environments on sensory perception of chocolate products: a preliminary study". *Foods*, 9, 515.

Korsgaard, D., Bjøner, T. and Nilsson, N.C. (2019) "Where would you like to eat? A formative evaluation of mixed-reality solitary meals in virtual environments for older adults with mobility impairments who live alone". *Food Research International*, 117, 30-39. https://doi.org/10.1016/j.foodres.2018.02.051

Koster, E.P. (2003) "The psychology of food choice: some encountered fallacies". *Food Quality and Preference*, 14, 359-373.

Kumar, J., Adhikari, K., Li, Y., Lindshield, E., Muturi, N. and Kidd, T. (2016) "Identifying barriers, perceptions and motivations related to healthy eating and physical activity among 6th to 8th grade, rural, limited-resource adolescents". *Health Education*, 116(2), 123-137. https://doi.org/10.1108/HE-03-2014-0035

Kwak, H.S., Jung, H.Y., Kim, M.J. and Kim, S.S. (2017) "Differences in consumer perception of Korean traditional soybean paste (Doenjang) between younger and older consumers by blind and informed tests". *Journal of Sensory Studies*, 32, e12302. https://doi.org/10.1111/joss.12302

Li, B.J. and Bailenson, J.N. (2018, 1 May) "Exploring the influence of haptic and olfactory cues of a virtual donut on satiation and eating behavior". *Presence*, 26 (3), 337-354. https://doi.org/10.1162/pres_a_00300

Lefebvre, S. and Orlowski, M. (2019) "Can, cup, or bottle? The influence of beverage vessel on taste and willingness to pay". *International Journal of Hospitality Management*, 76 A, 194-205. https://doi.org/10.1016/j.ijhm.2018.05.009

Lonbart, C., Millan, E., Normand, J.-M., Verhulst, A., Labbe-Pinlon, B. and Moreau, G. (2019) "Consumer perceptions and purchase behavior toward imperfect fruits and vegetables in an immersive virtual reality grocery store". *Journal of Retailing and Consumer Studies*, 48, 28-40.

Maynes-Aminzade, D. (2005) Edible Bits: Seamless Interfaces Between People, Data and Food. *Conference on Human Factors in Computing Systems (CHI'05)-Extended Abstracts. Citeseer, ACM, 2207-2210.*

Meiselman, H.L. (2003) "A three-factor approach to understanding food quality: the product, the person and the environment". *Food Service Technology*, 3, 99-105.

Meiselman, H.L. (2006) "The impact of context and environment on consumer food choice". In L. Frewer, H.C. van Trijp (eds). *Understanding Consumers of Food Products*. Woodhead Publishing, 67-92.

Meiselman, H.L. (2013) "The future in sensory/consumer research: evolving to a better science". *Food Quality and Preference*, 27, 208-214.

Miyashita, H. (2020). Norimaki Synthesiser: Taste Display Using Ion Electrophoresis in Five Gels. *CHI-EA'20: Extended Abstracts of the 2020 CHI Conference on Human Factors in Computing Systems*, 25-30 April 2020, 1-6. https://doi.org/10.1145/3334480.3382984

Narumi, T., Sato, M., Tanikawa, T. and Hirose, M. (2010) Evaluating Cross-Sensory Perception of Superimposing Virtual Color onto Real Drink: Toward Realization of Pseudo-Gustatory Displays. In *2010 Augmented Human Conference*. Megève France, 2-3 April 2010, Article No.:18.

Narumi, T., Nishizaka, S., Kajinami, T., Tanikawa, T. and Hirose, M. (2011) Meta Cookie: An Illusion-Based Gustatory Display. *International Conference on Virtual and Mixed Reality*. pp. 260-269. https://doi.org/10.1007/978-3-642-22021-0_29

Olabi, A.A., Lawless, H.T., Hunter, J.B., Levitsky, D.A. and B.P. Halpern. (2002) "The effect of microgravity and space flight on the chemical senses". *Journal of Food Science*, 67(2), 468-478.

Ong, S.K., Nee, A.Y.C. (2004) "A brief introduction of VR and AR applications in manufacturing". In S.K. Ong, A.Y.C. Nee (eds). *Virtual and Augmented Reality Applications in Manufacturing*. London: Springer. pp. 1-14.

Pennanen, K., Närväinen, J., Vanhatalo, S., Raisamo, R. and Sozer, N. (2020) "Effect of virtual eating environment on consumers' evaluations of healthy and unhealthy snacks". *Food Quality and Preference*, 82(June). https://doi.org/10.1016/j.foodqual.2020.103871

Peryam, D.R. and Pilgrim, F.J. (1957) "Hedonic scale method of measuring food preferences". *Food Technology*, 11, Suppl, 9-14.

Picket, B. and Dando, R. (2019) "Environmental immersion's influence on hedonics, perceived appropriateness, and willingness to pay in alcoholic beverages". *Foods*, 8, 42.

Pinto, V.R.A., Teixeira, C.G., Lima, T.L., De Almeida Prata, E.R.B., Vidigal, M.C.T.R., Martins, E., Perrone, I.T. and de Carvalho, A.F. (2020) "Health beliefs towards kefir correlate with emotion and attitude: a study using an emoji scale in Brazil". *Food Research International*, 129, 108833. https://doi.org/10.1016/j.foodres.2019.108833

Piqueras-Fiszman, B. and Jaeger, S.R. (2014) "The impact of evoked consumption contexts and appropriateness on emotion responses". *Food Quality and Preference*, 32, 277-288.

Pound, C., Duizer, L. and McDowell, K. (2000) "Improved consumer product development. Part one. Is a laboratory necessary to assess consumer opinion?". *British Food Journal*, 102, 810-820.

Ranasinghe, N. and Do, E.Y.-L. (2016) "Digital lollipop: studying electrical stimulation on the human tongue to simulate taste sensations". *ACM Transactions on multimedia Computing, Communication, & Application (TOMM)*, 13, 1-22, Article No: 5.

Ranasinghe, N., Nguyen, T.N.T., Liangkun, Y., Lin, L.-Y., Tolley, D. and Do E.Y.-L. (2017) Vocktail: A Virtual Cocktail for Pairing Digital Taste, Smell, and Color Sensations. In *Proceedings of the 25th ACM International Conference on Multimedia (MM '17)*. New York, NY, USA: Association for Computing Machinery. pp. 1139-1147. https://doi.org/10.1145/3123266.3123440

Redd, E.M. and de Castro, J.M. (1992) "Social facilitation of eating: effects of instructions to eat alone or with others". *Physiology and Behavior*, 52(4), 749-754.

Robinson, E., Aveyard, P., Daley, A., Jolly, K., Lewis, A., Lycett, D. and Higgs, S. (2013) "Eating attentively: a systematic review and meta-analysis of the effect of food intake on memory and awareness on eating". *American Journal of Clinical Nutrition*, 97, 728-742.

Robinson, T.E. and Berridge, K.C. (2003) "Addiction". *Annual Review of Psychology*, 54, 25-53.

Rollo, M.E., Bucher, T., Smith, S.P. and Collins, C.E. (2017) "ServAR: An augmented reality tool to guide the serving of food". *International Journal of Behavioral Nutrition and Physical Activity*, 14(1), 65.

Saleme, E.B., Covaci, A., Mesfin, G., Santos, C.A.S. and Ghinea, G. (2019) "Mulsemedia DIY: A survey of devices and a tutorial for building your own mulsemedia environment". *ACM Computing Surveys*, 52(3), 1-29, Article No. 58. https://doi.org/10.1145/3319853

Shilo, L., Sabbah, H., Hadari, R., Kowatz, S., Weinberg, U., Dolev, S., Dagan Y. and Stenkman, L. (2002) "The effects of coffee consumption on sleep and melatonin secretion". *Sleep Medicine*, 3(3), 271-273. https://doi.org/10.1016/S1389-9457(02)00015-1

Siegrist, M., Ung, C.-Y., Zank, M., Marinello, M., Kunz, A., Hartmann, C. and Menozzi, M. (2019) "Consumers' food selection behaviors in three-dimensional (3D) virtual reality". *Food Research International*, 117, 50-59. https://doi.org/10.1016/j.foodres.2018.02.033

Stafford L.D., Agobiani, F. and Fernandes, M. (2012) "Effects on noise and distraction on alcohol perception". *Food Quality and Preference*, 24, 218-224.

Stelick, A., Penano, A.G., Riak, A.C. and Dando, R. (2018) "Dynamic context sensory testing- a proof of concept study bringing virtual reality to the sensory booth". *Journal of Food Science*, 83, 2047-2051. https://doi.org/10.1111/1750-3841.14275

Stelmach-Mardas, M., Kleiser, C., Uzhova, I., Penalvo, J.L., La Torre, G., Palys, W., Lojko, D., Nimptsch, K., Suwalska, A., Linseinsen, J., Saulle, R., Colamesta, V., Boeing, H. and on behalf of the DEIDIPAC-Consortium. (2016) "Seasonality of food groups and total energy intake: a systematic review and meta-analysis". *European Journal of Clinical Nutrition*, 70, 700-708. https://doi.org/10.1038/ejcn.2015.224

Torrico, D.D., Fuentes, S., Gonzalez-Viejo, C., Ashman, H., Gunaratne, N.M., Gunaratne, T.M. and Dunshea, F.R. (2018) "Images and chocolate stimuli affect physiological and affective responses of consumers: a cross-cultural study". *Food Quality and Preference*, 65, 60-71.

Townshend, J.M. and Duka, T. (2001) "Attentional bias associated with alcohol cues: differences between heavy and occasional social drinkers". *Psychopharmacology*, 157 (1), 67-74.

Tuanquin, N.M.B., Hoersmann, S., Petersen, C.J. and Lindeman, R.W. (2018) The Effects Of Olfactory Stimulation and Active Participant on Food Cravings in Virtual Reality. *2018 IEEE Conference on Virtual Reality and 3D User Interfaces, VR, Reutlingen*. pp. 709-710. https://doi.org/10/1109/VR.2018.8446279

Van Herpen, E., Van Den Broek, E., Van Trijp, H.C. M. and Yu, T. (2016) "Can a virtual supermarket bring realism into the lab? Comparing shopping behavior using virtual and pictorial store representations to behavior in a physical store". *Appetite*, 107, 196–207.

Verastegui-Tena, L., van Trijp, H. and Piqueras-Fiszman, B. (2019) "Heart rate, skin conductance, and explicit responses to juice samples with varying levels of expectation disconfirmation". *Food Quality and Preference*, 71, 320–331.

Viswadhara Meenakshi, V. and Sowmya, K. (2003) ESTER: Experiencing Smell Through Electronic Reality. *2003 IEEE International Conference on Consumer Electronics*, Los Angeles, CA, USA, 17–19 June 2003. ICCE.

Whitmer, B.G. and Singer, M.J. (1998) "Measuring presense in virtual environments: a presence questionnaire". *Presence: Teleoperators and Virtual Environments*, 7, 225–240, Article 3.

de Wijk, R.A., Kooijman, V. Verhoeven, R.H.G., Holthuysen, N.T.E. and de Graaf, C. (2012) "Autonomic nervous system responses on and facial expressions to the sight, smell, and taste of liked and disliked foods". *Food Quality and Preference*, 26, 196–203.

Yan, K.S. and Dando, R. (2015) "A cross-modal role for audition in taste perception". *Journal of Experimental Psychology: Human Perception and Performance*, 41, 590–596.

Yoshimura, T. and Sakashita, Y. (2006) "Development of a perfume emission system via internet". *Journal of Computer Chemistry, Japan*, 5(4), 227–230. https://doi.org/10.2477/jccj.5.227

Zandstra, E.H. and Lion, R. (2019) "4 - In-home testing". In H.L. Meiselman (ed). Context: The Effects of Environment on Product Design and Evaluation. Duxford, UK Woodhead Publishing. pp. 67–85. https://doi.org/10.1016/B978-0-12-814495-4.00004-0

Zhou, P., Qin, M. and Dando, R. (2020) "A piña colada tastes better on a (virtual) beach". *Food Technology*, 74(4), 46–50.

Post Scriptum

I am writing this chapter in the time of COVID-19 social distancing, wearing masks while grocery shopping, Zoom calls with family and friends, Google Meet classrooms, and Skype cooking classes. I started this chapter with a quote I heard listening to NPR one day. Their story talked about a Zoom bar mitzvah ceremony held in New Jersey during quarantine, how one family and community were able to provide a meaningful and memorable experience for one of their members, even under highly unusual circumstances. In the words of one event participant, even though it was held via Zoom, the ceremony "felt real." The digital experience, while not the same, had a similar emotional feel to the real one.

By the time this book goes to print, more people may already have been vaccinated against COVID-19 and life would be under less-stringent circumstances. However, the situation will be different from preCOVID-19, before the masks and Zoom calls. The new technologies are here to stay, embrace them.

10

It ain't what you don't know that gets you in trouble.

It's what you know for sure that just ain't so.

Mark Twain

What this is about

This chapter highlights and provides efficient approaches in refining and screening product ideas for product developers to prioritize and classify insights in order to strategize their activities accordingly.

Product Innovation Toolbox: A Field Guide to Consumer Understanding and Research, Second Edition. Edited by Kannapon Lopetcharat, Dulce Paredes and Jacqueline H. Beckley.

Chapter 10

Tools to Refine and Screen Product Ideas in New Product Development

10.1

Contemporary Product Research Tools

Michele Foley

Key learnings

- ✓ The definition of a concept test
- ✓ The considerations in designing a concept test
- ✓ The considerations in interpreting a concept test

10.1.1 Introduction

Introducing a new product is expensive and risky, but it is critical to business growth. In most cases, the success of a new product cannot be known until after it is in the market. Although there are many factors that determine a product's success, the strength of the original idea in generating incremental growth is one of the most important. Deciding which ideas will be resourced is one of the most critical and most difficult business decisions in product innovation. Khurana and Rosenthal (1998), report that the greatest success comes to organizations that take a holistic process orientation to innovation.

Most businesses evaluate new ideas based on three broad criteria - technical feasibility, potential profitability, and consumer appeal. Scanlon (2009), in a *Business Week* article on innovation, reports that "For an idea to be considered

Product Innovation Toolbox: A Field Guide to Consumer Understanding and Research, Second Edition. Edited by Kannapon Lopetcharat, Dulce Paredes and Jacqueline H. Beckley.

for development, it has to meet Whirlpool's three-pronged definition of innovation: it must meet a consumer need in a fresh way; it must have the breadth to become a platform for related products; and it must lift earnings." For this discussion, we will consider only the consumer appeal - how relevant the idea is to the target consumer and how broad the appeal might be to a larger population.

Concept testing is a key element in successful, productive innovation efforts (Cooper and Edgett, 2008) because it provides a sense of potential and supports the definition of the product. Concept testing is a common practice in market research as indicated by the quantity of research suppliers found in the *Green Book*, the guide for buyers of marketing research services. In a recent search, 50 companies are listed for concept testing and 64 are listed for product testing (www.greenbook.org/market-research-firms/concept-testing). Product concept testing is an essential tool to ensure relevance and interest in the idea before significant investment is made in its development and commercialization.

10.1.2 What is a concept?

The Product Development and Management Association (PDMA, 2006) defines a concept as "a clearly written and possible visual description of the new product idea that includes its primary features and consumer benefits, combined with a broad understanding of the technology needed" (www.pdma.org/page/glossary).

Moskowitz et al. (2005) describe two types of concepts: a product concept and a positioning concept. Most concepts comprise product and positioning; both a description of a product or a service and benefits associated with the product. It provides a blueprint for the product.

There are many styles of concepts, from simple one-line descriptions of the product to complex descriptions with positioning, graphics, and price. The stage of development usually determines how complex or complete the idea. Early in concept development, simple single-line ideas might be screened to narrow down options. Ideas can come from a variety of sources - competitive intelligence, brainstorming sessions, or suggestion boxes. These ideas need to be put into context (consumer insight, benefit, or brand positioning) for testing. Then further refinement is used to develop the best positioning or selling orientation. The more complex concepts are used throughout product development to ensure the product design meets the expectations from the concept statement.

10.1.3 Elements of a concept

Because the concept is used in a variety of ways, it must be clearly communicated. In general, concepts are presented in a format shown in Figures 10.1.1 and 10.1.2 with the following five elements:

(1) Headline - summarizes the main idea of the new product.
(2) Accepted consumer belief - expression of a consumer need.
(3) Benefit - what's in it for the consumer.

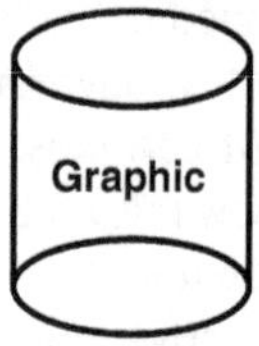

Introducing... "name and brief description"
It's important to me that... "accepted consumer belief"
Now, with "name", I get "benefit"
"Name" is "reason to believe"
Available in "store location" in "packaging" for "price"
(call to action)

Figure 10.1.1 Typical concept elements.

Introducing Michele's Oven Classics... Homestyle Cookie Favorites

Cooking from scratch is important to me but I don't always have the time .... especially when it comes to baking.

Now, with Michele's Oven Classics, I can provide cookies with that home-baked taste withot spending time baking in the kitchen.

Michele's Oven Classics are time-loved favorites made with the best ingredients, then frozen in individual packages to ensure freshness. We select the finest ingredients like creamery butter, real eggs and vanilla from Madagascar to create perfectly baked cookies. Then we flash freeze them in individual packs to ensure fresh-baked goodness. Simply thaw and eat or warm them up for that fresh-from-the oven experience.

Classic Favorites:

- Thin and crispy chocolate chip cookies
- Chewy oatmeal raisin cookies
- Rich and creamy sugar cookies

Available in the freezer in boxes of 12 cookies for $4

Figure 10.1.2 Example of a concept.

(4) Reason to believe - the product description, gives credibility to the promise, could include a graphic or image to help describe the idea.
(5) Call to action - where the consumer can buy it, price, etc.

This format is often referred to as the concept board because of the way it was traditionally shared in focus groups. Before revealing concepts to the target consumer, these basic elements of a good concept should be considered:

- Consistent voice; a first-person account can add personal relevance
- Neutral description of what the product experience would be in simple consumer language, not technical or business/marketing lingo
- Clear communication of relevant differentiated benefit
- Brand promise support.

10.1.4 What is a concept test?

Concept testing is the process of using quantitative and qualitative methods to evaluate consumer response to a product idea prior to the introduction of a product to the market (Jansen, 2019). The PDMA lists three types of tests in its glossary: concept optimization, concept screening, and concept testing.

Concept screening involves evaluating potential new product ideas during the discovery phase of a product development project based on their fit with business strategy, technical feasibility, manufacturability, and potential for financial success. In addition to screening ideas with consumers, internal assessments should be conducted with key functions in the organization based on their understanding of strategy, the market environment, and current capabilities to assess the potential for success.

Concept optimization with consumers is used to select the most appealing options of benefits and features to highlight in the concept test. The concept should be optimized to convey the attributes, benefits, and desired positioning adequately. Effective communication is important at this point to ensure a good idea is not eliminated because consumers did not understand it.

Concept testing provides insight into market potential for a product by identifying its potential success before investing in the time and resources to develop the product and marketing strategy. It can identify and prioritize ideas to meet new consumer needs as well as ideas to modify or reformulate existing products and services.

10.1.4.1 Concept screening tests

Concept screenings identify ideas that are sufficiently promising to merit further consideration. The concepts tested in this step are typically simple product descriptions and may include context (why this idea exists) and/or claims.

Concepts can be screened qualitatively or quantitatively. In this stage, it is possible to screen several ideas with the same consumers. If you know your target consumer, screening qualitatively is a rapid approach. Easy exercises such as sorting ideas into categories related to interest and uniqueness help identify priorities for further testing. Follow-up discussions can be used to identify areas for optimization and to explore elements of the final concept (accepted consumer belief, benefits, and reason to believe) before testing quantitatively.

Quantitative screening methods typically include testing multiple concepts online and some suppliers describe a tournament format in which top-rated ideas are then matched against each other to force winners. Sample sizes for quantitative screening vary depending on the number of concepts and knowledge of the consumer. The more concepts and/or scope of consumer targets drive larger sample sizes.

The analysis approach considers the interest or purchase intent scores and uniqueness. Higher potential products are high in both purchase intent and uniqueness. On the other extreme are low-purchase intent/low uniqueness rankings. These ideas are not worth pursuing either because there is low interest or there are other offerings in the marketplace. Ideas that are ranked high in purchase intent but low in uniqueness are not worth exploring further because there are probably many competitive offerings. Ideas that are ranked low in purchase interest but high in uniqueness may require further development in the optimization phase to communicate the idea and benefits more clearly.

Many organizations screen ideas through key internal business groups such as sales and manufacturing. With the ease and low cost of testing online, surveying both internal and external constituents provide a rich context for decision-making. While the consumers evaluate the ideas on purchase intent and uniqueness, the internal teams might evaluate the ideas on perceived relevance to customers or feasibility in the current systems.

10.1.4.2 Concept optimization tests

At the screening phase, the intent is to assess business potential and to decide which ideas to take to the next phase for optimization and further refinement. Ideas with higher potential from the screening phase enter optimization for further development. In this stage, qualitative interviews with the target audience can help explore usage, benefits, and positioning. They can be used to understand current solutions and drive differentiation from competitive offerings. In addition to general likes and dislikes, areas to explore include the consumer need and comparison to their current products or method of doing things. Discussion could also include the message and type of communications that would help them decide to buy the product or the service.

Following qualitative testing and before the concept test, conjoint analysis is recommended to optimize key elements of the positioning concept. When optimizing ideas, determining trade-offs can be very useful. Conjoint analysis is a technique that assesses the relative importance of individual components of an idea (see Chapter 8.2 for more information about conjoint analysis). This approach is recommended in the optimization phase to determine which elements are more compelling to consumers. It can involve the evaluation of rational or functional benefits, such as "a detergent that removes stains," nonrational or emotional benefits like "lets you be your best self," as well as, descriptions of product features, "available in your favorite flavors," store location, "you can find it in the dairy section" and price. This will ensure the final concepts are communicated clearly to consumers.

10.1.4.3 Concept tests

The final step in developing concepts is the quantitative test of the final concepts, potentially comparing them to other options and the competition. The most widely known method is an online survey to ask questions about what consumers think about the potential concept, and if they would be willing to purchase it. Typically, these concept tests are monadic, each consumer evaluates a single idea and, like the screening phase, interest, and uniqueness are primary measures. Beyond scales and open-ended responses, research suppliers have developed other techniques to better understand the ratings, such as highlighting in which consumers are able to indicate specific words or phrases that impact their reactions to the concept.

10.1.4.4 Concept product fit tests

Concept product fit testing is done as a final validation to ensure the product delivers on the concept message prior to product launch. It can be a central location test (CLT), home-use-testing (HUT), or test market (small-scale market launch) driven by business need and resources. The consumer metric for success is dependent on the type of innovation (incremental or disruptive) and can be derived from perception surveys (likely to buy) or behavioral approaches (pay for additional product) (Hanson, 2012).

10.1.5 Common measures

The key question to be answered prior to quantitative testing is "What criteria will be used to decide that the idea should go forward?" At this stage, most companies use benchmarks based on historical performance to set action standards. In establishing benchmarks, consider an appropriate context depending on the type of innovation under investigation. Different benchmarks should be set based on brand, target consumer, or type of innovation (extension vs. new-to-the-world). For more developed categories, volumetric assessments are used to determine whether the concept should proceed to product development. The following list includes some of the most typical measures:

- Appeal
- Uniqueness
- Believability
- Purchase intent
- Relevance, ability to meet consumer need
- Fit with brand.

Uniqueness and its ability to provoke interest measure a concept's novelty factor. Uniqueness reflects to what extent consumers see it as being different from products already in the market. Unique concepts lend themselves to more successful positioning efforts once they are introduced. The extent to which the concept is interesting measures a similar dimension. If consumers perceive the concept as interesting, they are more likely to be attracted to it.

Measures such as frequency of purchase and source of volume can be used to calculate potential based on the predicted size of the business. The outcome of the model is either an index of potential success or estimated volume or earnings in dollars.

10.1.6 Sampling: who do you test with?

In concept testing, at least two types of respondents are needed to understand the potential of the idea - a general population of potential users and the target consumer. A large sample of the general population is used early in the screening test to assess the breadth of acceptors and to identify subgroups who are most likely to be adopters. During optimization, the target consumer, either someone with the particular need or who is loyal to the brand, will be most helpful to guide the refinement of the idea. A common practice is to leverage lead users or consumer experts (Cooper and Edgett, 2008) in the idea creation and optimization stage. And at the concept testing phase, when the results provide an initial estimate of the size of the idea, information about the respondents' adoption orientation is relevant. Klink and Athaide (2006) report that purchase intent is an appropriate response for early adopters, a question such as "Would you purchase this product if a friend recommended it?" might be more appropriate for later adopters.

Prediction Markets are a recent alternative to testing with target consumers. A prediction market is a collection of people speculating on an event (for example, election results). In this wisdom-of-crowds approach, qualified participants build a portfolio (invest in concepts) that they believe best address a specific business question or consumer need. Instead of asking "Would you buy this concept?," participants respond to "Would others buy this concept?" This approach is a lower cost, faster alternative during the screening phase, especially if the target consumer is difficult to recruit. The approach also could be used with internal employees and category experts.

10.1.7 Biometrics applications

Neuroscience tools are described in Chapter 9.2. Biometric devices that track consumers' responses (galvanic skin response, eye tracking, and facial coding) provide insights in advertising, digital marketing, packaging, product development, and retailing (Farnsworth, 2019). Combining these techniques with traditional approaches can provide additional insights into the potential success of concepts. Respondents in neuromarketing studies are exposed to the same stimuli as in more traditional studies. One example is where the concept is created as a radio ad and the consumer listens to the verbalization of the concept (Pradeep, 2010). Responses are recorded continuously, allowing for detecting which of the elements generate the highest and lowest levels of effectiveness. Like looking at the integration of interest and uniqueness in the screening step, replacing uniqueness with emotional resonance provides compelling results. Higher- potential products are high in purchase intent and high resonance. On the other extreme are ideas with low purchase intent and low resonance. These ideas have low potential on both fronts. Ideas that are ranked high in purchase intent but low in emotional resonance, just as with uniqueness, are not worth exploring further because they are "me too" ideas. Ideas that require more optimization are ranked low in purchase interest but high in resonance. There is something in the idea that is compelling to the target but may not have broad appeal.

10.1.8 New developments in concept testing

10

Developments in digital marketing are driving innovations in concept testing. A relatively new approach to marketing called growth hacking is being applied to new product development (Holiday, 2014). In this approach, social media sites provide the platform for testing ideas. Using a website version of the idea, ads are created on social media platforms. Key measures such as engagement (how audiences interact with content) and conversion (for example, signing up on a website) are used to assess the concepts. Traditional tools such as online surveys can be incorporated into the website for additional understanding.

10.1.9 Conclusion: from winning idea to successful product

The process of concept testing is used to select, refine, and develop new ideas to increase their success. The newer or more disruptive the idea, the more refinement and development are needed to clearly communicate the idea.

Communication of the idea to the consumer can be challenging early in the idea stage because the developers may not have a way to describe the idea effectively. In addition, who you test with and how you measure success are critical. Using a blend of target consumers to evaluate the idea against the need for emotional resonance, brand loyalists to assess brand fit and a broader group to evaluate the potential size of the market can provide the most effective insights. The concept needs to be referenced throughout the development process to ensure the product design meets the expectations and promises described in the original idea. Referencing the concept during bench screening of prototypes provides a target to the project team and keeps them focused on key promises rather than their own personal preferences. The evaluation of the product with the concept by consumers validates the product experience and meets the expectations established by the original winning idea.

Discussion questions

- What makes an impactful product concept?
- How do you enhance and validate a concept through the research cycle?

References

Cooper, R.G. and Edgett, S.J. (2008) "Maximizing productivity in product innovation". *Research Technology Management*, 51(1) (March-April) for © Product Development Institute Inc. 2000-2010, 19 pages.

Farnsworth, B. (2019) *15 Powerful Examples of Neuromarketing in Action*. Retrieved from iMotions. www.imotions.com/blog/neuromarketing-examples (accessed 7 March 2022)

Hanson, J. (2012). "Product concept validation tests". In J. Beckley, D. Paredes and K. Lopetcharat (eds). *Product Innovation Toolbox*. Ames, Iowa: Wiley Blackwell, 317-324.

Jansen, B. (2019) *Concept Testing*. Retrieved from toolshero. www.toolshero.com/marketing/concept-testing (accessed 7 March 2022)

Khurana, A. and Rosenthal, S.R. (1998) "Towards holistic 'front ends' in new product development". *Journal of Product Innovation Management*, 15(1), 57-74.

Klink, R.P. and Athaide, G.A. (2006) "An illustration of potential sources of concept-test error". *Journal of Product Innovation Management*, 23, 359-370.

Moskowitz, H.R., Poretta, S. and Silcher, M. (2005) *Concept Research in Food Product Design and Development*. Ames, Iowa: Blackwell Publishing Professional.

Pradeep, A.K. (2010) *The Buying Brain, Secrets for Selling to the Unconscious Mind*. Hoboken, NJ: John Wiley & Sons.

Product Development & Management Association (2006) *The PDMA Glossary for New Product Development*. St. Paul, MN: Product Development Management Association (PDMA). https://www.pdma.org/page/glossary_access3?&hhsearchterms=%22glossary%22 (accessed 7 March 2022).

Holiday, R. (2014) *Growth Hacker Marketing: A Primer on the Future of PR, Marketing, and Advertising*. Penguin Random House LLC

Scanlon, J. (2009) *How Whirlpool Puts New Ideas Through the Wringer*. Business Week. www.businessweek.com/print/innovate/content/aug2009/id2009083_452757.htm (accessed 7 March 2022).

10.2

Insight Teams: An Adaptive, Self-directed Group to Discovery

Kannapon Lopetcharat, Dulce Paredes, and Jacqueline H. Beckley

The editors (adapted from the original chapter by Stacey Cox)

Key learnings

- ✓ How to utilize consumers differently for rapid development of innovative products
- ✓ Introduce a new "panel" format for development
- ✓ Discuss how insight teams differ from traditional tools used throughout the development process
- ✓ Identify what kind of information can be provided by Insight Teams

Note: What is described in this chapter makes certain assumptions:

(1) That some sort of "panel" may already exist
(2) You, the readers, are not starting from scratch
(3) If, you are starting from having no panels, you will use this chapter to tell you how to recruit and train the first panel and you or someone you consult with will play the role of the existing Insight Team to onboard the participants.

10

10.2.1 Insight Teams for discovery

In this chapter, we will focus on the use of a new type of consumer "panel" or team, as Stacey Cox, prefers to call them, which was used continuously at H.J. Heinz from 2000 until 2018 for product development discovery and direction (other companies and initiatives have embraced this approach but due to agreements, choose to not be recognized). The idea of an Insight Team was

Product Innovation Toolbox: A Field Guide to Consumer Understanding and Research, Second Edition. Edited by Kannapon Lopetcharat, Dulce Paredes and Jacqueline H. Beckley.

formed in 2000. At the time, Cox was beginning a product research department for part of the H.J. Heinz organization. The need for a descriptive panel as well as cost-effective and rapid methods for more consumer feedback earlier in the development process was identified. The need for creative ways to get all the needed work done was also identified, given this was a small (e.g. restrained resources) department. At times, as well, Cox, a new manager, would need another set of eyes, ears, and brains to help developers screen products against the consumer's expectations, provide feedback, attend meetings, survey the marketplace, and understand category dynamics.

In collaboration with a products research company (The Understanding & Insight Group), Cox evaluated tools she had used in other companies for driving product understanding, laid out the pros and cons, identified resource gaps and created a new tool, the Insight Team, to assist the mission of the department. With consultation, she first took a look at traditional descriptive methods. She had a number of years as a panel leader for several types of descriptive panels. She felt that descriptive panels provide good objective feedback on products. However, she felt there was a large gap in communication between the panel's work and the developer, in addition to the resource drain they could place on the organization. She then looked at other tools that Jackie Beckley from the U&I Group had at previous companies, narrative panels (a descriptive panel trained to be more elaborate in description around the story of the product) and proof-of-principle (POP) panels (perception and opinion of people) (Moskowitz et al., 2006).

In addition to descriptive panels, she evaluated the pros and cons of typical focus groups, groups lasting from 1.5 to 2 hours with developers trying to understand what product design parameters were of interest to the consumer. The main hurdle with this process was several clients indicated they were not getting enough useful direction to design more successful products. The traditional focus group was insufficient to really understand what that consumer expected from the promise of the concept and would truly purchase repeatedly. In an attempt to meet the client's needs, she introduced iterative consumer groups, focus groups where the consumer would return over several days, weeks, or months to provide more feedback to the developer as they proceeded through the development process, later titled a guidance panel.

10.2.2 Definition of an Insight Team

It appears that the term "panel" has many definitions in the consumer product goods industry. A basic definition is a group of consumers who are recruited and screened for participation in research. A panelist, according to Lawless and Heymann (1998), "connotes a participant as a member of a group that is often tested on more than one occasion." Participants on panels are recruited from some sort of source, most often a database of prerecruited individuals, who typically have been through some sort of screening process where profile information has been gathered (Resurreccion, 1998). The product research (sensory evaluation) field began using panels for research with the creation of trained or expert panels for descriptive analysis, that is, the quantitative descriptive analysis (QDA) method (Stone et al., 1974) and the Spectrum method (Meilgaard et al., 1987). Since then, panels of many types have emerged with the broad commercialization of database tools and do it yourself (DIY) approaches to all forms of marketplace research (ESOMAR, 2020).

Consumer and citizen panels can be divided into two types: those that are more focused on answering a product or package design question vs. those that would be set up for a marketing focused question. The Insight Team is an objectively trained "consumer panel," a group of individuals who have been recruited and coached and/or trained to provide iterative and objective, qualitative or quantitative feedback on anything from ideas, concepts or product/package designs. Over the years, the name Insight Panel, as originally called, became the Insight Team due to the dynamics of how they functioned as a collection of people. The tool functions as more than just a group of participants recruited for a task and providing "one-way" answers, but as a group of individuals/consumers trained to work as one team, all functioning together to drive not only development but the business forward. The Insight Team has been utilized to answer both types of questions, both business and technical. The Team has been utilized to ground the business in where the category is going, provide understanding on competitors' business models, language for concepts, product and package direction and design feedback, preparation insights, and screen suppliers.

The Insight Team at Heinz was a group of 10–12 individuals who started with an adaptive approach to looking at products and articulating observations around those products. They adaptively moved to wanting to rate and rank products using a more formalized approach and were eventually introduced to a classic scoring of attributes (the Heinz team used a Spectrum method since Cox was familiar with that approach [Meilgaard et al., 2007] though any method would have been selected), so they could function as a descriptive panel, in terms of evaluating products with attributes, aligning on ratings using the 15-point universal scale, and using references to guide the conversation, when need by a given task. ***But here is where they end as a traditional descriptive panel. As mentioned previously, Cox needed this team not only to provide the traditional descriptive output for some cross-functional partners, but she also needed them to be researchers and panel leaders at the same time!*** So the group was additionally trained in various aspects of team behavior, leadership, some business acumen, and research that would be necessary for them to succeed as a team (or a department of one) as well as individually.

10.2.3 When to apply the skills of an Insight Team

An Insight Team can be used anywhere in the development process (Figure 10.2.1).

At the beginning stage of the new product development (NPD) process, business assessment, and/or ideation, an Insight Team can be used to understand what is occurring in the marketplace. Heinz used them to identify white space or gaps in product design in the category prior to further ideation or concept work. An Insight Team can be used in the concept development and qualification

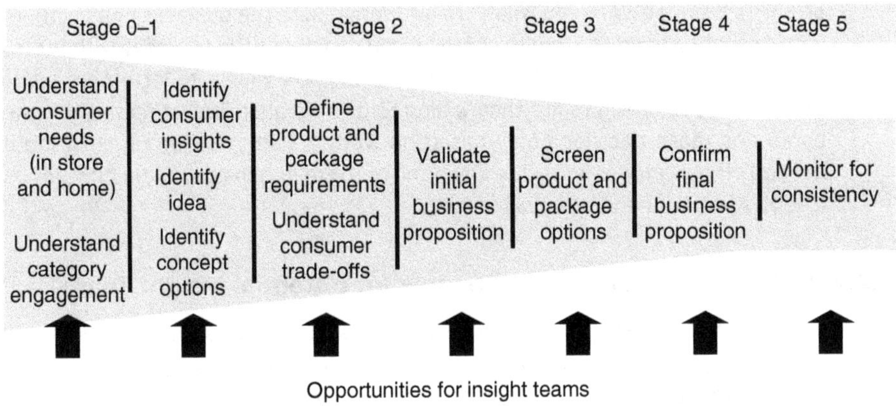

Figure 10.2.1 Utilization of Insight Teams in the development process.

process or Stage 2. Heinz used the Insight Team to evaluate initial concepts and to refine wording in the product description. In Stage 3, or product qualification, an Insight Team can be used in all sorts of ways. Cox used them to screen prototypes prior to quantitative consumer work; review the category to determine potential product and package drivers prior to qualitative consumer work; support the work of the developer or researcher with product purchase; lay out the potential consumer conversations for qualitative work; recommend stimuli for research; and to prepare as well as develop language for conjoint analysis (see Chapter 8.2 for details about conjoint analysis). Once the product is launched, the Insight Team can also be used for documenting the profile of the launched product, monitoring it over shelf-life, or anything else the cross-functional team might need throughout the process, such as data mining or report generation.

10.2.4 Implementing Insight Teams for development

10.2.4.1 Insight Team composition

The key to a successful Insight Team is having participants who:

(1) Love categories the company has businesses in
(2) Have background leading others, especially in teams if possible
(3) Have creative and innovative minds
(4) Are diverse in backgrounds and approaches to life.

So what is meant by this?

When embarking on an Insight Team, one should first take a hard look at your current descriptive panel and determine if they could do the work you need them to do. Can they move to the next level of understanding? Can they begin to integrate the business, development process, and marketplace using their objective training as a basis? Also, take a hard look at you and your organization. Are you willing to let go of some of the more traditional approaches to consumer panel or descriptive panel usage? Are you willing to let the panel see all the products branded? Are you willing to let them talk to the business units without you present? Are you willing to share all you can about the business with them? Then look at the organization, are your developers willing to let others guide them in product design? Are they willing to give up quantitative data (the numbers!) and make decisions only based on words? Can you give up shelf-life if needed? If the answer is "no" to any or all of these, the Insight Team may not be a good fit for you and/or your business.

10.2.4.2 Assessing your current panels for potential Insight Team

If you have decided to give it a try, then at the start of the recruiting process for the Insight Team, assuming a descriptive panel of some sort exists, Cox suggests to first look at the current members of your descriptive panel or an ongoing consumer panel that you have control over for projects. Begin by evaluating their styles of communication, personality styles, and learning styles. Cox had the team participate in the Myers Briggs test (Myers et al., 1998), as well as communicating and learning style assessments. She, at times, had the team participate in the team assessment from *The Five Dysfunctions of a Team* by Patrick Lencioni (2002). Through these tools, Cox has been able to determine where the gaps in a team are and what skill sets may be missing. It is helpful to talk to the team about their needs and the gaps they see at the team skills level, styles, or experiences that would be of value to increase the success of the team moving forward. Once the assessment is complete, a job description is written (see example below), and potential team members are recruited.

10.2.4.3 Recruiting and screening an Insight Team member

Cox's recommendation is not to use employees, if at all possible, for an Insight Team. This team needs to remain objective and needs to have the time available to do anything that is needed from them. Their job, in her experience, is to help the business with resources, not be a drain on them.

At Heinz, Cox worked through a temporary contractor firm to recruit consumers from various avenues of life. Once recruited by the firm, the firm does an initial prescreening. The initial pre-screening consists of questions relating to:

(1) Ability to provide objective evaluations, such as no dentures, no tongue, or nose piercings
(2) Ability to conduct work on a computer
(3) A certain level of business tool capacity such as Word, Excel, and PowerPoint knowledge and
(4) An ability to think and articulate creatively.

Figures 10.2.2 and 10.2.3 outline some of the skills and attributes to look for when recruiting new members to the Insight Team.

Table 10.2.1 is also an example of the job description for an Insight Team that is used in recruiting. Recruitment is conducted from both the public as well as family members/friends of company employees.

- [ ] Loves exploring, creating and learning about the category
- [ ] Has some business knowledge (this is a plus)
- [ ] Conducts himself or herself professionally
- [ ] Can work in the "grey areas"
- [ ] Can deal with change and can move rapidly from one area of the job to another - flexibility and adaptability are key
- [ ] Is Interested in assisting the company in developing innovative cutting edge products
- [ ] Is interested in a long-term part-time temporary position that has flexibility for additional hours if available
- [ ] Wants to work with a group that continually grows and learns
- [ ] Creativity is valuable: art, music, poetry, crafts, people-related things
- [ ] Demonstrates a balance between listening and leading
- [ ] Has basic writing, verbal and listening skills
- [ ] Is able to follow instructions well
- [ ] Computer skills are a plus: PowerPoint basics a must; Word and Excel nice to have
- [ ] Has their own transportation for shopping/restaurant excursions
- [ ] Is willing to work additional hours in evening and weekends if needed

Figure 10.2.2 An Insight Team member should possess the above 15 desired skills.

- [ ] Mature, well rounded individual with an open mind
- [ ] Someone who can think "outside the box"
- [ ] Good leadership and organizational skills
- [ ] Good personality and able to work in a group atmosphere
- [ ] Has patience and will take this job seriously
- [ ] Flexible, critical thinker, and can take constructive criticism
- [ ] Has excellent communication skills and is creative and articulate
- [ ] Follows directions and is detail oriented
- [ ] Not afraid to consume anything (food-wise)
- [ ] Adventurous
- [ ] Open to challenges

Figure 10.2.3 An Insight Team member should possess the above 11 desired attributes.

Table 10.2.1 An example of the job description for the Insight Team that is used in recruiting at Heinz.

The H.J. Heinz Company is currently recruiting persons interested in providing Heinz exceptional product insight through various product evaluations.

This exciting job entails the selection, purchase, preparation, and evaluation of a wide variety of food products, not limited to the Heinz products. The most critical piece of this position is the product evaluation. Products are evaluated daily, individually, and as a group. These evaluations are used to guide product development efforts at Heinz. A person in this job has many opportunities for interaction with various members of the business, especially the product developers.

A person applying to this job should be interested in a long-term, part-time commitment to the organization. The person should thoroughly enjoy working with others in a group setting to accomplish small to very large job assignments. The person should be open-minded and have a desire to listen to and share others' ideas and opinions.

The job requires a high degree of critical thinking, creativity, flexibility, self-motivation, and a drive to get the job done. Leaders will truly succeed in this type of work environment. The person should be willing to try new products and ideas on a daily basis.

Other job requirements:

- Flexibility to complete additional job assignments outside of the 9 a.m. to 1 p.m. work schedule: may include work in the evening or on the weekends from your home
- Ability to communicate through a wide variety of methods: writing, visual, talking, listening, presentation
- Organizational and time management skills
- Basic computer skills – Microsoft Office including: Word, Excel, and PowerPoint
- Transportation: Ability to travel to and from work, grocery stores, restaurants, etc.
- Loves to have fun!

Once prescreened, the first of three separate sessions are set up on site with the current Insight Team leading the session. For this session, basic sensory capabilities screening (taste, smell, visual, tactile, and auditory acuity), potential members are brought in for approximately 45 minutes. The Insight Team members provide a brief introduction and then the recruits are asked to complete several tasks. The introduction covers the purpose of the job, its roles and responsibilities, how the team interacts with the business and development cycles and the plans for the rest of the screening and training. The recruits then go through discrimination tests for basic taste thresholds, taste intensity ranking, aroma identification (of various herbs and spices), visual ranking of color intensity, and a visual exercise to determine their ability to estimate magnitudes. Participants have to pass at least four of the activities to move to day 2.

If recruits have made it to day 2, they are given an assignment for the next day to bring in a picture or two (or object) that describes a topic to them without, as in the case at Heinz, using food. For instance, they may be asked to bring in items that describe "genuine" without using food. They are asked to speak on their topic for five minutes in front of the group the next day. In the day 2 session,

8-10 participants are gathered in a group with the current team members watching or leading the discussion. The facilitator of the discussion reviews the day before and gives a little more information about the ins and outs of the Insight Team, its processes, project types, organizational structure, etc. The recruits are then asked to go around and share about themselves and the items they brought in and what the topic means to them. The purpose of the group is two-fold: (1) To see how articulate and creative they can be on short notice with an "out of the box" assignment, and (2) To see how they interact in a group setting, do they try to take over, do they only listen and don't add, or do they like to engage in the conversation?

Those recruits that pass the group discussion session are then asked back for a shadowing day. One or two recruits are asked to shadow one of the team members for the day. The team spends the day working on a current project and the recruits work right along with them. The recruits are taken to the store to understand how the team shops, prepares, and even evaluates products. Through this process, both the recruits and team members can determine who would be a good fit and the recruits can determine if they want to be on the team. The team is also able to provide the recruits even more insight into the team's roles and responsibilities in the development process. After day 3, the team determines the members with a final sign off from the manager.

10.2.4.4 Implementing the Insight Team process

The Insight Team's process is a comprehensive approach utilizing many different techniques to accomplish the required results. The training takes the team through four different sections:

(1) Team roles and responsibilities
(2) Parts of the process and tools
(3) Language (attributes) and
(4) Teaming skills.

For the training, Cox has been able to get new members contributing to business results within the first week of training. With some teams, due to the business needs and project timelines, she has been able to modify the training and have the new members acting as full contributing team members in three weeks.

10.2.4.5 Insight Team jobs

There are eight jobs within the Insight Team (Figure 10.2.4). Since the goal is to create a team of people who will be self-managed and self-directed, it is important to have the jobs of the team well understood by all participants and supervisors.

The person that leads the process for the project is the ***process facilitator***. Instead of having an outside panel leader or facilitator like many descriptive and consumer panels, the team rotates through the process facilitator position by project. The facilitator ensures the project makes its deadlines, and is point person for communication of issues with upper management, and is the key

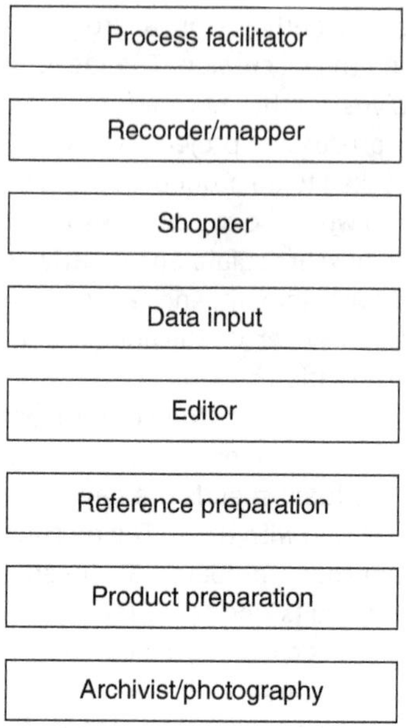

Figure 10.2.4 Insight Team roles and responsibilities.

contact between the team and the developer and/or requestor throughout the process. The other functions include ***recorder mapper***, who has responsibility for note taking and documenting all conversations within the team throughout the projects; ***shopper***, who is responsible for coordinating or conducting all shopping needed for the project; ***data input***, who inputs all notes, data, graphs and generates the initial report for review by the team; ***editor***, who is responsible for editing the report either individually or leading the team through the process; ***reference preparation***, who ensures all references for evaluations are available and prepped; ***product preparation***, who ensures all products needed for evaluation are available and prepped; and ***archivist/photographer***, who takes all photos needed for the report as well as ensuring all documents used during the project and the report are filed appropriately for future reference. The team rotates through all jobs for each project.

10.2.4.6 Steps in the Insight Team processes

The process for the Insight Team contains eight phases (Figure 10.2.5).

The first phase: Project initiation and planning provides the team with an opportunity to talk to the client, to understand their objectives and desired outcomes, and to plan how they may approach the project to get the best results and meet the timeline. This phase contains the following steps:

- Requestor initiates project with Insight Team manager to get on the schedule
- Manager reviews initial test request with the team
- Research proposal is developed, clarified, and agreed upon

Phase I	: Project initiation and planning
Phase II	: Experiential stories (history and experiences)
Phase III	: The marketplace
Phase IV	: The big picture
Phase V	: The details
Phase VI	: Report assembly
Phase VII	: Presentation of the findings
Phase VIII	: Team debrief

Figure 10.2.5 Phases of the Insight Team process.

- Process facilitator (with team if desired) develops a tentative agenda for all phases for team evaluation and alignment
- Jobs are assigned based on the job grid and job coordinators develop phase agendas for their jobs
- Phase agenda is reviewed client and finalized by the team
- Data input creates a folder for project information and inputs and files proposals electronically.

Phase 2: The experiential stories is an opportunity for the team to become grounded on what is already known about the topic. The team spends time discussing past experiences they or their friends and family have had related to the project topic. They spend time researching the topic on the Internet, books, journals, etc. This phase allows the team to refine their project direction and generate hypotheses when exploring the marketplace in phase 3. Phase steps include:

- Team conducts storytelling session on previous projects, stories, and experiences in the category
- Map of past experiences is developed and refined, providing context for the rest of the process
- Mapping software used to visually represent the experiences and refine throughout the project.

Phase 3: The marketplace, is designed to get the team grounded in what is occurring in the marketplace that may have an impact on the outcome of the project. Sometimes, the project stops here with the team reviewing the market-place and sharing findings. For instance, the Insight Team has had several projects where the client needed to understand how food service establishments menu and serve various products. The team went out and evaluated products in over one hundred restaurants and was able to provide a comprehensive understanding of product design across all types of establishments. For the marketplace phase, we spend time teaching the team how to shop differently, how to not only pick up what they need but to also see what else is going on in the categories they are familiar with, plus discovering what is happening in other noncompany related categories. How are consumers making decisions with all the product designs available to them? We have a shopping pyramid we use for teaching this marketplace phase of the project (Figure 10.2.6).

This phase includes the following:

- Team discusses how "deep" to go in the marketplace using Aha!'s from stories from phase 2
- Team refines what is needed for purchase and where the shopping is going to take place
- Shopper coordinates shopping trips throughout the process
- All products purchased are documented
- Marketplace experience is shared and informs the additional phases in the process.

Phase 4: Big picture is considered a rapid, at times "30,000 foot" review of the products. During this phase, the team becomes well versed in the world of this product experience or the category being reviewed. The team will

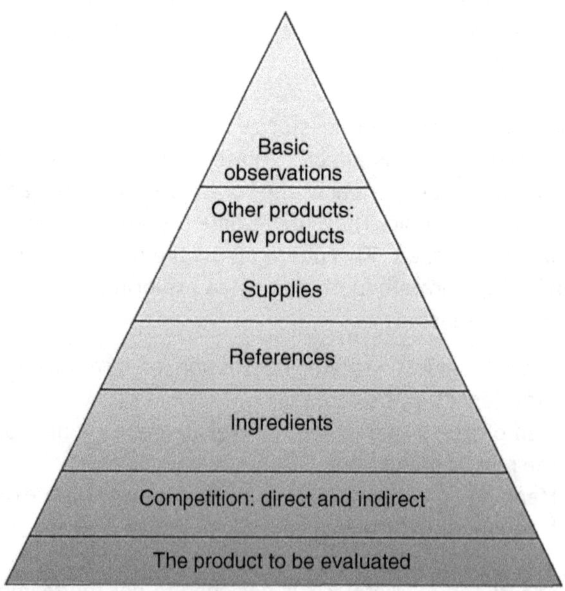

Figure 10.2.6 The shopping pyramid.

often spend time "powertasting" (Moskowitz et al., 2006) all the products to develop evaluation procedures and ballots, determine references that may be needed, etc. Steps include:

- Team develops a plan, procedures, and references (if needed) to be used for rapid product evaluations
- Team discusses what will be evaluated in the powertasting
- Product prep coordinates preparation of products
- Evaluation of the product during preparation is done by the preparation team
- Photographs are taken of the products
- Team evaluates all products and identifies key findings and determines if a detailed evaluation is needed
- Adjustments to the phase agenda may be made
- Overall category experience is documented electronically
- Process facilitator leads the discussions
- If a detailed evaluation is to occur, a ballot is developed including definitions and references
- Team decides if additional training is needed on specific terms.

In Phase 5: The details, the team focuses on more detailed evaluations of the product. In this phase, the traditional descriptive scales, references, and attributes are utilized in the process. Events which occur in this phase are:

- Team determines which (if any) key references will be needed for this phase
- Reference person coordinates shopping list and preparation of references as needed
- Reference person refines and documents preparation procedures electronically
- Product preparation obtains and documents products including preparation procedures electronically
- Team evaluates the products in context individually
- Process leader facilitates product evaluation discussion
- Recorder/mapper documents product discussion
- Team reviews and refines product evaluations prior to input
- Data input documents evaluations electronically (including maps/graphing).

During the last three phases of the process, Phases 6, 7, and 8: Report assembly, presentation of findings and team debrief, the Insight Team creates a report which contains:

- Project background
- Executive summary
- Key findings for the project and
- Recommended next steps (Figure 10.2.7).

The Insight Team then schedules the presentation with the client and other cross-functional team members. The Insight Team will then present the report often with samples of products they evaluated or recommend

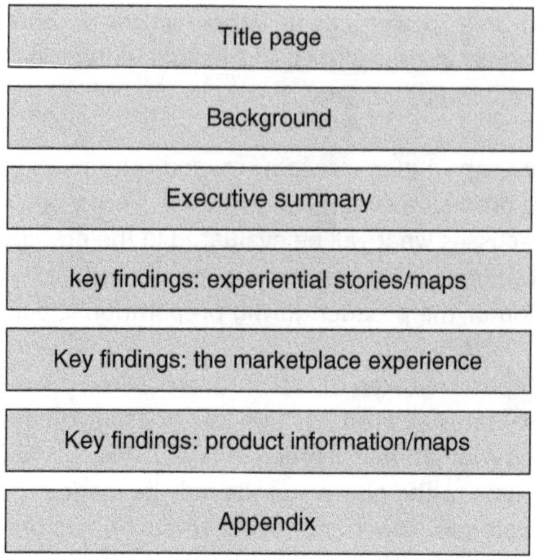

Figure 10.2.7 Insight Team report format.

products they have created based on key category drivers. It is during the presentation that one can often see the true leadership and innovation of all of the members of the team. The presentations are extremely engaging and insightful for the cross-functional teams. After the presentation, the Insight Team is then tasked with debriefing the outcome of the project. They review what worked well throughout not only the presentation but also the project in its entirety, how the team dynamics worked and how the team would like to change the process for the future. Key steps in these three final phases include

- Data input completes electronic documentation
- Editor coordinates the review of all electronic documentation
- Process facilitator coordinates team review of the final documentation
- Team prepares and gives the presentation to the requestor and cross-functional team
- Team debriefs on what went well and what could be better
- Final report and supporting documents filed electronically.

10.2.4.7 Maintaining the Insight Team

The key to maintaining a successful team is balance, a balance between creating a feeling of investment in the business among the team while ensuring they remain true to their job of providing objective, insightful feedback. Avenues Cox has found to keep this balance are twofold: (1) Including the team in as much of the business situation as possible, and (2) Ensuring they are a strong functioning team which can be self-directed and self-managed.

Motivating people to do their best, remain focused on the task at hand and stay on the team long term is a challenge. This topic has been debated over the years and can be seen still in online chats, at conferences and in training courses. Some people will recommend motivating team members through activities outside of work time. Cox has found having what is called a "red light day," a day to stop what the team is doing and engage in a more fun, light-hearted, team-building activity, such as visiting a museum or going out to breakfast, benefits the entire group. It is good practice to have an activity like this happen once every other month if possible. Another motivator people have tried is monetary or a similar type of incentive. This is provided, as well, yearly or depending on the success. There are also attendance incentives, thank you incentives, and other monetary incentives over the years. All of these levers help the team become a more unified force for the organization.

There are many, many potential motivators; however, the motivator that Cox found most useful is increasing the level of engagement of the team in the business and its strategies. She believes it is almost demotivating if the panel feels they just come in, do their job and don't really get to see the impact they are having on the business or products that are going to market. They want to feel they are a critical part of what makes the business successful. They want to know their input is of value and an integral part of what the business teams are using to make important strategic decisions.

But be careful! With the opening up of the communications lines with the business. Cox encountered teams displaying characteristics of overestimation of worth and entitlement when provided too much access to the daily business ins and outs. She has come across feelings of wanting to have involvement in all decisions being made, wanting to be able to have more influence on the direction of the business, and wanting to know more in areas that are not really integral to complete their work. Cox has seen teams get more caught up in the business or social aspect of the team and move away from the process and their role in the organization.

For a strong functioning team, Cox believes it is important that all team members, especially prospective recruits, understand that this job position is "the one job you will have in your lifetime where your performance as an individual doesn't matter as compared to that of the team. If the team succeeds then you all succeed, if the team fails, then you all fail." One of the ways to strengthen the team is through constant review of the teaming and individual skills training. The skills reviewed are:

- Communication, personality, and learning styles
- Listening and hearing
- Teaming and coaching
- Problem-solving and critical thinking
- Innovation and creativity.

Cox has used a multitude of resources throughout the team-building process. Books used are the following:

- *The Art of Possibility* by Rosamund Stone Zander and Benjamin Zander
- *A Theory of Shopping* by Daniel Miller
- *The Springboard* by Stephen Denning
- *Coaching for Commitment* by Dennis Kinlaw.

10.2.5 How to use the Insight Team

Cox has been able to utilize the Insight Team in so many different ways at Heinz to support the business. This list shows just a few examples of how to use this creative team in guiding business decisions and providing more insights into the consumer interaction with a product, package, or category:

- Reviewing various categories to determine potential product drivers and nuances
- Being trained observers in the marketplace or with consumer behavior
- Creating recipes with existing products and new ingredients for white space opportunities
- Creating the ideal prototype
- Developing names for new products based on flavor perception
- Confirming products cover a testable range in a design of experiments
- Providing concept wording
- Perceptual mapping of the category and prototypes
- Package design evaluation and recommendations
- Reviewing consumer data such as concept tests, central location tests, and segmentations to provide data interpretation support to the department
- Collecting and developing language for conjoint studies
- Supporting developer needs, such as checking product design at various cook times and ball park drain weights
- Traveling to customers to present Insight Team information
- Purchasing products and setting up category reviews (without evaluations) for developers.

The Insight Team can really be used to accomplish many different tasks for a research organization. However, Cox suggests three areas that she would not use their work for:

(1) Replace consumer acceptance or preference ratings from the target consumer
(2) Make any business decisions based on what they like or don't like or
(3) Make any type of volume forecasting prediction based on their findings.

10.2.6 Case study of using the Insight Team

One of the wins Cox shared when using the Insight Team was on a product that was not doing well in the marketplace; however, the category was growing. The developer needed to improve the product but was unsure where to start. The Insight Team reviewed the category utilizing the entire process. They "power-tasted" (Moskowitz et al., 2006) all of the category, conducted drain weights, analyzed potential ingredients from various marketplace sources, and created an ideal prototype that they felt delivered against the concept and its product promise and should beat the competition. The developer recreated the ideal formula utilizing ingredients in the Heinz system plus nine other options and tested them against the current product and lead competitor with the target consumers. The prototype that was created based on the Insight Team's ideal received: 73% top 2 box purchase intent, 7.4 in overall liking, and a 67% top 2 box concept/product fit, the highest of all the products evaluated.

10.2.7 The future of Insight Teams

In summary, Cox believes there will be many more types of opportunities for utilizing groups like the Insight Team to drive the business. More often than not, resources are limited, yet better insights and bigger innovations are still needed. This team can become an integral arm of the research organization increasing the reach of current resources, delivering true innovation, while utilizing a cost-effective approach. This team is able to function as objective researchers, descriptive experts, and insight gatherers for any consumer and product/package interaction.

The trend of using creative, highly articulate consumers for more impactful, insightful research has flourished in many organizations over the past 10–15 years. In fact, some say consumers today, according to Simon Chadwick, are even expecting to be more involved in the cocreation of products, communications, and brands (Chadwick, 2010). Cox has suggested in the future an increased opportunity for Insight Teams to interact even more closely with consumers to design and interpret. It is possible that target consumers could engage with the team in a process where consumers and the team are side by side creating a product, package, or idea. The idea of a highly trained group of people with the perspective of objectivity and detail vs. subjectivity and experience could be a powerhouse in developing relevant ideas rapidly. Cox's words proved prescient as a number of firms - among them information technology providers and consumer goods manufacturers - are already refining their co-creation strategies involving internal teams, consultants, and customers.

The future of teams like the Insight Team will require organizations to look hard at traditional approaches and ask: "Can they do more to drive the business than just provide descriptive profiles or point in time feedback? Can these creative and innovative minds be used to drive the organization faster through the development process and get to a better end result? Is it possible that these team members can become even more involved in the creation of the development process?" Insight Teams should become an even stronger partner with the business team, specifically the development team, leveraging their tools and working side by side on the bench, creating products that will sustain in the market.

Discussion questions

- Given the self-directed, self-managed nature of Insight Teams, why would they not be adopted as a substitute for other forms of consumer/trained panels?

- Why should companies trying to make offerings in today's world want to consider this approach within their organization?
- How might this approach be further adapted for the world we are living in?

References

Chadwick, S. (2010) *The Connectivity Revolution*. Research World, No. 23, September. ESOMAR magazine for marketing intelligence & decision making.

ESOMAR (2020) Global Market Research 2020 An ESOMAR Industry Report, digital pdf. www.esomar.org

Lawless, H. and Heymann, H. (1998) *Sensory Evaluation of Food: Principles and Practices*. Norwell, MA: Kluwer Academic/Plenum Publishers.

Lencioni, P. (2002) *The Five Dysfunctions of a Team*. San Francisco, CA: Jossey-Bass.

Meilgaard, M., Civille, G.V. and Carr, B.T. (1987) *Sensory Evaluation Techniques*. Boca-Raton, FL: CRC Press.

Meilgaard, M.C., Civille, G.V. and Carr, B.T. (2007) *Sensory Evaluation Techniques*. Boca Raton, FL: CRC Press, Taylor & Francis Group.

Moskowitz, H., Beckley, J. and Resurreccion, A. (2006) *Sensory and Consumer Research in Food Product Design and Development*. Ames, IA: IFT Press and Blackwell Publishing Professional.

Myers, I.B., McCaulley M.H., Quenk, N.L. and Hammer, A.L. (1998) *MBTI Manual (A Guide to the Development and Use of the Myers Briggs Type Indicator)* (3rd edition). Mountain View, CA: Consulting Psychologists Press.

Resurreccion, A. (1998) *Consumer Sensory Testing for Product Development*. Gaithersburg, MD: Aspen Publishers, Inc.

Stone, H., Sidel, J., Oliver, S., Woolsey, A. and Singleton, R.C. (1974) "Sensory evaluation by quantitative descriptive analysis". *Food Technology*, 28(11), 24-34.

10.3

Rapid Approaches in Defining the Product Space and Product Optimization

Jennifer Vahalik, Ratapol Teratanavat, Jennifer Lewis, Mary Sonnen, Melissa Jeltema, and Jacqueline H. Beckley

Key learnings

- ✓ Benefits of the rapid product navigation (RPN) and rapid product optimization (RPO) processes
- ✓ What to consider when designing RPN and RPO research
- ✓ How to develop the qualitative product space (QPS)
- ✓ How to execute the RPN and RPO processes

10.3.1 Doing more with less

What do you do when the marketing department wants to launch a new product or line extension in only a few months, and you have limited resources to do so? That's the situation that Company ABC's product researchers and product developers found themselves in when the marketing department identified an opportunity and wanted to launch a line extension product into an existing category, which has been dominated by its competitor. Another situation is when Company JKL wanted to revamp an existing product line to increase repeat purchase from their consumers. In both cases, both companies also faced resource constraints (time, money, etc.). The teams were asked to do more with less (e.g. less money, less time, and less people). So how can we most efficiently develop the best product possible? In this chapter, we will highlight new research approaches called rapid product navigation (RPN) and rapid product optimization (RPO) - these are rapid and cost-effective approaches in defining the product space so product developers can decide what to focus on RPN and how to optimize products RPO.

10

Product Innovation Toolbox: A Field Guide to Consumer Understanding and Research, Second Edition. Edited by Kannapon Lopetcharat, Dulce Paredes and Jacqueline H. Beckley.

10.3.2 Listening to understand

There are multiple approaches for developing new products. A product can be developed to match a predetermined descriptive profile. Or consumer acceptability tests (either central location or home use testing) can be used to screen and identify optimal products. Often, consumer acceptability testing requires iteration as it may take multiple rounds and a lot of money to reach an optimum product, if ever. A pure design of experiments modeling approach may lead to an optimum design that doesn't make sense, especially when there are many interactions between the factors. And when results don't make sense, we don't have any understanding of why people rated the products the way they did. What's missing from these approaches is the act of *listening*.

Both RPN (Lewis et al., 2010) and RPO rely heavily on consumer feedback to help guide where the product space is and whether we can achieve an optimal product, respectively. Both approaches consist of a series of qualitative discussion groups, so RPN/RPO is first and foremost, consumer-driven. By *listening to understand*, the project team can gain additional insights about the relative positioning of the prototypes, allowing them to make better decisions about product direction - as seen in this example of a product map (Figure 10.3.1), where after placement of an array of current market products, consumer research participants also indicated where their "optimum" product would be for these two attributes. RPN/RPO helps the product team identify critical product attributes and critical trade-offs consumers are willing to make. RPN/RPO is also, by definition, rapid. Because each product evaluation provides the consumer feedback that drives subsequent product trials, an acceptable product design can be reached quickly. RPN/RPO is a highly effective, powerful approach. The voice of the consumer is incorporated into all product decisions. Because

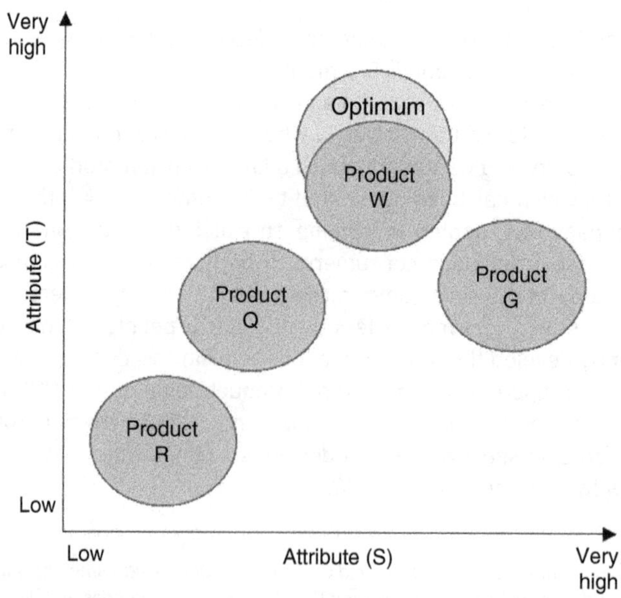

Figure 10.3.1 Example product map.

10

product developers are intimately involved with the research, they gain a deeper understanding of consumer wants and receive immediate, actionable direction. Compared to current methods, RPN/RPO provides better results, faster.

10.3.3 Defining rapid product navigation (RPN) and rapid product optimization (RPO)

RPN and RPO are consumer interactive approaches that allow us to quickly and effectively identify the product space to focus on and ultimately optimize products. Both approaches require stimuli, which could be market products, protocepts, or prototypes for consumers to react to. Having consumers interact with stimuli in the product space helps elicit real conversations about product attributes in context vs. in theory. Now, let us look into the similarities and differences between RPN and RPO.

> **RPN** is used during early stages of product development to explore the product space of a given category in order to identify what areas are important to consumers and what areas the company should focus on optimizing. This approach is suited for "brand new to the world" ideas for which there are no direct competitors - it helps *define the design space* of the new idea, using stimuli from existing market offerings within the broader category and/or adjacent categories. This approach can also be used for a company to review their existing product category to identify potential opportunities to optimize the product.
>
> **RPO** may follow RPN for a new idea or may be used to improve a product that is already on the market, with the goal of *optimizing the most important attributes* of the product. The attributes that are studied in RPO are either determined by previous RPNs or by other data that indicates the critical design components for that product. Stimuli include prototypes that are designed such that consumer perception of the prototypes covers the entire probable product space, along with current products and competitor products as benchmarks.

To further understand these approaches, we present an overview of best practices for setting up any RPN/RPO, followed by three case studies. The first case study represents an ideal (fictional) RPN, and the second and third case studies illustrate real-world adaptations of the RPN and RPO processes, respectively.

10.3.4 Recommended tools and "how to" implement RPO

10.3.4.1 Step 1: Building a foundation through product screening

One of the most important steps in the process is building the foundation for the RPN/RPO. Before conducting RPN/RPO, three things must be known: (1) the starting prototype for the exploration - this could be a new or existing product,

(2) the product developers must know the design elements (sensory attributes) that need to be explored and optimized, and (3) there must be some understanding of potential consumer segments. If consumer segments are thought to exist, the business must decide how to move forward: develop the product for only one segment, which meets the product design criteria for one segment, but may alienate another; or develop the product to have the broadest appeal across all segments, which may not be the best product for any of the segments. This information would come from the discovery and scoping stages of the product development process. If you do not know where you need to start, you should first do some preliminary product screening, or scoping.

Rapid screening of a broad array of product options can quickly determine which options are most acceptable to the intended audience and if different segments of consumers appear to prefer different product options. A broad array of differentiated prototypes for screening is essential because casting a wide net will reduce the possibility of missed product opportunities.

The method of screening depends on the product category and the number of prototypes to be screened. One way to select the best option to optimize is to conduct qualitative discussion groups that incorporate product trial RPN. The objective of these groups is to identify the best product direction to optimize, which product attributes need to be included in the optimization, and any consumer segments that might have different product sensory requirements. As demonstrated in Case Study 1 later in this chapter, RPN is an excellent method to gain the insights needed to design a successful RPO. However, an alternative screening method based on hedonics around attributes such as aroma or appearance may be necessary, if the product has a high propensity for carryover, or if only one product can be evaluated at a time when used as intended.

For example, Case Study 3 will demonstrate the use of aroma screening to identify the flavor direction for the smokeless tobacco (ST) product to be optimized. Research has shown that screening based on aroma is a good surrogate for screening based on taste because consumers expect products to taste like they smell (Simons et al., 2008). In addition, people can evaluate many aromas in a single session before boredom and fatigue set in, so screening by aroma only can expedite the research by very quickly reducing many flavor options down to a few that can be explored more deeply. Also, because the base product itself may have its own aroma, as with products such as coffee and tobacco, aromas should be screened in the presence of the base product. After an initial broad screening based on aroma, the remaining product options must be further narrowed down based on taste. While the aroma screening method discussed above works well for some products, e.g. coffee or tobacco, it may not suit all products. Whatever screening method you choose, you should come out of the screening process knowing which product you want to optimize, the product attributes that should be included in the optimization and an understanding of possible consumer segments that may have different sensory preferences.

10.3.4.2 Step 2: Conduct the sessions

The RPN/RPO process is a collaborative approach that requires the full participation of both product developers and product researchers in all aspects of the research, including defining the qualitative product space (QPS), defining the intended audience for the product, observing the group discussions, and deciding which products to present to the groups. The details for each of these areas are discussed below.

10.3.4.2.1 Step 2.1: Qualitative product space

During the RPN/RPO, the project team will need to be able to easily identify the prototype(s) that should be evaluated next, based on what they hear from the research participants. The way to do this is to translate the technical product design matrix into a sensory product space that is defined by the consumer's language - the QPS that is based on *design for perception* (DFP) vs. traditional design of experiments. For example, suppose the optimum balance of attributes P and M needs to be identified. So the QPS should be the space defined by those attributes (Figure 10.3.2). Prior to the RPN/RPO, the prototypes are arranged within the QPS by the project team. Placement decisions can be made based on input from multiple possible sources: product developers' expert opinions, descriptive panels, or findings from prior research. Sample selection should include a range of experiences that covers the entire qualitative space, rather than focusing too narrowly on what the project team believes will be the "winning" combinations.

Increasing the number of design elements or the number of levels will quickly increase the number of products in the QPS to an unmanageable level. For some product categories, it may be possible to create a new prototype by blending existing prototypes during the qualitative session based on feedback being gathered. If custom blending is not possible for a product category, an additional round of RPN/RPO may be required once additional prototypes can be made.

10.3.4.2.2 Step 2.2: Participant selection

Participants in the RPN/RPO research must be members of the intended audience for the product. Multiple groups of consumers are recruited to participate

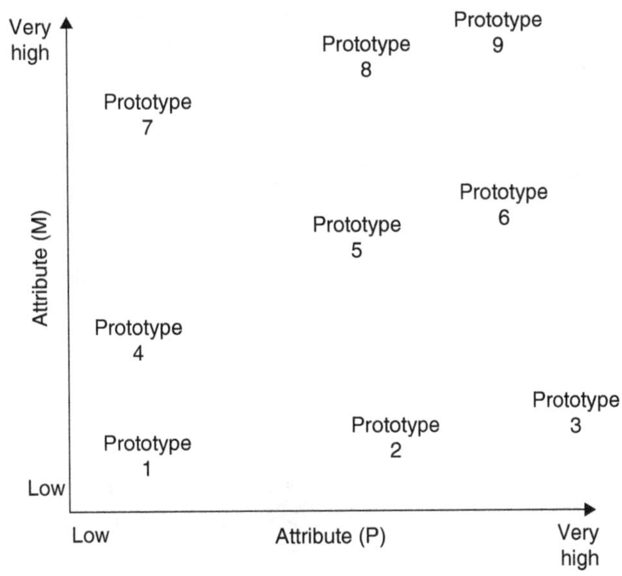

Figure 10.3.2 Simple qualitative product space (QPS).

in RPN/RPO discussion groups. If previous product screening or scoping research, such as RPN, has been conducted, those same participants may be recruited for the RPO. However, additional new recruits should also be included for a fresh perspective.

The number of discussion groups needed is determined based on the number of design elements to be explored RPN or optimized RPO, or the number of products that can be evaluated during one group session and the number of possible consumer segments. For smokeless tobacco in Case Study 3, the number of products per group is based on the normal duration of product use and on the potential for carry-over. For other types of products, the number of products per group depends on the complexity of the product being evaluated (i.e. for a sandwich, the bread, the toppings, and the condiments may all be part of the design), the ability to prepare multiple products if cooking is involved, and other factors that may be specific to the product type. If multiple consumer segments with different product preferences are believed to exist, additional consumers and groups must be recruited. These segments might be based on product usage, such as regular brand or frequency of use. They might also be based on differences in desired product experience - some consumers may desire a stronger flavor experience or a different texture experience than others, for example.

Groups with six to nine participants are recommended. Having more than nine consumers in a group is not only more difficult for the facilitator to manage but also limits the depth of discussion because more time must be spent to include all participants in the discussion. Fewer than six per group will require additional groups to ensure that enough "voices" are heard.

10.3.4.2.3 Step 2.3: Discussion flow

The most important thing to know about the RPN/RPO discussion groups is that they are simultaneous individual assessments. We are listening to understand the individual opinions, not come to a group consensus.

The discussion consists of three sections: introduction, stimulus-response (product trial), and summary (Figure 10.3.3). At the start of the session, perform

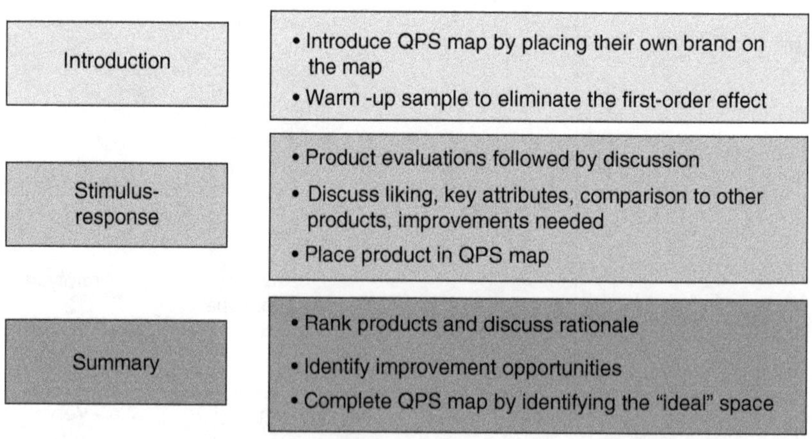

Figure 10.3.3 General discussion flow for rapid product optimization.

introductions to let participants get to know each other, and explain the objectives and the process for the discussion. They can get comfortable with the facilitator and each other by sharing their opinion of the product they currently use. Introduce the product map by asking the participants to place their own brand on the map. Also, a quick evaluation of a practice sample may be used to familiarize the participants with the process and to minimize the impact of order effects on the subsequent product evaluations.

During the stimulus-response phase of the discussion, ask the participants to try the product, and record their opinions on paper before any discussion takes place about that product. For some product categories, this does not take much time. The participants can try the product and evaluate it without much down time. In some cases, it may be possible for participants to evaluate a set of products during each evaluation period, rather than just one, as we will see in the first two case studies. However, some products, such as the smokeless tobacco in Case Study 3, require that the participants use a product for a period of time before doing the evaluation. It is important to give them ample time to think about the product they are evaluating and jot down any notes on their thoughts before beginning the discussion.

Once the evaluation of the first product (or product set) is complete, start the discussion to understand what they like or dislike about the product. How much do they like the flavor and how do they describe the flavor? Is the texture what they expect? What can we do to improve the product for them? How does the product compare to the product(s) they currently use? Ask the participants to place the prototype on the product map relative to the other products that were already placed on the map at the start of the session. Repeat the product trial, evaluation, discussion, and mapping for each product (or product set), cleansing the palate between each product as needed. Placement need not be unanimous since experiences differ with preference. Simply place the sample in multiple positions, noting how many individuals agree with each position.

After the evaluations are complete, ask the participants to rank the products from most liked to least liked and discuss the rationale for their decisions. Complete the product map by asking the participants to define their "ideal" space and what would need to be done to their highest-ranking product to enter that space, if it is not already there.

The discussion groups may be lengthy in duration to allow for product trial and in-depth discussion about each product. Groups extending beyond three hours should be designed to manage potential fatigue, boredom, and breaks. Again, the number of products presented during the group depends on the product category. For food product categories, such as the pasta that is described in Case Study 2, a good rule of thumb is around four to six products per half hour block during the session to allow for enough time for eating, discussion, and mapping. The number of products is highly dependent on the complexity of product and specificity of axes. In the case of products with high carry-over such as the smokeless tobacco that is described in Case Study 3, only four products were evaluated per session for two reasons. First, the product trial period during the sessions had to be long enough for the participants to feel comfortable that they could properly assess the product. Second, tobacco products have high carry-over, so there had to be a sufficient waiting period after the product was removed from the mouth before tasting the next product.

It is good practice to provide the participants with handouts, so they can rate their overall liking and record their thoughts on likes, dislikes, the product experience, or anything else they want to share.

10.3.4.2.4 Step 2.4: Conducting the discussion groups

During the RPN/RPO process, we are listening to the participants to understand several things. How much consensus is there? Are there groups whose liking is driven by different sensory factors? Are any of the prototypes polarizing? During the discussion groups, the product developers must be present and focused, listening to understand the participants' reactions to the prototypes, making decisions about the next prototypes to evaluate, and if possible, mixing prototypes on the fly to create new ones that better fit with the needs being described by the participants. Without the product developers' full engagement in the initial development of the QPS and in the navigation decisions, the project will not be successful. When mixing on the fly is not possible, it is important for product developers to create a number of hypotheses about *what they think they will hear* during the sessions, and develop a broader range than they think they will need in order to cover the sensory design space.

There are two approaches to navigating the QPS during RPN/RPO: navigating within a discussion group or navigating across discussion groups. For a straightforward situation with only two or three design elements being modified, the navigation can occur within each group. In RPN/RPO, the decision of which prototype to present next is ideally driven by the feedback from the previous product. This process can be repeated across multiple groups to build the research team's confidence in the results, by coming to consensus on the product design that best meets the product sensory criteria. For products that require time to prepare prior to evaluation, such as the pasta in Case Study 2, some decisions must be made in advance vs. truly "on the fly." In this case, deciding which *series* of stimuli should be shown next may be more feasible than making the decision sample by sample. Another approach is to decide what to show participants in one session based on feedback from those in prior sessions.

Because each group, and even each person, navigates to its own final product design, it is possible that this process can result in more than one product design that meets the product sensory criteria. In such a case, both products should be included in any resulting validation tests that may be conducted. If consumer segments are thought to exist, and the business wants to develop different products for each segment, the RPN/RPO process would need to be conducted separately for each segment to determine the final product design for that segment. While multiple segments can be seated in each group, it makes it more challenging to hear the voices separately, thus the recommendation to conduct separate groups when possible. When not possible for time or budgetary constraints, placement of samples on the map must be marked such that it is clear when segments have differing placement. In the postsession reporting, separate maps are often created for the segments to facilitate interpretation.

In an elementary example of conducting an RPN/RPO where sample order is determined "on the fly," assume that only two design elements require

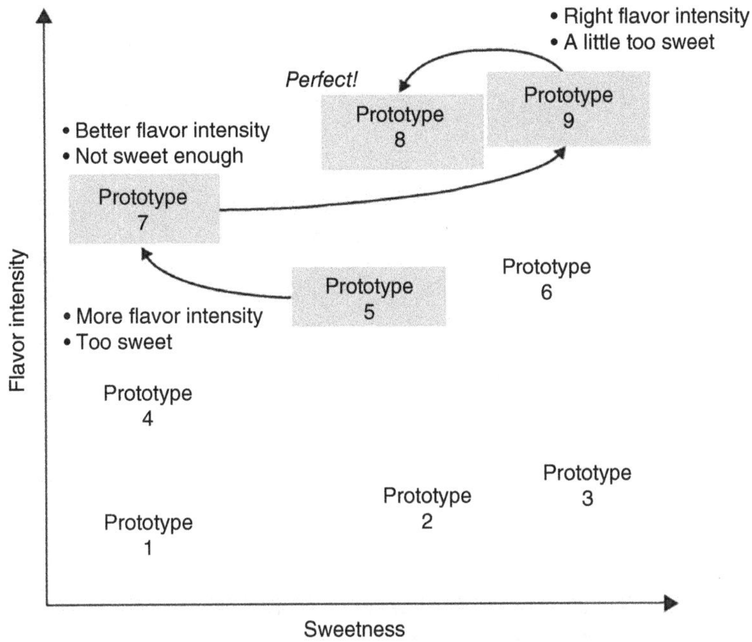

Figure 10.3.4 Simple two-factor evaluation path within a discussion group.

optimization, sweetness, and flavor intensity. Figure 10.3.4 shows a possible evaluation path from the baseline product to the product that best meets the product sensory criteria within a single discussion group. Consumers first evaluate and discuss the initial prototype, identifying the attribute changes needed to make it better, more sweetness for example. The next prototype for evaluation would then be one that offers more sweetness than the initial prototype. The evaluations and discussions continue in this way until the product design that best meets the product sensory criteria is reached. Figure 10.3.5 shows the final product map based on the consumer feedback. In this example, prototype 8 was clearly the winner, as it was within the ideal space identified by the participants.

However, product research is rarely that simple. When multiple design factors are involved, or when there are complicated interactions between the factors, the navigation must occur across the discussion groups, instead of within each group, because there is not enough time during a single group to evaluate enough prototypes to reach the final product design. This approach will be illustrated in the pasta RPN example in Case Study 2. Consider the multiple discussion groups as an iterative series of small-scale product trials where the feedback from the consumers in one group drives the focus of the product trials in the next group. In other words, at the conclusion of the first group, the project team determines the products to be presented to the next group based on the feedback from the first group. This decision is best made by the project team as a whole - the product researcher, product developers, and the facilitator. This adaptive approach allows the researcher to rapidly explore a greater number of factors than could be accomplished using a traditional experimental design methodology.

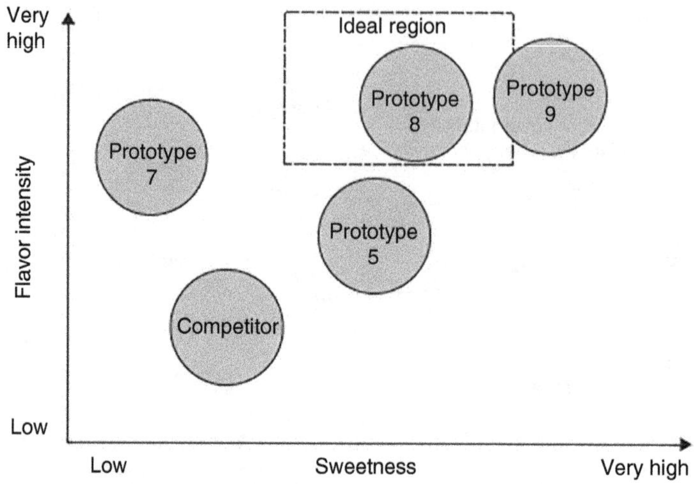

Figure 10.3.5 Example of a final product map.

When making the decision about the products to include in the next group, it is important to ensure that product design decisions are based on findings confirmed across multiple groups. For example, if the first group shows that ingredient 1 is better than ingredient 2 because ingredient 2 has an off-taste, then the two ingredients should be evaluated again in a future group, and the findings replicated, before eliminating ingredient 2 from consideration. This repetition across groups also allows the researcher to adjust the presentation orders to reduce the possibility of order effects biasing the results.

10.3.4.3 Step 3: Verify and validate the design

Depending on the product category and a company's comfort with risk, once a final prototype is selected, the acceptability of the product may need to be confirmed among members of the intended audience who did not participate in the RPN/RPO, ideally in market areas other than the market used during the RPN/RPO. The understanding of the key sensory attributes gained during the group discussions can drive the questionnaire development for this stage.

However, quantitative consumer tests are expensive and sometimes not necessary, so the sensory performance of the optimized prototype could be confirmed in a small-scale consumer test as a precursor to, or instead of, a larger study. Incorporating in-depth interviews following this test may provide additional information on minor changes to the product needed prior to the quantitative study (or prior to launch if this is the only validation step being performed).

Now that we have described the best practices for conducting any RPN/RPO, we present some specific examples of the process in action. Case Study 1 represents the "ideal" process where an RPN is followed by an RPO, while Case Studies 2 and 3 represent real-world adaptations where an RPN was not followed by an RPO and an RPO was conducted without being preceded by an RPN.

10.3.5 Three Case Studies that illustrate the uses of RPN/RPO

Case Study 1 Using rapid product navigation (RPN) to identify a new product opportunity space in the chocolate category and rapid product optimization (RPO) to finalize the product design

Confectionary company "Awesome Sweets" planned to enter the chocolate category. They knew from their current knowledge base that size and shape are important for confectionary products and had an established set of guardrails for their hard candies. Before committing resources to developing a range of chocolate prototypes for consumer testing, they wanted to understand if those same attributes are critical for a satisfying chocolate experience and if so, what the design space should be.

The product research team compiled a large array of products representing various sizes and shapes from their own product portfolio as well as likely competitors in the chocolate category. The product research team in conjunction with the cross-functional team (i.e. a team representing individuals with a broad range of responsibilities within the organization), narrowed down the array to eliminate duplicates while still covering the largest range possible, including those products they believed would be outside the bounds of acceptability. They also planned a *theoretical navigation path* for sample flights (in this case, groups of products that had similar shape/size characteristics) to follow during the sessions based on their hypotheses about consumers' preferences (Figure 10.3.6).

The company recruited consumers for the RPN who were current users of their products who also enjoy chocolate candy, with a subset who also use products from the likely competitor companies. Early in the discussion, it became clear that there were round vs. square preferences, so the second step of the

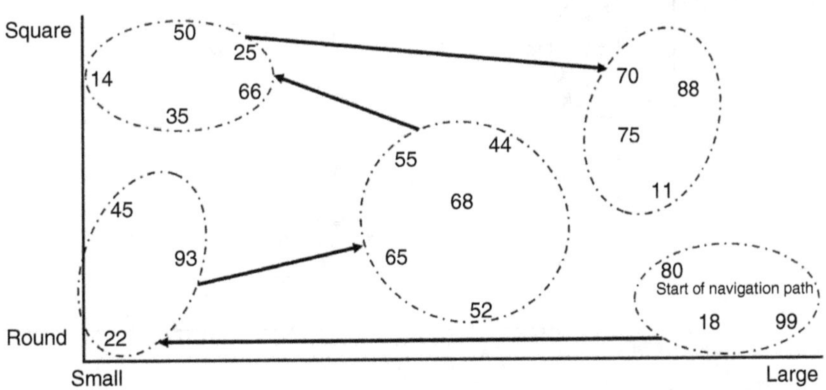

Figure 10.3.6 Landscape of shape and size for confections created by research team, with theoretical navigation path.

10

navigation path was changed and, subsequently, the entire path was modified as shown in Figure 10.3.7.

After three sessions of consumer feedback, during which the consumers placed their "ideal" product in the space after all samples were evaluated, the team discovered that there were two consumer segments, one that preferred medium, rounded shapes, and another that preferred small squares (Figure 10.3.8). Because the company did not have the capability to make square candies, they decided to conduct future RPO sessions using shapes ranging from flat ovals to round balls to determine the optimum design.

For the subsequent RPO, product developers created a DFP that included stimuli similar to the shapes and sizes that the "rounded" segment preferred in the RPN, and further pushed the boundaries of size (including thickness) and rounded shapes to pinpoint a clearer optimum for their new chocolate candy. In addition to the wide range of stimuli created by the team (Figure 10.3.9), potential competitive products (C1–C5) were also chosen that covered the space of thin to thick and "round disk" to "round ball."

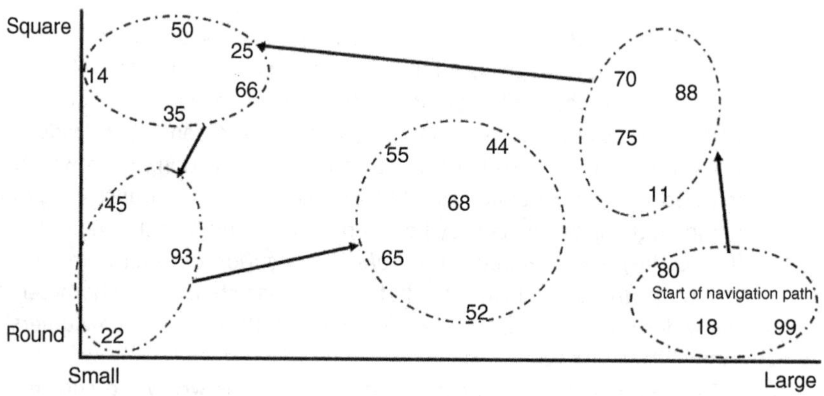

Figure 10.3.7 Actual navigation path based on conversation about size and shape preferences during session.

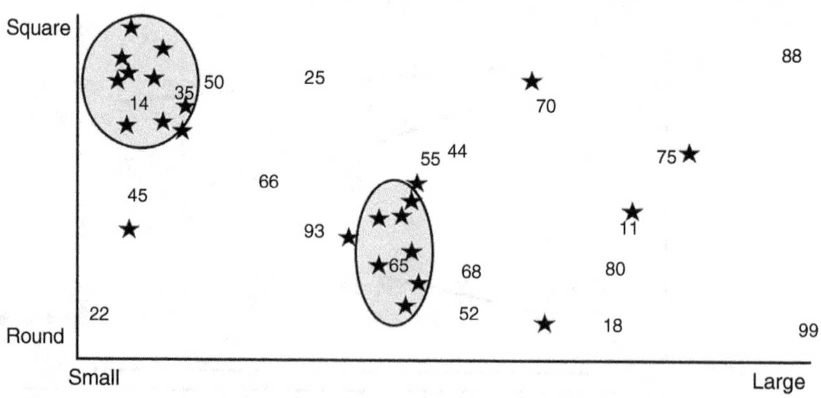

Figure 10.3.8 Landscape for shape and size created by consumers, with "ideals" noted as stars.

Three consumer workshops were again conducted, using recruiting specifications similar to those in the RPN. The team had planned a theoretical navigation path based on their learnings from the RPN, but ultimately followed a path that was guided by consumer feedback after the first flight of products. Participants placed each of the products on the map and again indicated where their ideal product would be (Figure 10.3.10).

After the three workshops, it was clear that shapes closer to a ball were preferred to round disks, but the thickness of the ball ranged from thin to thick depending how they consumed them. Those that desired a "poppable," easy to eat format preferred smaller balls, while those who wanted to eat one at a time and "savor it" preferred a larger ball. There were also differences based on preferences for dark vs. milk chocolate embedded in these behaviors, so the research team decided to do an additional follow-up RPO exploring the best combination of dark/milk chocolate vs. thickness of ball shapes to understand this further.

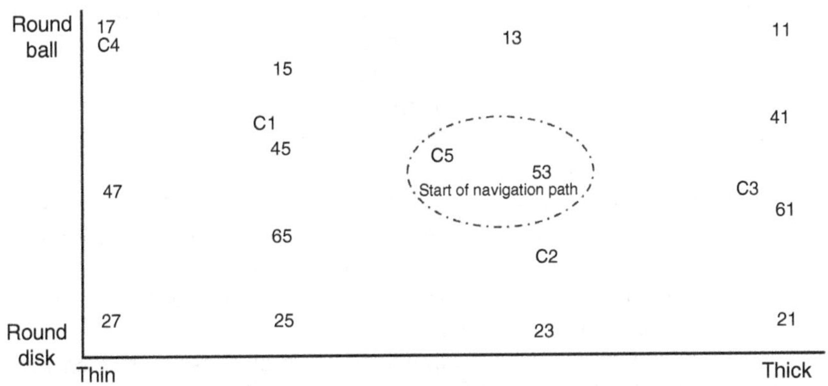

Figure 10.3.9 Design for perception of chocolate confections for RPO.

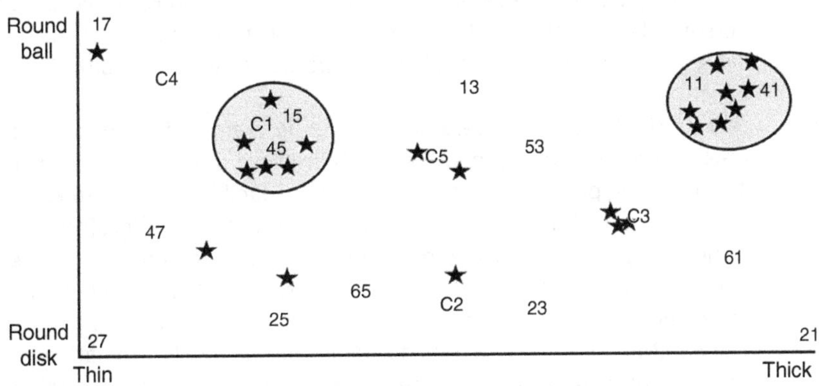

Figure 10.3.10 Landscape for shape and size created by consumers, with "ideals" noted as stars.

10

Case Study 2 Using rapid product navigation (RPN) to identify an opportunity space to improve a pasta product

Company JKL was looking to improve an in-market healthy pasta. Retailer data showed that this pasta had high initial purchase, but poor repeat purchase, indicating that the product concept and claims were intriguing to consumers, but the product itself was not delivering on consumer expectations. Marketing was open to formulation changes as long as the product could still make healthy claims. A cross-functional project team agreed that RPN would be the ideal method to have consumers evaluate a wide array of prototypes in order to significantly improve the product formulation and deliver on product expectations.

The first step was conducting workshops with current and target consumers. Prior to the workshops, participants were asked to purchase the in-market product and prepare it at home. They then shared their experiences during the workshop. This exercise revealed that the participants really liked the product concept and were excited to purchase and try it. However, after preparing the product at home, some participants found the flavor of the nontraditional ingredients in the product was too strong. Others noted that the product had off-flavors. Some participants liked the pasta texture while others did not. Additionally, they pointed out several issues with the product preparation. In addition to the sensory attributes, the participants gave feedback on what products they currently use, how the alternative pasta products could fit into their daily lives, and what their expectations would be of the products using the Game Board approach described in Chapter 6.3. This feedback was used to identify the critical product attributes.

After the initial round of workshops, the project team used marketing data, consumer feedback, and the workshop data to form hypotheses about how to improve the product. One hundred seventy-five hypotheses were generated that focused on the product formulation, product experience, and the impact on other parts of the business. These hypotheses were based on data, not personal beliefs, that had been gathered on the products up to this point. Table 10.3.1 is an example of one hypothesis around health and product appearance.

Once the hypotheses were generated, the team began prototyping for the RPN. Prototypes were generated to either prove or disprove each hypothesis. These prototypes were made in a pilot plant facility. In order to keep the texture consistent throughout the samples, the pasta in all prototypes was the same shape and size. Prototypes were made to represent the broadest product space possible. A wide range of ingredients was used, even if some ingredients were not technically feasible for scale-up due to processing capability and/or cost. The goal was not to design the *ideal* formulation in the prototypes but give participants the widest range of samples. In addition to the pilot plant-made prototypes, in-market products were used, which included both Company JKL's existing products and competitive products from both

Table 10.3.1 Hypothesis example for pasta product design.

Product lever	What we believe	Because/ so	Why we believe	Stimuli idea(s)
Product design	Naturally colorful ingredients are important in side dishes and meals	Because/ so	It signifies healthy	Colorful prototypes (red, green, orange, yellow, brown)

traditional and healthy categories. These in-market products provided benchmarks throughout the study.

Once all prototypes were produced, the RPN was conducted over two days with three separate consumer workshops. It was determined during the first phase of the research that flavor, texture, and product preparation were the three most critical characteristics of the product. Workshop 1 ($n = 9$) focused solely on product flavor, workshop 2 ($n = 9$) focused on product texture, and workshop 3 ($n = 9$) validated the flavor and texture results of the previous groups and focused on product preparation.

Since the overall flavor of pasta is relatively bland, many samples could be evaluated with very little carry over. Each workshop consisted of three sample rounds lasting 45–60 minutes with a total of 24 samples presented. This allowed participants enough time to taste samples, and thoroughly discuss their feedback with the facilitator. The first round was the longest, since participants had to define the flavor attributes and place them visually on an *X*-axis and *Y*-axis. After tasting the samples, participants defined flavor in the two ways that they felt were most important to them for a pasta product. The *X*-axis was defined as Plain/Flavorless/Bland → Overpowering and the *Y*-axis was defined as Flavor Belongs in Pasta → Flavor Does Not Belong in Pasta. These attributes are shown in Figure 10.3.11.

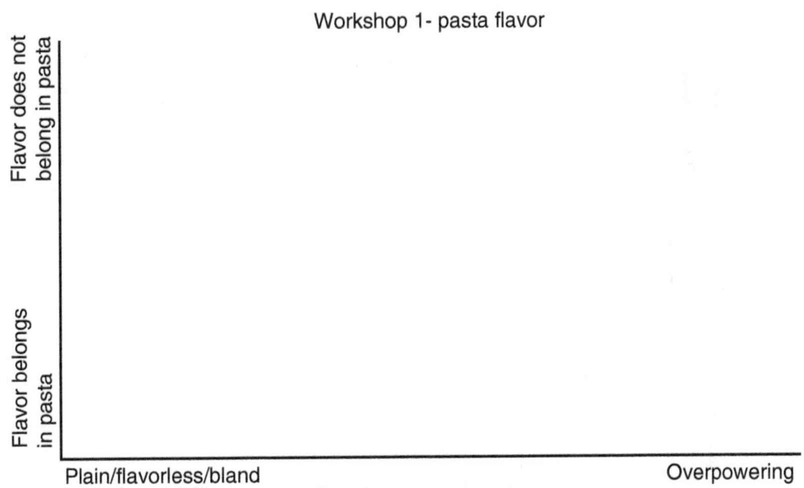

Figure 10.3.11 Consumer defined map – pasta flavor.

The samples in round 1 were preselected by the research team to have the widest range of flavor and were chosen during screening to have very little flavor to very intense flavor as evaluated by the product developers. Additionally, sample round 1 included several in-market products as benchmarks. Some samples, such as C, N, and T were repeat samples shown multiple times to evaluate any variation (Table 10.3.2). Due to the large number of products, stimuli in each round were presented in multiple, smaller flights.

Once the flavor descriptors were defined, participants placed the samples visually on the biplot in relation to one another (Figure 10.3.12). This graphic representation was critical to the project team because *they assumed the pasta should be as bland as possible*, just like the category leading product. However, some of the prototypes had strong flavors, and depending on the ingredients, that strong flavor was found to be either acceptable or not acceptable based on

Table 10.3.2 Workshop 1 flavor samples over three rounds.

Sample round	Samples	Notes
Round 1	A, B, C, D, E, F, G, H	C- current in-market product D- in-market competitor product E- in-market competitor product F- in-market competitor product H- current in-market product
Round 2	I, J, K, L, M, N, O, W	N- same as C K- in-market competitor
Round 3	P, Q, R, S, T, U, V, X	T- same as N and C R- same as E U- same as F

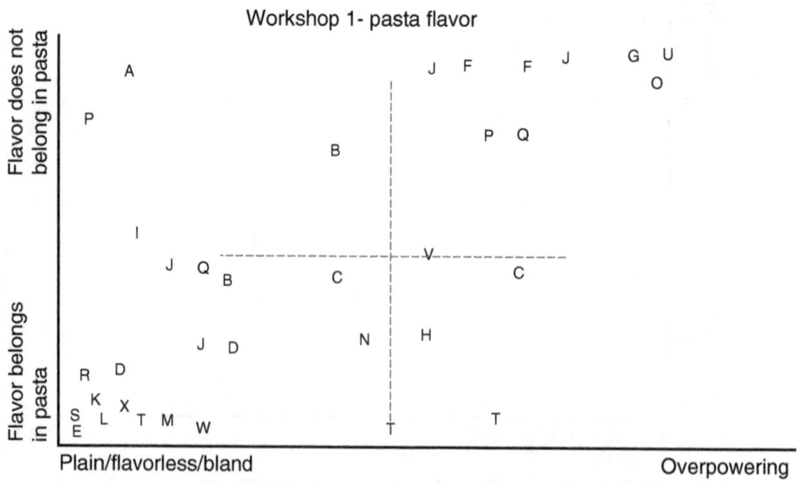

Figure 10.3.12 Pasta flavor mapping after three sample flights.

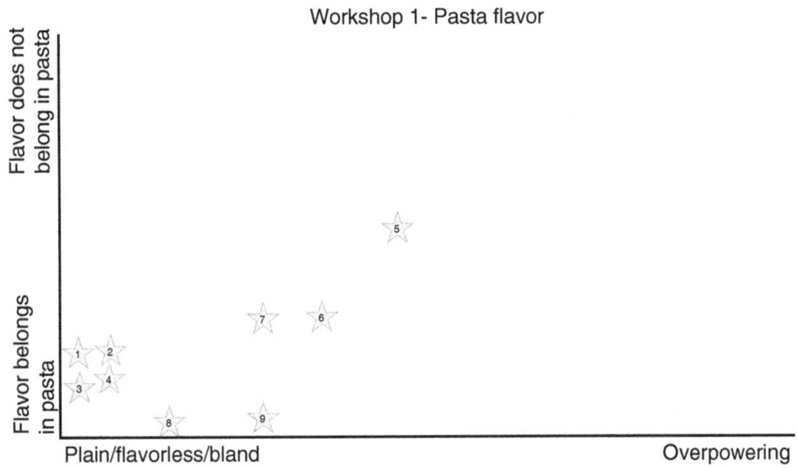

Figure 10.3.13 Ideal location of pasta flavor.

whether the participants felt the flavor belonged in pasta. Conversely, some of the samples with mild flavor were unacceptable because participants felt that flavor did not belong in pasta.

Many samples were placed on the biplot more than once to account for multiple experiences. For example, there were very different opinions on how strong the flavor was in sample J and whether this flavor belonged in pasta. This difference of opinion indicated that sample J could be polarizing to consumers (Figure 10.3.12).

After all the prototypes were plotted, participants were asked to indicate where their ideal flavor would be for pasta (Figure 10.3.13). About half of the participants wanted a very low flavor pasta with flavors that belong in pasta, which is what the project team expected. The other half of the group was open to more strongly flavored pasta, if the flavor was one that belonged in pasta.

The second consumer workshop focused exclusively on pasta texture and contained three 45–60-minute rounds. As in the previous workshop, participants defined two different dimensions for texture for the *X*- and *Y*-axes: Very Soft → Very Firm and Not at All Chewy → Very Chewy (Figure 10.3.14). These descriptors showed what pasta texture attributes were most important to these participants and how they described these textures. There are many ways to define texture, so using the consumers' definitions of firmness and chewiness aligned the project team on these terms, which were then used for future internal evaluations.

Similar to the flavor workshop, the samples in round 1 of the texture workshop were preselected by the research team to represent the widest range of textures. During screening the product developers identified these samples as ranging from soft to firm/crunchy. The first round included the current in-market products and in-market competitive products to establish benchmarks. The round 2 samples were selected to narrow the range of acceptable texture and were intended to represent various levels of chewiness in the products. The final

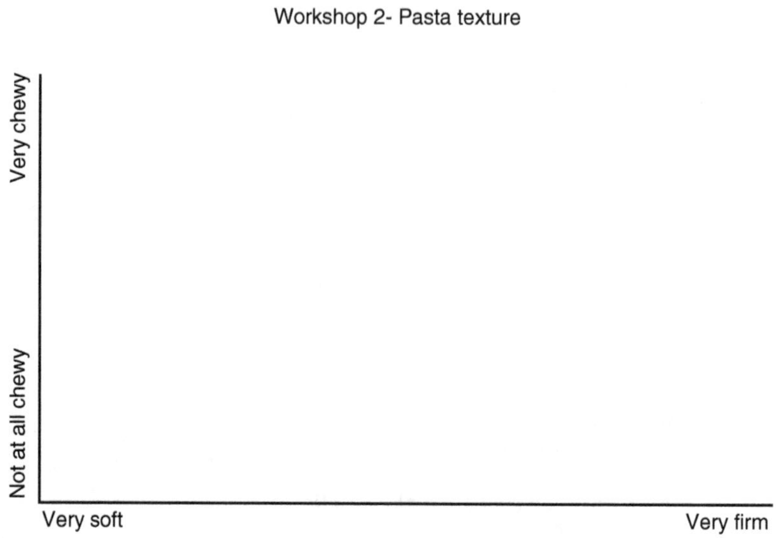

Figure 10.3.14 Consumer defined map – pasta texture.

Table 10.3.3 Workshop 2 texture samples over three rounds.

Sample round	Samples	Notes
Round 1	A, D, E, F, H, Q, S	D- in-market competitor product
		E- in-market competitor product
		F- in-market competitor
		H- current in-market product
Round 2	B, J, K, O, P, V, Y	Y- same as H
		K- in-market competitor
Round 3	E, L, M, U, W, X, Z	Z- same as Y and H
		U- same as F

round of samples was selected to narrow the range and identify those samples that were ideal to participants for either firmness and/or chewiness. Table 10.3.3 shows the samples evaluated in each round.

Across the three rounds, a total of 21 samples were evaluated by participants and placed on the biplot. Figure 10.3.15 illustrates that the prototypes selected were effective in representing a wide range of textures. Participants were able to easily differentiate between textures. Many samples are shown more than once on the biplot; as in the flavor workshop, these samples did not have consensus from participants about where they should be placed. This exercise showed that texture is very dependent on the individual's preference. For example, sample E is the in-market product leader and was considered too firm by some participants *and* too soft by others. This indicated to the team that the

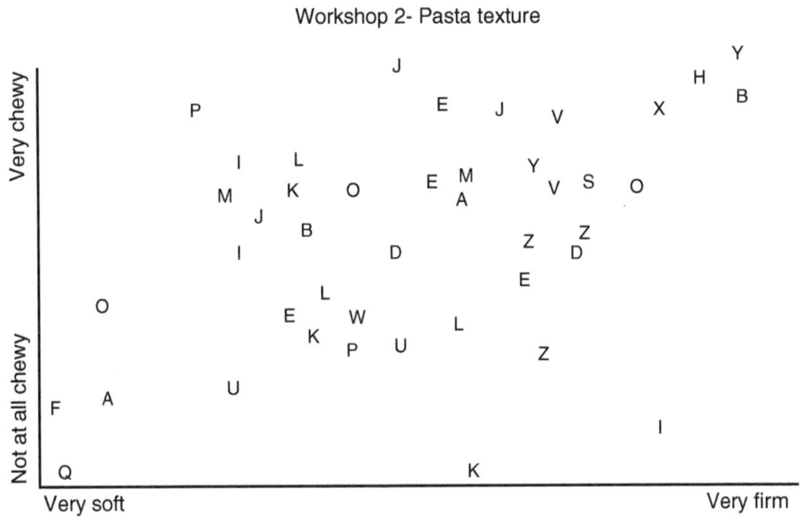

Figure 10.3.15 Pasta texture mapping after three sample rounds.

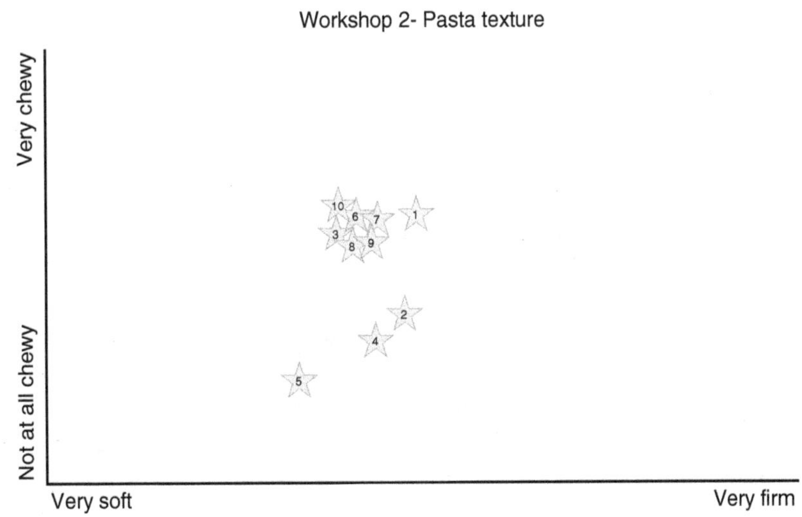

Figure 10.3.16 Ideal location of pasta texture.

product texture would not be ideal for every consumer. Therefore, the product preparation instructions would need to accommodate a wide range of texture preferences. The product should be prepared from firm to very soft, without sacrificing product quality, in order to meet the wide range of consumer preferences.

After three rounds of prototypes were evaluated, participants were asked to identify their ideal texture for the pasta (Figure 10.3.16) and, unlike flavor, there was a tighter range of what was acceptable. The main difference between participants' ideal textures was not the desired firmness, but rather how chewy the pasta should be.

The first objective of workshop 3 was to confirm the flavor and texture results from workshops 1 and 2. The second objective was to conduct an interactive exercise to examine hypotheses on product preparation. In workshop 3, participants did not identify critical flavor or texture attributes. Instead, the descriptors for both flavor and texture from workshops 1 and 2 were presented, along with a tighter range of samples chosen based on learnings from the first two workshops. The facilitator worked with participants in workshop 3 to ensure alignment and understanding of the flavor and texture biplots created by the individuals in workshops 1 and 2. In the first round, samples were evaluated only for flavor; in the second round, samples were evaluated only for texture. After both rounds of evaluations, participants identified their ideal range, which is circled on biplots in Figure 10.3.17.

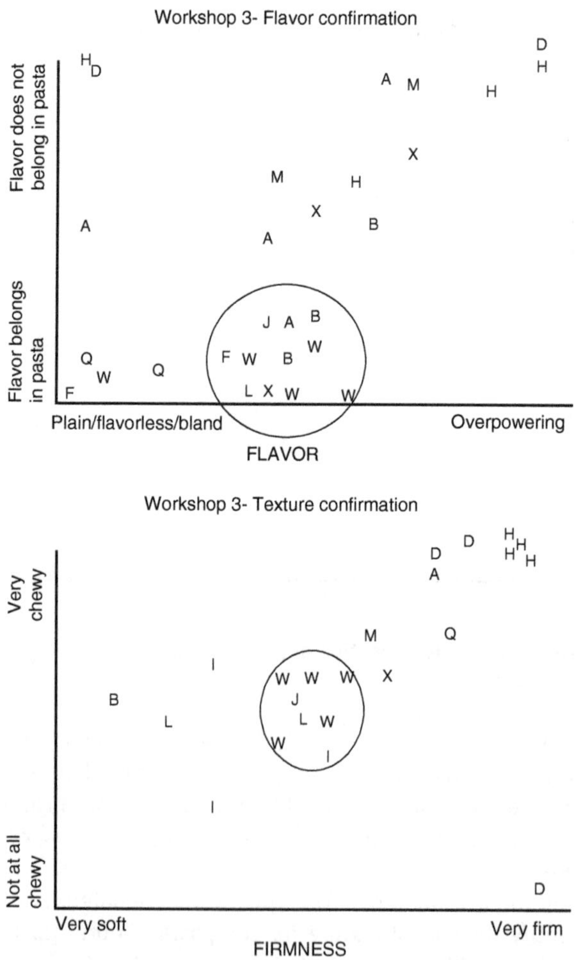

Figure 10.3.17 Pasta flavor and texture mapping validation.

The ideal flavor for pasta wasn't to be completely bland/flavorless. Instead, the pasta should have a mild flavor that belongs in the pasta. The desired flavor differentiated the pasta from traditional products and evoked the perception of "healthiness" to consumers. This finding eliminated several ingredients that product developers had been considering for the product reformulation. Despite the health claims that could be made with these ingredients, developers removed them from consideration due to their unacceptable type and strength of flavor.

The ideal product texture was found to be between soft and firm and between chewy and not chewy at all. This area is also generally where traditional pasta products fall on the biplot. Whether or not a prototype fits this description varies among individuals, since soft or chewy to one person may be different from soft or chewy to another. The most important learning on texture was that the product needed to be robust enough during preparation to be cooked to the "ideal" softness or firmness (without negatives such as mushiness or hard spots), depending on consumer preference.

After the flavor and texture evaluation rounds in workshop 3, the participants were brought to a kitchen where multiple stations were set up with in-market products as well as typical home cooking utensils and equipment. They were given a product and asked to follow the preparation instructions on the box. No additional instructions were provided. Participants made the pasta while the project team observed and asked questions. This exercise allowed the project team to draw several key conclusions about the product preparation instructions. These included the following:

(1) Preparation instructions needed to mimic traditional pasta. Consumers will assume that a healthy pasta that looks like traditional pasta will also be prepared the same way.
(2) Experienced cooks will not check the pot size and/or measure water to boil pasta.
(3) Some people will not wait for water to boil due to time constraints.
(4) Some people will follow the instructions exactly, without even testing the pasta for "doneness" before draining. Conversely, some people will not follow any of the instructions, including setting timers and measuring ingredients.

Looking at the data, the project team confirmed that existing in-market nontraditional pasta products, samples C, H, N, T, were not ideal for either flavor or texture for most consumers. This qualitative data confirmed that product reformulation was necessary for long-term product success. Additionally, the largest competitor, samples U and F, were also unacceptable to consumers. The flavor was too strong, and the texture was too firm. Lastly, results showed that traditional products in the category, samples E, D, K, and R, were very acceptable to consumers, which was expected by the research team. Using the RPN method, two new prototypes rose to the top. Samples L and W were not only within the ideal range for most consumers for both flavor and texture but also allowed for flexible preparation. These two prototype formulations quickly moved forward in the project and were evaluated for technical and market feasibility, with no follow-up RPO needed.

Case Study 3 Using rapid product optimization (RPO) to optimize a smokeless tobacco product[1]

Company ABC's marketing department identified an opportunity and wanted to launch a brand Y line extension into the smokeless tobacco (ST) flavor F segment. This flavor variety was dominated by the competitor's product X. Because this was a new segment for the company, some initial product scoping and screening was needed. Product development created 10 new prototype flavor variants for the flavor F line extension. However, some screening was required to identify the best flavor direction to take before a product could be developed. The company identified two important attributes (i.e. flavor intensity and sweetness level) that needed to be optimized. They had a tight timeline, and they opted for the RPO approach.

Instead of internally selecting which variation was the best to move forward, a central location test (CLT) was conducted among 50 adult ST consumers to evaluate the aroma liking of all 10 flavor variants plus two competitive products. Adult consumers were recruited who were users of competitor product X. They were recruited based on willingness and availability to participate in the research rather than for any specific expertise.

ST samples with the prototype flavors applied were placed in glass jars with screw-top lids, as were samples of the competitor's ST products. During the test, the participants removed the lid from the jar, sniffed the product, and rated their liking of the aroma on a 7-point hedonic scale (1 = dislike extremely, 7 = like extremely). The presentation order was balanced for order effects and first-order carry-over effects using a 12-product Williams design (Williams, 1949). The participants neutralized the odors in their nose by sniffing their own arms between each sample.

Since it was possible that not all of the adult consumers would like the same aromas, potential consumer segments were identified by clustering the consumers based on their aroma liking ratings of the 12 products. A k-means cluster analysis on the consumer/product matrix of mean liking scores indicated there might be five clusters and provided the mean aroma liking scores for each product by cluster (Table 10.3.4). Because of the small sample size, the results of the analysis were treated qualitatively, looking for direction only. Highlighting the two most liked products for each of the clusters showed which products most appealed to each potential consumer segment. This analysis helped to develop hypotheses about possible consumer segments, which were further explored during additional research to select the best option to move forward.

To further narrow the product options and select the best product to optimize, four discussion groups with product trials were conducted. The objective was to identify the best product direction to optimize, which product attributes needed to be included in the optimization, and any consumer segments that might have had different product sensory requirements.

Table 10.3.4 Mean aroma liking ratings by cluster (scale: 1 = dislike extremely, 7 = like extremely).

Cluster	n	P1	P2	P3	P4	P5	P6	P7	P8	P9	P10	C1	C2
1	11	4.78	4.33	3.33	5.56	1.89	5.00	5.11	5.11	4.44	5.56	2.33	2.22
2	14	4.77	4.62	4.46	4.00	3.77	4.92	3.92	4.62	3.31	4.08	4.69	5.69
3	11	2.80	4.10	5.20	4.90	1.50	3.60	4.30	3.80	4.40	2.00	4.30	2.90
4	6	3.40	3.80	5.20	2.40	2.40	4.20	3.40	2.80	2.80	6.40	4.20	2.60
5	8	5.29	4.86	6.43	5.86	5.43	6.57	5.86	5.14	4.86	6.71	6.29	3.43

In the aroma screening, the five potential clusters of consumers appeared to like different products. For this phase, those same consumers were asked to return to participate in the group discussions. They were assigned to groups based on which cluster they were in - those participants in cluster 1 were in group 1, and so on. Because cluster 4 and cluster 5 were small, they were combined into a single discussion group for efficiency, but the feedback from individuals in both clusters was considered separately.

Products 3, 4, 6, and 10 were the highest in aroma liking for the different clusters, and therefore were selected to be evaluated (tasted) and discussed in depth during each of the groups. Because the feedback on the products was similar across the groups, the project team decided to focus on the best overall option rather than different options for different segments. Products 3 and 6 had issues with the taste not meeting expectations based on the aroma or off-taste and were eliminated from consideration. Products 4 and 10 were both well liked, even though there were opportunities to improve the intensity and balance of the two flavors. Ultimately, product 4 was selected to move forward into the RPN process, because product 10 did not fit the concept of the flavor being developed, even though it was well-liked.

Product 4 was selected as the best starting point to optimize the ST line extension, but it still needed improvements to the balance of flavor intensity (ingredient 1) vs. sweetness (ingredient 2). The product development team also needed to select the most appropriate tobacco blend that would enhance rather than clash with the flavor and to determine if ingredient 3 was needed for improvements to pinch-ability and flavor duration. Figure 10.3.18 shows the QPS that was designed for this example. With two tobacco blends, three levels of ingredient 1, two levels of ingredient 2, and two levels of ingredient 3, there were 24 prototypes available. In the QPS, prototype E represents product 4, the product selected in the initial screening.

Because the potential line extension was in a segment dominated by the competitor product X, the objective of the RPO was to meet the product sensory criteria and to beat competitor X among adult consumers of competitor X, the intended audience. Six groups of eight adult competitor X consumers were recruited to participate in the RPO. Because the product screening research showed that product 4 (prototype E) appeared to have broad appeal, all groups were recruited from the same audience.

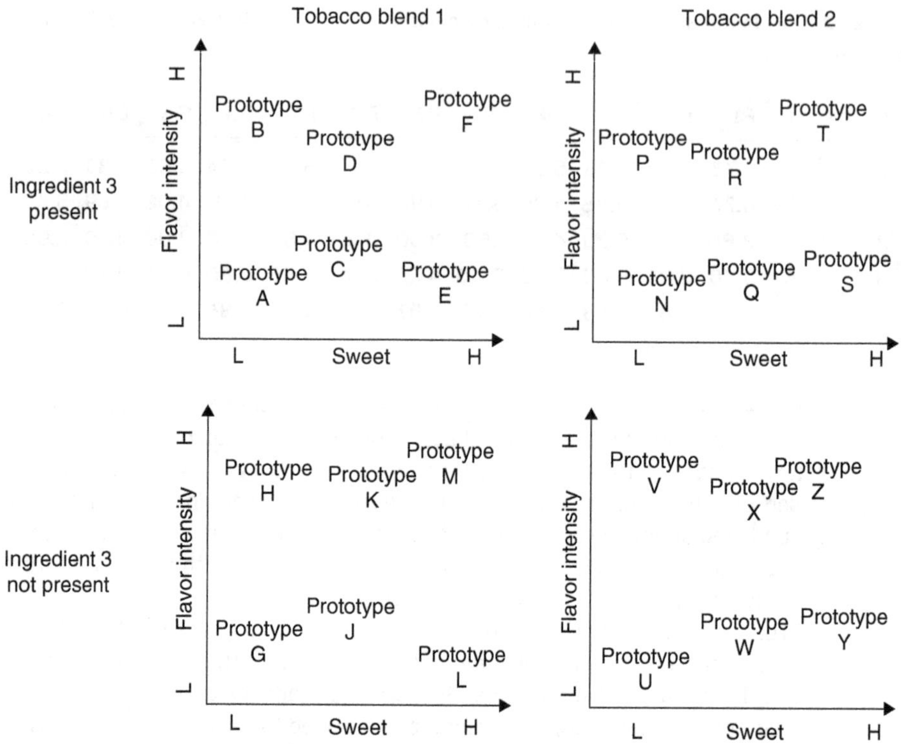

Figure 10.3.18 Smokeless tobacco line extension qualitative product space.

Because ST products are typically held in the mouth for an extended period, only four products could be evaluated during a single discussion group. Figure 10.3.19 shows the navigation path across the six discussion groups. The product selection decisions are discussed in detail below.

Group 1 explored the different tobacco blends. Since tobacco blend 2 had an off-taste, tobacco blend 1 appeared to be the best tobacco blend, but this needed to be verified in the next group. In addition, flavor intensity appeared to be too low, and sweetness appeared to be too high.

10

In group 2, the tobacco blend comparison was repeated at a different combination of flavor intensity and sweetness than in the first group, and varying levels of sweetness and flavor intensity were explored. Tobacco blend 2 again presented an off-taste, and because no interactions were expected between blend and varying levels of sweetness and flavor intensity and the presence/absence of ingredient 3, tobacco blend 2 was eliminated from further consideration.

For products with tobacco blend 1, a high level of flavor intensity with a low level of sweetness appeared to provide the desired flavor profile.

Groups 3–5 confirmed levels of flavor intensity (high) and sweetness (low) and also explored the need for ingredient 3. Since the prototype without ingredient 3 needed improvements to pinchability and flavor duration, it was decided to keep ingredient 3 in the product.

The final group discussion was used to verify the need for ingredient 3 and to verify the final prototype's improved performance over the baseline prototype

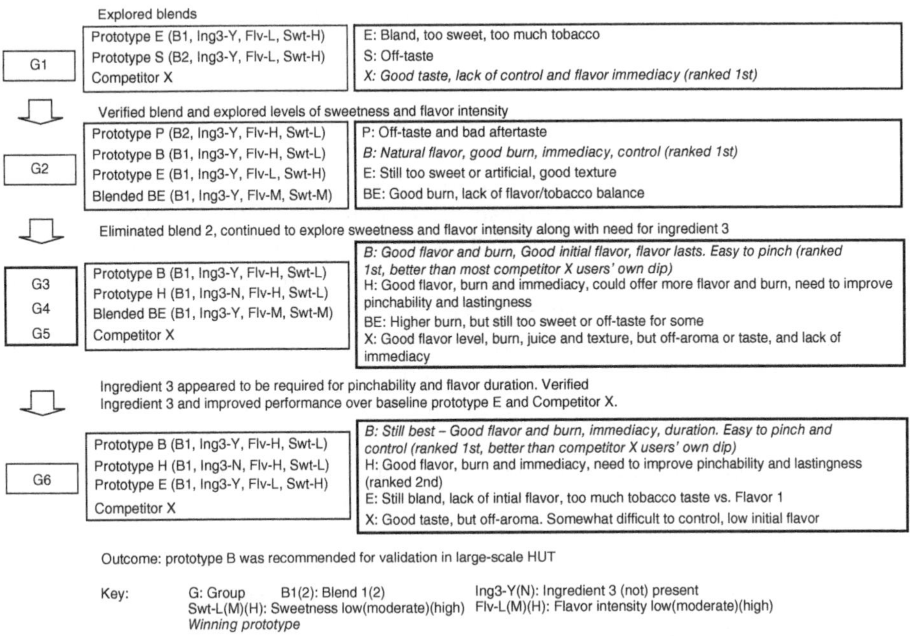

Figure 10.3.19 Smokeless tobacco line extension navigation path across discussion groups.

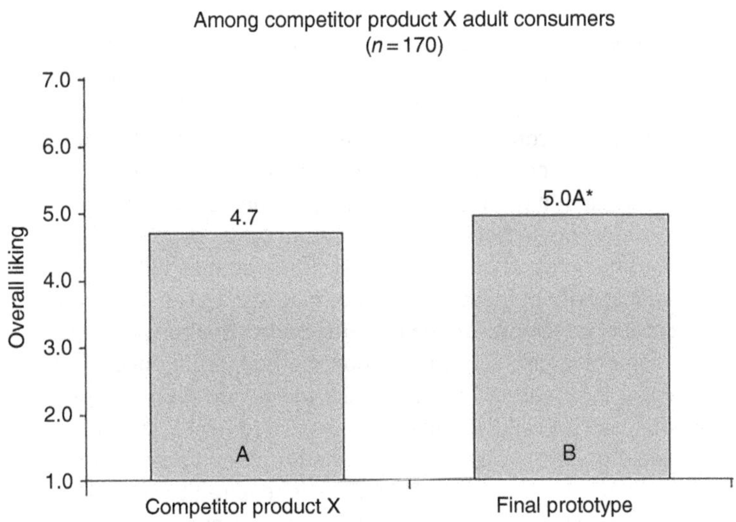

Figure 10.3.20 Blind home-use test results.

and competitor X. Prototype B with tobacco blend 1, high-flavor intensity, low sweetness, and ingredient 3 present was recommended for validation in a quantitative home-use test.

In a quantitative blind home-use test, the final prototype (prototype B) was found to be more acceptable to competitor product X adult consumers than their own brand (Figure 10.3.20).

10.3.6 Theoretical background of the tools

Over the years, multiple approaches for product optimization have been used; however, each has its own strengths and limitations. For example, a trial-and-error approach - develop a few prototypes, test for acceptability, reformulate, and repeat as necessary - can be very costly and time-consuming due to its iterative nature. Sometimes it can also be difficult to determine what direction to take from standard hedonic and descriptive measures. In another example, a prototype may receive lower liking scores than expected, and the descriptive attributes may not provide enough information to determine what needs to be improved in the product. Product developers must then make their best guess, and optimization may take several iterations, costing time and money. Product opportunities may also be missed by making assumptions too early in the process about which product direction will be most acceptable to the consumer.

Descriptive analysis using trained assessors may be used to understand how different factors impact sensory perception and to modify a product to meet certain sensory criteria. But the trained assessors cannot provide hedonic information (Stone and Sidel, 1993). Without consumer feedback, the product developer must assume that meeting the sensory criteria will result in an improved product.

Experimental design has long been used both for identifying the most important design factors and for modeling those factors to determine the optimum level of each factor (Moskowitz, 1995). Prototypes are produced that vary in certain factors according to the experimental design. The prototypes are then evaluated among consumers to gather hedonic ratings, and the results are analyzed to determine the optimum combination of factors. But because this approach can require a large number of prototypes when there are several factors to be considered, experimental design can be expensive and time-consuming, especially for products with high carry-over, such as tobacco products or for products with multiple components, such as frozen entrees. Also, modeling sensory responses based on an experimental design might not be feasible when there are extensive interactions between numerous factors. The complexity of these interactions may lead to results that are not actionable to product development because the models don't make sense.

While not a product optimization method, preference mapping can provide valuable information to the product developer about possible consumer segments, which products appeal to those segments, and where a new product could be placed to best appeal to the target audience (Yackinous et al., 1999). To do this, a broad range of products is required to span the sensory space to ensure sufficient differentiation between the products. If the products do not cover the sensory space, product opportunities may be missed. Also, a large number of consumers are required to be able to see potential consumer segments. As with the pure experimental design approach, this method can be very expensive.

To address the above issues, an adaptable, consumer-driven, RPN process for developing an optimal product was developed and refined. Consumers (users of the product) participate in qualitative discussion groups, during which prototypes are evaluated and discussed in depth, with participants' responses to each prototype determining the next prototype to be presented. The results from one group discussion may also be used to determine the products to be evaluated in subsequent groups. Within 4–12 iterative groups (2–4 days), the final product design can be reached. This adaptive, stimulus-response approach is consumer-centric in nature, which lends itself to understanding consumers' product sensory requirements. In addition, the business gains a deeper understanding of key design elements, which then feeds back into future product development projects.

10.3.7 Summary and future of the tools

This RPN process has demonstrated multiple successes for both product optimization and new product development projects across product categories. In the past, we performed multiple iterative loops of prototype development, screening, and consumer testing that took 18–24 months. RPN has greatly condensed that cycle time down to six months or less from scoping to commercialization, and thus has also greatly reduced costs. Also, starting with a broad scope and narrowing the product options down may reduce missed opportunities. The collaboration required for this approach to be successful helps to create a partnership between product researchers and product development.

The consumer-centric, qualitative nature of the research lends itself to understanding consumers' wants for the product sensory requirements and increases knowledge and understanding of the key design elements. The depth of discussion may also lead to more actionable insights than traditional quantitative tests because not only the "what" is understood but also the "why." It is possible to identify potential consumer segments that may not be obvious from traditional sensory evaluation and ensure that the resulting product design meets the product sensory requirements for the intended audience. Additionally, the real-time dialog with consumers provides a robust understanding of the consumer language that defines the product space and informs subsequent ballot development for product validation and other forms of traditional survey research.

This consumer-centric, RPN process not only benefits the current project, but also adds to existing organizational knowledge to benefit future projects.

Discussion questions

- What are the watchouts for designing an RPN/RPO study and what can you do to strengthen the learnings?
- What differentiates Design for Perception from Design of Experiment studies?
- What are the crucial roles of the cross-functional team during the entire RPN/RPO study?

Note

1. Due to proprietary concerns, actual research data and findings could not be presented in this case study. As a result, the following smokeless tobacco case study is fictional, but was devised based on actual research experiences with RPO in order to illustrate the concepts and procedures involved in the RPO process.

References

Lewis, J.N., Teratanavat, R., Beckley, J. and Jeltema, M.A. (2010) "Using a consumer-driven rapid product navigation process to develop an optimal product". *Food Quality and Preference*, 21(8), 1052-1058.

Moskowitz, H.R. (1995) "One practitioner's overview to applied product optimization". *Food Quality and Preference*, 6(2), 75-81.

Simons, C.T., Adam C., Kirkmeyer S., et al. (2008) Using Orthonasal Aroma Evaluation to Predict Consumer Liking. *Presented at the Society for Sensory Professionals*, Kentucky, 5-7 November 2008.

Stone, H. and Sidel, J. (1993) *Sensory Evaluation Practices* (2nd edition). San Diego, CA: Academic Press, Inc.

Williams, E.J. (1949) "Experimental designs balanced for the estimation of residual effects of treatments". *Australian Journal of Scientific Research, Series A: Physical Sciences*, 2, 149-168.

Yackinous, C., Wee, C. and Guinard, J.X. (1999) "Internal preference mapping of hedonic ratings for ranch salad dressings varying in fat and garlic flavor". *Food Quality and Preference*, 10(4-5), 401-409.

10.4

Free-Choice in Context Preference Ranking: A New Approach for Portfolio Assessment

Ratapol Teratanavat, James Mwai, Melissa Jeltema, and Jennifer Vahalik

Key learnings

- ✓ A new approach to assess overall product performance and identify optimal product portfolio selection (e.g. what is the optimum assortment of product offerings for a target consumer group)
- ✓ Rationale and benefits of assessing preference ranking through free-choice and in-context method
- ✓ How to apply this approach (research set-up and data analysis) to answer business questions around category/competitive benchmarking and product portfolio optimization.

10.4.1 Want to offer more. . .but how many is too many?

Consider a situation when a consumer packaged goods (CPG) company, called company X, is looking to expand its product portfolio. The company currently offers four stock keeping unit (SKUs) and is considering adding some new products into this product line. Within this product category, there exists a competitor - assume the key competitor is company Y which offers two SKUs, see Figure 10.4.1. Working in the consumer product research group for company X, you are asked to team up with marketing and product development to help answer the following questions:

- Question 1: With the current offering, from the pure product standpoint, what is company X's reach vs. competitor Y's?

Product Innovation Toolbox: A Field Guide to Consumer Understanding and Research, Second Edition. Edited by Kannapon Lopetcharat, Dulce Paredes and Jacqueline H. Beckley.

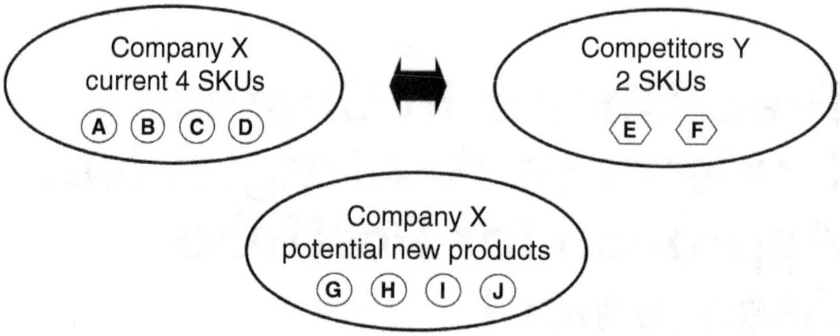

Figure 10.4.1 The case scenario for the company X.

- Question 2: For company X, do they need all four SKUs? Should they streamline and keep only the two best performing SKUs?
- Question 3: If company X wants to change its product offerings, what is the best mix of their products? How many do they need to beat the competition?

Driven by the need to remain competitive and relevant, many CPG companies need continuous innovation. The introduction of new products or varieties to a product line raises the need to determine the optimal product with the goal of building product portfolios that are profitable and different enough to generate additional revenue. An understanding of the competitive landscape and the performance of the company's portfolio against competing product lines is also of importance in measuring the impact of innovation and providing information on potential product opportunities. This innovation may take the form of product line extensions or the development of a completely new line of products, for a given target consumer audience.

10.4.2 Current approaches on product line extension

10

In the case of product line extensions, the company needs to first understand its position as compared to its competition. The questions that marketing and product development teams get asked by the executives are as follows: "How do our products stack up vs. competitors? And are our current products better than the competition?" Then, the team needs to address the scenario question: "What if we add line extensions or replace the existing SKUs with new products? How does this impact the overall portfolio?" For the completely new line of products, the questions become: "What is the optimum mix of products to launch and how many and which ones should be included?"

From a product performance standpoint, one approach is to rely on hedonic ratings such as liking or purchase interest (Lawless and Heymann, 1998). Generally, by comparing mean liking or TB/T2B on purchase interest, this would indicate whether a product is more or less appealing than the other, but it does not provide information on whether one *product bundle* is more appealing than another bundle. More importantly, with hedonic measures, many companies generally rely on test statistics such as average or TB/T2B ratings (Sambandam and Hausser, 1998). This only tells us whether the product has broad appeal, but doesn't necessarily reveal consumer segments, where one product appeals to one group and another product appeals to a different consumer group. Also, generally, the average scores tend to be similar; it is hard to differentiate one from the other, see Figure 10.4.2.

Another approach is a paired preference test (O'Mahony, 2007). This approach can be used when we have a small set of products, generally for head-to-head comparison, for example own vs. competitor, see Figure 10.4.3. However, it does not work when we have a large set of products we want to compare. In addition, the preference question is often asked after the hedonic rating, which could create a potential consistency bias from respondents.

Perrot et al. (2018) applied an internal preference mapping to define the best product portfolio study. They conducted a study to identify the optimal mix of soluble/instant coffee in three different markets, where 400-plus consumers evaluated a variety of products, including their own current market products, their prototypes, and their competitive market products in the mix. They applied a cluster analysis to identify consumer segments and determine the best performing product, based on liking rating, for each consumer segment. The suggestion for optimal product portfolio can be derived by combining the best-performing product from each consumer segment. This combination may be one of the

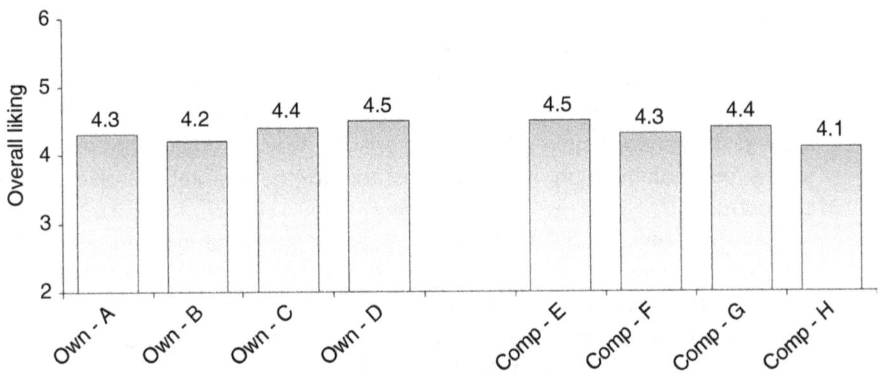

Figure 10.4.2 Using hedonic measure (e.g. mean overall liking) to compare how one product performs in comparison to others.

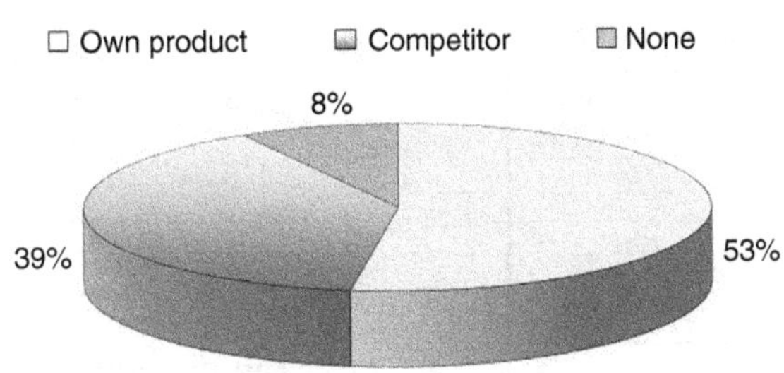

Figure 10.4.3 Preference ranking measure.

10

potential solutions, but it doesn't consider a synergistic effect that could take place when we consider other possible combinations.

From the product optimization standpoint, the commonly known approach is total unduplicated reach and frequency or TURF (Kreiger and Green, 2000; Miaoulis et al., 1990). It identifies the optimal mix of products that maximizes consumer reach by selecting the combination that complements well within the mix and gains highest reach, regardless of individual product appeal. The example in Figure 10.4.4 illustrates how TURF works. In this example, the company wants to select two out of three products to launch. The solution is to go with products A and C. The combination of product A and B is not selected even though each individually has high appeal. However, they overlap - they appeal to the same group of consumers and as a result, the total reach is less than when products A and C are offered.

While TURF has been commonly used, its limitations have been discussed in the literature (Conklin and Lipovetsky, 1999; Cui, 2004; Lipovetsky, 2008). It offers multiple solutions that are close to one another. In addition, it shows the potential mix for various scenarios, but it doesn't recommend *how many* to include. Fayle and Ennis (2010) also indicated that TURF is a flexible technique and easy to understand and explain; nonetheless, its visualization is not possible for larger numbers of concepts, and there is no exact solution for large problems - only approximate solutions exist. An alternative approach called efficient total unduplicated reach and frequency (eTURF) was introduced to help provide optimal solutions in those cases (Ennis et al., 2012). Their study showed that eTURF helped determine optimal solutions when there were lots of options to consider within reasonable time. Nonetheless, for this case study, we find some additional limitations with TURF. This may work when there is no competition, but when competitive offerings exist, it does not indicate how many products are required to be in a superior position. Also, it doesn't tell us where our combination is likely to stand in the market relative to the competition.

Given that these current approaches have some limitations, an alternative approach was developed to help deal with business questions regarding competitive assessment and portfolio optimization. This chapter will explain in detail this new alternative approach - how to execute and analyze data, and how to use information to help identify the optimal product portfolio and determine how well this optimal portfolio mix compares to the relevant competition. In order to differentiate the output collected from this approach from traditional

Ex: Select 2 products to the market

	TB purchase interest (%)
Product A	30
Product B	20
Product C	15

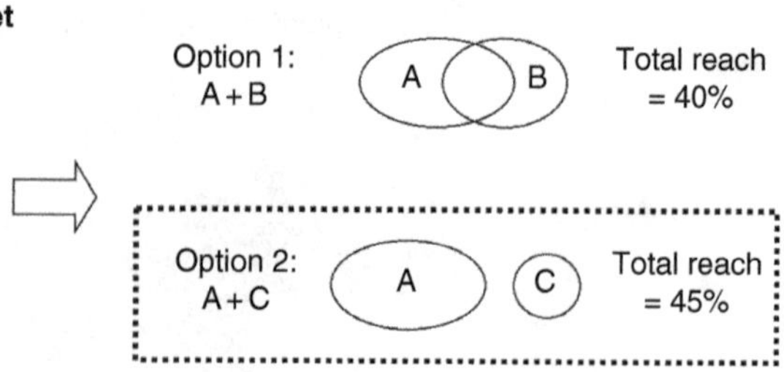

Figure 10.4.4 An example illustrates how TURF approach works.

TURF Analysis as well as to avoid confusion among team members about the source of data, we refer to "reach" as "potential market share," or PMS.

10.4.3 Free-choice in context preference ranking

This alternative new approach is called "free-choice in context preference ranking." This method measures preference based on multiple use/free choice by allowing consumers to use products freely in their natural setting. It allows researchers to consider a large set of products simultaneously and include competitive products whenever possible. This preference ranking information is used to help address business questions around the product portfolio comparison and optimization.

10.4.3.1 Methodology/set-up

The "free-choice in context preference ranking" method involves three steps (see Figure 10.4.5).

Step 1: Recruit consumers who are the target audience of these products. Then, assign a set of products (competitive products can be included) to everyone. Generally, products should be grouped in a small set of 4-5 products. Grouping can be done randomly or arranged based on product similarity. For instance, one set may have products with mint flavors (wintergreen, peppermint, and spearmint) and the other set may have products with other non-mint flavors.

Step 2: Within each set, ask consumer research participants to use products and indicate, for each product, whether or not it is in their consideration set. They can be asked whether it is something they would consider using in the

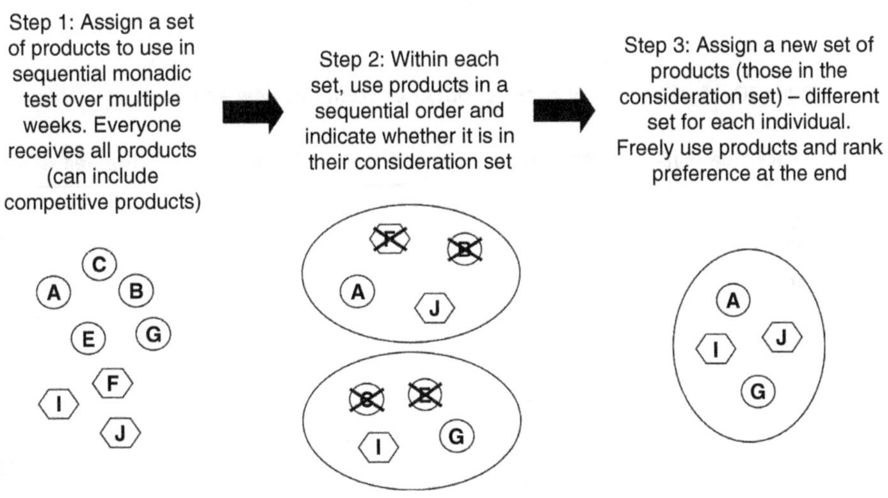

Figure 10.4.5 Free-choice in -context preference ranking set up.

10

future, or whether they would see themselves regularly purchasing and using the product. Consumers are instructed to go through all sets of products. Instructions are provided to consumers on the order of product evaluation according to a balanced, randomized design across product samples. During this sequential monadic test, traditional product attributes can still be gathered as well.

Step 3: Based on the information collected at step 2, the same consumers receive a new set of products that includes only products that were in their consideration set from the previous step. This set of products will vary by person as individuals will naturally have different products in their consideration set. During this step, they can freely use products; however, they want to in their own setting. They can go back and forth or use each again to help form their opinion. At the end of this step, they report their preference ranking.

Now, let us go back to the case study to see how this new approach helps address important business questions with regard to competitive assessment and product portfolio optimization. Due to proprietary concerns, actual research data and findings are not presented. However, a hypothetical illustrative case is developed to demonstrate the concepts and procedures involved.

This case study is based on research that was conducted with approximately 200 adult consumers. There were a total of ten products included in the study – four products (A, B, C, and D) from company X's current offerings, two products (E and F) from competitor Y's offerings, and four new products (G, H, I, and J) that company X considered as line extensions. These ten products were grouped into two sets (five products each) based on their flavor profile and were evaluated over three weeks. During the first two weeks, consumers were asked to use each set of the products. At week 3, they received a new set of products consisting of those that they indicated were part of their consideration set during the first two weeks. Table 10.4.1 illustrates an example of preference ranking data obtained from this study.

Table 10.4.1 Example of data set obtained from this new approach.

Participant ID	Ranked 1st	Ranked 2nd	Ranked 3rd	Ranked 4th
101	C	A	E	–
102	H	D	F	E
103	E	F	B	C
104	G	E	–	–
...	...	...	...	...
...	...	...	...	...
...	...	...	...	...
...	...	...	...	...

Note: Product not shown means it is not in the consideration set for that participant.

10.4.3.2 Data analysis and interpretation

10.4.3.2.1 Question 1: With the current offering, what is company X's potential market share (PMS) vs. competitor Y?

The first question focused on the assessment of the current offerings - from a pure product standpoint, did company X's product mix beat the competition (Figure 10.4.6)?

The first step in analyzing the data was to count the number of times each of these six products was ranked first (note: although ten products were included in this study, to answer this question, only six currently available products, that is, A, B, C, and D from company X vs. E and F from competitor Y, were included in the analysis). In the case where other products than these six products (i.e. G, H, I, and J) were ranked first, we went to the next rank until we found one of these six products. For instance, in Table 10.4.2, participant ID 102 ranked product H first, but this product was not included in the analysis for this question; as a result, product D was considered a first rank for this consumer (the same approach was taken for participant ID 104).

Figure 10.4.7 shows the output from this analysis, where company X received approximately 60% PMS vs. 30% PMS from competitor Y. Based on this analysis, one would conclude that company X was in a good position with these four product offerings, within the context of two product offerings from competitor Y.

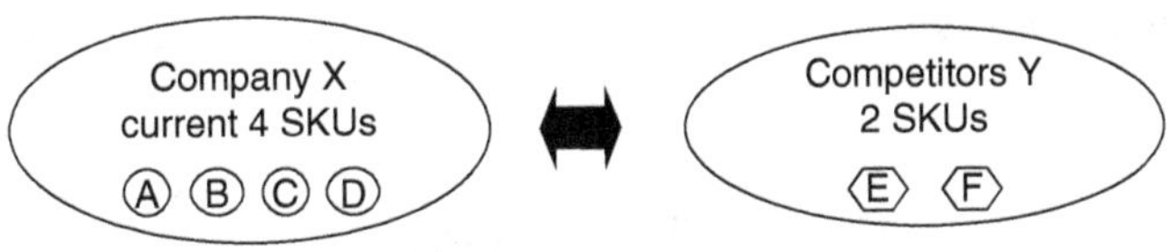

Figure 10.4.6 Number of products currently offered by company X and Y in the same market.

Table 10.4.2 Identify order of product preference for each consumer.

Participant ID	Ranked 1st	Ranked 2nd	Ranked 3rd	Ranked 4th
101	C	A	E	–
102	H	D	F	E
103	E	F	B	C
104	G	E	–	–
...	...	...	...	...
...	...	...	...	...
...	...	...	...	...
...	...	...	...	...

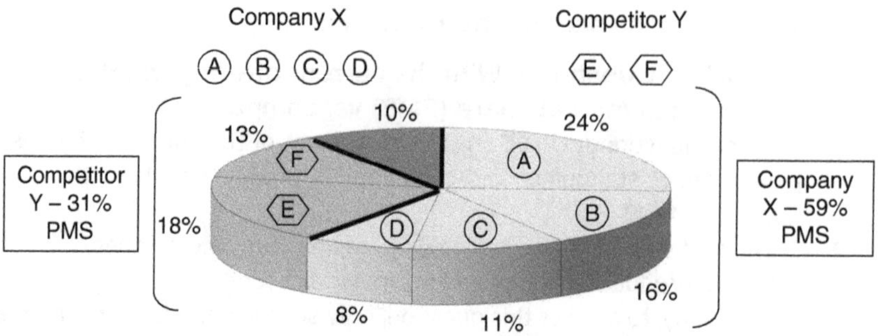

Figure 10.4.7 Preference ranking data with four SKUs from company X and two SKUs from the competitor Y.

10.4.3.2.2 Question 2: For company X, is there a way to streamline the current offering? Do they need all four SKUs? What if they keep only the two best-performing SKUs?

Now, company X wanted to know if they needed to have all four products. Could they streamline their portfolio? What if they offered only two leading products, the same as the competitor (Figure 10.4.8)? Is this enough to maintain the competitive advantage? Without having to conduct another study to answer this question, we can just simply go back to the same ranking data and run a new analysis.

To answer this particular question, only four products were included in the analysis – the two current best-selling products from company X (assume for this example A and B) and the two products from competitor Y (E and F), see above. We went through the same process as we did previously, see Table 10.4.3:

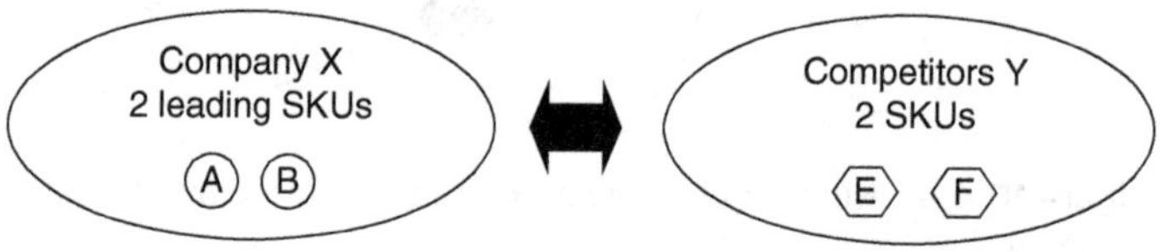

Figure 10.4.8 Depiction if company X offers only its two leading products (A and B) to compete with company Y's products (E and F).

Table 10.4.3 Identify number rank first on four products (A, B vs. E, F).

Participant ID	Ranked 1st	Ranked 2nd	Ranked 3rd	Ranked 4th
101	Ⓒ	Ⓐ	⟨E⟩	–
102	Ⓗ	Ⓓ	⟨F⟩	⟨E⟩
103	⟨E⟩	⟨F⟩	Ⓑ	Ⓒ
104	Ⓖ	⟨E⟩	–	–
...	...	...	...	...
...	...	...	...	...
...	...	...	...	...
...	...	...	...	...

Note: Product not shown means it is not in the consideration set for that participant.

10

- Identify the frequency with which each product gets ranked first
- If other products, other than these four products, were ranked first, go to the next rank until one of these four products was chosen and an additional analysis step:
- Identify the number of participants who have none of these four products in the consideration set.

Figure 10.4.9 shows that if company X decided to keep only two products (A and B), they would actually be worse off than with four products (A, B, C, and D). In this scenario, competitor Y would have a higher PMS, which could be because consumers who liked the other two products (i.e. C and D) from company X switched to competitor Y as their product was no longer available. Based on this analysis, company X should not streamline its offerings; rather, they should revisit their own lineup and try to understand why they would need more products in their portfolio than the competition has in order to be in a superior position.

10.4.3.2.3 Question 3: If company X wants to change its product offerings, what is the best mix of their products? How many do they need to beat the competition?

Now, if company X wants to rearrange their product line - they want to revisit their current four product offerings (A, B, C, and D) along with four new products (G, H, I, and J) that product developers have been working on (Figure 10.4.10).

The question becomes *"How many products should they offer and what are those products?"* To answer these questions, we conducted the scenario

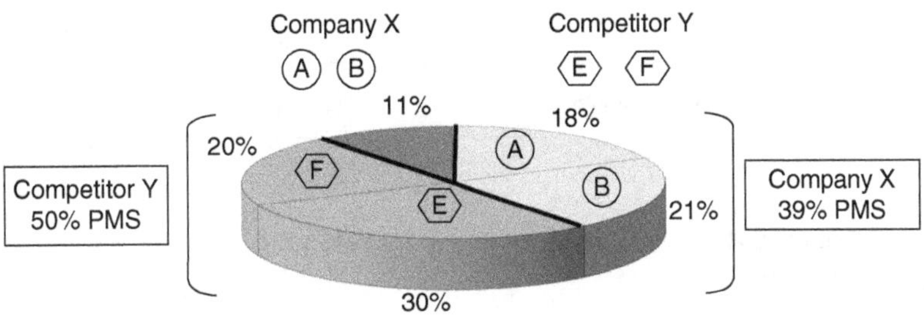

Figure 10.4.9 Preference ranking data with two leading SKUs from company X and two SKUs from competitor Y.

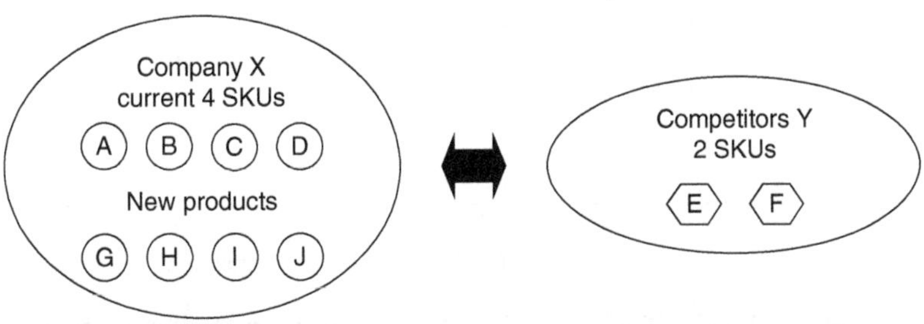

Figure 10.4.10 Depiction if company X offers additional products from four new options (G, H, I, and J) to compete with company Y's products (E and F).

10

analysis by running multiple analyses based on the number of products the company could potentially offer.

For each scenario, the best product mix was identified and company X's PMS vs. competitor Y's PMS was estimated, see Figure 10.4.11. For instance, if company X were to put out only one product, they should select product B - in this case company X would receive approximately 30% PMS, whereas competitor Y would receive 50% PMS. The analysis was conducted for all possible scenarios - from a single product that company X offers to all eight products together.

The next step was to plot the percentage PMS in each scenario, see Figure 10.4.12. It shows that company X would need at least three products to be most competitive with company Y. If this was the option to go forward with, we could further identify the best mix of three products for company X to offer. Based on this result, again, company X should strongly reconsider its own

Scenario 1: Company X offers 1 product

	% PMS
Company X: (B)	30%
Competitor Y: ⟨E⟩ ⟨F⟩	55%

Scenario 2: Company X offers 2 products

	% PMS
Company X: (B) (A)	39%
Competitor Y: ⟨E⟩ ⟨F⟩	50%

⋮

Scenario 8: Company X offers 8 products

	% PMS
Company X: (A) (B) (C) (D) (G) (H) (I) (J)	60%
Competitor Y: ⟨E⟩ ⟨F⟩	30%

Figure 10.4.11 Comparing company X's PMS vs. competitor Y's PMS for each scenario.

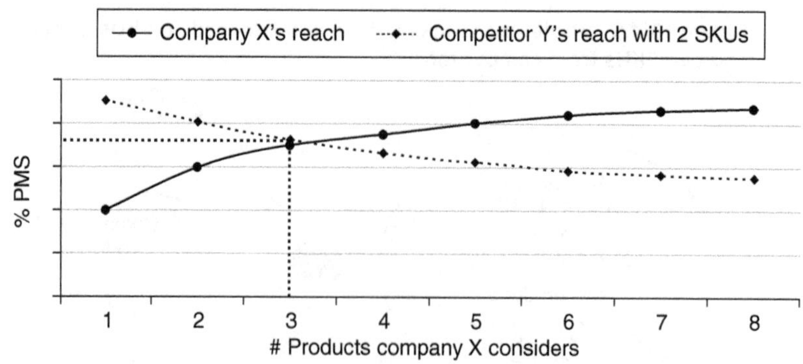

Figure 10.4.12 Identify number product mix using equilibrium approach.

product lineup and try to understand why they would need more products in their portfolio than Company Y in order to have a competitive advantage.

It is worth mentioning that this approach can be extended when new products are introduced at a later stage without having to start everything all over again. For instance, what if competitor Y decided, later on, to launch new SKUs? In this case, there is no need to begin from scratch. The researchers can simply invite the same group of consumers back and give them a set of new products along with the product they most preferred from the previous study. The analysis can be rerun to assess the impact of the competitor's new offerings on company X's PMS. With the new competitive set and analyses, we can answer the following: What is the defensive product strategy for company X? Should they include some new products? How much will they gain back? etc.

10.4.4 Practical considerations

- What can you do if products cannot be picked up by research participants multiple times (or even once)?
- What if your company or brand dominates the category (and you may not need to consider your competition) and still wants to expand your product offerings in the market?

These two questions are likely not that unusual in today's environment. The free choice profiling approach described earlier, where the researchers were able to place prototypes and competitive products with participants more than once is highly adaptable, since situations exist where this is simply not possible. The following example shows how this approach was adapted for a research project that required shipping products to participants for a brand that had no competitive set. The business question to be answered for this study was "Among these eight options, what is the right *number and combination* of products for the brand to offer going forward?"

In this study, eight complex products that required preparation by the consumer were evaluated: four current products (K, L, M, N) and four potential line extensions (P, Q, R, S). This line of products was "new to the world" and had no competitors at initial launch, and still none at the time the potential line extensions were being formulated. Therefore, there were no competitive products available to include in the study. Additionally, in order to ensure representation of a specific target demographic, a highly geographically dispersed population of 250 individuals was used, which required shipping of all products vs. pickup at facilities.

Given that re-distribution of only those products that were in their consideration set to the participants was not possible logistically, a modified approach was adopted. Participants received two full sets of eight products: one set marked "open now" and one set marked "open later." Products in the "open now" set were used to collect traditional hedonic and purchase interest measures and for determining which products, if any, they would buy regularly ("in consideration set" as described previously in Step 2). After all samples were prepared and evaluated (in randomized, sequential monadic order), participants were instructed to open the "open later" set and remove the specific products that were in their consideration set and prepare and evaluate them again, in any order they wished. Preference ranking was then collected on this smaller subset of "re-taste" samples as described previously in Step 3. See Appendix 10.4.A for example questionnaire.

Since the client team was open to replacing some of the existing SKUs with the new line extensions, two scenarios were possible as outcomes based on insights from data analysis. (1) Add the highest performing new SKUs to the existing 4-product line, and (2) Revamp the entire line to include the highest performing products in the study. In either scenario, we still needed to understand both WHICH products and HOW MANY to include in the new product lineup.

10.4.4.1 Question #1: Which SKUs should be added if the original 4 product line is retained?

To answer this question, the original 4 SKUs were forced into the analyses, regardless of their ranking position across individuals. The line extensions were then added to the analyses, one at a time, to determine their preference relative to the originals. Based on preference data and other measures gathered in the study, a 5 or 6 product line was recommended - adding product P, or products P and Q to the original 4 SKU line. The company could choose to add two additional flavors to their current lineup and gain 2% more PMS. If only 1 flavor was added, no consumers would be lost overall - Figure 10.4.13 shows that with both

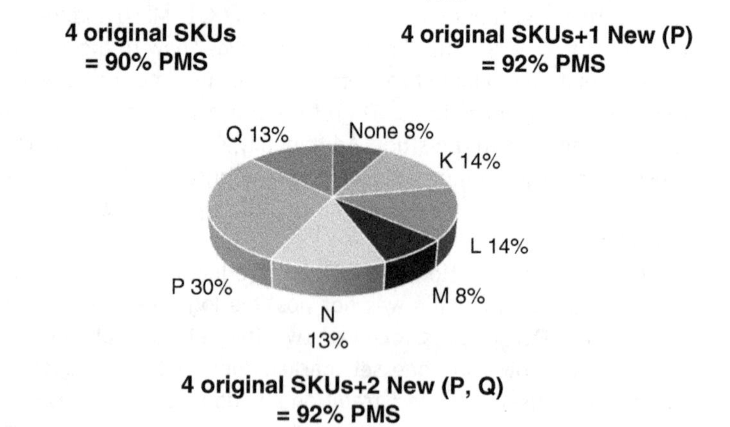

Figure 10.4.13 First-choice preference ranking data for a 4, 5, or 6 product lineups depicted as PMS when all 4 original SKUs are retained.

10

5 and 6 products in the lineup, only 8% of consumers would not have a product in their consideration set. Although adding Product P only added 2% to PMS, it was actually the most preferred in the 5-product mix. While no additional gains would be realized, overall, by adding product Q along with product P, the flavor of product Q was thought at the time to be more in line with the interests of the initial target consumer. This niche target represented a subset of the overall population in this study and gave product Q higher ratings overall than others. Additionally, as you will see in Question 2, product Q was a better performer than one of the existing 4 SKUs, and this information was also taken into consideration when making recommendations. What the company would ultimately choose to do would come down to a business decision of whether the benefits of adding both P and Q outweighed the additional fees associated with carrying more products in the line (manufacturing, packaging, marketing, slotting, etc.).

10.4.4.2 Question #2: Which SKUs should be carried if the whole line is revamped?

To answer question #2, all eight products are analyzed in the mix. Again, a five or six product lineup was recommended, but since it is possible to drop one of the original SKUs in favor of one that was a better performer in the study, we see that the combination has changed for the five-sample array. Table 10.4.4 shows that, all things considered, new product "P" outperforms all other products and is again included in a 5-product array, but original product "M" is now replaced by new product "Q". Carrying the analyses through to include all eight products, we can see that there is nothing to be gained by including more than six total products as the percentage of people with no products in their consideration set levels off at 8%, and adding a 7th or 8th product merely takes potential market share away from one or more of the other products in the line.

Table 10.4.4 First choice preference data for a 5, 6, 7, or 8 product lineup when all products are included in analysis.

		% first choice based on # of products in array			
		5	6	7	8
K	Existing	15%	14%	14%	13%
L		15%	14%	13%	12%
M		–	8%	7%	6%
N		15%	13%	10%	10%
P	New	32%	30%	28%	28%
Q		14%	13%	13%	12%
R		–	–	–	5%
S		–	–	6%	6%
None		9%	8%	8%	8%

10

Beyond the numbers: As with any research study, one or two data points in isolation should never paint the entire picture that a company uses for critical business decisions. The data in this example study, along with other factors, including weekly velocity of the existing SKUs, allowed the company to make the most informed decision for their brand.

10.4.5 Theoretical backgrounds of free-choice in context preference ranking

In developing a new product line, companies sometimes face the challenge of identifying the right mix of product varieties for a given target consumer audience. For instance, in developing a new sports drink line, the company needs to determine how many flavors should be offered. Generally, offering multiple flavors tends to increase appeal to a broader audience; however, this can be costly for marketers. The company needs to determine the right number of flavors to offer that gain the broadest PMS (i.e. not appeal to the same audience) and the composition of that flavor mix. In addition, the company will want to gain insights on how its optimal flavor mix performs compared to its competition in order to guide its business decision process.

The procedures outlined in this chapter present a new approach called the free-choice in context preference ranking method for optimal product portfolio selection. The traditional approach, such as TURF analysis, relies on purchase interest or liking measures, whereas this approach relies on the placement of products into a consideration set and a comparative measure, that is, preference ranking. First, participants evaluate all products and indicate those that fall into their consideration set. Then, they rank the products in their consideration set based on preference. With this approach, competitive products are also included whenever available to help guide the portfolio selection in greater context and allow head-to-head comparisons on overall appeal. Scenario analysis is conducted and the optimal product portfolio is identified as one that has the highest PMS <u>and</u> is higher than the competitive set. It has been shown that this new approach provides insightful information to help identify the optimal product portfolio and to determine how well this optimal portfolio mix performs from a product standpoint compared to the competitors.

10.4.6 Summary and future

This alternative approach, the free-choice in context preference ranking method, has been shown to help address complex business questions when it comes to portfolio assessment or optimization. It is based on preference ranking data and the preference is based on multiple uses in a natural setting. This method also allows you to include a large set of products, including competitive products. And from what we experienced, the output from this is intuitive and easy to understand and communicate to the business.

Note that the approach discussed in this chapter focuses on the pure product performance (i.e. a blind product test). It does not take into account the effect of branding and other marketing factors that could impact product choice and purchase decision. As a result, the estimated PMS should be used cautiously; it should not be simply equated to potential market share. Nonetheless, this method can be applied to branded products as well - it is interesting to see how consumers react to products under two different contexts - one in the absence of the branding and the other with all under the brand. The area one may want to consider as a future study is to compare results from this approach to TURF, which is traditionally used in market research, as well as other approaches that have been introduced such as SURF - structural unduplicated reach and frequency (Lipovetsky, 2008), Shapley value analysis (Conklin and Lipovetsky, 1999; Cui, 2004) or landscape segmentation analysis (LSA) maximum coverage/portfolio optimization (Fayle and Ennis, 2010).

Discussion questions

- What is the value of Free-Choice in Context Preference Ranking in current category assessment and line extension opportunities?
- Can it be used for creating a totally new category?

10.4.A Appendix

Examples of questions to collect free choice in-context preference ranking data

(1) **Product in consideration set**
Is this a product that you would choose to purchase on a regular basis?
[ ] Yes
[ ] No
Why or why not?
(Note: only products with "Yes" should be included in their final ranking).

(2) **In-context preference ranking**
Please rank the products you tried from most liked to least liked. Only those that you said you would use on a regular basis are included in this list.
1_________ Product Most Liked
2_________
...._________
...._________ Product Least Liked
If we were to offer you 4 additional products, which would you choose? You can choose any combination of products. E.g. all 4 of one product, 2 of 1, 2 of another, 4 different products, etc. Which ones would you choose? Remember, you have 4 total products to choose.
Product__________ How many?
Product___________ How many?
Product____________ How many?
Product____________ How many?

10

References

Conklin, M. and Lipovetsky, S. (1999) "A winning tool for CPG - the Shapley value game theory method brings advantages to marketing managers evaluating product line flavor decisions". *Marketing Research*, 11(4) (Winter 1999/Spring 2000), 23-27.

Cui, D. (2004) Appropriate Application of TURF and Shapley Value for Product Line Optimization. A White Paper from the Ipsos Group.

Ennis, J.M., Fayle, C.M. and Ennis, D.M. (2012) "eTURF: a competitive TURF algorithm for large datasets". *Food Quality and Preference*, 23(1), 44-48.

Fayle, C.M. and Ennis, J. (2010) An Efficient Approach to Solving Complex Market Research Problems. *Institute for Perception. Paper Presented at the 2010 Sensometric Conference*, Rotterdam, The Netherlands.

Kreiger, A.M. and Green, P.E. (2000) "TURF revisited: enhancements to total unduplicated reach and frequency analysis". *Marketing Research*, 12(4) (Winter), 30-36.

Lawless, H.T. and Heymann, H. (1998) *Sensory Evaluation of Food - Principles and Practices*. New York: Springer Science.

Lipovetsky, S. (2008) "SURF - structural unduplicated reach and frequency: latent class TURF and Shapley value analysis". *International Journal of Information and Technology & Decision Making*, 7(2), 203-216.

Miaoulis, G., Free, V. and Parsons, H. (1990) "TURF: a new planning approach for product line extensions". *Marketing Research* (March), 28-40.

O'Mahony, M. (2007) "Conducting difference testing and preference trials properly for food product development". In H. MacFie (ed.) *Consumer-Led Food Product Development*. Cambridge, UK: Woodhead Publishing in Food Science, Technology, and Nutrition, 434-452.

Perrot, M., Pineau, N., Antille, N., Moser, M., Lepage, M., Thaler, T., Voirin, A. and Rytz, A. (2018) "Use of multi-market preference mapping to design efficient product portfolio". *Food Quality Preference*, 64(March), 238-244.

Sambandam, R. and Hausser, G. (1998) An Alternative Method of Reporting Customer Satisfaction Scores. White Paper from TRC Research Insight Direction. Available at https://trcmarketresearch.com/whitepaper/an-alternative-method-of-reporting-customer-satisfaction-scores-2/ (accessed 8 March 2022).

10.5

Extended Use Product Research for Predicting Market Success

Ratapol Teratanavat, Melissa Jeltema, Stephanie Plunkett, and Jennifer Vahalik

Key learnings

- ✓ Why extended use product research (EUPR): Rationale and benefits over traditional sensory research
- ✓ How to conduct extended product use: Factors, research flow, key measures
- ✓ What you gain from product use research: Insights that provide a deeper understanding of consumer product experience that go beyond measuring product performance
- ✓ When to use extended product use research: Desired approach for solving business problems regarding product viability and understanding potential market success.

10.5.1 Challenges in validating and predicting the success of new product launch

10

A leading consumer packaged goods (CPG) company wanted to explore new product opportunities to market under their major brand. These products were unlike any currently on the market. The brand team and product development faced two immediate challenges: First, they needed to determine which, if any, of the new products were viable options, and second, determine the optimal portfolio of products to achieve broad appeal. They also wanted to avoid new product launch pitfalls as the majority of new products fail within the first year and subsequently are removed from the shelf

Product Innovation Toolbox: A Field Guide to Consumer Understanding and Research, Second Edition. Edited by Kannapon Lopetcharat, Dulce Paredes and Jacqueline H. Beckley.

(Beckley et al., 2017; Stevens and Burley, 1997). While there are a host of reasons why new products may fail, we contend that the variability in predicting new product success can be improved through a better understanding of two important questions. First, do we have the right product, and second, what is the market potential?

For new product ideas, traditional market research concept testing is used to assess the idea viability. It may have suggested that people are interested in these ideas; however, while concept testing of novel products provides a good indication of the general consumer need, it often is unable to predict how the consumer will react to the actual product. Additionally, neither concept testing nor single use concept/product testing can provide an indication of whether consumers will adopt new and novel products that require a change in behavior. Ultimately, companies want to know how the new products will affect their bottom line. Will they gain market share? Through enhancement of the traditional sensory research approaches, we might better understand the consumer in context and ultimately the real marketplace potential.

In the conventional approaches, the consumer and product research group supports product development by utilizing well-known approaches to define overall product performance (see Figure 10.5.1 for the commonly used approaches by sensory professionals) (Beckley, 2007; Moskowitz et al., 2012). Discrimination and descriptive measures provide direction for product development to make improvements against specific attributes measured (Stone and Sidel, 2007). Additionally, affective measures of liking are utilized, sometimes exclusively, as an indicator of product performance (Lawless and Heymann, 1998). Finally, traditional approaches may extend to qualitative testing which allows for a more in-depth understanding of product performance. While these conventional approaches work for specific questions of interest, their use has limitations in application to broader business strategic questions like new product viability and product portfolio optimization.

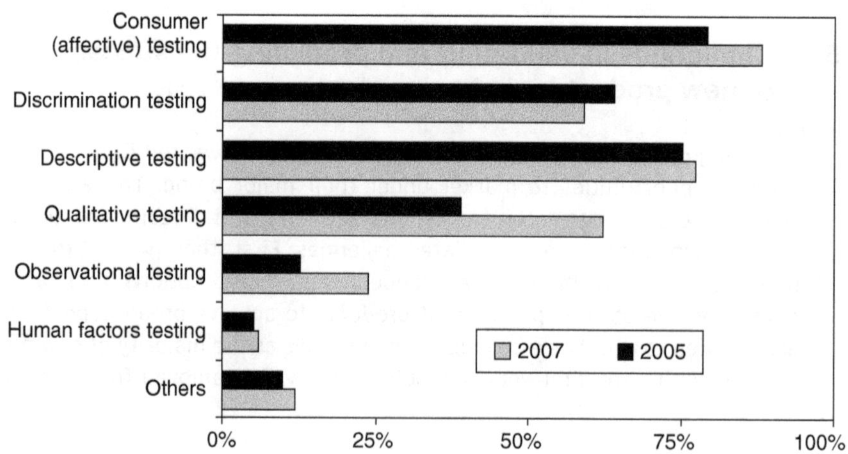

Figure 10.5.1 Commonly used sensory research methods, surveyed among sensory professionals.

10

10.5.2 Limitations of traditional approaches

Traditionally, the criterion used to determine whether a product is ready to launch focused primarily on standard product performance measures of liking, and/or other hedonic analogs, purchase interest, and preference. Metrics of success were established as a comparative assessment. That is, hedonic measures have to be at or above an established benchmark. Similarly, a preference or liking measure of one product would be compared with another product (usually one or more competitors). But rarely are consumer decisions confined to these comparative assessments. Thus, these traditional approaches are useful only in the context of the conditions of consumer choice they are measuring. More specifically, the comparative assessments may inform product potential, but they are incomplete for determining how consumers will leverage the comparative assessments in their overall product choice. Despite their popularity and usefulness, we believe that traditional methods do not provide adequate information for novel products to answer whether or not a product is ready to launch.

The complexity of consumer decision-making is a topic garnering much attention and stems from the vast array of product choices that are available to consumers. Large companies, like Google, are actively investing in the understanding of the consumer decision-making process to help with product launch decisions (Rennie et al., 2020). In the case of new product launches, it is important to understand whether this new product will be adopted by consumers over their current products, and why or why not? This includes understanding the trade-offs that someone may be willing to make. People may not like the taste of a particular product, like wheat bread (this can be reflected by low hedonic ratings), but they still choose to use or buy the product over white bread because it offers other benefits. Taking a more holistic view of the consumer-decision-making process builds upon consumer product preferences to include preferences in context of lifestyle, attitudes, and values. The type of trade-offs and degree to which they are made depends upon what the individual values most.

As we have strived to improve our approaches to product research, we realized that we needed to go beyond an understanding of liking or preference for our products. In essence, we needed to go beyond just the sensory attributes and understand why consumers want certain products and what benefits they are looking for. Using traditional sensory research approaches and relying on hedonic measures does not help us answer these questions. Therefore, we knew we needed to develop an alternative approach which provides specific information to the business.

Another key component for estimating a successful product launch is knowing how hedonic measures are reflected in actual behavior (i.e., in the context of the natural environment, what drivers of liking are most influential to product choice?). The concern we have with traditional sensory approaches is that research findings may not generalize to the natural setting or the marketplace (Moskowitz et al., 2012). Thus, the traditional approaches have the same deficits as laboratory research when compared with research conducted in the natural environment. For example, in laboratory settings, palatability has been shown to be a main driver of intake. Additionally, in these same settings, the time of day has not been shown to influence the amount of food consumed.

However, in the natural environment, time of day is significantly more influential on the amount of food consumed, with little effect on palatability (deCastro, 2000). Interestingly, this has been an ongoing concern within the sensory and consumer research discipline, and recently, Halim et al. (2019) published an article discussing the validity concern of the laboratory preference test vs. the real-time choice behavior of consumers. While it was indicated that preference predictions from the laboratory environment were good predictors of real preference over time, the study did not address the fact that consumers could change their product choice over an extended period because they might want to "switch things up," not because they had inconsistency in their state of preference.

Traditional approaches do not provide an understanding of the decision-making process in the context of the real world. This understanding, however, is a relevant outcome of extended product use testing. This is because extended product use testing considers the context of the marketplace landscape, and within that, how the product measures up against an appropriately determined benchmark. Furthermore, we include the trade-off question as part of the extended use method as we recognize the importance of individual differences in how consumers make choices as they weigh experience, preferences, attitudes and values, and individual needs. In addition, when using extended use testing, product viability is more in focus as we understand the drivers that maintain the relationship between x and y in order to achieve product loyalty. In short, extended product use testing allows a deeper understanding of the decision-making process as it relates to the consumer/product relationship in the context of everyday life.

The foundation from which this alternative approach is developed is simply a thorough understanding of consumer behavior. In order to develop successful products, we need to truly understand people (Schieffer, 2005), and sensory attributes and liking are only a small part of that understanding. To be consumer-centric, we must understand the consumer decision-making process and assess product performance in the context of free choice and behavior. We have been successful internally in demonstrating that go/no-go decisions should be rooted in understanding the consumer-product connection in order to be able to determine whether the product is likely to be successful. In other words, it is not just whether and how much consumers like the product. It is important to understand the driving factors - what is important to them, why and how often they will be using products, and whether and how this product fits into their life vs. other products that are available for them.

10.5.3 An alternative: extended use product research

The approach we describe here is called "extended use product research" (EUPR). This approach combines traditional methods (i.e., ballot measures) with behavioral measures such as product choice and usage as well as in-depth, in-context interviews (i.e., consumer decision-making criteria) evaluated over extended periods. It offers insights beyond traditional sensory research approaches (e.g., descriptive, discrimination, and affective) to help understand consumer perspective on products and assess the potential market success. It enables us to understand the consumer interaction with the product (decision-making process and product design criteria) and to obtain behavioral measures that help determine market potential.

This approach addresses the following important questions that will ultimately assess market potential:

- What factors influence consumer choice decisions, beyond sensory attributes?
- What are the product design criteria? How important are they?
- Does the product response remain the same or change over time?
- How well does the product fit into consumers' lives?
- What is the usage pattern? How does it change over time?
- Is the product good as it is, or what changes need to be made to be good enough?

Now, we will walk through the steps of this method. At each step, we will review how to execute the approach, the type of information provided and how to utilize the information to help make business decisions.

10.5.4 Steps in conducting extended use product research (EUPR)

EUPR testing steps begin after a proof of concept has been established. That is, the products used must have been sufficiently optimized for go/no go decisions for this method to be successful. There are four critical steps to the process that produce the insights around the consumer decision-making process, knowledge about the consumer and product interaction, and assessment of market potential. Figure 10.5.2 outlines the general steps in conducting an EUPR.

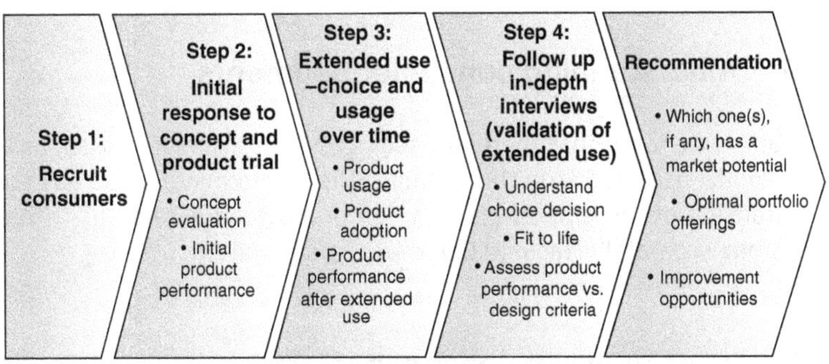

Figure 10.5.2 Steps in conducting an extended use product research (EUPR) to gauge new products' potential in marketplace.

10

Step 1 is to recruit the consumers for the study. In general, we would recruit consumers who currently use or are familiar with the product. For those "new to the world" products, we would screen consumers who express interest in the concept. Also depending on project objectives, we may want to screen consumers based on form and flavor interest.

Step 2 is to introduce concept and product to consumers. This step allows consumers to become aware of the product idea and to try the product so they can familiarize themselves with the product and any product options (e.g., flavors or forms). *Tip*: For products that are very new and where we anticipate initial barriers to entry, a group setting can be used to encourage interaction and sharing of product experiences. In our research utilizing this process, we found group interaction very helpful to get consumers past any reservations they may have had with the product, as well as helping them to understand what the product is about and how to use it. This group setting appears to be particularly useful for a "new to the world" product since consumers will be unfamiliar with the concept. During Step 2, we can also assess the initial response to the concept and products.

Step 3 is to offer research participants the option to take home a product(s) to use. They could choose from the products in the choice set provided to them. The number of products to be included in the choice set as well as the number of products each person is allowed to take home depends on your research objectives. We tracked each consumer's choice selection and also gave them a journal to track their usage and write any thoughts and comments about each product.

Step 4 is in-depth discussions with consumers at selected intervals to understand product experiences and the product adoption process. We also collected information to understand usage and product experience and to assess product performance after extended use. This provides understanding of any changes in consumer experience, behavior, and usage over time. At the end, we conducted in-depth interviews to help understand consumers' decision-making processes and validate the product performance against the design criteria.

10.5.5 Understanding consumer segments

Often, we learn that there are segments of consumers based on interest and product acceptance. In this example, the in-depth interviews gave us an understanding of the differences between groups and allowed us to evaluate each group separately (Table 10.5.1).

10.5.6 Assessment of sensory performance

Similar to traditional sensory research approaches, for each consumer segment, we observe and compare how consumers rate one product vs. others.

Table 10.5.1 Different responses from different segments – they were identified based on interest/acceptance.

High acceptance	Audience segment	% Acceptance	Rationale based on in-depth interviews
⇧	Segment 1	60%	• More open/accepting to concepts • More familiar with products • High enjoyment
	Segment 2	40%	• Busy lifestyle; social interaction • Have interest but do not enjoy using these products
Moderate acceptance	Segment 3	20%	• Don't really have any interest for these new products

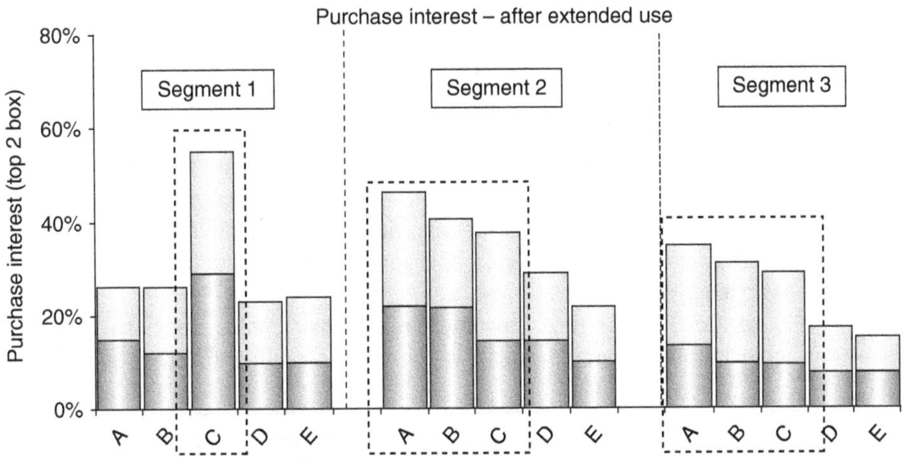

Figure 10.5.3 Product response by consumer segments. Purchase interest (top 2 box) refers to % Definitely/Probably would buy.

In Figure 10.5.3, different segments reacted differently to these products (i.e. one product did not fit all), so we needed to assess product performance separately by consumer segments. We also reviewed the rating on sensory attributes and tried to explain the drivers of product acceptance.

Extended product use testing allows us to compare product performance over time. Figure 10.5.4 shows product response varying at different stages. In this example, some products were well received at the beginning but performed poorly after an extended period, whereas other products did not do well at the beginning but performed better over time. It is important to understand the reasons behind this variation in performance. It could be because of sensory attributes, but could equally be due to other reasons, such as fit to life. Sensory is only one aspect – there are so many other factors that drive product acceptance and rejection at different stages.

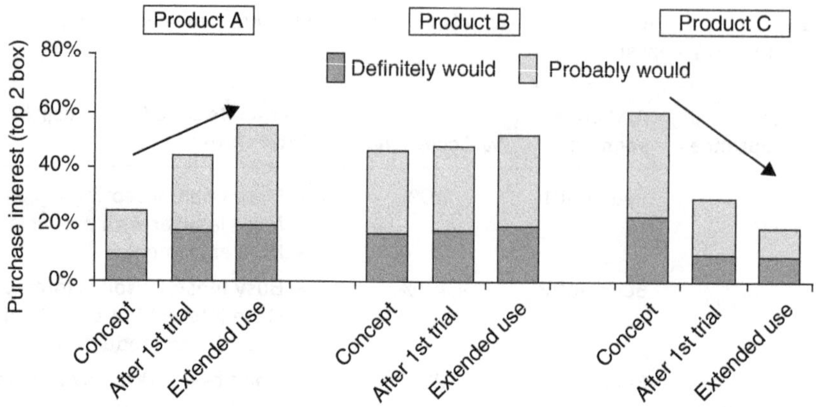

Figure 10.5.4 Extended product use research allows us to compare product performance over time. Purchase interest (top 2 box) refers to % Definitely/ Probably would buy.

10.5.7 Understanding how consumers make choice decisions

In-depth interviews allow us to understand how consumers made their choices. We find that several factors, *not just sensory attributes*, influence product choice decisions before, during and after experience with the products. This helps us understand why consumers accept or reject products. Understanding factors that influence their choice decisions allows us to determine what challenges each product faces.

Figure 10.5.5 shows how consumers made their choice decisions from this example. In this research example, we provided consumers with three different

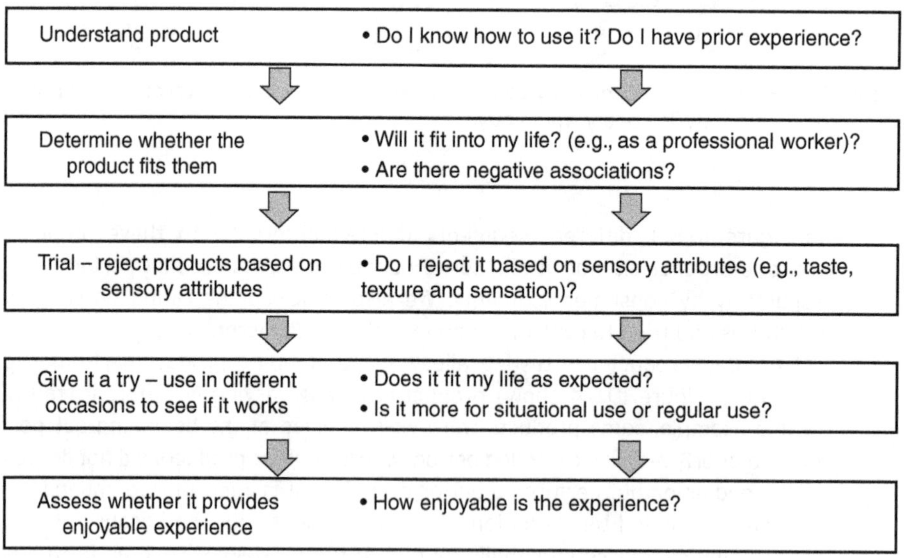

Figure 10.5.5 An example of output from in-depth interview in extended use product research.

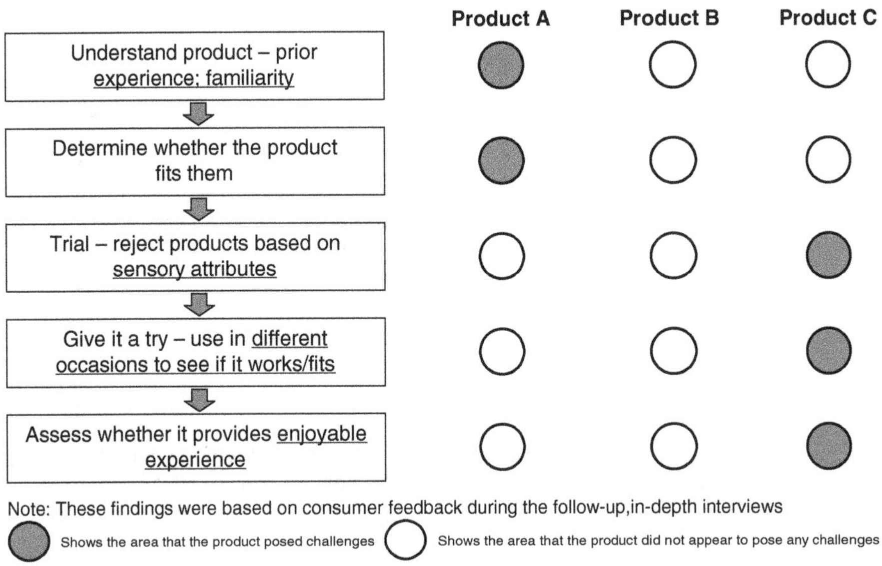

Figure 10.5.6 Understanding factors that influence their choice decision allows us to determine what challenges each product faces.

options (products A, B, and C), all of which were developed against the same consumer need. As you can see from Figure 10.5.6, the challenges, or pain points, varied by product. While the initial reaction to product A was problematic, after use, the product was actually well received. The opposite was true of product C. While the issues with product A might be overcome with adequate marketing and communication, the issues with product C made it a poor option for commercialization. Product B was actually the best option of the three, as it had no pain points.

10.5.8 Using behavioral measures to help assess product viability

Throughout the extended use period, we determine how well consumers adopt each product. We are able to observe the usage patterns through their choice selections. These are important behavioral measures that go beyond what traditional approaches provide. Tracking product usage helps assess short-term and long-term product performance. It helps us understand what it will take for the new products to be acceptable and adopted. Also, we can determine the impact of new product introductions on the usage or consumption of existing products.

Understanding product experience is critical, particularly for new products. This is important because, for a new product to be successful, it is not just about product attributes anymore. It is about understanding the value consumers gain from the product. This is very critical, particularly for new products. Not only do we want to assess whether the new products are acceptable, we also need to

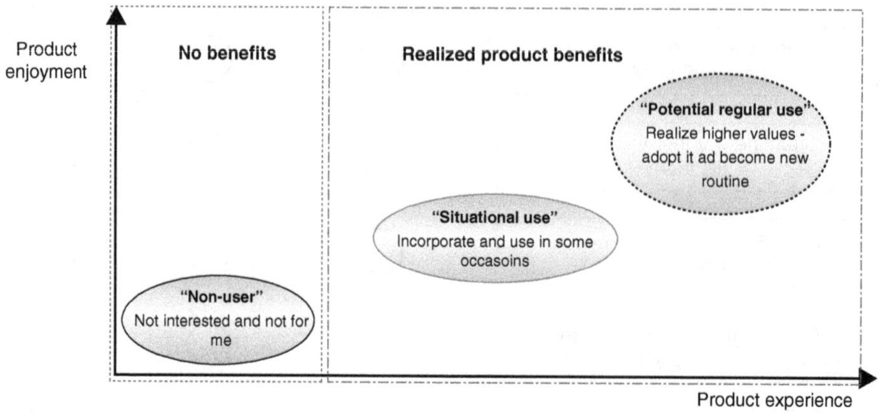

Figure 10.5.7 Illustration of consumer segments based on behavioral measures. Through this research, we were able to classify each person as a "user" or "nonuser" of each product. Among users, they were also segmented into situational users and regular users.

know why they would want to use these products, or not. Why would they choose to use these products instead of something else, and what benefits do consumers find from using these products?

Through listening to individuals' stories about their product experiences, we are able to identify consumer segments. In this example case, we were able to identify a group of consumers who were not interested in the product. Once they tried it, they realized that these products were not for them, so they were not really using the products. Then, we had another group of consumers who saw benefits of these products during the extended usage period and had adopted them. There were segments within this group as well. Some saw limited benefits and therefore used the products only in certain situations. Others realized more benefits. Those who really loved one or more of the products truly adopted them and the new product(s) became their new routine. The degree of adoption was related to the amount of benefits realized (how large was the need) as well as the length of the product experience (see Figure 10.5.7). This is particularly true with new products that require a behavioral change.

10.5.9 Among users, there were also segments of situational users and regular users

In order to adequately understand how well these products perform, secondary measures in addition to product usage are valuable. At the beginning of the research example, we gathered measures to quantify the degree to which each consumer "needed" the benefits that these products were designed to deliver. Toward the end of the research, we then measured how well the products delivered against these needs (Table 10.5.2). We also gathered information on the critical sensory parameters and measured, for each product, how well each delivered against those sensory requirements, using a KANO diagram

Table 10.5.2 Comparison of product benefits consumers gain from using various products.

Ranking-importance of benefits

Benefit/gap relevancy of benefit[1]	Product A (%)	Product B (%)	Product C (%)
• I like that I am not being judged by other people when I use these products	15	65	45
• It fits better with my lifestyle	20	60	30
• I don't have to go out of my way to enjoy these products	30	70	40
• I can use these products anywhere anytime	35	50	15
• I enjoy the taste and flavor of these products	50	85	65

[1] Numbers are % consumers who feel that the product fulfills a benefit.

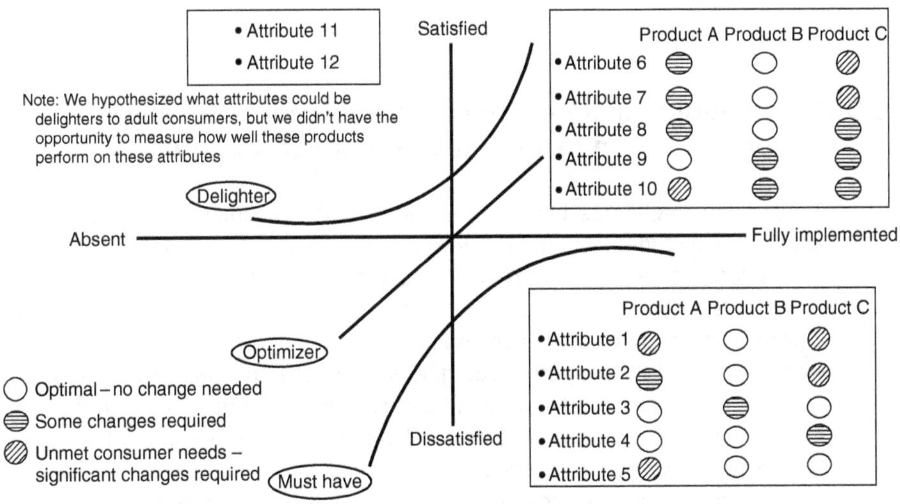

Figure 10.5.8 Using KANO diagram to help assess how well the product meets the sensory requirement.

(Figure 10.5.8) (see Chapter 8.1 for additional tools based on KANO philosophy). These measures, as well as the in-depth interviews, allowed us to understand at a deeper level the performance and potential of each product.

We then quantified and predicted the market potential to determine if the product was a viable option and thus should be pursued further. Figure 10.5.9 shows the comparison of data from the extended product use research with

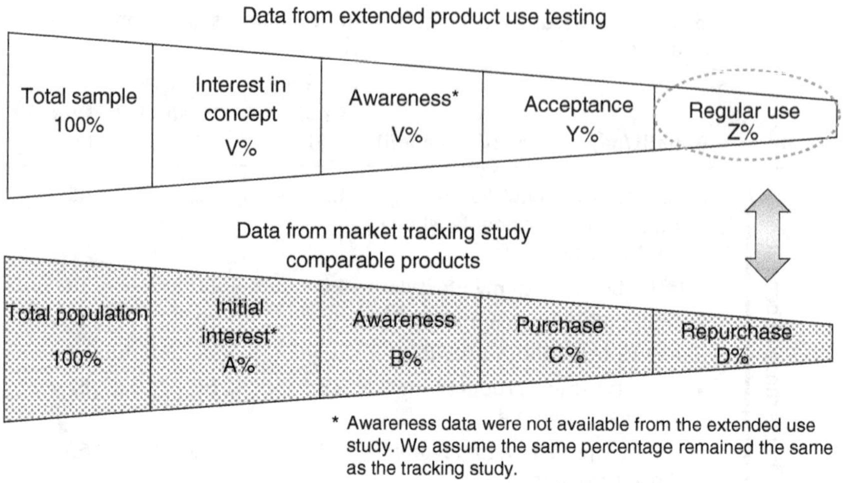

Figure 10.5.9 Comparison of data from extended product use research with market tracking data.

market tracking data; this allowed us to determine if the product is a viable option and thus should be pursued further. The final outcome was the ability to obtain an initial estimate on marketplace performance by providing volume projections.

10.5.10 Additional insights on consumer choice selection – learning from extended use product research

This section highlights two key insights about consumer choices based on another case study, using the EUPR Approach (Teratanavat and Paredes, 2015). In this case, consumer and product researchers would like to understand how consumers respond to and/or make choice selection on a given product with different flavors. Each year, food and beverage companies launch new products with a variety of flavors across categories. Generally, they fall into two types of flavors – mainstream and emerging flavors. Mainstream flavors are those commonly found in any given category and are those that consumers tend to be familiar with. Emerging flavors are less familiar to consumers, yet trendy, and these could help create a point of differentiation within the category. It is important for a company that considers launching new products in a well-established category to understand consumer behavior when it comes to flavor choice selection to determine what flavor(s) to launch.

10.5.10.1 Insights 1: Familiarity is the key. Mainstream flavors generally do better as consumers are more used to these flavors

Figure 10.5.10 shows the average overall liking of a particular product with different flavors during a traditional home use test. As you can see, mainstream flavor options generally received higher liking ratings than emerging flavors.

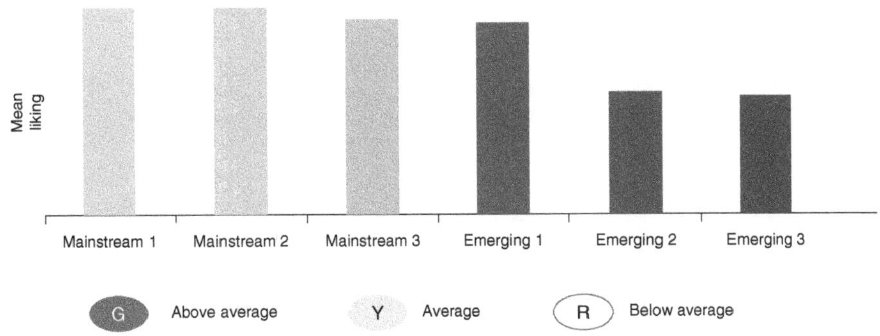

Figure 10.5.10 Overall liking data comparison between mainstream flavors and emerging flavors during home use test.

This finding suggests that consumers like what they are familiar with and therefore tend to gravitate toward mainstream flavors that they are used to. Some emerging flavors (e.g., Emerging 1) may also receive a good liking score, but in general, they do not do as well as mainstream flavors. Some feedback we received about emerging flavors include *"It tastes different from other flavors I tried. Very good but caught me off guard when I first tried it."* or *"As much as I enjoyed this one I'd stick to my current flavor. I feel like I'm getting a bit more out of my current brand."* We consider this a "Novelty Hurdle" that companies need to address and try to overcome if they consider launching new flavors that are less familiar to consumers.

To help identify which flavors really do well, we look into the extended use data (see Figure 10.5.11). During this extended use period, consumers are encouraged to experiment with different flavors, and after a few weeks, a clear pattern of preference starts to emerge. It helps validate that mainstream flavors perform well not just at the initial use but also over an extended period. Some emerging flavors (e.g., Emerging 1) did well initially but did not sustain. Based on this learning, we would consider data from both

Products	Week 0 consideration set	Week 1 choice selection	Week 2 choice selection	Week 3 choice selection
Mainstream 1	G	G	G	G
Mainstream 2	G	G	Y	G
Mainstream 3	G	Y	G	G
Emerging 1	G	G	Y	Y
Emerging 2	Y	R	Y	R
Emerging 3	R	Y	R	R

Figure 10.5.11 Overall product performance comparison between mainstream flavors and emerging flavors over extended use period.

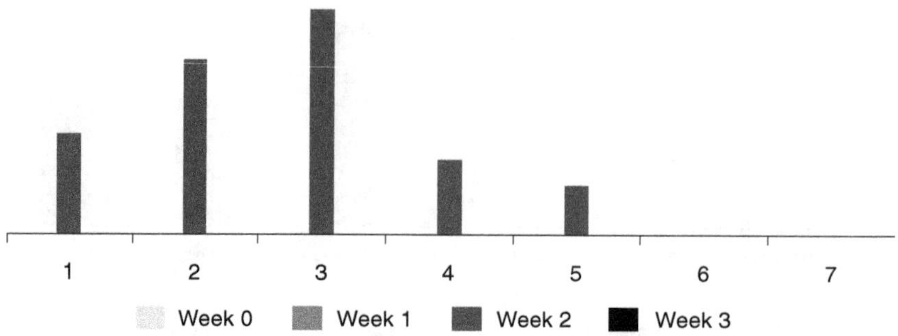

Figure 10.5.12 Number of products that consumers put in their consideration set over time.

initial use and extended use and winning flavors should be those that perform well consistently *over time*.

10.5.10.2 Insights 2: Consumers have a limited number of products in their consideration/choice set. They may try new and different flavors, but they tend to go back to their limited preferred choices.

The EUPR allows us to better understand how many products consumers have in their choice set. In the real world, when consumers go to a grocery store they are faced with many flavor or product options. For instance, when consumers go to the yogurt aisle and see 8–10 different flavors, do they tend to stick with their favorite flavor(s), or do they keep trying new flavors? Figure 10.5.12 shows the distribution of products in consumers' consideration set over time. Initially (week 0), many people have between three and five products in the consideration set, but the number declines over time. By week 3, most have only one to three products in their consideration set. This learning suggests that consumers may want to try or experience new flavors at the beginning, but over time, they will narrow down to a few favorite flavors.

10.5.11 Philosophy behind extended use product research

During new product development, sensory, and consumer research is generally conducted to determine how new products appeal to consumers and how new products perform compared to competitive products (Moskowitz et al., 2012). Several measures have been used to measure overall product performance, such as liking, preference and purchase interest. These measures are generally collected from consumers before and after product evaluation (i.e., pretrial and posttrial) through mall intercepts, central location testing, or home-use testing. While these measures have been widely used, how well these measures can accurately predict product success in the marketplace is often questioned. A more behavioral measure of product adoption can eliminate some of the limitations of other methods.

One approach that can be used is to develop an extended use consumer panel. Researchers can observe initial responses on concept and product trial from potential consumers and measure how their reactions with products change over time. Researchers can also measure long-term behaviors such as product choice selection patterns in different scenarios. These behavioral measures may help identify consumer segments based on product preference. Understanding sensory segments is important for product development and marketing to determine product mix strategies. These behavioral measures may help researchers better understand actual consumer behavior, as compared to other stated measures such as overall liking, stated preference, or purchase interest. We think the approach described in this chapter is very useful because it not only provides a snapshot of how acceptable the product is at the evaluation point but also allows an assessment and understanding of product performance both in the short term and long term. It also enables us to bridge hedonic measures to actual behavior. Additionally, it allows us to gain a deeper understanding of consumer product experience not only around product attributes but also the functional and emotional benefits that are important to consumers.

10.5.12 Summary and future

EUPR is an approach that integrates behavioral measures (e.g., product choice selection and usage) in addition to hedonic measures (liking, stated preference, or purchase interest). It allows researchers to gain a deeper understanding of consumer product experience and uncover the "why" behind how all the pieces of the product experience fit together. It offers different views to assess product acceptance, short-term, vs. long-term performance. These behavioral measures provide more solid ground than one-time measures (liking, purchase interest) to assess likelihood of product success and product viability to help businesses make informed decisions.

Discussion questions

- What are the advantages of an Extended Use Product Research over traditional market testing approaches?
- How do we assess the advantages - and watch-outs - of creating an Extended Use Product Panel?

10

References

Beckley, J.H. (2007) A biennial survey conducted by It! Ventures LLC and The Understanding & Insight Group LLC to understand the changing role of the Sensory Professional.

Beckley, J.H., Herzog, L.H. and Foley, M.M. (2017) *Accelerating New Food Product Design and Development* (2nd edition). Hoboken, NJ: John Wiley & Sons Ltd and the Institute of Food Technologists.

deCastro, J.M. (2000) "Eating behavior: lessons from the real world of humans". *Nutrition*, 16(10), 800-813.

Halim, J., Sinaga, W.S.L. and O'Mahony, M. (2019) "What do laboratory preference tests tell us about real life (operational) preferences: a preliminary investigation". *Food Quality and Preference*, 76(September), 60-70.

Lawless, H.T. and Heymann, H. (1998) *Sensory Evaluation of Food - Principles and Practices*. New York: Springer Science.

Moskowitz, H.R., Beckley, J.H. and Resurreccion, A.V.A. (2012) *Sensory and Consumer Research in Food Product Design and Development* (2nd edition). Ames, IA: Wiley-Blackwell Publishing.

Rennie, A., Protheroe, J, Charron, C. and Breatnach, G. (2020) *Decoding Decision - Making Sense of the Messy Middle*. Think with Google [white paper]. Available at https://www.thinkwithgoogle.com/_qs/documents/9998/Decoding_Decisions_The_Messy_Middle_of_Purchase_Behavior.pdf (accessed 8 March 2022).

Schieffer, R. (2005) *Unlocking the Mind of the Market - Ten Key Customer Insights*. Mason, OH: Thomson/South-Western.

Stevens, G.A. and Burley, J. (1997, May/June) "3,000 raw ideas = 1 commercial success!". *Research Technology Management*, 40(3), 16-27.

Stone, H. and Sidel, J.L. (2007) "Sensory research and consumer-led food product development". In H. MacFie (ed.) *Consumer-Led Food Product Development*. Boca Raton, FL: CRC Press. pp. 307-320.

Teratanavat, R. and Paredes, D. (2015) "New and Unique Flavors: Do Consumer Really Want Them? Insights From Extended Home Use and Behavioral Measures". The IFT Meeting, 13 July. Chicago, IL.

Part III

Word of the Wise: Wisdom From Experienced Explorers

11

We rarely start off with the traditional business question, which is, 'Where is the white space?' What we look at first is, where's the human problem, and what is the burgeoning science that supports a valid solution for addressing the problem?

Ben Goodwin, CEO – Olipop, 2021

What this is about

Two seasoned Consumer Explorers share their journey from the beginning of their careers and what inspired them to see consumers as the center of innovation to the processes they followed, cocreated, and/or built for robust innovation. In tandem, Stacey Cox and Anthony Jackel share real-life experiences and advice on how to move from a Data Gatherer to a Strategic Research Partner and Insights Leader with a seat at the decision table.

Product Innovation Toolbox: A Field Guide to Consumer Understanding and Research, Second Edition. Edited by Kannapon Lopetcharat, Dulce Paredes and Jacqueline H. Beckley.

Chapter 11

Putting It All Together: Driving Consumer-Centric Innovation in an Organization

Stacey Cox and Anthony Jackel

Key learnings

✓ Putting the consumer story at the front and center of your innovation

✓ How the path of successful innovation looks like

✓ Harnessing the power of the past and using tools to set up for success

✓ Transforming yourself from a research runner to the insight influencer

11.1 For successful innovation, the consumer story must be front and center

Being consumer centric is the goal of any business whether you are a Fortune 500 company or a pop-up store on the corner. Every business wants to sell something that someone wants or needs and not just once, but every time that need or want appears. So why does every business need to be consumer-centric in their daily approach? At the end of the day, it is the individual that needs to be excited by the offering if it is to sell successfully. The way to ensure success is for the consumer story to be front and center of the innovation process.

11

What does being consumer centric mean? It is not as easy as it may often seem to others in an organization. In our years of business experience, we have seen too many in industry say, "well we know what the consumer wants even better than they do," with brand teams saying, "our employees are consumers so they can decide," or even "let's just look at the trends and take a risk that we

Product Innovation Toolbox: A Field Guide to Consumer Understanding and Research, Second Edition. Edited by Kannapon Lopetcharat, Dulce Paredes and Jacqueline H. Beckley.

know what that consumer will want." Yet the flaw as we see is that we are not asking the person who will be paying the money for the product or service what they want, need, and will spend their money on during a transaction.

Putting the consumer as your focal point is as simple as knowing your consumer better than anyone else, or as a colleague once said, "be able to talk about these people like they are your best friend." Too often, we have sat in research and a colleague has said, "So what do you think of the idea, the product, etc." The fundamental issue is "I am not that person/consumer." None of us creating the product or service in an industry is really a true representation of the consumer. As researchers or business owners/inventors, we can be a spark of inspiration, but we still need to understand if others, who we hope will engage with our product, are genuinely interested in the idea. We have the innate bias to want to sell our product or service; therefore, we are not like our consumer. We do not have the same needs as they do, so what we create has to be about solving their problem, not ours.

Making the consumer the center of the innovation process can be accomplished in three simple steps:

(1) find your love of human behavior
(2) know your biases and "leave them at the door" during consumer engagement, and
(3) build or use a simple framework to monitor, refine, and tell the consumer's story as the basis of your innovation process.

11.1.1 Step One: Finding your love of human behavior

Growing up as a "minister's kid," one of us developed a love for watching, understanding, and empathizing with human behavior early on in our life. Putting the consumer in the center of your innovation cannot be something you only do from 8 to 5. One must love watching and understanding all day, every day. It should be a focus everywhere you go. Watching and loving human behavior is an essential tool toward telling that consumer story, you must want to tell it with their ups and downs, their successes and failures, and within the context of everything going on around them. And developing this deep-rooted love of telling their story does not have to cost a fortune. It can be done as you go about your daily life. For instance, we have become good at "eavesdropping" at the train station or in a busy restaurant. Also, we have picked one social media platform to follow daily for months. On that platform, we try to follow new industries we are not currently working in and follow completely new influencers we have not heard of, to understand how other people around the globe navigate their lives. We are constantly listening and watching what or could matter to "my" consumer, what makes them happy, scared, nervous, excited. With this knowledge, we then apply our learnings to what it means for the goods and services we work on in my organization.

All these actions give us a deeper appreciation of how all humans, and ideally my consumers, are trying to move through their life with their needs and wants. The Pandemic year of 2020, for all of its horrors, was an excellent time to practice or hone skills. Many have asked how we feel as researchers during the historical events of 2020. Our response, "couldn't be a better time to watch and understand human behavior." 2020 provided moments where we could observe the rawness of consumer beliefs and emotions, observe major shifts in consumer habits, and get to be a part of whole new consumer stories being written.

11.1.2 Step Two: Know your biases

Succeeding at being consumer centric is knowing your biases and being able to "leave them at the door" when you enter the consumer's world. In our experiences, we see many in all industries talking with their biases when they discuss consumer needs *versus* what the consumer is really telling them what they need. However, to be successful with innovation, having an accurate idea of what your biases are, where they are coming from, and not allowing them to enter your mind or being during your research is essential. Biases bring our opinion as business owners into the conversation and muddy the truth in the consumer story we are viewing. It sounds easy, yet this is the hardest thing for individuals to do, especially in the innovation process. Yet these biases cause us not to hear and fully appreciate the consumer and their lives. Having deep self-awareness, practicing this awareness every day and also finding research partners to hold you accountable for not bringing the biases into your thinking while you are engaging in research are tools that we use to try to separate our biases from our thinking.

11.1.3 Step Three: Knowing the consumer better than anyone else

Have a simple way to build and continually refine what the consumer story is for yourself and the organization. As researchers in the consumer packaged good (CPG) industry for over 25 years, we know there is no shortage of data on the consumer. The challenge more often than not is knowing which data to watch, use, or is truthful in the consumer story *versus* being noise to distract or mislead. Several years ago, in partnership with The Understanding & Insight Group, one of us developed a simple framework, the Eating Behavior Framework© (Figure 11.1), as a way for our organization to understand the complex, yet entire story of the people we hope will purchase our products, our consumer.

This framework was created out of the need for more successful product and package innovation. As is not uncommon to the CPG industry, we often went into research with one view of the consumer, either screening them demographically, with one attitudinal segmentation algorithm or using their use of the category or brand. The issue, however, was once we would develop against that "lens" of the consumer, inevitably, we would miss some signal from the consumer, and the product would not stick on the shelf past the first year. In 2006, we started to understand the consumer from a more holistic vantage point by putting together the many "consumer lenses" we had in the company, i.e., attitudinal segments, by creating a few more lenses, i.e., Product Preference Segments and Internal and External Influencers and incorporating lens from other strategic partners, Mindset Segments *from Moskowitz Jacobs, Inc. and The Understanding & Insight Group* and Need States from *Synovate, Censidym*, which became *IPSOS Inc.*

The compiling of these multiple consumer lenses into an easy to understand consumer framework allowed us to listen to, comprehend and piece together,

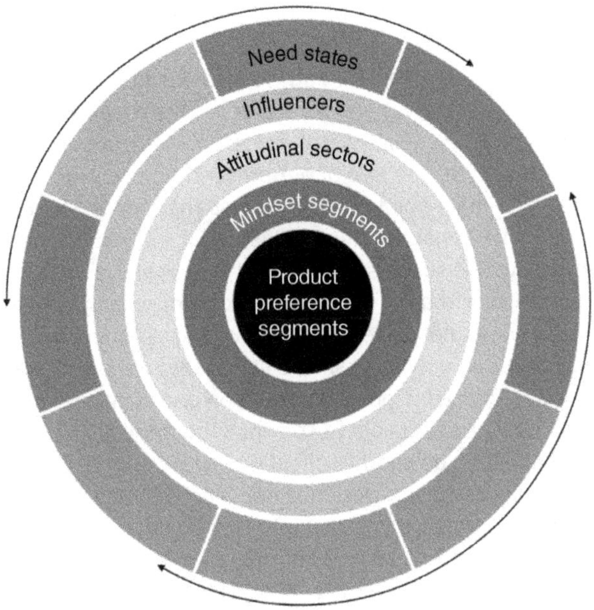

Figure 11.1 Eating Behavior Framework – Stacey Cox, The Understanding & Insight Group (2006).

rapidly, all facets of the consumer to identify whitespace opportunities in their lives. A **framework to listen** allows us to identify what is true and real and be able to monitor and refine the story quickly as the consumer's world evolves. A listening framework such as the Eating Behavior Framework also allows one to sift through and categorize new data or new ways to look at the consumer quickly to identify what is fitting and what is not fitting into the consumer story. For instance, in the Eating Behavior Framework we utilize 20 filters, i.e. Internal and External Influencers, to rapidly identify new information and use it to be more successful in predicting new or changing consumer behavior. A listening framework is a great tool to enable business teams to know when and how to listen to the world outside the organization, while maintaining the consumer front and center of our thinking. It enables teams to not have to focus on how to listen but on the **signals** and more easily **ignore the noise** in the consumer conversation, thereby allowing a faster path to breakthrough innovation and sales driving communication.

It is also important to ensure your consumer story or framework allows you to capture all aspects of the consumer's life. Too often, companies or businesses focus on just understanding the piece of the consumer story that is relevant to the innovation to be launched, yet this is not the whole picture of the individual or group. The choices consumers make in one aspect of their life is impacted by all the other aspects. Therefore, as a business stakeholder, ensure that you are driving the organization to listen to and understand the entirety of the consumer's story, stretching beyond your industry or business.

11.2 What does the path to successful innovation look like?

Before we discuss what the path to success may look like, let us think about what successful innovation may mean in our respective industries. Many years ago, we were asked to speak at a conference on our perspectives of what is successful innovation. The title of our presentation was *Innovating for Success and Survival*. The main point of our talk was that successful innovation is really in the eye of the beholder. Many industries have hurdles, benchmarks, norms to call an innovation successful. However, it is what the business considers success to the company that is most important to use as the key metric. Successful innovation does not have to be flashy and create a huge buzz. A successful innovation, to us, can be as simple as figuring out a new way to get home when there is new construction or blockage on one's typical travel route. For publicly held companies, being able to convince the market that your metric is an innovation metric can be another challenge.

For us, successful innovation is one where the consumer has adopted a new solution to a problem and sticks with that solution vs. reverting back to their old habits. The product or service offering must be one that provides a new answer to the consumer's problem without adding any new anxiety or uncertainty. When the consumer tries the new innovation, if they think, even for a minute, it will not work as well as their old one or the new solution will introduce new problems, they will revert back to old habits and the innovation will fail. Business teams who think through how the new product or service will address current problems without creating new problems will have more success in launching their innovation. Thinking through the potential anxiety a new solution will create in a consumer is not always easy and at times teams need to be pushed to think more broadly about how their new solution could create new problems for the consumer. In Figure 11.2, we see the path to successful innovation from *The Garage Group*. The path to getting a consumer to try or use and then stick with the new solution or innovation is to reduce the anxiety or uncertainty of change in the new solution so that the consumer does not revert to existing habits and allegiances.

So how do you ensure you are on the right path with the consumer in the driver's seat for your innovation to ensure long-term consumer adoption?

11

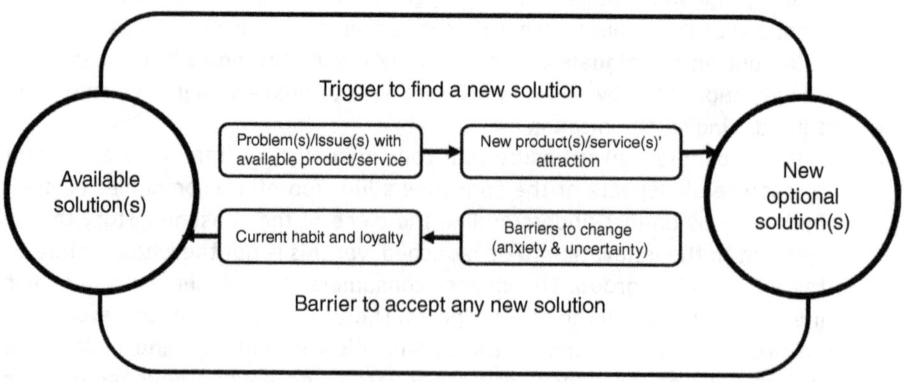

Figure 11.2 What Needs to be True to Change Behavior, The Garage Group (2018).

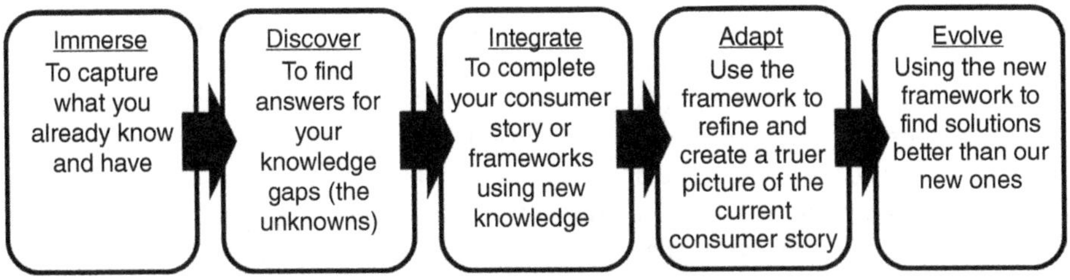

Figure 11.3 Five steps to successful consumer understanding.

We recommend a simple five-step process that allows teams to build a deep consumer knowledge (Figure 11.3). The steps take a team through immersing themselves in the knowledge the organization has about the consumer, discovering where the gaps in knowledge exist and moving to discover the answers, integrating this new knowledge into the team's current framework, adapting the framework by incorporating the new understanding and then evolving the framework to be the current true articulation of the consumer story.

The first step of the process is **Immerse**, capture what is already known about the consumer within the organization, immerse oneself in past successful and failed innovations, and then most importantly, identify how the team will immerse themselves in the consumer's world. The team needs to go and be a part of the consumer's world, not just sitting in a back room of a focus facility, be where people are either in the physical sense (ethnographies), in simulated venues or online, and to watch, watch, watch. In our opinion, to truly feel immersed in the consumer's life, one set of focus groups or one central location test is never enough. Being able to immerse oneself in our consumer's lives daily for months, 24×7 is the best. Only then can you truly articulate the consumer story and identify true opportunities. While this may not be possible for certain projects, keeping the perspective and mindset of that "consumer" in mind as you go through a day or days as many of us experienced during the 2020 pandemic, can augment deep, constant immersions. When COVID-19 hit, businesses needed to have quick responses to a widespread disruption. Business teams who had been studying consumers deeply were able to use what they already knew and react to the shifts in purchase and consumption behaviors, almost anticipating how people who were studied previously and deeply would respond.

Second step is **Discover**. *Based* on one's learnings in Immerse, set out to Discover the answers to the unknown or the yet to be confirmed fully known about the consumer story. In our experience, the best way to approach discovery is through hypothesis testing, just like we learned early on in beginning science, applying the hypotheses to either qualitative or quantitative methodologies. Create hypotheses statements with the team on knowledge gaps, statements that you believe will be true and set out to prove or disprove your theories with consumers. This method has proven to be effective for teams to fill in their knowledge gaps of the consumer's story and hear parts of the story they might have missed if they approached the Discover phase with a simple question and answer approach. This approach has been well written by organizational psychologist Adam Grant in his book, *Think Again* (Grant, 2021).

Once the team has gathered new knowledge and filled in the gaps, the next step is to **Integrate** this knowledge into the consumer story or framework that was created from previous work or a hypothetical consumer story that was created before the work. Take the new insights and determine where and how they fit into the framework you are using. Determine if the insight is completely new or just a better way of articulating a preexisting understanding, or if it does not fit into your framework at all. Throwing out data is always an option. Not all data is bad data, but it may not always be the right data for the story we are telling at the time. Taking listening frameworks into research allows teams to filter out data that is noise to the story and show where and how to integrate critical data into the right places for a more authentic picture of the consumer, at the current point in time. Frameworks allow us to have a solid starting point of understanding to judge new data and determine whether we have a breakthrough Aha! or just a new view of the same information.

Once the data collection has happened and it is integrated into the framework, then **Adapt** the framework for a refined view demonstrating a truer picture of the current consumer story. Keep what is still true from the past and adapt the pieces of the framework that have shifted with new learning. For example, when the Eating Behavior Framework was created back in 2006, it was developed out of the understanding of consumer eating patterns from studying over 300+ years of healthy eating. Since then, we have continued to adapt the model as a new or evolved understanding of the consumer has emerged. Any framework should be treated as a living, breathing story. However, even 14 years later, with each new bit of learning, we reevaluate the framework's articulation of the consumer story and adapt the articulation to ensure it is truly conveying the consumer today.

Last step in the process is to **Evolve**. It is this step that often gets ignored, especially by brand teams and researchers alike. In most industries, we are all too fast to put the framework of understanding on the "shelf" once the research is done or the innovation is launched. We think we have learned all that we need to for a successful launch and move onto the next task at hand. Yet the last step is the most important. We have invested time and resources into discovering consumer needs and creating breakthrough solutions based on these Aha!'s! that we hope address consumers' needs the best. So, why stop there, the consumer's story doesn't. Their story continues to evolve and solutions that may be better than ours continue to come into their view for consideration. The most successful path to the creation of breakthrough innovation is one grounded in the belief that we are never done evolving our consumer story and that listening frameworks are critical tools to enable this rapid, insightful evolution.

Harnessing the power of the past and using tools to set up for success

11.3 Learning from the past before jumping to collect new information

Whenever encountering a new problem from your internal brand team, sometimes your first inclination is to reach out to research agency partners to help you field research to find the answer. This is certainly an option for solving the problem and bringing information to the organization to move forward, but we would argue this is not the first step you should take as an experienced insights leader.

One of the first things you should do when taking on a new project or encountering a question from your internal teams is deep dive into past experiences to help guide your thinking. Often, a previous insights team has explored the very same question or worked on a project that was focused on the same topic. We can never stress it enough with our team that you can become such a powerful insights leader within the organization by grounding yourself in what has been explored in the past. By doing this, you not only become better versed in the history of your brands but you also can then be the voice in the room that can help influence the direction that the brand should go based on what you know has worked or not in the past. It would take many hands to count the number of times a team brought up a "new idea" to explore and by deep diving into old files, we were able to show it was not all that new of an idea and that it had been explored previously. Successfully driving consumer-centric innovation starts with educating yourself on the past and harnessing that knowledge as power. Plus, think about the power you would have in your company by answering questions the teams have without spending a dime on running new research!

If you are currently thinking to yourself, "I wish our company had a database and repository of old research to explore!", then you officially have your next project cut out for you! Do yourself and your company a favor and sign up to lead an initiative to build an internal research database system. There are plenty of companies that specialize in research database management systems. All of them can certainly help you get started and provide you with a solution to fit your needs. However, if you do not have the budget to partner with one of these companies to set a repository system up, something as simple as a protected shared folder system within your organization will work just fine to start housing your research. Knowledge is power, and a research repository is the conduit for that knowledge to flow to your company's insights for many years to come.

11.4 Creating the critical internal contract of the research plan

There is a great quote by Albert Einstein that goes, "If I had an hour to solve a problem, I'd spend 55 minutes thinking about the problem and 5 minutes thinking about solutions." This quote could not apply any better for how every insights leader should think about setting up a plan before running new primary research. Far too often we resort to jumping right in to run new research only to get the results back and find out that we have more questions than answers. Worse is when we move quickly to run consumer research to then find that the results do not actually solve the questions that the team really is asking. That is why it is critical to spend a significant amount of time with your internal stakeholders drafting up what we will refer to as the "Research Plan," before jumping to what tools and methods you will use.

A Research Plan is essentially what you will draft between your brand teams and yourself to serve as a contract to ensure everyone is on the same page with the project background, consumer target, and success criteria of the overall project. Before you can go into your toolbox to recommend what methodologies to use to solve the team's problems, you must first spend the much-needed time

Consumer Insights & Strategy

To: Date/Year

Project Name

The purpose of this document is to outline a clear research plan to address the key questions and objectives relating to ________

Brand objective/JTBD:

Consumer target:

Project business background & tier:

Project success criteria: what must be true at the end of the project to call it a success:

Project research background:

Figure 11.4 Research Plan Template example.

aligning on these items upfront. Figure 11.4 is an example of the upfront section of the research plan template we have been using on our team to gain the background and goals of the project. It has all of key project information that can quickly ground everyone on the project: the objectives and/or "Job to be Done" (JTBD) of the brand, the research consumer target, the background for why this project exists, how we will classify this project as successful, and any past research that has been conducted on the brand or topic. All sections are important, but I would argue the most critical piece that we find ourselves going back and forth on is the Project Success Criteria. We cannot stress enough how important it is for the team to really spend time on making sure it is clear to everyone what "success" looks like at the end of the long innovation journey.

Even though this is an internal "contract," it is essential for you as the research leader to make certain that all of the studies you plan later on in this document ("contract") help the team achieve that success criteria.

This part of the process is hardly ever quick and easy; rather, it is the one that tends to require the most discussion and debate. The silver lining of this arduous process is that once you scribe all this information on paper and align upfront, putting together the plan for how your innovation project will achieve success is simple! Einstein has been quoted to say, "Spend the time making sure you and the team really understand the problems you are trying to solve before jumping to the tools that will give you the solutions."

11.5 Gathering the data to help influence the direction of innovation and conversation

At this point in the book, you know that there are countless ways in which you can conduct research to help collect data for your projects. Obviously, the ability to gather information is not the problem we usually face as research professionals. What tends to be the bigger hurdle is knowing what tools to use based on what phase in the project you are in and what questions you are trying to answer with your team. Whether you are at the point of front-end ideation or the whole way to the final validation stage, your role as the consumer centric person in the organization is choosing the right tools and methods to successfully steer the direction of a project and influence the conversation with leadership.

As mentioned earlier in Section 11.4 in this chapter, the importance of defining the problem and understanding what it is you are trying to solve is something that should not be understated. Once you figure out the core questions you are trying to shine light on, you can move to searching through your toolbox of methods to figure out which is the right one to use to further the innovation design. As we have learned, there are so many incredible studies you can conduct. It can honestly be overwhelming for a new insight professional to enter the field and try to navigate and understand all the different ways in which they can run research. Our best advice to any person working in an insight's role is to spend as much time as possible learning all about the different methodologies at your disposal. The collection of tools that you have gathered from this book can make you such a powerful leader in your organization if you really take the time to truly understand them, know exactly how they work, why they are designed that way, and when, in your research plan, they become best utilized. Once you master the tools and fully understand the team's objectives, success criteria, and what they need to answer, drafting up the remaining part of the research plan is relatively fast and simple! Figure 11.5 captures the latter half of

Research Plan Summary:

Total Estimated Timing:
Total Estimated Cost: $

Research Plan Detail:

Phase 1: Consumer Targeting, Category Understanding, Brand Positioning, etc.

1. **Tool**
   Objective:
   Method:
   Success Criteria:
   Vendor:
   Cost:
   Timing:

Phase 2: Idea Scoping, Concept Development and Validation

1. **Tool**
   Objective:
   Method:
   Success Criteria:
   Vendor:
   Cost:
   Timing:

Phase 3: Product & Package Development and Validation

1. **Tool**
   Objective:
   Method:
   Success Criteria:
   Vendor:
   Cost:
   Timing:

Phase 4: Communication Development & Validation

1. **Tool**
   Objective:
   Method:
   Success Criteria:
   Vendor:
   Cost:
   Timing:

Figure 11.5 Key steps in the research plan.

the Research Plan document we use in our teams. Our teams use this template to house the key information about the project and map out all of the research we will conduct across each phase of the innovation workstream from foundational consumer and category understanding to product validation and even communication work. Having one document that captures the project plan on a page helps keep you and your teams focused and forces your internal clients to strategically think through all of the questions they will have along the way, rather than running and planning research as the team progresses. The ability to think longer term and plan through all of the questions the team will have down the line is not an easy task, but we promise that spending due diligence in doing that up front will allow you to have a much more strategic and well thought out innovation plan.

A challenge to any insights lead is to work to become just as much of an expert in the different methods as the agencies that you partner with to field them. That does not just mean asking for a PowerPoint that provides an overview of the tool and reading it in your spare time. It involves sitting down with your agency partner, having them walk you through the tool step by step, and asking questions along the way. You cannot just have surface-level understanding of how the research works. If you only have a surface-level understanding of how the research works, you should expect only surface-level insights. A good agency should love to sit down and talk in detail about the research that they can conduct with you. If they are not willing to take the time to go through the method in detail, then you probably should not invest your time and company's budget to work with them. Also, do not just accept the methods for what they are and run with it. We suggest that you investigate and question the way the tool works, not just to make sure you feel it will work for what you need but also to be ready for any of the questions that your internal team is going to have. Finally, we would suggest that before you field any research with your consumers that you run through the survey or method yourself. You and your team would be surprised how much you can learn about the research that you are running and become better entrenched in how the tool works by going through what the consumer will see themselves. Learn the tool, go through the tool yourself, become an expert on the tool, and then you can use the tool to help gather the data you need to help drive your projects.

Synthesize and simplify: designing and utilizing analytical structures and constructs

11.6 Connecting the dots of multiple pieces of data and research

Fast forward to the point in which you finished up a few different pieces of your research plan. By this time, you have spent a solid chunk of your resources on harvesting a lot of data for you to use to help drive your innovation project forward. The typical pattern that exists in most research plans usually follows the cadence of fielding a specific piece of the plan, reporting out on the results, then moving onto the next phase of the research. While that certainly is the purpose of a planned-out approach, what you might miss if you do not think about the lessons in totality are the insights that bubble together in a pattern when reflected by you across studies. In fact, we would argue that the results of just one specific study should only be considered "data" and/or simply "findings." You do not tend to discover true insights and powerful, motivating nuggets until you look across all the research you conducted and find the patterns that exist throughout the whole discovery journey. Connecting the dots of multiple data points allows you to pull out critical learnings that help you have impactful influence in your organization.

We will give you an example of this from a recent project that we have been tackling with one of our teams this past year. We had been working on a larger strategic platform for our company, which was all about trying to uncover the next level of convenience for one of our product categories. Because this was such a big initiative, it had a lot of resources allocated for staff and budget. We essentially were given the freedom as the insights team to craft a full research plan that we deemed was necessary to accomplish the goals and objectives of the project. We have always operated with the mindset that a great insights leader achieves respect not by how complicated and robust his or her plan is, but how focused and efficient it is to help drive the project without irresponsibly burning through money. While this may seem like a random aside, we feel it is necessary to give the advice that when your company gives you what looks like a blank check for your research quest, do not take advantage of it. You will be met with way more applause if you bring the needed insight to the organization and drive innovation through a strategic and efficient plan *versus* an overly robust and complicated plan that costs the company more than what it pays back in returns. We are beginning to come full circle, because in order to have a well thought out plan that even the finance or accounting department will respect you for, you must know your methods like the back of your hand (e.g. very well and familiar)!

Back to our story: The team and us were at the point of embarking on a long journey of uncovering what the next level of convenience looked like for our consumers. We find that these sort of bigger strategic platform projects tend to start you out in a very cloudy setting with not much direction and sometimes not even an existing "map" to help you navigate through the project. What needs to be done prior to taking your first steps is harnessing the power of the past to set yourself up to know what you know currently and what you still need to explore. The process of knowledge mapping what your team already knows from previous research and old projects is imperative to realizing future success. Once we had laid out all of the "knowns" and "unknowns", we were ready to draw out our map for how we were going to discover all that we **needed to discover** in order to drive our project forward. The knowledge map kept us from spending unnecessary resources (time, money, mental energy) on areas we knew well (Beckley et al., 2017).

There are so many different tools that you can grab from your research toolbox to use throughout the many phases of your innovation projects. You will discover that more complex projects will lead to needing multiple tools to finish the job. Just as a house requires more than just a hammer and nails, driving consumer centric innovation requires a variety of research tools to "build the house" and launch successful products into the market. As the son of a carpenter, I can confidently say that the strongest and longest standing buildings are those that are built using more than just a single hammer and nails. Successful innovation requires having a full toolbox and the knowledge and confidence to use many different tools. Using multiple tools acknowledges that when the topic is complex, expecting any single solutions to provide all of the knowledge is foolhardy.

Back to our project: We were at the point where we had planned out and fielded a large amount of our research including ethnographies, a category lab, ideations, co-creations, and so on (Harte and Jeltema, 2017; Moskowitz, 2017; Vahalik and Alfieri, 2017). We could have easily used each of these studies to just answer the specific questions at hand and move through the process one study at a time. However, the true power came from being able to take all the many different pieces of research and "put it all together" to discover the game-changing insights that permeated across the studies. Like the idea of compounding interest with saved funds, we propose that there is something in research called "**compounding insights**." Just as a financial advisor would promote compounding interest as a way for any investor to be wealthy in the long term, we suggest that some of the richest insights come from the compounding learning you gather from study after study, not by one single report. The ability to put all those insights together and discover the patterns that no one else took the time to find is how you can really show your true value as an insight leader.

NOTE: One of the authors (Cox) has created a consumer centric program over more than 20 years that involves the use of "labs." Depending on the needs of the project, the labs may focus on products or categories or other factors important to the business. The use of the term "category lab" above designates to the authors a particular test design, a series of actions that are known to have specific outcomes. These labs are usually elongated in length (three to five hours) and are considered in-depth qualitative sessions.

11.7 Creative listening frameworks to help navigate the consumer conversation

One of the best ways to put together an array of information for your organization to understand is creating **listening frameworks**. Think about the framework to tell the organization how to listen and understand everything you have learned about your consumer. You have no doubt learned so much through all the research you have conducted that you should be able to talk about your consumer like your best friend or even beyond your best friend because you understand so much. Not only can you talk about these people but also you should be able to listen to them and understand what they truly mean by what they say during qualitative research. Creating one of these listening frameworks will help you show your organization everything you have learned so far about the project by drafting up principles for your team to know before "entering the conversation," and then know how to "navigate the conversation" once they are listening behind the reflective glass mirror or in the homes of consumers.

First, you need to lay out all the key principles you have learned about your consumer thus far. Our advice is to go back to your knowledge mapping report and remind yourself of your knowns and unknowns. By now, you should have already begun to move some of those unknowns to the known section of the report because your research should have been designed to answer those outstanding unknown gaps or questions. As an example, during our project, one of the biggest unknowns was how consumers define "convenience" in food and then how they prioritize all those factors that exist in that broad definition. While we had many other unknowns that we needed to solve, this was a big, essential item that we had to spend a lot of time researching and unpacking. As you might expect based on what we discussed so far, the answer did not come from one single isolated study. To define and decode "convenience," we had to take everything we learned across all of our research touchpoints (ethnographies, category lab, consumer co-creation, etc.) and find the consistent truth that allowed us to deeply understand this topic. By creating a listening framework, we were able to make sense of all of the information across the studies and provide our organization with a way to better understand what consumers mean by "convenience" and how we navigate that conversation across our businesses.

Defining and laying out the principles of the framework allows you to first ground everyone in your organizations about the facts of the topic that you know to be true before starting to teach them how to listen to the consumers. This is an important first step because it allows the stakeholders to (hopefully) set aside their own biases and understand what is absolute about the consumer that they are about to listen to and understand. As insight professionals, we all should know how important it is to stay objective and avoid having any of our own personal bias cloud our consumer understanding. All of us will no doubt work with incredibly intelligent cross-functional partners. Each of those team members will need to learn to also set aside their ongoing biases related to product, people, behaviors, etc., to clearly listen to what people are saying and what they genuinely mean by their conversations. One of the best ways to guide your team through the process of setting aside their dispositions is to have them own up to every one of them in the form of hypothesis generation. Before going into research, have your full cross-functional team discuss and lay out all their hypotheses about the project in order for you to test out, confirm, and disconfirm all of them along the way. This process not only helps to identify each person's ingoing beliefs but it also helps you as a researcher understand what you are going to need to prove or disprove as truth before being able to better understand your consumers. An example from our convenience project was criticality for our insights team to discuss all of our hypotheses about what convenience meant for consumers and then show how, through our many phases of research, we learned we were completely wrong. Had we clung to those initial biases and went straight into innovation development before first establishing the principles of what convenience really means to our target, we would have completely missed critical consumer conversations that lead to some of our bigger product insights and innovations.

After you have laid the foundation of your framework with the key principles, you are ready to teach your teams how to listen to consumers. The exact listening notes you pull out very much depend on each project, but you will learn about what is important to know and listen for in all your upfront research. In our work, we realized it was critical to understand the consumer's life stage and the mealtime occasion they were thinking about before talking about convenience. It might seem like a minor detail to know where the consumers in your front room are in their life as far as family, marriage, and living situation goes, but we found through all of our research that this was a key guide for us to successfully navigate the conversations. Additionally, something as simple as which mealtime occasion they were thinking about had a huge impact on the kind of solutions they needed based on their pain points and what they are dealing with during that specific time. These were just two of the pieces that we laid out for our team to understand as a part of our listening framework. There were about six pieces in total that we felt were important to think about and listen for while we were in consumer research. Without these guides, it would have made navigating the conversation much more difficult because our teams would not have known what to be listening for in all the conversation. As you could imagine, trying to keep up with the many voices in the front room can be an arduous task that could challenge even the best listener. By creating a listening framework, you can set you and your teams up to better navigate the conversation and pull out the key insights to successfully drive forward your innovation projects.

A word to the reader: most of us have never been taught how to listen. It is important to realize this when you embark on a series of research activities in which listening clearly is paramount. The actions we have described above assume that while our cross-functional partners are all highly intelligent, most come to our research not knowing that they actually might not listen well! (Murphy, 2019).

11.8 Operationalizing your learnings with visual product models

The topic of model building was discussed in-depth in other chapters of this book (Chapter 6.3), so this section is not meant to go into extreme detail of how to build models, but rather help give you a real-life example of when models come into place and how they can drive monumental change in an organization. We firmly believe that the best way to teach someone about a topic and make it memorable is through visualization. The best tool we find to bring to life and distill an overly complex and detailed consumer conversation is through a product model. One of the difficulties we tend to have as researchers is that we are the closest people in our organization to the consumers and the ones who spend the most time with the research we collect. Being that close puts us into an incredibly unique (and exciting if you ask us!) position. We have the daunting opportunity to bring to life the learnings of all our research in an easily understood way so that the insights can permeate and proliferate across the organization shaping strategy, business decisions, and ultimately, sales.

Imagine now that you have successfully conducted all your needed foundational research for your innovation project. In our example, this would have been the point at which we finished up our multiple studies and were ready to ground the organization in what we learned from our consumers about convenience to tee up the whitespace opportunities from an innovation standpoint. Additionally, we had taken the time to put together a listening framework to help our teams better understand the consumer conversation of our convenience platform. There were many facets of convenience that we unpacked through the journey, and we needed an easy way to put it all together and operationalize our learnings for the organization to clearly understand how to innovate against certain spaces. That is when our decision to put it all into a visual model came into play. Creating product models can be intimidating at first because you tend to be starting from nothing and need to use your knowledge and research to help put together something tangible for your team to quickly and easily understand. One of the best pieces of research that we like to utilize to help inform the creation of the model is an **in-depth qualitative lab**. By going deep on a topic with consumers for hours, you can essentially hire the consumers to serve as your contractors to help build your model visualization (see NOTE: above).

In our case, we brought in consumers to explore the world of convenience products that exist for them today in the grocery store. We spent two full days diagnosing how they used the different products, where and when they were used in their repertoire, and why they turned to each option throughout the week. The numerous hours of groups had our cross-functional team walking away from the research feeling like we learned a ton of information. The challenge for our insights team was to figure out how we could distill everything we learned into an actionable output for our teams. All of the discussion and product investigation allowed us to clearly see that there was a fascinating relationship between how *functionally convenient* an item was and the *resulting feeling someone reported they experienced* after using it to make a meal for themselves and their family. We will not be able to go too much in-depth with the specifics, but just seeing the blind model itself can help you to begin to visualize how we were able to funnel the complex consumer conversation into a digestible model.

With any model, you are trying to understand and establish the relationship that exists in individuals' minds as they think about a category or collection of options they have at their disposal. Whether you are exploring a very specific, tangible category like condiments or investigating a broader, intangible idea like "tastes great," you can create a model to explain the relationships that exist between products and how consumers really think about the category. We always tend to find that the most easily explained models come in the form of a two-dimensional plane. One great approach that is often used within our company is Qualitative Multivariate Analysis (QMA), which is an approach developed by The U&I Group LLC and U&I Collaboration LLC. The approach works well in getting teams to quickly visualize and understand how the consumer is making sense of a "space." (For more specifics on QMA, please see Chapter 6.2 in this book.)

An important consideration as you create and build out the experience model is what you establish as your axes (Figure 11.6). Typically, these come directly from the mouths of your consumers or are derived from past research. We discovered our key axes through all of our foundational research and the creation of our listening framework. What our lab allowed us to do was successfully put our framework and learnings into action by visualizing the world of convenience products. In our model, we ended up with eight of these spaces (Figure 11.7), which we found were either extremely crowded and occupied by competitors or

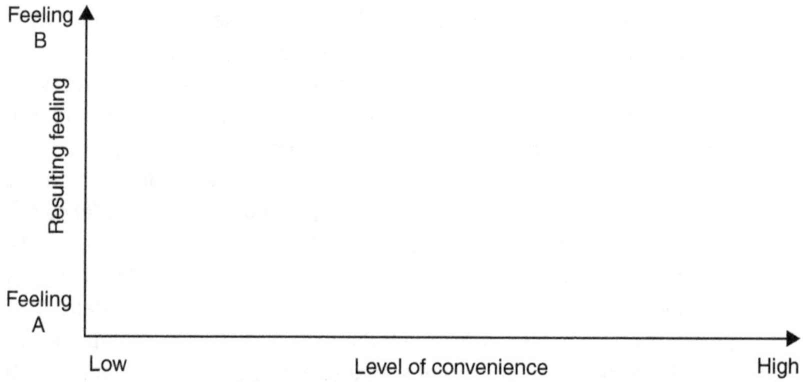

Figure 11.6 **Creating axes for Qualitative Multivariate Analysis (QMA)**
The Understanding & Insight Group, LLC.

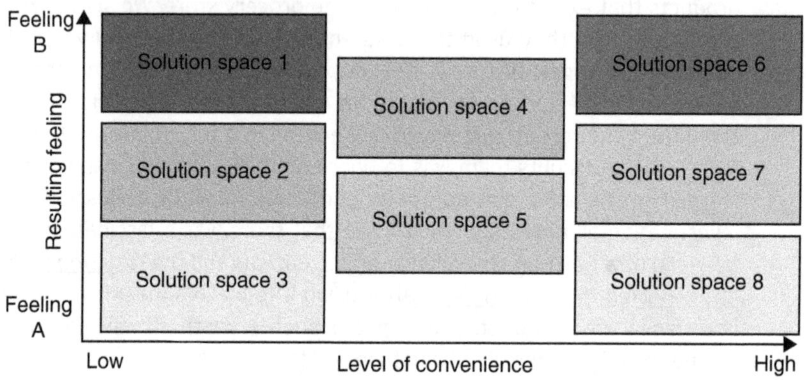

Figure 11.7 **Building a listening framework to track the consumer conversation.**

had a lot of opportunity and whitespace. This not only helped us understand the roles of certain products in people's lives, but it also showed us all the Solution Spaces that existed in the world of convenience where we could innovate. From here, our team was able to develop concepts and continue down our research plan through concept iterations, testing, and product development. Just as he would tell us to spend most of our time on understanding the problem, Einstein would likely agree with us that spending a disproportionate amount of our time and resources on conducting all of the needed foundational research before jumping straight to concepts and solutions is an appropriate consumer centric design approach.

We will be the first to admit that as researchers we can sometimes get a little caught up in all our data. Sometimes all of those PowerPoint slides filled with plenty of metrics, stat testing, and charts can feel like the best security blanket for knowing you will have enough data to back up any point you want to make or answer any possible question. However, we have learned through our insights career that the most influential and strategic reports tend to be more powerful if they can be distilled down to just a few meaningful pages. This topic will be discussed in more detail in Section 11.9, but the point here is ***that there is no***

correlation between the number of slides you have in a report and the influence you can have with leadership on a project. To successfully drive consumer-centric innovation and be turned to as the expert in the room, you need to show the organization how you can synthesize complicated and complex research into an easily understandable story. Your CEO and senior leaders do not have enough time (or may choose to not have the time) to read through a 20-plus page report. For your hard work to be recognized and used by the most senior people in your organization, you will need to challenge yourself to tell the story of your data in a concise and simplified way. One of the best feelings you can experience as an insights lead is seeing your leadership team present a strategy or company update using slides you created yourself.

11.9 Crafting the influential strategic conversation to make sense of it all for action

Now that we know how to create frameworks to organize our knowledge and understand the nuances of the consumer for developing successful innovation, we need to convince our organization to act or at the very least, modify their mindset or understanding of the consumer. To create movement or change in an organization, you have to hone your skills in the art of crafting a strategic conversation that leverages the frameworks of understanding to influence the change in behavior or gain the alignment you need to move forward.

What is an influential strategic conversation? An **Influential Strategic Conversation** is one which taps both the individual and collective knowledge of the team and uses it to produce a shared understanding and vision to move forward and act. To be effective, it must utilize the collective understanding of the team or the insight knowledge base, not just one person's thinking or "gut." We want to use this knowledge to evolve the understanding of the team we have so we can rapidly gain alignment on a plan of new action. Per Dr. Andrew V. Abela's, The Extreme Presentation™ Method, we want to cause our audience to go from one way of thinking and doing, to a new way of thinking and doing, ideally aligning with the action we want the organization to take with our information (Abela, 2010).

To better understand what "Crafting an influential Strategic Conversation" entails, we should have a common understanding of the parts: (1) influential, (2) strategic, and (3) conversation (Figure 11.8).

Let us start with why "influential?" In many organizations, conversations (or presentations) are a daily occurrence, even hourly. However, the intent to influence gets lost in too much data, not being clear on what we are trying to do in the conversation, or lack of understanding of the problem we are trying to solve in the conversation. Many conversations, when not **crafted** correctly, become only informative conversations with no action or alignment on a path forward. By the end of the presentation, we are often back where we started, not having decided or our audience is more confused than ever. From our perspective, conversations can either be participatory or influential. Before embarking on a conversation, it is important that the crafter of the conversation is clear in their own heads which one they want to be pursued with their audience. A participatory conversation is

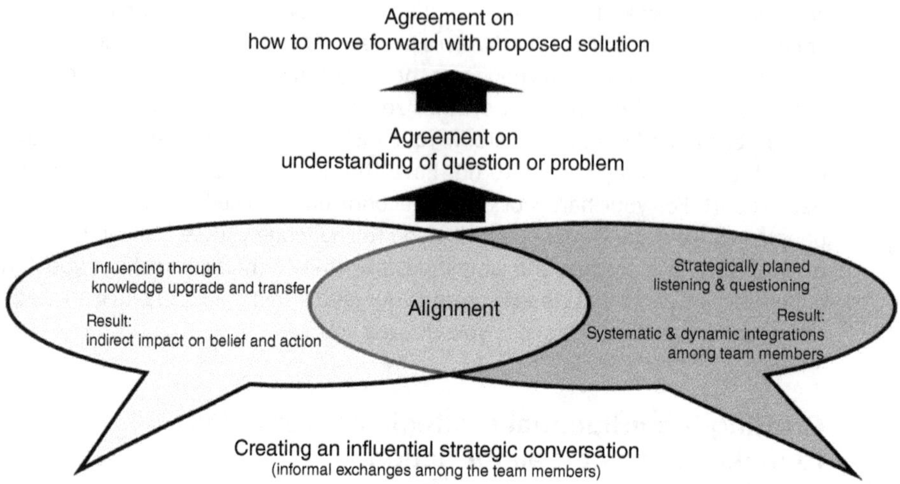

Figure 11.8 An Influential Strategic Conversation. Source: Adapted from Abela (2010).

one of participating by listening, asking clarifying questions, and acknowledging the hearing of the content, but there is no expectation of resulting change in behavior, intent, or action. An **influential conversation** is one that is crafted with the intent to change a person's or team's mindset or behavior through the transfer of the knowledge that we hold through our frameworks or models.

How do we approach this in our crafting of a conversation that influences? Abela suggests we must start with understanding who we are as people as well as their audience that we need to influence. Thinking through the members of the audience in the room and where one might get easier alignment vs. not is important before starting the process of crafting how one will navigate to decision or action. In addition, if the goal is to influence, you must stay focused on solving their problem whatever that may be, vs. solving your own issues. Focusing on the audience's problem, not yours, allows them to remain engaged throughout the conversation, opening the opportunity for them to be influenced through your strategic process of knowledge transfer.

Once we have outlined what behavior or mindset we want to influence through the transfer of the knowledge, we then need to think through the navigation or the **strategy of the conversation.** What is the story we need to guide them in listening with the result of alignment and take to action? We need to lay out the steps on how we are going to move our audience from Point A to Point B quickly and efficiently with the knowledge that we possess even before we make slides on our computers. We want to use our knowledge and frameworks as tools to guide the listening and resulting conversation with the audience. We want to start thinking through the questions that we need to answer as part of the strategic conversation and ensure that our frameworks provide the depth of data needed for our audience behind those answers. What do we know or not know, is that data qualitative or quantitative, is it new data or something we have already known? We need to consider how the knowledge we have, and the framework of the consumer story can be leveraged to systematically move the audience to alignment.

An approach we have used for many years to craft a conversation strategically is Abela's S.Co.R.E. Method (S.Co.R.E. stands for a checklist that screens for: **S**ituation, **Co**mplication, **R**esolution, **E**xample). It is an approach where the conversation begins with the current situation or the state of the business and moves the audience through conflict or problem identification followed immediately with resolution of the problem through your solution supported by examples, i.e. data or often in my case, a framework of the consumer's story. Figure 11.9 shows us how a crafter of the conversation can strategically move the audience from the problem of not being able to grow consumer preference to a potential resolution of test and learn, without a lot of extraneous data noise.

We have seen the most success in the process of crafting strategic conversations when the Eating Behavior framework (discussed above) is used as the basis for examples. An assumption is that since the framework is already a culmination of knowledge gained and updated throughout listening and observing the consumer and/or an evolution of past models of understanding. This is really the "secret sauce strategy" for telling the story of the consumer in a conversation. It is an effective way to navigate stakeholder alignment to solutions in our conversation. Using a focused, evolving framework allows for rapid, easy transfer of knowledge for executives who must make decisions based on short, quick top-level meetings.

The last piece of crafting an Influential Strategic Conversation, is ensuring it is a **conversation**, not a one-way presentation, dialogue, or pitch. We don't know if the reader has been in any meetings where the person kicks off the presentation with "this is informal, please ask lots of questions and then proceeds to have 59 slides for a 60-minute time slot" or spends the entire time with their 100 slides solving their problem, not yours. The most productive presentations we encoun-

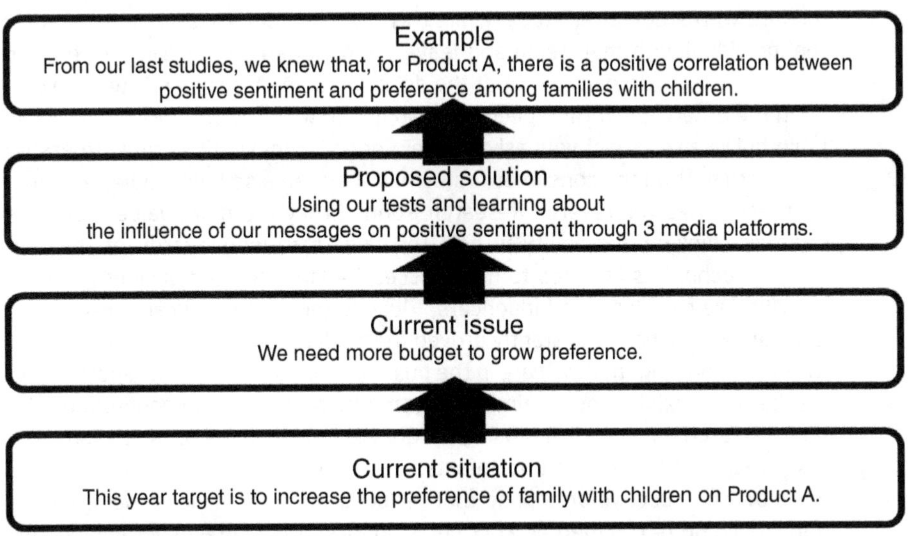

Figure 11.9 Example of utilizing The Extreme Presentation™ S.Co.R.E. Method to outline the conversation strategy. Source: Adapted from Abela (2010).

ter are ones where there is an informal exchange of thought and perspective among the audience, but it is enabled by having a quarter of the slides to fill the time slot or ensuring slides are set up to anticipate many questions ahead of time, leaving the audience time to ask gap issues we hadn't considered ahead of time. One cannot be successful in changing mindset or behavior if the audience is not allowed to engage in the conversation because there is no time left to discuss the solutions. Again, Abela has some good tips and tricks we have used over the years like eliminating chart junk, crafting slides, instead, that move the audience's eyes through the story we are telling and ideally creating a whole Influential Strategic Conversation into one slide for a 60-minute time slot.

Evolving from a research runner to an insights influencer

11.10 Moving from a transactional relationship to an integral strategic partner

Believe it or not, there is a spectrum within the insight profession to explain the different kinds of styles each of us exhibit in our organization. Regardless of whether you are on the client side or agency side, we believe these different designations hold true.[1] On one end of the spectrum is what we call the "Research Runner." A person operating with this type of mindset is the one who is an expert in scoping out research, fielding studies, and turning around reports. They are undoubtedly experts when it comes to knowing which tool in their toolbox to pull out depending on what stage in the process the team is in, and their internal clients, like the brand team, know that they can turn to this person at any point when they need quick answers. Teams will praise this type of research based on their speed, dependability, and ability to drive the projects forward with study after study. While this style can quickly earn accolades for the individual with their team and leave them wowed by the "skill" of turning around research, to earn a seat at the decision table, this style has to evolve.

On the other side of this spectrum is someone who not only has a seat at the table but is also proactively asked to join decision-making meetings to assure teams are putting the consumer first. While they can also field studies and analyze results like the revered Research Runner, they are more valued for their strategic thinking and ability to help the team plan longer term for the consumer touchpoints required to have successful launches and businesses. They are what we call "Insights Influencers." Just as the name suggests, this person is someone who has consistently proved themselves to be of value to the organization by showing that they are in the business for more than just reporting out results. They exhibit some skin in the game by driving recommendations with their research and are always using insights to influence the organization.

At first, it can feel a bit uncomfortable and even intimidating to be putting one's own perspective on a page that includes what you believe the organization should do next. Whether working as an insight lead in a major conglomerate, a small start-up, or on the agency side delivering the results to clients, the recommendations you make can have big implications. To help illustrate this, here is an example that many of us working on whitespace innovation have or will encounter in our time:

Example: Put yourself in the shoes of an insight lead who has been working on a team who were tasked to create the next generation of innovation for a long-established brand.

(1) This project is coming at the time where small insurgent brands have consistently been popping up in the category that your brand has owned for decades. Because there is a lot of heart for bringing new life to the brand, there are plenty of resources that have been designated to the project including people, time, and money.
(2) Fast forward to the point in which you have already spent months conducting all of the foundational research to understand the pain points in the category, what consumers are yearning for, and where the most appealing Solution Spaces are that your team can focus upon.
(3) To this point, there has been plenty of work on figuring out the potential ideas, but they are still very much in their infancy and not yet fully baked. The team finally has enough information to work through rough costing to figure out the price they are going to have to charge to make money.
(4) While the ideas are exciting and certainly deliver on the uncovered pain points, the cost for solving them is rather high. The team finds out they need to price the innovations at about triple the cost of the typical items on the shelf.
(5) At this point, there has already been months of work, a lot of money, and plenty of time spent on the project, so there is a great deal of sweat equity by all the cross-functional teams into these ideas.
(6) Through the research plan drafted with the team, it is known that the next step is some sort of concept testing in order to more fully size and measure the potential of the idea. This is when things become a little uncomfortable. After much anticipation, you finally receive the results back on your concept testing from which you find out that the proposition has potential but has major red flags from a pricing standpoint.
(7) Let us explore how our two different types of researchers would approach reporting out on the results.

Approach One: For this example, the Research Runner would quickly pull together to review the data, make sure it looks correct, and send off the final slides and data tables to the team. The Research Runner certainly feels invested in the project from all the time and energy put into the project, but at this point, they feel it is up to the team to decide what they want to do based on the results.

Approach Two: The ballad of the Insights Influencer would sound quite a bit different. While they would also review the results with a fine-tooth comb to check to make sure everything seems intact with the data, the Insights Influencer's biggest proportion of their time is spent mulling over the results and considering what the team should do. They take not only this one study into consideration but also all learnings gathered up until this point. The past learnings show that the team is sitting on some strong ideas; however, the value proposition for consumers just does not seem like it is going to work. As the well-established Insights Influencer, the organization fully trusts and depends on them to occupy their seat at the table with a strong point of view rooted in the consumer truths they find in the research. Will they recommend killing the project and funneling the company's resources elsewhere? Will they use all the research to craft a story showing there still some potential for the team to keep working on?

These parallel stories of the Research Runner and Insights Influencer illustrate just how different of a mindset you can exhibit depending on how you position yourself as a leader. While it might sound a bit simpler and less stressful to just hand over the results, we promise you that the excitement and respect that comes from being a strategic partner who operates with some skin in the game is worth it.

Whether one is just starting out in an insights career or several years in, you need to take the time to do some introspection and think about what kind of researcher you are operating as and if that matches with who you want to be. Depending on how you want to operate, there might not be a right or wrong answer. There are plenty of people who do very well being expert Research Runners. You can become very sought after by different teams within an organization as a strong researcher who can turn around results very quickly. On the other hand, if you want to be considered a strategic partner who has that seat at the table and a say in decision-making, you will need to evolve into an Insights Influencer. As you take hold of this responsibility, you will learn that your job is to truly drive consumer centric innovation by using the tools you learned throughout this book to deliver results, influence decisions, and help build a better tomorrow for your consumers.

11.11 What does the future hold for the world of insights?

We have covered a great deal of information in the book so far, and we would love to leave you with a few thoughts and an outlook of our wonderful world of consumer research. If you ask us, we think it is a remarkably interesting time to be in the field. Technology has become a huge game-changer in the way we do research. In just the past few years, we have seen countless research companies offering solutions that not only allow us to field research quicker but also do some of the more complex analyses in a "Do It Yourself" (DIY) manner. Technology is certainly allowing us to find shiny new tools to put into our toolbox. Most of these savvy solutions come advertised with benefits like "quick turn," "agile," or "DIY." These tools that promise results overnight or even instantaneously might tempt us to adopt more of a Research Runner style where we are valued for the speed at which we can turn around research. The challenge for us is to figure out how to use these new tools and techniques and not become the new normal of conducting one-off, overnight studies as a replacement for the tried and true methods. You can prove the insights function to be incredibly valuable to an organization by marrying the speed at which you can run robust research with strategic, critical thinking. Proving value is more important than ever in today's business world. The idea of cost cutting is certainly not a new topic. Companies have and will always be trying to figure out ways to be efficient with marketing budgets in order to make the most out of the bottom line. One of the areas of the budget that tends to receive scrutiny are any activities that are deemed as "Non-Working," which essentially refers to any spend that is not considered consumer facing, like advertising, coupons, etc. In most organizations, the money spent to conduct research falls into that non-working bucket. That means it is becoming even more imperative to prove the worth of the insights department and the research we conduct. As Insights Influencers, we need to prove that we are a critical component of the company that they could not operate or live without.

We would love to close out this chapter of the book, which is all about "putting it all together" by telling you about a short story that really stuck with me after attending a recent research conference. The keynote speaker gave an incredible and very humbling presentation on the insights function today that exists in most companies. He used the human body as a metaphor for the way of talking about how most organizations operate today. As he went through the presentation, he revealed how each function and/or role relates to the body. For example, CEOs are considered the eyes since they set the vision for the organization, while R&D would be the feet since they create the products that are the base of the company onto which the rest of the company is built. It then became the point at which we were about to reveal what part of the body the insight function was considered. Guesses came out from the crowd, but none of them were close. The speaker went on to reveal that he believed that most insights functions today operate as if they are the appendix of the human body: the organization knows that they exist, but they do not necessarily know what exactly they do. As sad as that might sound, it should light a fire into any researcher reading this book to know that they can have a bigger impact in an organization and should not be thought of as a noncritical organ. A final push to everyone still reading at this point is to prove to the organization that you work with that you are critical. We need to evolve as researchers to become something more imperative for survival. We should strive to be considered something like the gut, which the company listens to when they need some reassurance or certainly before moving forward or the brain that holds all of the consumer knowledge that the body needs to know in order to survive in today's dynamic environment. With our toolbox in hand and the wherewithal to make a difference, we have no doubt that we will be able to prove value by driving consumer-centric innovation that improves consumers' lives and helps our organizations flourish for many years to come.

Discussion questions

- Why do authors emphasize the use of "the consumer story"? What does that matter to innovation?
- What would it take for your organization to implement the compounding of insights approach recommended by the authors?
- Cox and Jackel speak to a range of professional behaviors, like the Research Runner and Insights Influencer, what do we think allows people to feel comfortable working in one role or another?

Note

1. For purposes of this chapter, we have only described the ends of the spectrum, not all of the insight and behaviors between the Research Runner and the Insight Influencer.

References

Abela, A.V. (2010) *The Presentation: A Story About Communicating Successfully with Very Few Slides*. Create Space Publishing.

Beckley, J.H., Herzog, L.J. and Foley, M. (2017) *Accelerating New Food Product Design and Development* (2nd edition). Hoboken, NJ: Wiley/Blackwell.

Grant, A. (2021) *Think Again*. NY, NY: Viking Penguin Random House, LLC.

Harte, B and Jeltema, M. (2017) "Chapter 14 design of packaging that fills the promise and meets/exceeds consumer expectations". In J.H. Beckley, L.J. Herzog, M. Foley (eds). *Accelerating New Food Product Design and Development* (2nd edition). Hoboken, NJ: Wiley/Blackwell.

Moskowitz, D. (2017) "Chapter 17 Category appraisal and ingredient search: identifying key sensory factors and product features at the early development stage". In J.H. Beckley, L.J. Herzog, M. Foley (eds). *Accelerating New Food Product Design and Development* (2nd edition). Hoboken, NJ: Wiley/Blackwell.

Murphy, K. (2019) *You're Not Listening*. NY, NY: Celadon Books.

Vahalik, J. and Alfieri, L. (2017) "Chapter 13 Applying processes that accelerate new product development". In J.H. Beckley, L.J. Herzog, M. Foley (eds). *Accelerating New Food Product Design and Development* (2nd edition). Hoboken, NJ: Wiley Blackwell.

12

It matters who is included in the average, and who is excluded.

David Freedman
author of *On Types of Scientific Enquiry: The Role of Qualitative Reasoning (2008)*

What this is about

The use of concepts around experimentation, design of experiments, and application of statistics is important to innovation and design. The chapter provides guidance on how to apply these methods for consumer exploration and insight discovery.

Product Innovation Toolbox: A Field Guide to Consumer Understanding and Research, Second Edition. Edited by Kannapon Lopetcharat, Dulce Paredes and Jacqueline H. Beckley.

Chapter 12

Above Averages: Use of Statistics and Design of Experiments in Product Innovation Applications

Kannapon Lopetcharat, Dulce Paredes, and Jacqueline H. Beckley

(updated original chapter by Frank Rossi)

Key learnings

- ✓ Experimentation vs. Testing
- ✓ The statistical design of experiments approach
- ✓ When and where it can be applied in product innovation studies
- ✓ The statistician as a valuable member of product innovation teams
- ✓ Big data and its applications

Throughout this field guide, defining objectives for research has been a critical step in using any tool described in this book. Having and setting clear objectives are the practical manifestation of applying the "*experimentation*" concept that has been an integral part of all chapters in the field guide.

For trained scientists, experimentation is the breath of science, and hypothesis testing is the breath of all research. By having a hypothesis (a specific, testable prediction about what you expect to happen in your study), a scientist can formulate a research objective and design an experiment. But one does not need to be a trained scientist to create hypotheses and design research. Unless you are creating a study that is exploratory in nature, your hypothesis should always explain what you expect to happen during your experiment or research. A hypothesis does not have to be right. While the hypothesis predicts what the researchers expect to see, the goal of research

Product Innovation Toolbox: A Field Guide to Consumer Understanding and Research, Second Edition. Edited by Kannapon Lopetcharat, Dulce Paredes and Jacqueline H. Beckley.

is to determine whether this assumption is right or wrong. When conducting an experiment, researchers might explore several different factors to determine which ones might contribute to the ultimate outcome. However, for many business research practitioners, hypotheses and research designs may sound foreign. Coupled with statistics, the experimentation concept may turn out to be elusive. It is usually confused with testing and, worse, data collection.

12.1 Experimentation vs. testing

Testing and experimentation share common traits as both need a person/team to follow specific procedures to collect and analyze the data to obtain a summary of the information. However, testing and experimentation are different from each other because of the reasons to perform them. The objective of testing is "to evaluate or validate" something; meanwhile, the objective of experimentation is "to discover or to test a hypothesis." Testing is a specific form of experimentation.

For example, when you want to choose a product concept, a study is usually conducted to help you choose the best concepts among the candidates. The objective of the study is "to evaluate" the concepts and select the best one; hence, this study is called concept testing and not concept experimentation (you can learn more about concept testing and screening in Chapter 10.1). The same logic can be applied to product testing when the objective is to screen the prototypes or to select the best prototypes to move forward in new product development. Therefore, to create and run a test, one must have a set of criteria (a.k.a. key performance indicator [KPI]) or a benchmark to allow one to decide to move forward or not. Without KPIs and action standards, the test will be wasted if not used to guide the next steps. *In general, testing is a special form of experimentation where the discovery or hypothesis is to prove that your product or concept meets your criteria or not.*

Since the beginning of the 2010s, experimentation has been popularized among business researchers due to the advancement of digital technologies and the booming of internet-based businesses. Many companies have adopted and advocated an experimentation culture such as Amazon, booking.com, Uber, and Netflix. These companies created an experimentation culture named "data democratization" where running experiments, sharing results and making results-based decisions. One of the most popular experimental designs used in business research is A/B testing (ironically but appropriately, the name uses "testing" and not A/B experimentation) (Thomke, 2020).

A/B testing is one of the most basic designs of an experiment where it involves only two factors of interest, A and B. It has an objective to discover which option is better by evaluating a set of KPIs. From 30 000+ Internet-based startups, Koning et al. (2020) reported that A/B testing helped companies to free up time and resources from thinking about the small stuff and to focus on more important activities such as generating more ideas. Consequently, the companies that adopted and implemented A/B testing, on average, had higher engagement from users, had greater likelihood to be funded by venture capitals and launched more products.

To discover insights, especially high impact ones, running a series of A/B tests may not be appropriate in many situations. A/B testing, formally called two-factor design, is appropriate when the two factors (in this case options A and B) can be manipulated independently. Real-world applications of such conditions are changing colors on a website, adding a button on a web-interface, comparing two different messages in written/verbal promotions, etc. However, the A/B testing design is inappropriate in many situations, where things are not discrete as A and B and where factors interact with each other. In those cases, more efficient and accurate approaches are needed. Statistical Experimental Design is more appropriate when one wants to optimize a product where different formulation or processing factors are involved.

We feel that Consumer Explorers (CEs) should consider this quote from Peter Drucker, a well-regarded management consultant:

> *Efficiency is doing things right; effectiveness is doing the right things.*

We recommend continuing to read about experimental design and statistics.

12.2 Statistical experimental design

Many of you are familiar with product optimization studies where different sets of variables are studied to identify optimal product performance.

Published examples of product optimization studies include Griffin and Stauffer (1990) and Moskowitz (1987). In these studies, ingredient levels and/or processing conditions (factors) are systematically varied according to a specific experimental design. Often experimenters vary just one factor, identify the optimal level and repeat with the remaining factors. This experimental approach is less desirable as it fails to account for the potential interacting effects of the factors, resulting in missed global optimums. It also often requires more experimental runs.

For example, in the development of a new cream-based smoothie, factors in the experimental design could be cream level and sugar level. Developers identify ranges for each of these factors, and an experimental design is used to determine prototype products to be created and tested with consumers (Figure 12.1 and Table 12.1).

An important aspect of this experimental design is that the prototypes define a space, and the resulting performance scores will be used to create a mathematical model that can be used to predict performance within the space (Figure 12.2).

The consumer research study at this stage could be conducted in several ways. Consumers could evaluate all 11 prototypes across a few sessions or could evaluate a subset of the products. The questionnaire could ask just a liking question or a few questions in addition to liking. The specifics will depend on the research objectives and aspects particular to the product category, such as fatiguing and carry-over effects. Additionally, consumers will evaluate products in different orders to minimize the effect of position bias across the study.

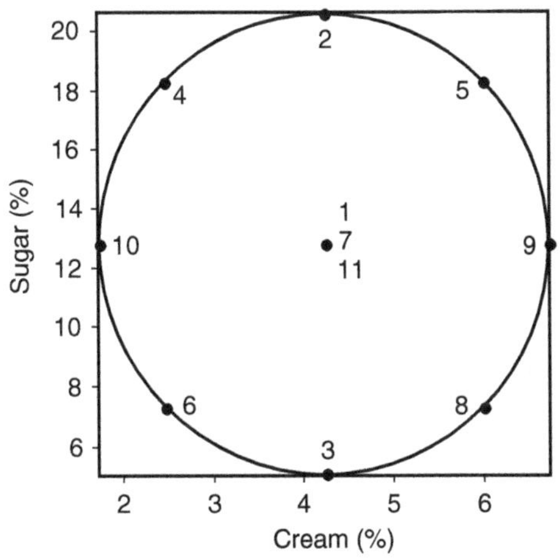

Figure 12.1 The design space for the smoothie, which is in this case a circle. Note how the design points are equidistant from each other on the circle and the center of the circle.

Table 12.1 A design created for the cream level range of 1.75–6.73% and a sugar level range of 5.03–20.53%.

Product	Cream (%)	Sugar (%)	Liking
1	4.24	12.78	6.73
2	4.24	20.53	7.20
3	4.24	5.03	4.90
4	2.48	18.26	7.21
5	6.00	18.26	6.82
6	2.48	7.30	5.47
7	4.24	12.78	6.71
8	6.00	7.30	5.86
9	6.73	12.78	6.62
10	1.75	12.78	6.64
11	4.24	12.78	6.78

For the smoothie study, consumers evaluated all prototypes, and the average liking scores were computed. These became the responses in the mathematical modeling (Figure 12.2).

Predictions outside the circle are not considered reliable since no data was collected beyond the space. The northwestern part of the circle is the area with the highest liking, indicating that higher levels of sugar with lower levels of cream are most liked. Cost information could be overlaid to cost optimize well-liked sugar and cream combinations. Note that the predicted optimum sugar- and cream-level combinations may be ones that have never been created or tested and would need some type of verification testing. This plot can be a

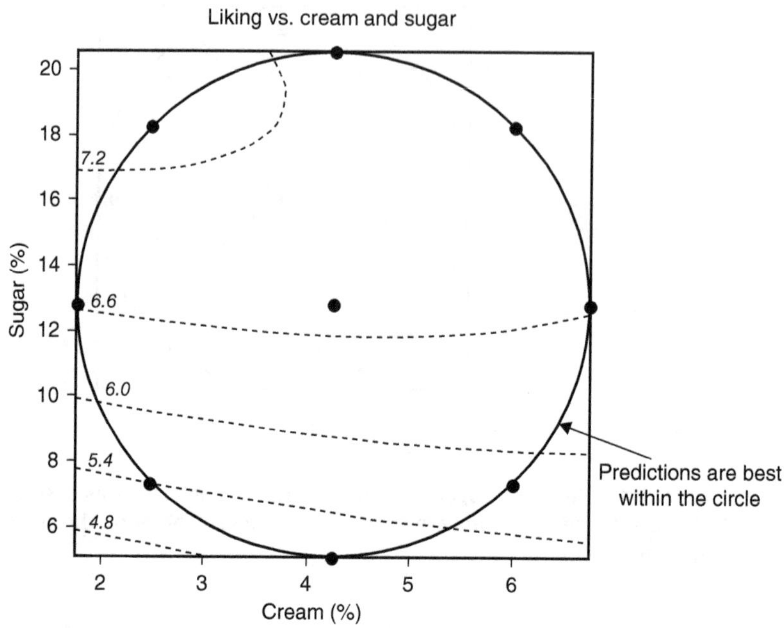

Figure 12.2 A contour plot of the modeling results, with levels of liking depicted as curves.

powerful tool in identifying the "sweet spots" where the ideal level of several responses can be achieved.

Statistical experimental designs fall into several categories, whose use depends on a specific experimental objective. They can be categorized into four key types:

(1) **Screening** designs are used to collect data to quantify the effect of factors with a goal of either reducing the number of factors under investigation or identifying the factor ranges for further study (a.k.a. Testing). For example, many starch and gum types could be studied to see which ones are most promising in making a low-fat cheese sauce perform most like a full fat sauce.
(2) **Response surface** designs are used to collect data so that a detailed model of the space defined by the factor ranges can be developed. The smoothie optimization study described above is an example of a response surface design.
(3) **Mixture** designs are special forms of screening or response surface designs where the factor levels are constrained to a sum total. For example, in developing a fruit juice blend, the ratios of three fruit juices (cherry, apple, and berry) are varied but will always sum to 100%.
(4) **Robust (Taguchi)** designs are used to determine the factor settings that reduce product variability due to difficult or impossible to control noise variation. An example would be an experiment to determine ingredient levels that make a dry mix pudding thickness robust to the fat level of the milk used in its preparation (skim, 2% or whole milk).

The factors and factor levels for each of these design types are dictated by the specific research objectives in the innovation process and the nature of the factors studied at this stage.

The smoothie example demonstrates the statistical experimental design approach applied to consumer research applied late in the innovation process. Product optimization studies are also used in research with existing products. But the statistical experimental design approach can be used much earlier in the innovation process. Most quantitative consumer studies can benefit from the approach.

An example of the statistical experimental design approach used earlier in the innovation process is a conjoint study. Factors in the statistical design are elements of a product concept or features of a product. Consumers evaluate concepts that are combinations of the elements that are determined by the factorial combinations of the elements. In the development of a concept for mini muffins, three factors under consideration could be the dome height, surface texture cracks, and serving size (Table 12.2).

Table 12.2 A potential experimental design for these three factors, the full set of factorial combinations of the level of the three factors.

Serial #	Size (in.)	Cracks	Dome
001	1.7	Smooth - no cracks	Flat - no dome
002	1.7	Medium cracks	Flat - no dome
003	1.7	Big cracks	Flat - no dome
004	1.7	Smooth - no cracks	Medium dome
005	1.7	Medium cracks	Medium dome
006	1.7	Big cracks	Medium dome
007	1.7	Smooth - no cracks	High dome
008	1.7	Medium cracks	High dome
009	1.7	Big cracks	High dome
010	2.1	Smooth - no cracks	Flat - no dome
011	2.1	Medium cracks	Flat - no dome
012	2.1	Big cracks	Flat - no dome
013	2.1	Smooth - no cracks	Medium dome
014	2.1	Medium cracks	Medium dome
015	2.1	Big cracks	Medium dome
016	2.1	Smooth - no Cracks	High dome
017	2.1	Medium cracks	High dome
018	2.1	Big cracks	High dome
019	3.0	Smooth - no cracks	Flat - no dome
020	3.0	Medium cracks	Flat - no dome
021	3.0	Big cracks	Flat - no dome
022	3.0	Smooth - no cracks	Medium dome
023	3.0	Medium cracks	Medium dome
024	3.0	Big cracks	Medium dome
025	3.0	Smooth - no cracks	High dome
026	3.0	Medium cracks	High dome
027	3.0	Big cracks	High dome

In the conjoint study, the consumers typically evaluate visual depictions of the factor level combinations. As in the optimization study, the goal is to develop a mathematical model of the performance scores based on the factors to predict desirable factor level combinations. A detailed description of conjoint analysis can be found in Chapter 8.2.

Another example of where the statistical experimental design approach can be used earlier in the innovation process is in Kansei engineering studies (Nagamachi and Lokman, 2010). Through the Kansei engineering methodology, potential product design elements are identified that can represent the factors in a statistical experimental design. While this approach is not taught as part of the Kansei engineering methodology, it is ideally suited to this step in the methodology. Consumers would then evaluate the object images, real, drawn, or computer generated, that represent combinations of design elements determined by the statistical experimental design.

The concept of experimentation is a requisite for CEs to discover insights. Unlike testing, where a very specific hypothesis is to evaluate or validate, experimentation's main purpose is discovering insights through hypothesis testing (in the scientific sense and not strictly the statistical sense). Conducting a series of A/B testing or statistical experimental designs can lead to insight discovery. Restrictions in creating stimuli (products and/or concepts) and contexts will dictate which design should be used in which context. With an experimentation guided research program, the data gathered from one study to the next synergizes with each other, allows CEs to learn continuously and increases the likelihood to create great ideas.

12.3 Brief history of experimental design

The theoretical background for statistical experimental designs dates to R.A. Fisher's proposed methodology in the 1935 textbook *The Design of Experiments*. While many of the early applications of the methodology were in agricultural research, advancements of the theory to applications in industry have been documented in work by Box et al. (2005) and Myers et al. (2009).

Many statistical software programs have incorporated the techniques into their application offerings, and several software packages have been created specifically for the industrial application of statistically designed experiments.

The statistical experimental design approach has gained increasing popularity in recent years in many fields. Availability of software tools to implement the techniques has made the implementation of the techniques easier and more time efficient. Computational advancements have made the design creation more flexible to practical research constraints. Research in the field continues to make statistical experimental design approach a critical tool in the product development process in a widening list of industries.

12.4 The age of big data and data science

Since the early 2000s, big data has become an important part of businesses, especially in the research field (SAS, 2021). Companies have large amounts of data from many sources from within or outside their organizations and the advent of online platforms has caused the size of databases to grow exponentially. In the beginning, big data was understood as a large volume of data. However, the defining characteristic of big data is not really its size, but how it is organized. Big data combines many data sets from different studies and sources with different data structures. This creates a large amount of data and a lot of analytical challenges. In the pre-2000s, a similar type of data was metadata. However, metadata sets are more prevalent in Internet-based businesses and societies and the data sets have now become known as big data.

Big data has at least three major characteristics: volume, velocity, and variety (SAS, 2021).

(1) **Volume:** In the 21st century, businesses collect data from many sources such as business transactions, smart digital devices (internet of things, IoT), records from equipment. In research, social media and digital devices are very popular modes of data collection and enable researchers to gather large amounts of data from a study. With a conventional relational database, storing, managing, analyzing, and relating data from different studies, sources, and modes was a big problem. However, adopting a new kind of database called a nonrelational database allows cheaper and more efficient data storage. This nonrelational database is usually called graph data or network data.

(2) **Velocity:** The growth of IoT makes it easy to collect large amounts of data. In fact, a new word, "streaming" is used instead of "collecting" for data movement. Velocity refers to the "rate of growth of the database."

(3) **Variety:** Data, in traditional databases, is usually referred to as a structured and numerical set of information. However, data, in the 21st century, comes in many forms, such as text documents, e-mails, videos, audios, stickers, emoji. Big data can store and handle all these forms of information.

Graph data or network data is the foundation of big data where the database can be expanded indefinitely. In traditional data, each number in a cell contains a meaning with/without the degree of intensity of this particular meaning; therefore, to find a relationship between two variables or more, a statistical method must be performed, and the relationship can be "derived" from a statistical analysis. On the contrary, in a graph data set, each number in a cell is an "actual relationship" observed directly (not derived from any mathematical calculation). Therefore, this nonstructured data set can be expanded. More details, descriptions, and applications of graph data or network data can be found in Chapters 7.1 and 7.2.

For example, the data in Table 12.1 can be converted into three graph data sets based on three relationships between the 11 products (cream% difference, sugar% differences, and liking score differences). The three data sets can be expanded in the future with different studies that contain the same relationships (Table 12.3).

Table 12.3 Converting data from Table 12.1 to graph data sets based on three relationships in the data: (1) cream% differences, (2) sugar% differences, and (3) liking score differences for the 11 products (P).

	P01	P02	P03	P04	P05	P06	P07	P08	P09	P10	P11
	Cream % differences										
P01	0.0	0.0	0.0	1.8	1.8	1.8	0.0	1.8	2.5	2.5	0.0
P02	0.0	0.0	0.0	1.8	1.8	1.8	0.0	1.8	2.5	2.5	0.0
P03	0.0	0.0	0.0	1.8	1.8	1.8	0.0	1.8	2.5	2.5	0.0
P04	1.8	1.8	1.8	0.0	3.5	0.0	1.8	3.5	4.3	0.7	1.8
P05	1.8	1.8	1.8	3.5	0.0	3.5	1.8	0.0	0.7	4.3	1.8
P06	1.8	1.8	1.8	0.0	3.5	0.0	1.8	3.5	4.3	0.7	1.8
P07	0.0	0.0	0.0	1.8	1.8	1.8	0.0	1.8	2.5	2.5	0.0
P08	1.8	1.8	1.8	3.5	0.0	3.5	1.8	0.0	0.7	4.3	1.8
P09	2.5	2.5	2.5	4.3	0.7	4.3	2.5	0.7	0.0	5.0	2.5
P10	2.5	2.5	2.5	0.7	4.3	0.7	2.5	4.3	5.0	0.0	2.5
P11	0.0	0.0	0.0	1.8	1.8	1.8	0.0	1.8	2.5	2.5	0.0
	Sugar % differences										
P01	0.0	7.8	7.8	5.5	5.5	5.5	0.0	5.5	0.0	0.0	0.0
P02	7.8	0.0	15.5	2.3	2.3	13.2	7.8	13.2	7.8	7.8	7.8
P03	7.8	15.5	0.0	13.2	13.2	2.3	7.8	2.3	7.8	7.8	7.8
P04	5.5	2.3	13.2	0.0	0.0	11.0	5.5	11.0	5.5	5.5	5.5
P05	5.5	2.3	13.2	0.0	0.0	11.0	5.5	11.0	5.5	5.5	5.5
P06	5.5	13.2	2.3	11.0	11.0	0.0	5.5	0.0	5.5	5.5	5.5
P07	0.0	7.8	7.8	5.5	5.5	5.5	0.0	5.5	0.0	0.0	0.0
P08	5.5	13.2	2.3	11.0	11.0	0.0	5.5	0.0	5.5	5.5	5.5
P09	0.0	7.8	7.8	5.5	5.5	5.5	0.0	5.5	0.0	0.0	0.0
P10	0.0	7.8	7.8	5.5	5.5	5.5	0.0	5.5	0.0	0.0	0.0
P11	0.0	7.8	7.8	5.5	5.5	5.5	0.0	5.5	0.0	0.0	0.0
	Liking differences										
P01	0.00	0.47	1.83	0.48	0.09	1.26	0.02	0.87	0.11	0.09	0.05
P02	0.47	0.0	2.30	0.01	0.38	1.73	0.49	1.34	0.58	0.56	0.42
P03	1.83	2.30	0.00	2.31	1.92	0.57	1.81	0.96	1.72	1.74	1.88
P04	0.48	0.01	2.31	0.00	0.39	1.74	0.50	1.35	0.59	0.57	0.43
P05	0.09	0.38	1.92	0.39	0.00	1.35	0.11	0.96	0.20	0.18	0.04
P06	1.26	1.73	0.57	1.74	1.35	0.00	1.24	0.39	1.15	1.17	1.31
P07	0.02	0.49	1.81	0.50	0.11	1.24	0.00	0.85	0.09	0.07	0.07
P08	0.87	1.34	0.96	1.35	0.96	0.39	0.85	0.00	0.76	0.78	0.92
P09	0.11	0.58	1.72	0.59	0.20	1.15	0.09	0.76	0.00	0.02	0.16
P10	0.09	0.56	1.74	0.57	0.18	1.17	0.07	0.78	0.02	0.00	0.14
P11	0.05	0.42	1.88	0.43	0.04	1.31	0.07	0.92	0.16	0.14	0.00

Because of the expandable nature of big data, it allows CEs to look through a vast amount of data efficiently and identify patterns in the stream of data quickly. Applying graph theory and network analysis on the big data opens new points of views for CEs to identify insights.

12.5 Managing experimentation

Today, getting data is not difficult anymore, but the stream of big data can be overwhelming for any nondata scientists or nondata specialists. Below are management recommendations for data managers based on Lucy Luo (2020) and Rouke (2016).

(1) **Championing experimentation through delegation and empowerment:** Experimentation is the foundation of discovering insights. To persuade the whole organization to embrace experimentation needs managers who can delegate and empower employees to select the best ideas based on the results of the experiments.

(2) **Ensuring that every study has at least a clear objective:** This is to prevent turning experimentation or testing into a data feed center because running studies is now easy to do. It is very easy to conduct an online A/B test now, but it does not mean one can neglect "carefully designed experiment" practice. Knowing why A won over B is much better than knowing that A won over B. In addition, knowing many whys will lead to the discovery of high-impact insights.

(3) **Be patient:** Implementing an experimentation strategy does not provide an "instant" impact on organization performance. Long-term commitment is needed to see its impact. Koning et al. (2020) reported that it took about six months for companies to notice the changes after adopting experimentation strategy using A/B testing.

(4) **Give an appropriate amount of time for your experiments:** Even though many organizations manage to streamline many processes in conducting experiments such as A/B testing online, it is still very important to give an appropriate amount of time to think through and carefully design the series of experiments. Quick A/B testing is just a tool; the experiment is the thought behind those actions and that is the reason for conducting any test. And although multivariate testing (simultaneous testing of multiple factors) is more complex than A/B testing, it can provide greater efficiency and learning about interactions than running a series of A/B tests.

(5) **Outsource your responsibilities carefully:** Research agencies, specialists, and strategic research partners are a big help for efficiently conducting experiments since they can share many responsibilities. However, certain activities are very critical and should be done by CEs and internal team members only. For example, deciding on a study objective is the CE and core team's responsibility. Strategic partners can help by providing recommendations and cautions for appropriate execution steps to achieve the objective. In addition, outsourcing design components for your experiments

is acceptable but the thought process and "designing of experiments" is your responsibility.

(6) **Planning your analysis before conducting any experiment:** Analysis paralysis happens quite often in many studies because of overthinking the results and failure to set clear action standards or guardrails on using the research outcome. It means researchers interpret the results beyond its objectives. What we recommend to CEs is allotting time and procedure for data analysis before conducting an experiment, planning proactively for decisions that will be made once the results come out (allot more time for an irreversible decision than a reversible decision) and letting the result speak for itself by avoiding debates of opinions.

(7) **Running a lot of experiments to produce strong evidence:** Most experiments fail but learning why the experiment fails provides valuable lessons. For any important hypothesis, it is better to run many experiments and seek convergence of evidence than to run few experiments and make a guess because of weak evidence. Ability to differentiate weak from strong evidence is important and setting context in your experiments can help. For example, testing five different prototypes in a laboratory will give a lot of significant differences among the samples. Then pick the best three and run them against each other in a home-use test in the next round. This will provide strong evidence and validate the previous results. Higher variability context produces stronger evidence in experimentation.

(8) **Flawless execution is your goal:** Avoid messy experiments when you can. This can be achieved by providing a well-defined hypothesis, experimental procedures, and metrics. If many experiments will be conducted, ensure that the data sets obtained are comparable. This can be achieved by (1) using a standardized form in managing the experiment, (2) explicitly describing all details in the experiments, and (3) including everyone in the team at the design stage to ensure thorough understanding of the experiments.

(9) **Having multiple metrics to measure performances:** To gain deeper insights, many points of views are needed. First, "read between the lines" when people answer your questions. Most of the time people answer because they are asked, and the important messages are not obvious to see or hear directly. Similarly, learnings can be buried in behavioral data especially if it was fleeting or an unarticulated frustration. Having a combination of survey and behavioral data in context is recommended.

(10) **Avoiding confirmation bias:** A good experiment is an experiment that you will get results that agree with you 50% of the time. This means, the results can contradict with your prior belief 50% of the time. Confirmation bias happens when one decides toward certain outcomes because the outcomes agree with one's prior assumption. Additionally, creating many competing hypotheses will help elevate this bias. Lastly, conducting many experiments can lead to converging evidence for each hypothesis. To avoid this bias, it is recommended to form a cross-category team to bring in different perspectives.

(11) **Learning from experiments and bringing it to action:** The goal of conducting an experiment is not just to test and learn but to decide based on

actual results and insights to improve the business. To do so, it is recommended to (1) allot time for synthesizing and ideating after getting the results, (2) keep your eyes on patterns and avoid nitpicking on minute details of experiments, and (3) create the habit of ideating from the results of your experiments to meet and, ultimately, move business goals.

12.6 Summary and future

Experimentation is the foundation of learning and discovering insights. Having hypotheses for all experiments will help you learn, make good decisions and discover efficiently. Currently, this concept is on the rise as many processes can be digitized and are easier to conduct. For more complex situations, sophisticated statistical design of experiments will be needed. This will depend on the nature of goals, situations, and product categories.

The statistical design of experiments provides several important benefits over less rigorous experimental approaches. The ability to isolate the effects of the different factors makes for unambiguous learning. The focus on model development and prediction identifies opportunities that may not have been considered beforehand. Innovation research would benefit in utilizing the approach in the many areas where it can be applied.

As has been demonstrated, the statistical experimental design approach can be applied throughout many stages of the innovation process. Therefore, a statistician can be a valuable member of a product innovation team. The role of a statistician in the innovation process is not limited to data analysis. Inclusion of the statistician on the team ensures that designs of all aspects of the studies are efficient and effective in collecting data to reach the goal at that stage in the process. When the team ultimately needs to make a decision on the introduction of the fruits of the innovation process, the statistician can ensure that the data collected will be sufficient to manage the associated business risks to appropriate levels.

Working with diverse forms of big data from structured databases to unstructured sources is inevitable for the CEs. New approaches for collecting, processing, managing, and analyzing will continue to advance but so will the challenge of dealing with big data. Schmelzer (2021) reiterated the challenging V's in big data, namely massive volume: velocity/speed of generating new data, variety of types/sources and he added veracity or varying levels of quality. Computing speed, advanced analytics, artificial intelligence (AI), and machine learning (ML) technologies are all projected to grow rapidly in business operations from planning to predictions. These digital technologies can free humans from repetitive work and potentially enable them to create new platforms for more meaningful work. Defining the hypothesis will remain the cornerstone of research and will still be a human responsibility along with the intent of the ensuing design. In addition, spotting "weak patterns" (a.k.a. budding insights), with the aid of ML and Augmented Intelligence (ML + human intelligence), will still be in CEs' hands. Hypothesis building, learning, and practicing experimentation and synthesizing its results will continue to be very useful and critical skills for efficiently and meaningfully discovering high potential insights.

Discussion questions

- Why is it important to consider experimentation vs. testing?
- Must all research start with a hypothesis?
- How can you convert big data to deep data?

References

Box, G., Hunter, J. and Hunter, W. (2005) *Statistics for Experimenters: Design, Innovation and Discovery*. Hoboken, NJ: John Wiley & Sons.

Griffin, R. and Stauffer, L. (1990) "Product optimization in central location testing and subsequent validation and calibration in home-use testing". *Journal of Sensory Studies*, 5(4), 231-240.

Koning, R., Hasan, S. and Chatterji, A. (2020) Digital Experimentation and Startup Performance: Evidence from A/B Testing. *Harvard Business School Working Paper*, No. 20-018, August 2019. (Revised September 2020. SSRN Working Paper Series, No. 3440291, August 2019).

Luo, L. (2020) The Most Common Experiment Pitfalls and How to Avoid Them. strategyzer.com (accessed 01 May 2021).

Moskowitz, H. (1987) "Optimizing consumer acceptance and perceived product quality". In J.G. Kapsalis (ed.) *Objective Methods in Food Quality Assessment*. Boca Raton, FL: CRC Press. pp. 183-229.

Myers, R., Montgomery, D. and Anderson-Cook, C. (2009) *Response Surface Methodology: Process and Product Optimization Using Designed Experiments*. New York: John Wiley & Sons.

Nagamachi, M. and Lokman, A. (2010) *Kansei Engineering (Industrial Innovation)*. Boca Raton, FL: CRC Press.

Rouke, P. (2016) The Growth Strategy That's Being Ignored: A Story of Untapped Potential. UK. PRWD.

SAS (2021) Big Data: What it is and why it matters. https://www.sas.com/en_th/insights/big-data/what-is-big-data.html (accessed 07 March 2022).

Schmelzer, R. (2021). Top trends in big data for 2021 and beyond. Published 20 April 2021. techtarget.com

Thomke, S.H. (2020) *Experimentation Works: The Surprising Power of Business Experiments*. Boston, MA: Harvard Business Review Press.

13

Online, word of mouth changes. In the old world, someone might tell five people, but in the new world, you can tell 5,000. But be careful because it works both ways. If you make a customer happy, he can tell 5,000 people, too.

Jeff Bezos
American industrialist & founder of Amazon.com

What this is about

Experts and social media influencers have been used by brands to push their products. The author shares her experiences and observations on how they can be engaged in innovation from inspiration to product delivery. She highlights the disruptive potentials of the social-media incubator influencer model for categories rooted in personalization trends. While influencer marketing has been proved successful in marketing and sales, using influencer's power for innovation is an emerging challenge. The author shares her experiences and observations - plus a caveat - on how influencers can be involved in innovation - from inspiration to product delivery, especially for categories rooted in personalization trends.

Product Innovation Toolbox: A Field Guide to Consumer Understanding and Research, Second Edition. Edited by Kannapon Lopetcharat, Dulce Paredes and Jacqueline H. Beckley.

Chapter 13

How to Work with Industry Experts and Influencers for Innovation

Veronica Symon

Key learnings

- ✓ Social-media incubator influencer model
- ✓ Influencer marketing
- ✓ Leveraging influencer's power for innovation
- ✓ Managing influencers and their influences

13.1 Introduction

Influencer marketing has become more important than ever. The best influencers are able to generate high-quality content that's on brand without appearing overproduced or inauthentic. Creating high-quality content without the support of a full in-house team is more valuable now than ever before.

According to *Business Insider*,[1] the influencer marketing industry is on track to be worth up to US$15 billion by 2022, up from as much as US$8 billion in 2019. Businesses are making US$5.20 for every US$1 spent on influencer marketing. Forty percent of consumers said they are more likely to trust a brand that features an influencer they know, and 25% said influencers who endorse brands shape their perceptions according to the said report.

While influencer marketing has been proved successful in marketing and sales, using influencer's power for innovation is an emerging challenge. In this chapter, I will discuss how to work with influencers to support innovation, and the areas to keep an eye on when using this approach.

What does an influencer have to do with innovation? Let us' review the diffusion of innovations theory first. It is a theory that seeks to explain how, why, and at what rate new ideas and technology spread. Everett Rogers who was a professor of communication studies, explained the theory in his book *Diffusion of*

Product Innovation Toolbox: A Field Guide to Consumer Understanding and Research, Second Edition. Edited by Kannapon Lopetcharat, Dulce Paredes and Jacqueline H. Beckley.

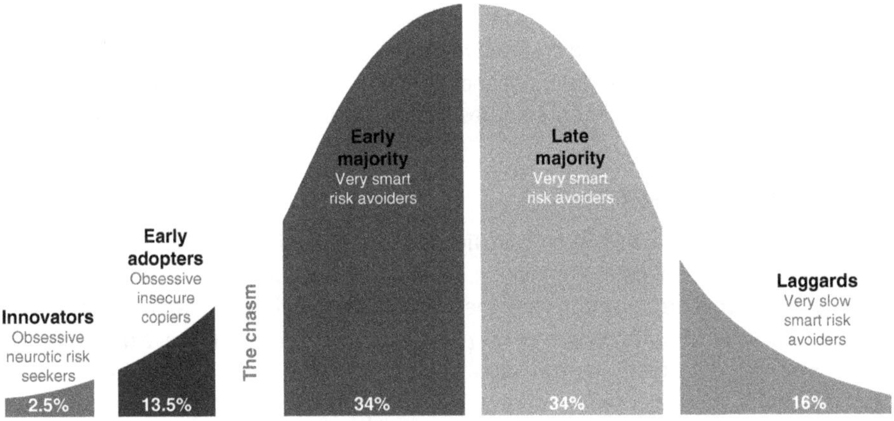

Figure 13.1 Adopter categorization on the basis of innovativeness. Source: Adapted from Rogers (2003).

Innovations, first published in 1962, and now in its fifth edition (2003). Rogers states that diffusion is the process by which an innovation is communicated over time among the participants in a social system. The origins of the diffusion of innovations theory are varied across multiple disciplines.

Rogers proposes that four main elements influence the spread of a new idea: the innovation itself, communication channels, time, and a social system (Rogers, 2003). This process relies heavily on human capital. The innovation must be widely adopted to self-sustain. Within the rate of adoption, there is a point at which an innovation reaches critical mass.

The categories of adopters are innovators, early adopters, early majority, late majority, and laggards. Diffusion manifests in different ways and is highly dependent on the type of adopters and innovation-decision process. The criterion for the adopter categorization is innovativeness, defined as the degree to which an individual adopts an idea (Figure 13.1).

Innovators have a huge importance with their leading roles. Influencers fit into the innovators' category. They invent and adapt to newness and influence others. Together with their wide network, their position in the adoption system creates a great influential power on consumer groups. A product life cycle mainly includes four main stages: introduction, growth, maturity, and decline (Shankar, 2008). Diffusion theory suggests that the word of mouth communication from previous adopters strongly determines a new product's sales growth (Krishnan et al., 2000). Overall, influencers would contribute to innovation's introduction and growth significantly. "So, who are these influencers and how could they impact innovation?"

13.2 Meet the influencers

There are several related definitions out there. Influencers are people who affect or change the way other people behave by (Cambridge Dictionary). Key opinion leaders (KOL) are people who are trusted for their expertise in specific areas ("reliable and trustworthy image"). Most recently, a Chinese word "Wanghong" means someone focused on creative content and wielding their influence on social media (https://tenbagroup.com/the-rise-of-the-wanghong-economy-in-china/).

In general, there are three types of influencers based on general definitions.

13.2.1 Technical experts

They have expertise in certain fields such as baking, distilling and blending, beauty care, culinary art, fashion design, and bartending. They have their own professional organizations and their opinions are well respected by their respective trades, corporations, and savvy consumers. They usually have gone through accredited professional training and have acquired many years of experience. Many of them work in corporate R&D organizations and contribute to product design and creation. Others act as independent consultants for large corporations and business owners.

These experts usually are briefed by marketing with a specific concept, then they are free to create and bring the concept to life with their technical expertise. Once the initial prototypes are done, they will then present to stakeholders such as marketing and management teams for screening and review. After collecting feedback, they will further improve the prototypes until internal stakeholders are satisfied. Another scenario could be using external experts to screen internal generated prototypes and provide feedback to project teams. Both pathways are beneficial in screening prototypes to a couple of team's favorites for final confirmation testing. However, there might be potential risk with these methods. A major concern is that the consumer's opinion is missing from the process. Experts' opinions may not always match what the consumer wants, especially for disruptive innovation that is new to the market and widely accepted by the consumers. It is also possible that without consumers' validation, an expert's opinion has limited influencing power, especially when a group of experts don't agree with each other.

I recommend enhancing these noted experts' influencing power by connecting them with consumers through social media. Today's consumers tend to demand more and more transparency on products from ingredients to processing. They appreciate being involved in the innovation process either for ideation or product design. We as Consumer Explorers (CE) and/or the Brand Team could work with e-commerce and public relation firms to create social media content for experts, then broadcast their creation process to consumers using social media platforms. There are several potential benefits for doing this such as making brands more personal and approachable, enhancing consumer engagement with transparency and collecting feedback from consumers at an early stage.

Another approach is to use artificial intelligence (AI) data analytics to quantify the expert's opinion. Currently, companies usually work with a handful of experts for a specific innovation project. Their feedback is often qualitative. To have better confidence, we could consider using a research platform with built-in AI tools such as Immerse by MetrixLab (www.metrixlab.com). This platform can reach out to 80-plus people from their expert-database such as chefs or bartenders to generate feedback on a specific concept or product. The platform processes the experts' responses and generates a rating ballot based on most common statements. In turn, the experts are asked to rate "Agree" or "Disagree" on these statements. At the end, the most agreeable positioning or product attributes are generated in a ranked order in real time. This can be a powerful and rapid tool to identify an optimal concept and product benefit/characteristic. Since the insights generated are both qualitative and quantitative, the project team could make a decision with higher confidence.

13.2.2 Key opinion leaders (KOLs) and celebrities

Using socially influential people such as politicians, athletes, and celebrities to promote brands is quite common. Brands usually have various types of partnerships with KOL and celebrities. Some brands just partner with them for marketing and sales promotion, while others have a much closer relationship with them. In fact, some KOL and celebrities are coowners of the brands. In this case, they take more responsibilities including participating in brand strategy building, innovation ideation, product design, and becoming marketing champions after launch. The benefits of these partnerships are to use the KOL and celebrities' established reputation and fan base. Since their personal reputations are tied to the brands, their avid followers would trust the brands more (Web, 2021). One of the successful examples is Kanye West's Yeezy Season line. He partnered with Adidas to design and produce Yeezy Season 1 - a collection of clothing and footwear that he described as the world's "first solutions-based" clothing line. He was able to bring the concept to life with a professional designer's help. It was such a hit and later on he launched seven more Yeezy seasons.

Although some KOL and celebrities are involved in innovation ideation and product design, most are not. Often, they could only allocate limited time on a specific project due to their tight schedules. In addition, recent consumer trends show that consumers' value less on KOL and celebrities' opinions than their peers, family, and friends (Tam and Menendez, 2016). This has brought a new focus to engage accessible, credible, and relatable social influencers for our target consumers.

13.2.3 Social media influencers

Digital technology has integrated social media into many people's lives. Social media influencers have emerged as an important marketing vehicle for brands. Who are the social media influencers? They are people who have built a reputation for their knowledge and expertise on a specific topic. They make regular posts about that topic on their preferred social media channels and generate large followings of enthusiastic, engaged people who pay close attention to their views. In the United States, there are four major categories of the social media influencers:

(1) **The mega influencer:** According to Gil Eyal, CEO and founder of influence marketing platform HYPR Brands, a mega influence is defined as being more famous than influential. They aren't necessarily subject matter experts, but they definitely provide a lot of reach in one hit. Usually, they have over a million followers in their social media accounts.

(2) **The macro influencer:** Macro influencers are similar to mega influencers, with the main difference being macro influencers usually rise to fame through the web, as opposed to the genuine celebrities that make up mega influencers. Macro influencers can be podcasters, vloggers, social media stars, and influential bloggers. Their audience size would normally be between 100 000 and a million followers.

(3) **The micro influencer:** When you're ready to move your influence marketing campaign from an awareness tactic to a more defined one, you need to start looking for more effective methods of communication and promotion. While mega- and macro- influencers can still impact results at this stage, this is the time when you would work with micro-influencers. This type of influencer's audience ranges from 10 000 to 100 000 followers. While their audience may be smaller, they are more invested in the influencer and their niche.

(4) **The nano influencer:** Nano influencers are a relatively newer category in influencer marketing. These influencers are often relatable because they are not household names with millions of followers. They usually have between 1000 and 10 000 followers. They are much more like your family and friends, which brings authenticity and trustworthiness to what they share.

Currently, social media influencers have taken a more important role in marketing. Influencer marketing is when a business works with an influential person to promote something - a product, service, or campaign. In today's digital world, more than two-thirds of North American retailers use some form of influencer marketing (https://blog.hootsuite.com/influencer-marketing/). And almost half of US and UK digital marketers spend at least 10% of their marketing communication budget on influencer marketing. A survey found that 34% of daily US. Instagram users bought something because a blogger or influencer recommended it. Instagram is the platform of choice for social influencers, though Tiktok and other apps are making inroads in this arena. Indeed, 89% of marketers identify it as one of the most important channels for influencer marketing. This number is projected to further grow this year, according to Christina Newberry from Hootsuite (https://blog.hootsuite.com/influencer-marketing/).

13.3 Could we go a step further, leverage social media influencers for innovation?

Like KOL and celebrities, social media influencers are mainly used for marketing and brand endorsement. How we could use this booming venue for innovation remains an interesting challenge. However, there have been some promising explorations in China that set up models for some industries. China's influencer marketing industry is three to five years ahead of the rest of the world (such as https://tenbagroup.com/working-with-chinese-influencers-to-market-your-business-in-china/). The advanced functionality and integration of social media, e-commerce, and digital payment in China have enabled business models that are yet to be explored in other countries. They have created an influencer incubator.

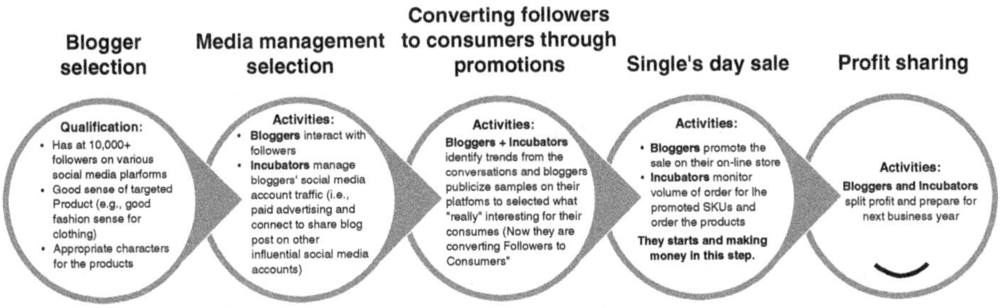

Figure 13.2 The influencer incubator model. Source: Adapted from Whaley (2019).

What is an influencer incubator? Unlike influencers in the West who rely on the traditional sponsored-post monetization model, many influencers in China have created their own incredibly successful brands, manufacturing, and selling their products specifically designed for their audience (Figure 13.2).

Influencer incubators use a fast-fashion model to compress the concept-to-consumer timeline. Every step of the product development life cycle is designed to be shared on social media by the influencer. With product development to marketing under one roof and brand, and advocates interacting throughout the process, culturally relevant products aim to birth new trends on social media. Social platforms are influencer brand tools used to ask consumers what they want, rather than telling them (Whaley, 2019).

With this model and tech-enabled supply chain improvement, the influencer co-designs the product with consumers and quickly manufactures and delivers it directly to consumers. The innovation process has been compressed from [product design → produce → distribute → display] to [product design → consumers]. The benefits are obvious: direct interaction with consumers for feedback, enhanced consumer engagement, and transparency from the code signing process, shortened timelines and a big cost saving from skipping distributor/wholesalers and stores. All of these would benefit both consumers and brands. Consumers receive personalized products with most likely lower prices; brands become much more competitive as they customize products for targeted consumers faster and cheaper.

This model specially works well for certain categories, such as fashion and accessories, beauty cares and household goods where personalization/customization is key for success. Two tech-driven functions contribute significantly to the success of Chinese social influencer models. One is a live streaming format for broadcasting the influencers, and another is one-click purchase for the consumers. Both technologies have enabled high levels of engagement for potential consumers. Google, Facebook, and Instagram have been following China's model closely. Once they develop similar technologies, brands outside of Asian markets could partner with social influencers to carry on their innovation projects or cocreate new brands with them. Looking into the future, social influencers have a huge potential as they could take multiple tasks, such as voice of brands, research and design partners, retail stores, consumer complaint departments, and future trendsetters. Are we ready for the transition?

13.4 Practical tips

I recommend individual brands do their own analysis on what influencer models best fit their business needs and consumers. You could use various types of influencers to benefit your long and short-term brand strategies. Below are some tips:

First: Choose the right type of influencers you would like to partner with. I am not surprised to hear that the majority of consumer packaged goods (CPG) companies still want to work with traditional KOL and celebrities as we all have had lots of experience and successful cases with them already. But I would advocate integrating their campaigns with the other two types of influencers - *KOLs/celebrities* and *social media influencers*. Technical experts' opinions could really appeal to product savvy consumers who are interested in experience and continuous learning in a specific category, such as wine enthusiasts. Having direct conversations with the experts and getting invited to product creation would certainly motivate these consumers to stay loyal. Brands can also create a premium personalized line for these consumers. On the other hand, the social media influencers could be extremely appealing to certain segments of consumers, such as those following specific diets (e.g. vegan, vegetarian, or keto). They are looking for niche voices that match with their beliefs and disciplines. A regular person sharing a similar lifestyle and challenges in life would be more impactful and relevant than a mega celebrity living a very different life from their own.

No matter which type of influencers are chosen by a brand to partner with, it is essential to use social media platforms. With fast development of technology, people routinely use social media for information search and social connections. Many types of sophisticated technologies have been applied into the platforms to ensure a pleasant, functional and fast consumer experience. Some functions are key to attract consumers and collect feedback from them, such as live streaming and instant messaging with multiple people. Social Search Optimization (SSO) has also emerged as a new generation of search engines that allow people to search not only the web but social media posts, private conversations, and miniprograms. What is especially important is that social media influencers hold a unique advantage in the next generation of search because social search algorithms take follower count, comments, reposts, and likes into account when calculating a post's rank. Their influence power can only grow with the development of new search engines in the future.

Second: I want to emphasize that it is also crucial to build and maintain a good relationship with consumer communities with influencers. Brands and influencers should know very well about their consumers including their demographics, beliefs, concerns, emotion, lifestyles, and expectations of the brands. It is encouraged to have periodical live streaming events to enhance engagement, familiarity to the brands and gain trust of consumers. Brands and influencers should tailor their program for specific consumer communities with an authentic and fun vibe. As an influencer, she or he is not only an advisor but also an entertainer, a host of a party and a community leader. Constant growing and nurturing the consumer community is needed. As brand ambassadors, celebrities, and influencers attitudes and behaviors must align with brand persona and target communities' values at all times. Any mishaps or fallouts of the celebrities and influencers from the communities can backfire to the brand and cause negative impressions to the brand and, worst, destructive behavior toward the brand.

Third: I would like to challenge brands to consider the social media influencer incubator model, especially, if you are in the business of fashion, beauty care, and household goods where customization is a key for success. You will need a dedicated team including marketing, creative, product design, and supply professionals to work with your influencers. This team should also be constantly monitoring the changes in the influencer's image and be prepared to answer queries or amend/pull out the campaign when the influencer is involved in a scandal or social mishap. The team should also be aware of current events to monitor that the social influencer is not aligned with extremist views that could potentially damage the brand. The activities they create need to be more inclusive rather than divisive and related to their targeted community current sentiments.

13.5 Conclusion

Despite upfront investment on time and resource, potential payback is balanced with shortened product launch cycle and lower cost. Best of all, you could be a trendsetter as you out beat your competitors on consumer insight, timing, and cost. In order to do so, a creative, learn and grow mindset with an agile and action-oriented culture are needed. Creating this shorter pathway of design and delivering products to consumers with influencers is an innovation by itself. We need the same energy, creativity, and disciplines as we innovate a new product. It's challenging ... but also rewarding.

Note

1. https://www.businessinsider.com/influencer-marketing-report

References

Krishnan, T.V., Bass, F.M. and Kumar, V. (2000) "Impact of a late entrant on the diffusion of a new product/service". *Journal of Marketing Research*, 37(2), 269-278.

Rogers, E.M. (2003) *Diffusion of Innovations*, (5th ed.). New York: Free Press.

Shankar, V. (2008) "The evolution of markets: innovation adoption, diffusion, market growth, new product entry, and competitor responses". In S. Shane (ed.) *Handbook of Technology and Innovation Management*. Great Britain: Wiley. pp. 57-112.

Tam, D.N. and Menendez, S. (2016) Consumers trust social media stars more than celebrities or ads. https://www.marketplace.org/2016/03/31/consumers-trust-social-media-stars-more-celebrities-or-ads/ (accessed 13 April 2021).

Web, A. (2021) Jay-Z's Luxury Champagne Deal with Moet Is a Celebrity Winner: sometimes it's more lucrative to keep your name off the bottle. https://www.bloomberg.com/opinion/articles/2021-02-23/jay-z-and-lvmh-champagne-show-the-winning-formula-for-celebrity-collabs (accessed 13 April 2021).

Whaley, E. (2019) The future of influencer marketing in China. https://www.luxurysociety.com/en/articles/2019/06/the-future-of-influencer-marketing-is-in-china (accessed 08 March 2022).

14

If you don't seek out allies and helpers, then you will be isolated and weak.
Sun Tzu The Art of War

What this is about

Supplementing internal capabilities with virtual staff enables a company to compete effectively and expand innovation. This mutually beneficial partnership is grounded on having common goals, building trust, working out the kinks in the process while celebrating and learning from both successes and failures.

Product Innovation Toolbox: A Field Guide to Consumer Understanding and Research, Second Edition. Edited by Kannapon Lopetcharat, Dulce Paredes and Jacqueline H. Beckley.

14

Chapter 14

Words of the Wise – Virtual Staff

Carter Green, Ratapol Teratanavat, and Dulce Paredes

Key learnings

- ✓ Foundations for building a strong working relationship
- ✓ Steps to creating a virtual staff
- ✓ Successful collaborations

14.1 Why a virtual staff?

Innovation does not have to be a tangible product. It can be a process that has commercial value and benefit. For us, that process is the creation of a virtual staff that covers talent, flexibility, and efficiency. More importantly, our virtual staffs share foundational values such as strong collaboration; mutual trust; respect for confidentiality and intellectual property; curiosity for new approaches; purposeful design thinking and efficient project activation. The concept of reinforcing current capabilities (resources) is not new as we have seen the rise of open innovation platforms and individual contractors that fueled the Gig economy in recent years. Although the concept of virtual staff started with people resources, it applies as well to shared technologies and manufacturing. The concept is especially relevant in today's hybrid work environment where having a trusted, engaged, productive, agile, flexible, and technology competent workforce is invaluable amidst limited resources and staff trained in accelerated technologies.

In 2011, our leadership team (at Takasago, International) created a Consumer Insights and Market Research global group dedicated to the flavor division and independent from its original affiliation with the fragrance group (another division in the company). The challenge was to quickly build the global group using existing resources, new hires, and collaborations. Acknowledging there are traditional outsourcing routes for finding suppliers and temporary workers to work on specific projects and support functions, the desire for a long-term solution became an opportunity. For this global group, we needed a cost-effective remote or virtual

Product Innovation Toolbox: A Field Guide to Consumer Understanding and Research, Second Edition. Edited by Kannapon Lopetcharat, Dulce Paredes and Jacqueline H. Beckley.

staff that was not limited by physical office space or other constraints. Besides, the work will be across borders/continents so it should not be limited to specific work hours/time zones. Building a dedicated remote team using external resources still requires an experienced internal team to efficiently screen, manage, and integrate new team members to ensure that everyone shares the key values mentioned above. Finally, we knew we needed to approach our resources differently with the expectation that everyone will have our shared commitment and vision for success. We/the internal team must create a path to (1) sustainable virtual staff organizational growth with the business and (2) customer project deliverables with a clear partnership strategy/vision for success on a business scale.

Now 10 years into the creation of our virtual staff concept, we can reflect on how it allowed us to build our team capabilities, effectively function as a global group and provide "return on investment" with global and regional business wins. We have refined the concept, buttressed by insights from recent global pandemic events that impacted how companies manage employees, and manufacturing processes during a Force Majeure event.

We will share, from experience, what is necessary to develop a virtual staff that can boost a company's commercial capabilities to facilitate efficient interactions with a business partner, a client, or a customer.

14.2 What is virtual staff and what is required to be one?

Virtual staff are defined as an individual, a group, or a company from outside of your organization that work to support you and your team. It allows your company to leverage external resources with specific expertise and skill sets that you need and/or that may not exist within your company. The concept of a virtual staff is similar to a virtual team as discussed by Ferrazzi (2014) as those who work independently and are not part of the company payroll; however, they are available and provide project support as if they were employees within your company. This is important because the financial commitment necessary for a large internal team with all necessary skill sets may be cost prohibitive for some companies. There are other independent workers or support functions that one can tap into such as gig workers, contractors, consultants, suppliers, or even strategic partners, but virtual staffs have unique characteristics (see Figure 14.1).

On the surface, virtual staff may be perceived as gig workers or contractors. Gig workers are defined as independent contractors that have formal agreements with your company to support various types of projects (Petriglieri et al., 2018). The gig economy has been booming in the past several years because many companies can tap external resources with various skill sets to help address their business challenges or work on projects without paying for benefits or training (Iacurci, 2020). However, virtual staff are different from gig workers or contractors in the level of trust and shared accountabilities. With virtual staff, the goal is to develop a strong relationship to the point that your company becomes willing to share corporate internal knowledge, strategies, and resources. The same level of trust or long-term relationship commitment may not be possible or desirable with gig workers or contractors. It is also possible for a gig worker or contractor relationship to evolve into a virtual staff

14

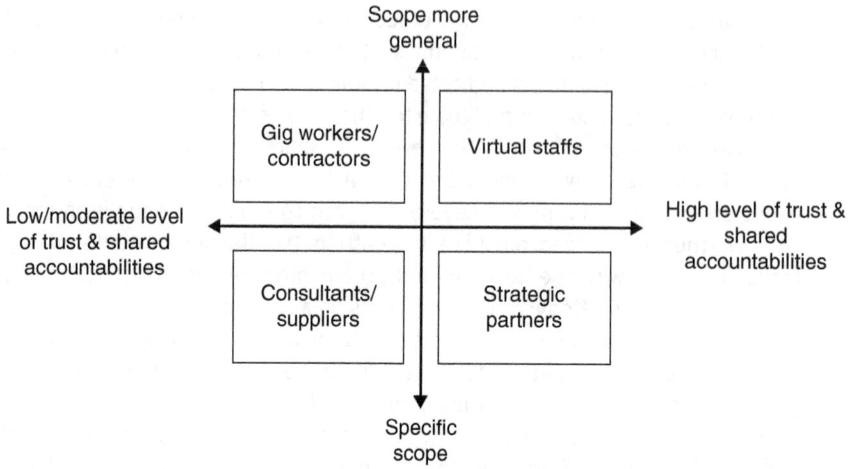

Figure 14.1 How virtual staff differ from gig workers, consultants, and strategic partners.

member. In addition, these virtual staff members have their "skin in the game" as they share responsibilities and accountabilities in the success and failure of the projects. This mindset is different from gig workers as they may not be entrusted and/or be expected to have full accountability due to their limited view of the big picture.

The same rule of engagement applies to relationships with consultants and suppliers/agencies. Typically, consultants and suppliers work on specific defined projects and provide knowledge and expertise to support a project or contract specification. Nonetheless, they are not considered part of the team as there is no expectation to share internal knowledge and resources as if they are your company staff. With virtual staff members, a company has more open-minded flexibility to engage on broader terms. The virtual staff scope of project work often starts more general and then evolves in a natural manner through project concept ideation like working with an internal team. In marked contrast, often working with strategic partners, consultants, and suppliers requires a clearly defined scope up-front based on individual project objectives, budget, and timeline.

14.3 How do you go about building and utilizing a virtual staff?

Step 1: Have all involved parties start on the same page with the end goal of getting a business win.

A full business partnership with virtual staff members typically requires a strong relationship built on trust, a common goal, and a reason to believe the goal is achievable. The parties should acknowledge individual strengths and capabilities, and both believe that their shared expertise and strategizing can be synergistic and accelerate the project timeline. It is important that your virtual staff possess the knowledge, experience, and capabilities necessary to complete projects. It is a bonus if your virtual staff have diverse business networking and strong professional relationships with team members and/or customers that facilitate day one familiarity and trust.

Following a year of shared COVID-19 experience, the necessity of virtual interactions has never been more apparent. The need to move out of our comfort zones and work together remotely is no longer a choice but essential for ongoing business interactions. During this time, the importance of business networking and strong professional relationships/friendships has never been more key to efficient and effective management of projects to achieve goals and overcome business challenges. Developing the mutual trust between you and your virtual staff is required for the effective implementation of the true partnership approach. This will facilitate the sharing of relevant information and materials so the team (you and your virtual staff) can develop shared common goals. Given that companies continue to be challenged with tight timelines and limited resources, it is essential that the team driving customer business opportunities has a clear vision of the goal for all parties to strive toward.

Step 2: Have the proper legal documents signed at the get go and agree with an endpoint in mind (Think backwards from the goal to the beginning of the project.).

All successful collaborations are built on mutual trust and a clear understanding of the assets (e.g. intellectual property) each party brings to the relationship. Each party obligates, via legal agreements, the other to protect the information that is valued by the company (e.g. intellectual property). There are different types of legal agreement such as confidential disclosure agreement (CDA), material transfer agreement (MTA), and joint development agreement (JDA) (see Figure 14.2). A detailed discussion of legal documents needed to form a cooperative and mutually beneficial relationship can be found in Paredes et al. (2017). In the case of virtual staff development, CDA and JDA are relevant agreements to consider.

Generally, most companies will start with a CDA. The CDA provides a framework within the Standard Process for the disclosure of proprietary information between parties (i.e. your company and virtual staff) related to a defined

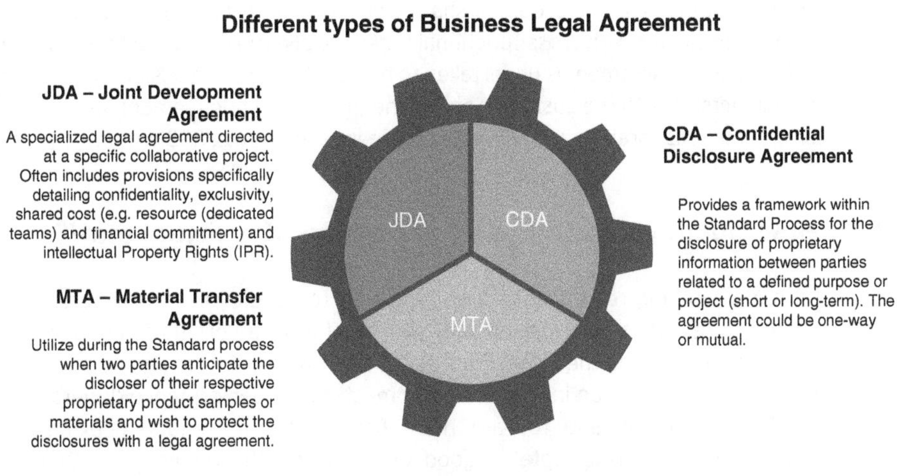

Figure 14.2 Different types of business legal agreement.

purpose or project (short or long-term). The CDA typically includes provisions defining the term for maintaining the confidentiality of the information and is often perpetual for information defined as a Trade Secret.

There are two types of CDAs that can be defined simply as One-Way and Mutual. Both types are often necessary during the Standard Process product development cycle. The one-way CDA is used to protect the disclosure of proprietary information to your virtual staff. Often, the information is disclosed for safety and regulatory purposes such as formula or other trade secret information as well as the disclosure of proprietary information related to a product base (e.g. formulation and/or processing conditions) to a flavor house (receiver). The agreement term is typically perpetual or until the information is disclosed or available in the public domain. In contrast, a mutual CDA is utilized during the Standard Process when both parties (e.g. you and your virtual staff member) anticipate the disclosure of their respective proprietary information and wish to protect the disclosures with a legal agreement.

Over the course of time, the level of trust and shared accountabilities between you and your virtual staff grow to the point that you want to develop a stronger collaboration. At this stage, a JDA should be considered. The JDA is a specialized legal agreement between two parties directed at a specific collaborative project. The JDA often includes provisions specifically detailing confidentiality, exclusivity, shared cost (e.g. resource [dedicated teams] and financial commitment) and intellectual property rights (IPRs). Both parties invest the time and effort to negotiate an equitable JDA based on the belief that product development success will be fast-tracked and imminent. In short, both parties are highly motivated toward a shared success as both have "skin-in-the-game."

Step 3: Design a pilot proof of principle (POP) study with virtual staff members to build trust, work out the kinks in the process, and especially if both parties are new collaborators.

A Pilot or POP study should be considered as an initial "test drive" of the working relationship between the team (i.e. internal staff and virtual staff members) and a customer team. A POP study is typically short term (e.g. three months) and it can include providing limited resources, sharing of information, having access to a variety of team players with cross-functional (interdisciplinary) responsibilities. After a POP, a review/assessment is undertaken to make a "go/no go" decision to continue the partnership with the customer and refine the rules of engagement, as needed, for greater collaboration and expedited business success (i.e. $$$ = money).

14.4 How would you rate the performance of a virtual staff?

As in any working relationship, it is important to have a set of attributes and metrics to assess the performance of your staff after a period of engagement. We recommend using the KANO framework (see Chapter 8.1 for details on applying KANO framework on innovation) with must-haves, performance drivers, and delighter attributes and to rate them based on the frequency exhibited. A criterion of what is acceptable, good, or superior with frequency can then be created for each of these attributes. Table 14.1 is a sample framework that we

Table 14.1 Assessment factors for virtual staff.

	Attribute	Always	Most of the time	Sometimes	Rarely	Never
Must haves	Curiosity	☐	☐	☐	☐	☐
	Individual initiative	☐	☐	☐	☐	☐
	Action oriented	☐	☐	☐	☐	☐
	Have demonstrated analytical capacity	☐	☐	☐	☐	☐
	Willingness to practice what has been taught until one "succeeds"	☐	☐	☐	☐	☐
	Ability to demonstrate sustained energy for a given project (focus)	☐	☐	☐	☐	☐
	Willingness to reach out and connect with internal and external individuals	☐	☐	☐	☐	☐
	Ability to take an insight and craft a compelling and engaging story	☐	☐	☐	☐	☐
	Ability to transform questions into testable hypotheses	☐	☐	☐	☐	☐
Performance drivers	Bringing others along	☐	☐	☐	☐	☐
	Demonstrate ability for abstract thinking	☐	☐	☐	☐	☐
	Create a plan that has all necessary product/communication/marketing design elements	☐	☐	☐	☐	☐
	Execute the above plan	☐	☐	☐	☐	☐
	Create a clear concise report with necessary journalistic standards (e.g., Start with outline, develop content, move to conclusions, be able to edit self)	☐	☐	☐	☐	☐
	Understands how to take reasonable risks to advantage business	☐	☐	☐	☐	☐
Delighters	Insatiable curiosity	☐	☐	☐	☐	☐
	Self-starter-can create new to the world approaches to captivate customers and advantage business	☐	☐	☐	☐	☐
	Engages broadly with the expertise community that benefits business	☐	☐	☐	☐	☐

created with our first virtual staff from The Understanding & Insights Group. The attributes listed cover the same capabilities that we would expect our staff to possess with appropriate training and coaching as we go through projects.

14.5 How does virtual staff work based on success case studies?

When our global consumer insights program was formed, we needed to build broad knowledge across consumer product categories defined by our ongoing flavor business needs. It was a huge undertaking to build the internal infrastructure and conduct research globally at the same time. Our approach was to present the business opportunity to a handful of strategic research partners vetted from previous or ongoing focused project interactions. We broadened the scope of our working relationships from working on a specific project to covering various aspects of customer project support formerly delegated to internal staff/team members (see Figure 14.3) (e.g. designing and running consumer research studies, training and onboarding new team members).

As noted earlier, the first virtual staff that we engaged was the Understanding & Insights Group. We had a strategic meeting to discuss the vision of a Consumer Insights and Market Research Team that is seen as a leader of insights so compelling that everyone would want to use them. We created training programs for onboarding new staff and initiated major applied consumer and product research in major categories. While the team leader (author D. Paredes) focused efforts on building the internal infrastructure, the virtual staff team was delegated the responsibility to conduct global research initiative pilot studies that were later replicated in multiple countries. We had a client manager as a key contact who keeps track and immediately identifies drifts in schedules and plans that will cause resource and timing issues. Within the first three years, we were able to complete multiple category research projects in over 20 countries.

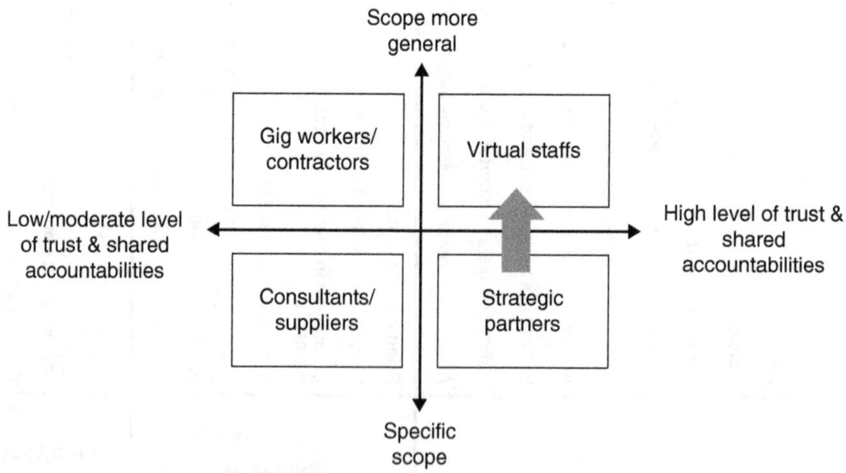

Figure 14.3 From strategic partner to virtual staffs.

Our business relationships have strengthened and with ongoing learning experiences we continue to work with the same virtual staff team to this day. We expanded our virtual staff arrangement to other groups as new global initiatives and technical scope are created.

As we reflect on the success of this virtual staff program and the forged collaborative relationships, the most important aspect is the level of trust that we have developed with each other. From the beginning of this global corporate initiative, we have been open and concise in describing our vision and willingness to share internal corporate knowledge, which empowered our virtual staff to be fully committed or "all in." As a direct result, our virtual staff has taken on responsibilities and accountabilities as if the success and failure of our initiatives were their own. We have truly embraced our virtual staff as part of our core team and 10 years later, a handful of our virtual staff ended up having more continuity than internal staff. Our virtual staff continues to be part of our onboarding program to train new staff.

By way of example and in general terms, we can discuss the evolution of a successful business partnership with one of our key customers, Company A (due to existing agreements, only generalities will be disclosed). As is common in the flavor industry, Company A typically provides a competitive project brief to several flavor suppliers detailing a new business opportunity and guidelines for flavor sample submissions. Over the course of several years, we demonstrate our commitment and willingness to work together and provide distinct technical capabilities and services. In addition, Company A took note of our willingness to share knowledge and technical expertise during the various stages of the commercial product development. As a result, the trust between our two companies was heightened, and we embarked on Joint Development Agreements for technology-based projects directed toward commercial product development. The evolution of this business relationship was facilitated by empowering a true partnership team (i.e. internal staff/virtual staff members and a customer team). Our collective "boots on the ground" and "hands on" cross-functional team demonstrated its expertise with effective ideation, development, and validation of commercial targets. With this elevated working relationship from supplier to virtual staff (Figure 14.4), we were able to jointly develop several new products and successfully launch them in the marketplace.

It has been our experience that the collaborative sharing of high-value information and materials by a food company and a flavor house can lead to a successful joint product development effort. Acknowledging that a flavor is a complex mixture of ingredients that may not behave predictably in a complex consumer product base necessitates the need for close collaborative discussions to ensure expedited and successful product development. The business relationship between food companies and flavor houses will evolve with time as they continue to build trust and joint wins (i.e. $$$) along the way.

14.6 Conclusion[1]

During the COVID-19 pandemic, the consumer product development timeline and all manufacturing related issues have been challenged. During this time, the importance of business networking and strong professional relationships has

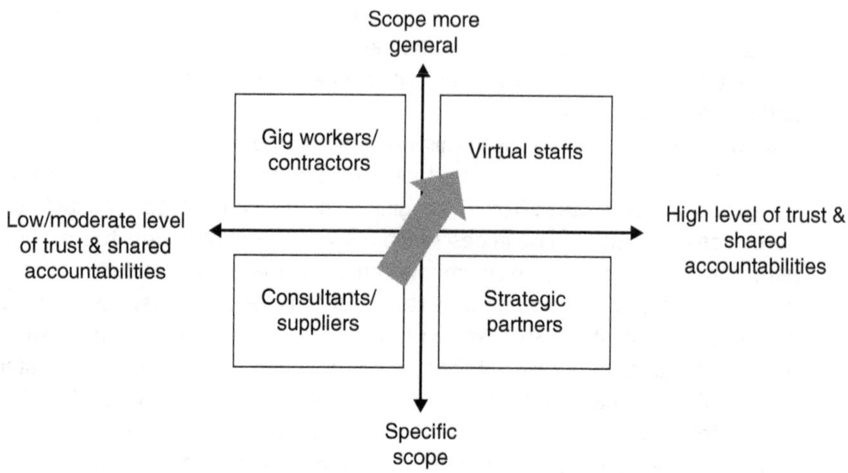

Figure 14.4 From consultants/suppliers to virtual staff.

never been more key to efficient and effective management of business opportunities and goals in the global market.

Perhaps the most relevant example delineating the power of working together toward a defined goal is the BioNTech and Pfizer collaborative COVID-19 vaccine development. They had an existing robust innovation process and trusted research partnership in place which enabled them to react at breakthrough speed and technical rigor on the development and delivery of one of the first COVID-19 vaccines. In a Financial Times Weekend Festival interview (2021) of the two BioNTech co-founders, Ungur Sahin and Ozlem Turici, the panel moderator and FT science editor, Clive Cookson asked them on how it was to partner with the Big Pharma company. Dr. Sahin said, "When we started the project, we understood very early on that developing a vaccine and delivering it worldwide will need international collaboration. This is nothing that we can do alone. Therefore, we were interested to partner with a company, bringing complimentary skills and sharing the same values and same way of scientific focused development as we do. Pfizer was an ideal partner for that. We already had a collaboration and we trusted each other. Pfizer knew already about our technology and so it was a very clear decision."

Our discussion of how to supplement internal capabilities with virtual staff for enhanced business interactions provides experiential insights into how a company can compete effectively in business and expand innovation. Mutually beneficial virtual partnerships grounded with common defined business goals and trust can lead to long-term success.

Discussion questions

- Where can you see virtual staff being used in your organization?
- What are the metrics for a successful virtual staff program?

Note

1. The authors acknowledge the contribution of Brian Buck, Senior Vice President and General Manager, Flavor Division, Takasago International Corporation (United States) for championing our first Virtual Staff program.

References

Ferrazzi, K. (2014) *Getting Virtual team Right*. Harvard Business Review (December 2014). https://hbr.org/2014/12/getting-virtual-teams-right (accessed 08 March 2022).

Financial Times Weekend Festival (2021, March 20) *The Vaccine Visionaries* [Video file]. https://ftweekend.live.ft.com/agenda/session/490312 (accessed 20 March 2021).

Iacurci, G. (2020) *The Gig Economy Has Ballooned by 6 Million People Since 2010. Financial Worries May Follow*. CNBC Personal Finance. https://www.cnbc.com/2020/02/04/gig-economy-grows-15percent-over-past-decade-adp-report.html (accessed 08 March 2022).

Paredes, D., Teratanavat, R. and Green, C. (2017) "A flavor supplier perspective - let's cut to the chase and win together!". In J. Beckley, L. Herzog and M. Foley (eds). *Accelerating New Food Product Design and Development* (2nd edition). Hoboke, NJ: John Wiley & Sons, Inc.

Petriglieri, G., Ashford, S.J. and Wrzesniewski, A. (2018, March-April) *Thriving in the Gig Economy*. Harvard Business Review.

15

"Translation is that which transforms everything so that nothing changes."

Günter Grass
1999 Nobel Prize laureate in Literature

What this is about

Over the years, a lot has been published on cross-cultural consumer research (for extensive reviews see [Ares, 2018; Buil et al., 2012; Camacho, 2012; Lee and Lopetcharat, 2017; Sin et al., 1999] among others). What you will find in this chapter are our learnings and insider tips, accumulated from years of experience as practitioners having to design and manage projects from the customer side and not from the research agency perspective. Our extensive and diverse focus is field and reality based. In each section, we provide a series of important considerations to take into account while conducting multicultural research.

Product Innovation Toolbox: A Field Guide to Consumer Understanding and Research, Second Edition. Edited by Kannapon Lopetcharat, Dulce Paredes and Jacqueline H. Beckley.

15

Chapter 15

Found in Translation: The Adventure of Conducting Multicultural Consumer Research

Vanessa Zuccoli and Paulina Morquecho-Campos

Key learnings

- ✓ Cross-cultural research
- ✓ Logistic preparation for multinational studies
- ✓ Testing product with multinational consumers in one site

15.1 Setting the scene: plan ahead

Multinational corporations frequently expect products to work across several markets. Yet, every country is unique. When engaging in consumer research in a country other than yours, and especially in the case of having to field in several different countries, it is crucial to identify and understand your target consumers and their cultural background to conduct a successful research. Research methods, questionnaires, and scaling techniques might all need to be significantly changed and adapted.

First and foremost, it is crucial to choose a local fielding partner with a deep knowledge of all aspects of the local culture and who has an extensive experience in conducting research locally. The objective and context of the research must be carefully explained and shared with the partner in order for them to provide suggestions on how to customize all steps of the research. Invest some time to screen different options when it comes to fielding partners and consult colleagues or acquaintances in the industry to select the most suitable one. The quality of the service provided, especially in emerging markets, might vary very significantly.

Set aside enough time beforehand for this exercise and make sure you have some flexibility or have built buffer time in case of logistical problems. It is your job to explain to all stakeholders, both internal and external, the situation and gain alignment for a suitable timeline. It is not realistic to expect to keep a

Product Innovation Toolbox: A Field Guide to Consumer Understanding and Research, Second Edition. Edited by Kannapon Lopetcharat, Dulce Paredes and Jacqueline H. Beckley.

similar timeframe for a multicountry research to be simply translated from English to German, as an example, and when it needs to be adapted (not just translated) for a Middle Eastern or African market.

The importance of considering cultural differences is critical when it comes to international market and consumer research. Differences can be either within or across culture and both aspects must be carefully considered upfront.

Tylor (1871) defined culture as a "complex whole which includes knowledge, belief, art, morals, law, customs, and any other capabilities and habits acquired by man as a member of society." Models have been elaborated to try to estimate which cultures are similar and which are dissimilar. Hofstede's and Schwartz's theories were useful to have a first basic understanding of which cultures are closer and which ones are different in terms of several parameters such as power distance, uncertainty avoidance, individualism, and masculinity (for further review see (Hofstede, 2011; Schwartz, 2006)). Having a simple look at existing data based on these parameters will make it evident that Japan and France are very different culturally, as an example. Whereas in other cases, another cases another country can work as a proxy thanks to cultural similarities(this will be addressed in section 15.2.3). Hence, we most likely have to give special attention on how to approach research questions and how to design and field in these countries if we have to work in both for the same project.

Response bias is when participants tend to systematically respond to questions in a particular manner (Baumgartner and Steenkamp, 2001). Response style differs widely across different geographies. Several researches have been conducted to understand the response style among different countries and some response styles have been identified (Baumgartner and Steenkamp, 2001; Harzing, 2006; van Herk et al., 2004; Yao et al., 2003). People from countries such as the United States, some European or Latin American countries tend to use the extreme part of the scale (extreme response style), while people from countries such as Japan or China prefer to stay close to the middle part (middle response style) (Harzing, 2006). To avoid cultural differences in response style, some researchers recommended the use of other methodologies instead of scales such as ranking, best-worst scaling, or an extra procedure known as data equivalence (or comparability for further information see Ares, 2018; Buil et al., 2012; Harzing, 2006; Sin et al., 1999).

The models mentioned above provide early watch-outs and consequently, some heads-ups to the researcher. However, a deeper insider gaze is needed to avoid major missteps. For example, one aspect that needs careful consideration when it comes to research in the Middle East and Africa is gender. Several years ago, the authors were in the process of conducting research on chewing gums in multiple countries. The research included visual stimuli for the respondents to make the size and shape of the chewing gum apparent. What we considered to be a perfectly harmless way of presenting visual stimuli to respondents in Europe or United States, a female hand holding a piece of chewing gum (Figure 15.1), is considered inappropriate when fielding in countries such as Saudi Arabia, hence, the research visual was customized by changing the visual of the same piece of gum lying next to a ruler instead (Figure 15.2).

Quota sampling balancing genders is standard in any European country. However, it would be meaningless for any research questions related to shopping behaviors for food staple or cooking ingredients in most African and Latin

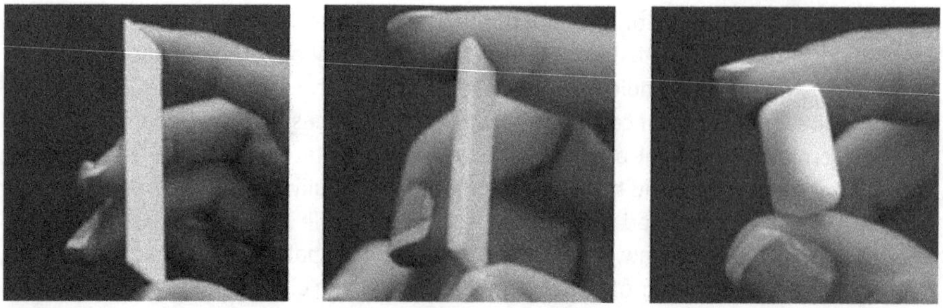

Figure 15.1 Visuals for chewing gum study for Europe and United States.

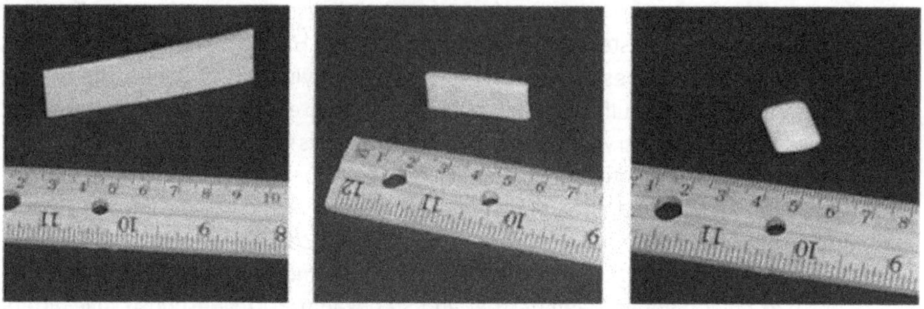

Figure 15.2 Visuals for chewing gum study for Saudi Arabia.

American countries where women are usually the sole shoppers and uniquely in charge of cooking, hence, sample selection will be skewed in that direction.

Looking at intra-cultural diversity, one factor to always keep in mind is that many countries consist of several cultural groups (or subcultures). These subcultures may be people from different ethnic groups within the same country or migrants who developed a new subculture (Sobal, 1998). For instance, densely populated countries such as India and China have different languages, religions, traditional foods, and customs within the same country.

Generally speaking, quota sampling is suggested when running cross-cultural research. In working out the quotas, the sociodemographic characteristics should be carefully considered to not only be relevant to the research question but also suitable to each specific country. Some countries have a larger distribution in their socioeconomic status which might influence their food accessibility, selection, and consumption. It is important to stabilize your target group to recruit your consumers, and first considering obviously your research objective.

A good example of the need to consider cultural differences in quota sampling comes from a research done in Nigeria. Nigeria is an incredibly complex country ethnically and culturally. The country has more than 500 languages currently spoken, and a thousand dialects and ethnic groups. Fortunately you can conduct your research in English which the vast majority of population understands. The largest ethnic groups are the Hausa, the Fulani, the Igbo, the Yoruba, the Tiv people, and the Efik-Ibibio. Each one has very specific culinary habits, linked to radically different traditions, habits, and religions. But how to approach research on such a complex environment?. Evidently, we need to make sure we are first of all aware of these circumstances. Once again, plan ahead and budget time to discuss with local partners and for desk research. In the sampling phase, we made sure our sample was as much as possible balanced for the different largest ethnic groups. Second, as our study included central location tests (CLTs) as well as consumer workshops, we had to make sure that the different groups were comfortable to interact with each other. If any problem was present in between any of those test scenarios, we ensured that groupings were homogeneous.

15

It is also important to consider the age and gender distribution in all the study sites and identify historical taste preferences and usage in the specific categories of interest For example, children in Latin America are looking for an explosion of colors and flavors in their candies. In particular, in Mexico, the candies can be sweet, bittersweet, and/or sour, but mainly with different levels of spiciness, while in other countries, candies are mostly sweet.

Lastly but not least important, you should identify the difference in uses, aims and ways of interacting with one product, which might vary in several countries. If you want to test the same product in different study sites across the globe, make sure that product is important and familiar for all the population groups. Otherwise, provide enough information to the participants to understand your goal. A good example is when the authors were fielding a global tea research in several extremely different markets, from Russia to Latin America, through Japan and the United States. We were testing different varieties of tea from different countries. In Mexico, we decided also to include mate tea, which is a traditional South American drink. After some market research, we believed it could perfectly work as an innovative idea in the emerging tea market in Mexico. Mexican tea consumers did not like this profile, most of them associated the product with a fresh/menthol note, which was related to dental products. After some market research, we believe it could perfectly work as an innovative idea in the emerging tea market in Mexico. Mexican tea consumers did not like this profile, most of them associated the product with a fresh /menthol note, which are related to dental products. We learned from this experience that it was necessary to do a first stage of research to test our hypothesis before going to product test. Testing the unfamiliar product with the concept of an iconic product with special benefits could have yielded different results.

15.2 Infrastructure, logistics and company: everything you take for granted, DON'T!

As we might have understood already, conducting international research is much more complex than running domestic research and the type of research method itself might have to be adapted to each geography in scope.

15.2.1 Data Collection Mode

In developing countries, mail surveys are not ideal due to widespread illiteracy rate and the near impossibility of reaching rural areas. Telephone interviews are also very hard to accomplish, not only due to low penetration of phones (landlines) but also for cultural habits which might prevent people, especially women, from talking to strangers.

In-home personal interviews with a researcher of a comparable cultural and socioeconomic background are then most likely the best medium to conduct questionnaires in developing countries. Mall intercept can work in some Middle Eastern countries such as the United Arab Emirates where modern retailing is very well developed, but also in this case the research must carefully consider the local social norms for approaching women. This latter technique would be less suitable to countries in Africa where traditional retailing prevails. Paper diaries and one-on-one interviews are effective and relatively easy methods to gather data in most developing countries.

Running an online research, simply translated to local languages, could result in the difficulty of reaching sample quota, and second, lead to unreliable results if they are not representative of the different social and economic realities of the country. The rise of mobile research has evolved and guidelines for the best practice, data protection, sample quality, and other guidelines have been developed (ESOMAR/GRBN, 2017). However, accessibility and cost of Internet access are some limiting factors in many countries. For example, the cost of 1 GB of Internet is higher than US$20 in Chad, Benin, and Malawi, while in India, Israel, and Kyrgyzstan, the cost is lower than US$0.25 (cable.co.uk, 2020). Use of technology might not always be the convenient solution. Even though you can provide the site with tablets or laptops, your segment or population might not be used to it or might not feel comfortable using it. Moreover, consider that the Internet connection of every country may differ, producing some technical problems or restraining the type of questionnaire to be performed. Finally, some groups of people such as older consumers might feel more comfortable answering a questionnaire on paper or even an interview than on a tablet.

The use of mobile phones, both feature phones and increasingly smartphones, is growing at an incredible pace in many developing countries who have bypassed the use of landlines. Mobile phone research is already the most suitable medium especially among the the young and tech-savvy population. Another interesting opportunity to reach these groups in social media.

15.2.2 The logistics

While designing a CLT, it is important to start the research of state-of-the-art facilities as soon as possible and, if feasible, visiting more than one venue in person. It is fundamental to be able to trust a very experienced local fielding partner but sometimes reality differs from the pictures in the brochures. Skilled local personnel - agencies leads, moderators, staff to prepare, and to distribute samples, etc., - might also be very hard to find on short notice. We have mentioned this before, but we cannot emphasize this enough: **it is indispensable to find and partner with a very experienced and trusted local field agency to help prepare all that is needed on the logistics side.**

Structural and environment problems with transportation are also extremely common in developing countries. Buses to reach the facilities and allowing for an ample margin of delay in starting tests is indispensable while running CLTs in Lagos, Nigeria, where transportation challenges can be daunting including massive traffic jams and inability for some of the participants to access public transportation. For this study, we learned that housewives from the lower economic strata could not afford to reach the venue independently so transportation was arranged beforehand and several test locations within the city were booked, to minimize the commute.

15.2.3 Legal and political environment

Knowledge and understanding of the legal and political environment of each country is fundamentally important. Understanding if not only any barrier is imposed to competition as an example but also any framework in relation to product quality and packaging. A good example is customs regulations in Ecuador that are quite strict for the imports of (food) products to the country. For instance, they have a tax for every product with a logo.

The choice of the country or countries in which to run a research depends obviously on strategic reasons such as the business relevance of that area for the company, institution, or agency designing it. There are some concrete factors to be considered at the very early stages of the research conception. The true relevance of testing the product in that country needs to be considered beforehand. When we ran chewing gum results mentioned in 15.1, Latin American countries (Mexico and Columbia) were included as a part of company strategy. However, while checking the results, we realized that chewing gum consumers in those countries were not willing to explore new ideas in their well-known products.

It is sometimes possible to identify countries in which it is easier or cheaper to do field research that can act as a proxy for other areas. Once, the authors were interested in understanding ready-to-drink coffee consumption in Europe. While running research in France, Germany, and Italy would have been ideal, budget considerations prevented us from fielding in three countries. A preliminary study of existing products on the market and a review of secondary research showed us that we could run in Switzerland instead. There was convincingly enough evidence that taste preference and consumption patterns were similar in this country to the other three countries. A careful selection of demographics and respondent selection, spoken languages, and different cultures in a geo-diverse area of the country ensured reliable results. Questionnaires were translated in several local languages and brands or shops names mentioned in the research were adapted to locally available ones.

15.2.4 Preventing “lost in translation”

We express ourselves better in our mother tongue, where we know the terms and expressions to communicate in our daily life. The same applies when we are answering a survey or giving our opinion about a product. Therefore, the questionnaire to be used in the multicultural study should be translated in each of the languages of the countries of interest. Although this task could seem like an easy one, it certainly implies crucial challenges such as preserving the same meaning and lack of terms in other languages (Buil et al., 2012). There are different translation techniques, but the most common is the back-translation (Ares, 2018; Buil et al., 2012). This procedure consists of the translation from the original language to the target one made by a bilingual person, followed by the back translation from the target language to the original one performed by another bilingual person (Brislin, 1970). The goal of this procedure is to have equivalent questionnaires for an adequate comparison of the results.

This also brings us to an essential part of the preparation: piloting. Pilot studies are crucial to ensure the clarity of the questionnaire/survey (instructions, questions, etc.), and familiarity of the staff with the procedure and logistics. Detection and adjustment of any potential area of improvement can be performed before the real data collection. It can also verify if the procedure is adequate for all the location centers to have an equivalent data collection.

Seasons are an important factor which determines the food choice based on availability of certain food products such as fruits and vegetables, taste and cost (Connors et al., 2001; Sobal et al., 2006). Some product usage is very seasonal, as we quickly realized while studying ready-to-drink tea. This product category usage is almost entirely skewed in the summer season for many countries. We noticed that running research on this category in the summer was indeed getting us higher liking scores for most concepts, especially the ones with summer flavors. People were more likely to express positive judgment of teas as they were thirsty for them while it was hot outside! Likewise, running stories on refreshment might yield very different results in winter compared to summer.

15.3 Multicultural background in just one site

The challenges mentioned above have to be considered when you are performing a global study. However, we have also identified other challenges when a study was conducted in one site with international consumers.

15.3.1 Food attributes

When you test products that are not only well known in the country but is popular among the international consumers, it is important to pilot asking the familiarity of these foods with your study population. If you want to introduce a product that is not familiar with your test population, make sure to give them more background information about the product. Furthermore, not all the population groups will like or be familiar with certain attributes or products. For example, Dutch people are not very familiar with the texture of a gelatin (such as Jell-O).

15.3.2 Associations and expectations

Everyone might have different experiences with several food products, and it might depend on their background, encounter with different (international) food products, liking, etc. A good example of this is the different perception and association of odors. Once, we were looking for odor that were associated with savory among international people. During the pilot stage, one of the available odors was a corn odor. The group of people that were piloting the odors were Europeans and Latin Americans. The Europeans agreed that the odor evoked savory memories, while the Latin Americans associated with sweet corn and categorized it as a sweet product. This odor was not used for the final experiment to avoid any further association bias.

15.3.3 Use of English

Although participants are screened to be fluent in a common language such as English, it is good practice to use common words or define any word or attribute that might not be easy to understand. An example is while asking for "How *savory* is this food product?" on an online questionnaire. Although the "*savory*" term is a common term in the food industry, we have several questions regarding this term. Some consumers translated it to their native language and the translation was "*tasty*," confusing the consumers and getting biased answers.

After receiving the feedback from some consumers, we decided to add a short explanation (savory/salty; savory/not sweet) to give a clearer idea.

15.3.4 Differences among the same language

Different words used to define the same object. Even in countries sharing the same language, it is very common to use different words to refer to the same object. Therefore, it is crucial to identify these word differences. For example, two popular fruits may have several words in Spanish-speaking countries. Let us start with the word "banana" which it is "plátano" in Mexico, Chile, and Spain; "banana" in Peru, Argentina, Ecuador, Uruguay, also used in Spain but for a different type of banana; "banano" in Colombia and "Cambur" in Venezuela. The second example is the word "strawberry" which is "fresa" in Mexico, Spain, Peru, Colombia, Uruguay, etc., but "frutilla" in Ecuador, Chile, and Argentina. Therefore, piloting your questionnaire should be done before starting the data collection. Share it with all the study sites for feedback and/or final approval.

15.4 Conclusions: found in translation

Conducting research across different countries and target demographics is a challenging but exciting adventure, and we want to summarize some of the key points we think are important to consider not to have bad surprises - but only good ones:

- Budget enough time to carefully prepare your research, negotiating upfront with all stakeholders.
- Consider possible proxy countries in case this is possible to save time and money.
- Especially when researching in developing markets, evaluate different partners and choose a local fielding agency with a long experience and a deep knowledge of the local culture. Check facilities in person if possible, in case of product research.
- Define the research question carefully and adapt them if needed. Be particularly mindful of product usage in different countries.
- Select the demographics of interest based on the research question - use quota sampling. Consider gender and age specificities.
- Check with your local partners any issue in terms of interactions between different genders, ethnic groups, or socioeconomic groups in case of qualitative research.
- Consider differences in response styles in the different countries and consider the possibility of changing scales from one country to the other.
- Prepare your questionnaire's translation carefully, adapting not only words but also any brand mentioned or any question on product usage. Do not forget to back-translate it if time allows it.
- When research is conducted in English but respondents are not native speakers, check competency, and word usage upfront.
- When research is conducted in one language across different geographies, check for differences in word usage.
- Always pilot to ensure or detect if your questionnaire (instructions, questions, etc.), is understandable and makes sense, and if the procedure is adequate for all the location centers.

15

Discussion questions

- What do you need to consider when conducting cross-cultural or multinational consumer research?
- Thinking back to the entire chapter, what is your key take-away regarding good implementation of a cross-culture study?
- In addition to language used, what other factors that you need to pay attention to when executing a multinational or cross-culture study?

References

Ares, G. (2018) "Methodological issues in cross-cultural sensory and consumer research". *Food Quality and Preference*, 64(March), 253-263. https://doi.org/10.1016/j.foodqual.2016.10.007

Baumgartner, H. and Steenkamp, J. (2001) "Response styles in marketing research: a cross-national investigation". *Journal of Marketing Research*, 38(2), 143-156.

Brislin, R. (1970) "Back translation for cross-cultural research". *Journal of Cross Cultural Psychology*, 1(3), 185-216.

Buil, I., De Chernatony, L. and Martínez, E. (2012) "Methodological issues in cross-cultural research: an overview and recommendations". *Journal of Targeting, Measurement and Analysis for Marketing*, 20(3-4), 223-234. https://doi.org/10.1057/jt.2012.18

Cable.co.uk (2020) *Worldwide Mobile Data Pricing 2020*. https://www.cable.co.uk/mobiles/worldwide-data-pricing/ (accessed 8 March 2022).

Camacho, A. (2012) "Best practices in global testing and multi-cultural consumer research". In J. Beckley, D. Paredes and K. Lopetcharats (eds.) *Product Innovation Toolbox: A Field Guide to Consumer Understanding and Research* (1st edition). Ames, Iowa: Wiley-Blackwell. pp. 365-371. https://doi.org/10.1002/9781118229248.ch11c

Connors, M., Bisogni, C.A., Sobal, J. and Devine, C.M. (2001) "Managing values in personal food systems". *Appetite*, 36(3), 189-200. https://doi.org/10.1006/appe.2001.0400

ESOMAR/GRBN (2017) *Guideline on Mobile Research*. https://www.esomar.org/what-we-do/code-guidelines/esomargrbn-guideline-on-mobile-research (accessed 8 March 2022).

Harzing, A.W. (2006) "Response styles in cross-national survey research: a 26-country study". *International Journal of Cross Cultural Management*, 6(2), 243-266. https://doi.org/10.1177/1470595806066332

Hofstede, G. (2011) "Dimensionalizing cultures: the hofstede model in context". *Online Readings in Psychology and Culture*, 2(1). https://doi.org/10.9707/2307-0919.1014

Lee, H. and Lopetcharat, K. (2017) "Effect of culture on sensory and consumer research: Asian perspectives". *Current Opinion in Food Science*, 15(June), 22-29. https://doi.org/10.1016/j.cofs.2017.04.003

Schwartz, S.H. (2006) "A theory of cultural value orientations: explication and applications". *Comparative Sociology*, 5(2-3), 136-182.

Sin, L.Y.M., Cheung, G.W.H. and Lee, R. (1999) "Methodology in cross-cultural consumer research". *Journal of International Consumer Marketing*, 11(4), 75-96. https://doi.org/10.1300/j046v11n04_05

Sobal, J. (1998) "Cultural comparison research designs in food, eating, and nutrition". *Food Quality and Preference*, 9(6), 385-392. https://doi.org/10.1016/s0950-3293(98)00029-9

Sobal, J., Bisogni, C.A., Devine, C.M. and Jastran, M. (2006) "A conceptual model of the food choice process over the life course". In R. Shepherd and M. Raats (eds.) *The Psychology of Food Choice*. Cambridge, MA: CABI Publishing. pp. 1-18. https://doi.org/10.1079/9780851990323.0001

Tylor, E.B. (1871) J. Murray (ed.) *Primitive Culture: Researches Into the Development of Mythology, Philosophy, Religion, Art, and Custom*. London, UK: Bradbery, Evans and Co. https://books.google.nl/books?id=AucLAAAAIAAJ&pg=PA1&hl=es&source=gbs_toc_r&cad=4#v=onepage&q&f=false (accessed 8 March 2022).

van Herk, H., Poortinga, Y.H. and Verhallen, T.M.M. (2004) "Response styles in rating scales: evidence of method bias in data from six EU countries". *Journal of Cross-Cultural Psychology*, 35(3), 346-360. https://doi.org/10.1177/0022022104264126

Yao, E., Lim, J., Tamaki, K., Ishii, R., Kim, K.O. and O'Mahony, M. (2003) "Structured and unstructured 9-point hedonic scales: a cross cultural study with American, Japanese and Korean consumers". *Journal of Sensory Studies*, 18(2), 115-139. https://doi.org/10.1111/j.1745-459X.2003.tb00379.x

16

Data can reveal patterns from the past.
It takes judgment to predict how these patterns will evolve in the future

Adam Grant

What this is about

The authors summarize predictions and weak signals on future directions of sensory and consumer research using text analytics to summarize notes from industry and technical conferences and papers presented the past two years. The authors show how text analytics partnered with content expertise can be used to transform large sets of unstructured data to structured data, find connections and patterns, generate insights, and create sturdier bridges to future understanding.

Product Innovation Toolbox: A Field Guide to Consumer Understanding and Research, Second Edition. Edited by Kannapon Lopetcharat, Dulce Paredes and Jacqueline H. Beckley.

16

Chapter 16

Sturdy Bridges to Future Trends

Katherine C. S. Rhodes, Dulce Paredes, and Jacqueline H. Beckley

Key learnings

- ✓ Definition of Text Mining
- ✓ Redefining Qual-Quant Data
- ✓ Text Mining to identify emerging trends in consumer research
- ✓ Importance of human expertise in using AI/ML Tools

16.1 Introduction

Sensory and consumer research, tools and methods are constantly evolving. The year 2020 was no exception as researchers and consumers alike were forced to adapt when traditional research methods, like in person research, were temporarily halted due to pandemic restrictions. To harness insights from the already available seas of data, many technologically savvy researchers adopted rapid insights generation tools. Data mining tools, such as text analytical software, prove especially helpful in the digestion of large amounts of qualitative data for review, tracking, and gathering of new insights when partnered with content expertise.

Text mining, a subfield of data mining that uses text-based algorithms, is used by researchers to analyze large amounts of unstructured text data or even audio and video when combined with text-to-speech technology. While text-mining tools are useful for analysis, it is important to note that the accurate generation of insights goes beyond machine learning. Early text mining explorer, Hearst (1999), stated that text analysis requires ". . . a mixture of computationally-driven and user-guided analysis," which, according to Kotu and Deshpande (2018), is at "the heart of supervised models used in predictive analysis." A combination of analyst content expertise and software expertise is required to make this approach both possible and meaningful when capturing data patterns and insights.

Product Innovation Toolbox: A Field Guide to Consumer Understanding and Research, Second Edition. Edited by Kannapon Lopetcharat, Dulce Paredes and Jacqueline H. Beckley.

In a review of sensory and consumer research data, we demonstrate the dynamic and resource effective nature of text mining by highlighting topics, tools, and methods of interest in the field. In this example, 185 discussions around new tools and methods from publications and conferences in the realm of sensory and consumer science throughout 2020 and up to March 2021 were mined.

From this analysis, discussion around data, experience, machine technology, and humans surfaced to the top of several of the results. In this chapter, the context of these topics as well as the changing definition of data is explored further.

16.2 Redefining data

Data is defined as information gathered through observation for reference and analysis. Consumer Explorers (CEs) further define it as quantitative and qualitative data to differentiate those expressed in numbers vs. text, visual, or voice. Text data outputs from qualitative research such as ethnographies, blogs, and interviews are referred to as unstructured or messy data since they cannot easily be relegated to columns and rows during data collection. Unstructured data must be transformed manually or through computer-assisted labeling often called "tagging." As CEs, we often hear, "quantitative research gives you the 'what' while qualitative research gives you the 'why.'" Quantitative data provides the comfort of numbers and statistics while text rich data seems more nebulous and is usually associated with a word cloud or an interesting quote to latch on to an insight. Text mining tools shifted this paradigm by allowing the transformation of unstructured data to structured data, enabling further statistical analysis. In tandem with quantitative data from topic categories, text mining bridges the quantitative and qualitative divide to reveal deeper information (e.g., more data, see Figure 16.1).

In a world with copious amounts of data at their disposal, many researchers find themselves in a "data rich, insights poor" dichotomy. The challenge for sensory and consumer researchers is to identify meaningful emotional connections within this data to humanize the insights on a level that engages shareholders for strategic activation.

To harness meaningful insights from the plethora of data available, many researchers have adopted modern data mining tools. For this chapter, we used text mining software (Provalis QDA Miner 6 and WordStat 8) to investigate the direction of sensory and consumer research tools and methods by exploring the findings from technical and market research communities in the field of sensory and consumer research. We mined 185 sources from 2020-Q1 2021 conferences, talks, and articles over the period of 2020 to Q1 2021 using TF-IDF1, topic modeling, dendrograms, proximity plots, and link analyses (although there are other methods available to explore the data as well).

Before analyzing the data, a categorization dictionary was applied to allow for selection of relevant words and grouping of similar terms. Dictionary building is a critical step in the process as it allows for selection of relevant words and grouping of similar terms for analyses. Categorization dictionaries help to group variant forms of a word or terms with similar meaning under a single category for analyses. For example, both "VR" and "virtual reality" were grouped under "Virtual Reality." This step can easily be one of the most time-consuming parts

16

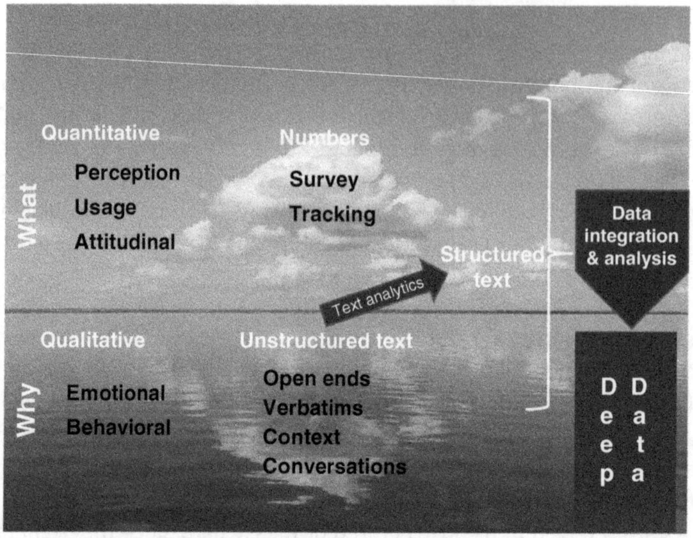

Figure 16.1 Text Analytics bridge qualitative and quantitative data analysis for deep data insights. Source: Adapted from Paredes et al. (2018).

of the process as it requires filtering through existing words and categorizing like-terms. In this study, the dictionary was built around sensory terms of interest from the term frequencies lists and was supplemented with synonyms from online resources (multiple sensory and market research websites). After applying the dictionary to the project, only words included in the dictionary appeared in the analyses. A frequency criterion was also set in some analyses to capture terms occurring more than a defined number of times or in a defined number of cases. This allows inclusion of potentially relevant terms from slipping past the text analysis. Several methods and best practices for dictionary building are already in existence and are also a good point of reference (Deng et al., 2017; Xu and Bengston, 1997).

Findings from this dive were compared to previous analyses of sensory and consumer research data from 2000 to 2019 (Speight et al., 2019) where several longitudinal trends in tools and methods were observed, including a decline in discussion around older sensory methods and increase in conversation around temporal and experiential methods.

Overall, we saw three main areas of focus in the 2020-Q1 2021 data dive:

- **Experience:** The rise of X (experience), user experience (UX) and consumer experience (CX)
- **Machine Technology:** Augmented and Artificial Intelligence (AI), Machine Learning (ML)
- **Humans:** Consumer focused, Purposeful branding, and connecting with consumers on a personal and emotional level.

16.2.1 Experience

Both the terms "**experience**" and "**environment**" appear in the top 50 words from the 2020-Q1 2021 data when investigating words using TF-IDF. **Experience** has always been an important driver for consumers when it comes to product liking, but it has **especially grown during the pandemic** as consumers crave new experiences in and outside of the home. Furthermore, it is important to researchers to deliver and capture accurate consumer experiences with products in an increasingly virtual world. When diving into a link analysis of "experience" and "environment" (Figure 16.2), we discover that the connected words include "**immersive**," "**virtual reality**," "**storytelling**," "**familiar**," "**empathy**," "**human**," and "**emotions**." These findings show us how connected technology and emotions are to human experience in the world of sensory and consumer research.

In recent years, the acronym "X" has become shorthand for the term "experience" seen most prevalently in "CX", and "UX". While "CX" and "UX" do not appear in the top 50 words from 2020-Q1 2021 data, they do have numerous mentions and have grown in recent years. When we look at connections to "UX" and "CX" using link analysis (Figure 16.3), we see links to "data collection," "artificial intelligence," and "machine learning" further tying experience with tech tools in research.

When looking at the topics that were extracted from the use of topic modeling on the 2020-Q1 2021 data, we see the topic "Immersive Technology" which further confirms the importance of technology in studying consumer perception, behavior, and experience in various contexts. Keywords that topic modeling for "Immersive Technology" include the following: *immersive, virtual reality, lab, real, environment, real pub, immersive techniques, real life, immersive room, laboratory environment, real environment, immersive technologies, immersive virtual, real time, and virtual environment*. Some of these words can also be seen in the connections in Figures 16.2 and 16.3.

In a dendrogram (agglomeration order) (Figure 16.4) of dictionary words from 2020-Q1 2021, we see "experience," "UX," and "CX" have similar word linkages to the findings above. This time, however, we pick up on the relationship between the word "feedback" and "experience." The word "immersive" was strongly correlated to "experience" in our previous analysis so we explored its connection in the dendrogram and saw links to "augmented reality," "virtual," and "CLTs (central location tests)." Since many CLTs were halted during the pandemic, the relationship with virtual methods and CLTs suggests the consideration of virtual alternatives to the in-person methods. When looking at "CX" and "UX," we once again see relationships to machine technology and pick up on the word "clusters" suggesting the application of CX and UX with consumer segmentation.

16.2.2 Machine technology

Several contemporary sensory and consumer research tools involve the use of machine learning and artificial and/or augmented intelligence. Commercial tools using machine learning have been developed to collect and identify top consumer responses and iterate results in real time; identify quotes of interest in curated datasets; and mine and quantify social media and other consumer and market data. So it comes as no surprise that among the top 50 words from our 2020-Q1 2021 data, "technology," "social media," "digital/online," and "automation" all appear.

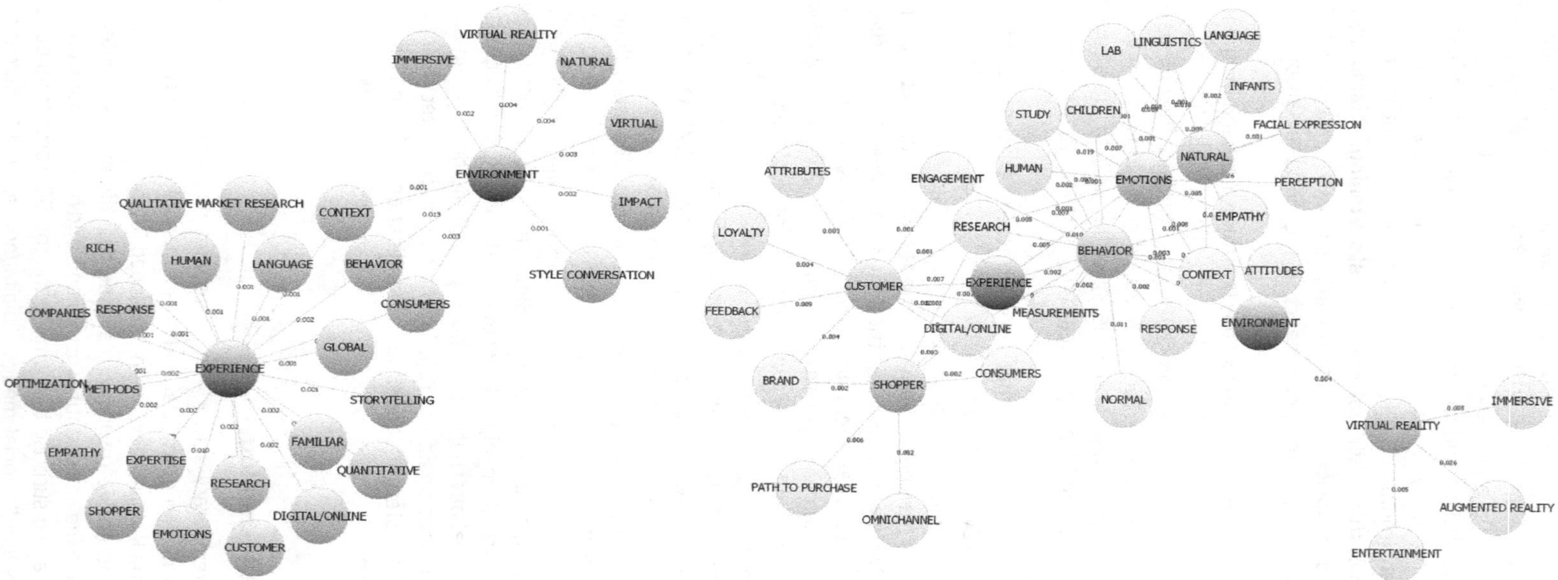

Figure 16.2 Primary (left) and secondary (right) link analysis of dictionary words related to "experience" and "environment" from data mining of 2020-Q1 2021 sensory and consumer science data.

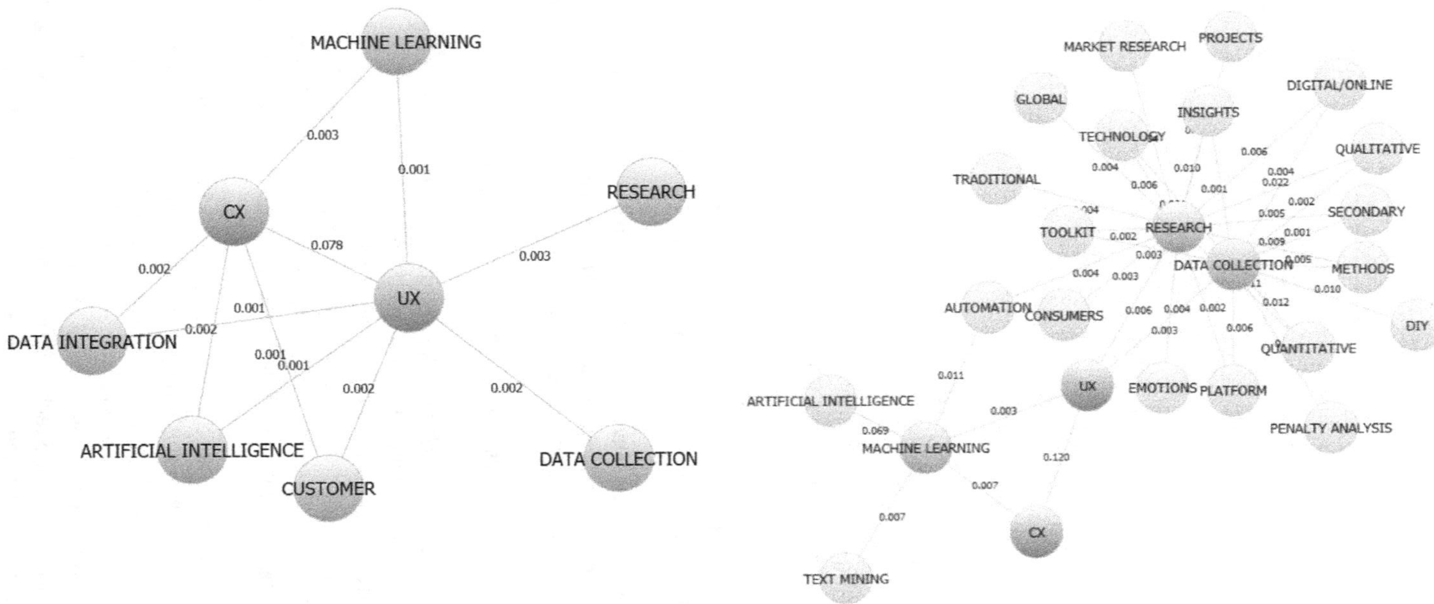

Figure 16.3 Primary (left) and secondary (right) link analysis of dictionary words related to "UX" and "CX" from data mining of 2020-Q1 2021 sensory and consumer science data.

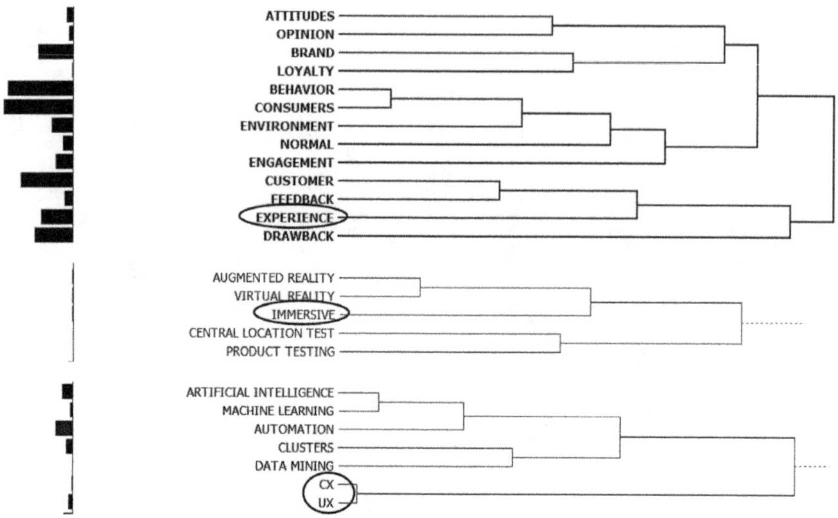

Figure 16.4 Dendrogram (agglomeration order) excerpts from data mining of 2020-Q1 2021 sensory and consumer science data.

When investigating the data through topic modeling, several themes that involve machine technology emerged, namely:

- ***Agile research*** (keywords: agile market research & insights, DIY, quick)
- ***Machine learning, Text mining, Automation*** (keywords: machine learning, text mining, verbatim analysis, artificial intelligence, automation, pathfinder)
- ***Comment analysis*** (keywords: comment analysis, mixed models)
- ***Consumer neuroscience data integration*** (keywords: eye tracking, neuroscience, data integration, big data, storytelling, consumer neuroscience, consumer experience)
- ***Chatbots*** (keywords: borderless, chatbots, messaging, contemporary, popular)
- ***Digital listening platforms*** (keywords: platform, digital/online, real time, community, intelligence, listening platform)
- ***Face reader*** (keywords: face reader, oral processing)
- ***Implicit and Explicit measurements*** (keywords: implicit, explicit, measurements, implicit methods, explicit and implicit, implicit and explicit, emotion measurements, implicit measures, implicit association test, emotion measures)

When we examine the connections to "artificial intelligence" and "machine learning" using link analysis (Figure 16.5), we see shared connections around "text mining," "automation," "data mining," and "CX." These are the areas with the strongest machine tech connections when thinking about tools and agile research. Some other connections of interest are "dirty" and "DIY" showing the messy nature of machine tech in insights generation. It is important to remember that technology alone is not what makes the insights great. The machine must be partnered with well thought-out datasets, strong categorization dictionaries, and subject matter experts for results interpretation. The connection

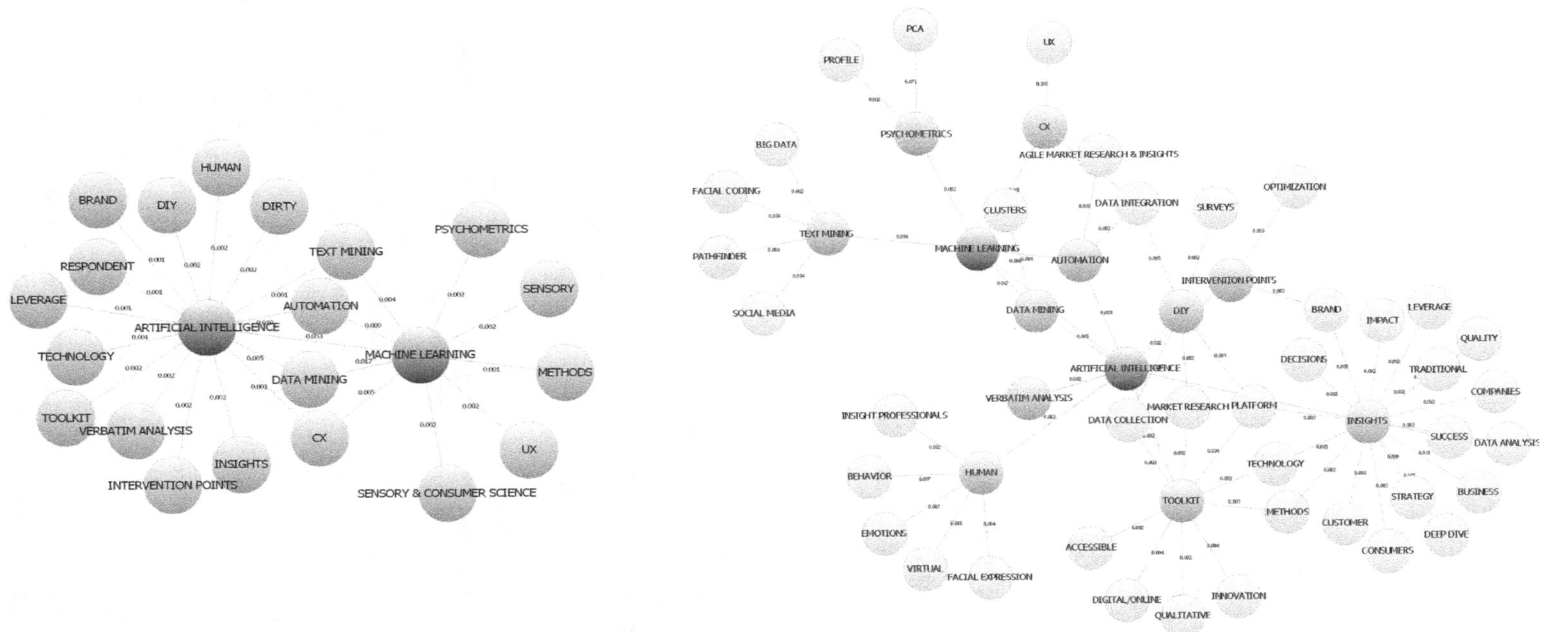

Figure 16.5 Primary (left) and secondary (right) link analysis of dictionary words related to "Artificial Intelligence" and "Machine Learning" from data mining of 2020-Q1 2021 sensory and consumer science data.

to "humans" is especially important in driving home the point that machine technology still needs an operator. When looking at the secondary connections to "humans," we see "insights professionals" as a top connection.

16.2.3 Human

"Human" is a word of importance in "experience," "machine technology," and overall, when looking at analyses of the 2020-Q1 2021 data. Nowadays, it is all too easy for researchers to get lost in the sea of data and forget to find meaningful human connections which ultimately lead to stakeholder support and market success. More than ever, there is a push to make data more relatable to "humans" and not just a number. The word "emotions" (which ranks #1 in the top 50 list) is especially important in relating data to humans. The phrase "facial expressions" also appears in the top 50 and is related to "human." It is a topic of much debate as many researchers attempt to understand human emotions through expression. Data bias from historical collections, algorithms, and cultural backgrounds influence facial and emotional expressions. When looking at other links to "human," words that stand out are "insights," "intelligence," "experience," "context," and "measurements" (Figure 16.6).

In the topic-modeling portion of this analysis, several topics revolving around "human" and understanding emotions surfaced, including the following:

- ***Consumer empathy*** (keywords: loyalty, connects program, customer, super fans, empathy, customer insight, cognitive empathy, emotional empathy)
- ***Emotion lexicon*** (keywords: lexicon, clusters, projective mapping, emotions, language, linguistics, positive emotions, negative emotions, emotion lexicon, emotions elicited)
- ***Emoji scales*** (keywords: emoji, language, CATA (Check All That Apply) terms and questions, RATA (Rate All That Apply) questions, emoji questionnaires, facial emoji, emoji meanings, frequency of emoji, grimacing face)
- ***Consumer behavior*** (keywords: behavior, consumers, attitudes, normal, environment, consumer behavior, behavior change, moral behavior, sustainable consumer behavior, sustainable behavior, sustainable consumer)

Looking at keyword-in-context around "human" reveals some interesting quotes: "the power of human" (Hanover and McCune, 2020), "virtual humans are vivid" (Barrett et al., 2019), "human talent" (Information Tools Ltd., 2020), "advances in human computer interaction" (Birchfield et al., 2008), and "human coders" (Barrett et al., 2019). Here, we see a lot of juxtaposition with machine and human as well as an emphasis on the importance of human. Often, researchers forget that tools like text mining software still require a human interface to make sense of the data. Humans are very much an important part of the technology advancement process and the future of research is a human-machine hybrid.

In a dendrogram excerpt from the 2020-2021 data (Figure 16.7), we can also see the relationship between "human" and "virtual."

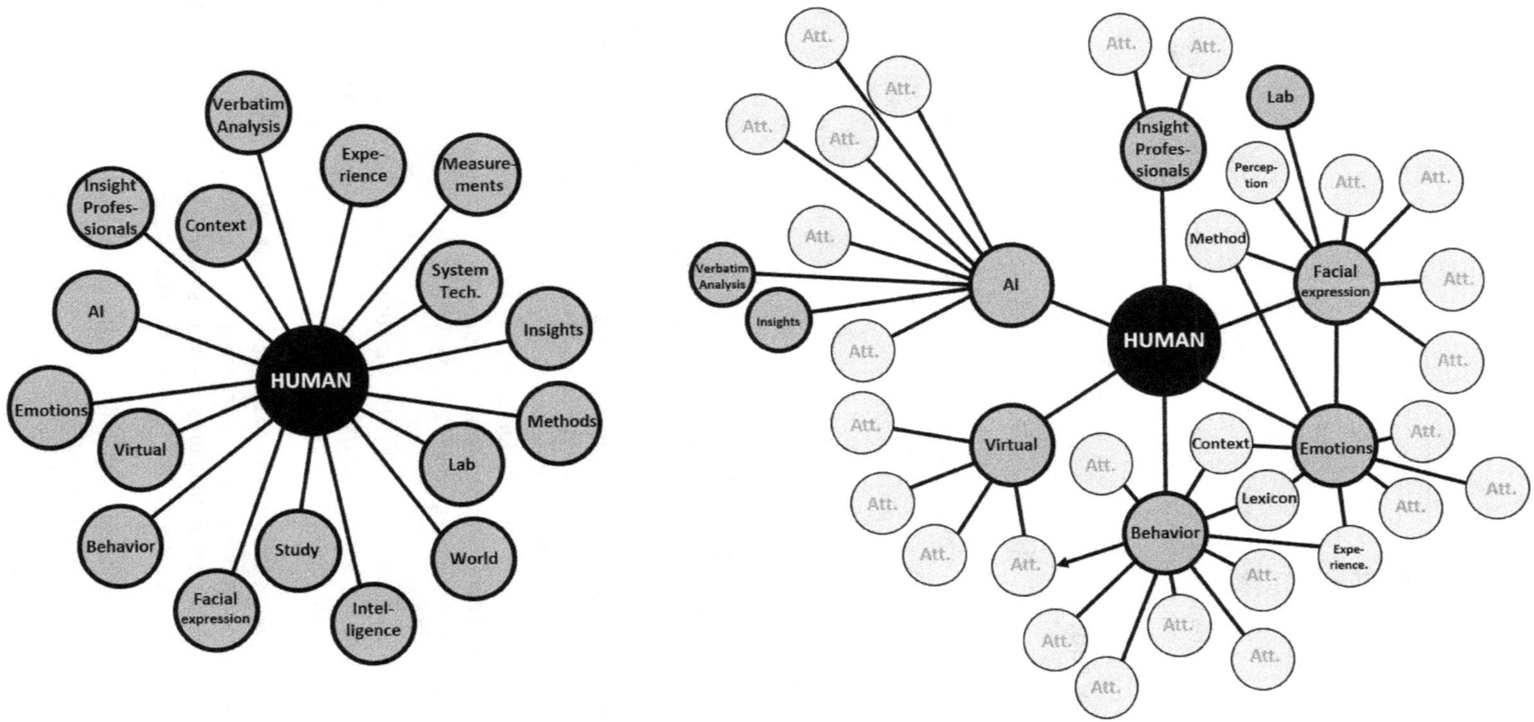

Figure 16.6 Primary (left) and secondary (right) link analysis of dictionary words related to "Human" from data mining of 2020-Q1 2021 sensory and consumer science data.

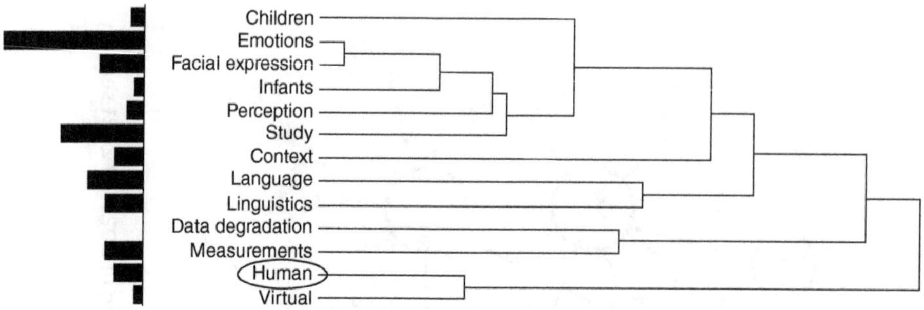

Figure 16.7 Dendrogram (agglomeration order) excerpts from data mining of 2020-2021 sensory and consumer science data.

16.3 Legacy tools

While the main topics of interest from this data dive revolve around experience, machine technology, and human, there were still several legacy topics extracted from the dive that are worth mentioning. For example, several topics revolving around traditional tools were also picked up and mostly revolve around surveys, CLTs, statistical analyses, and lab research:

- ***Qualitative and Quantitative research*** (keywords: qualitative, quantitative, methods, research, qualitative research, market research, qualitative and quantitative, qualitative methods, research methods, quantitative research, international journal of qualitative methods)
- ***Idea screening*** (keywords: idea screening, expertise, high-quality)
- ***Sensory evaluation*** (keywords: sensory & consumer science, sensory, consumer science, sensory booth, sensory and consumer, sensory properties, sensory attributes, sensory perception, sensory quality, sensory science, sensory testing, sensory characteristics, sensory evaluation, sensory perceptions, sensory and consumer research)
- ***Hedonic scales*** (keywords: hedonic, taste intensity, taste perception, hedonic scale, point hedonic scale, hedonic rating, hedonic acceptance, hedonic recognition, hedonic scores, dislike extremely)

Note: these words were also linked to "late fusion," "deep learning," and "facial expression analysis" suggesting that there is a hybrid approach of combining hedonic scales with these methods.

- ***ANOVA*** (keywords: ANOVA, paired comparison, MANOVA, discrimination tests)
- ***Linear regression*** (keywords: linear regression, logistic regression)
- ***Shelf life instrumental measures*** (keywords: shelf life, sensory vs. instrumental measures, safe, trained sensory experts)

We also found people groups of interest in the world of sensory and consumer research from the topic modeling:

- Gen Z & Millennials
- Diversity/Representation

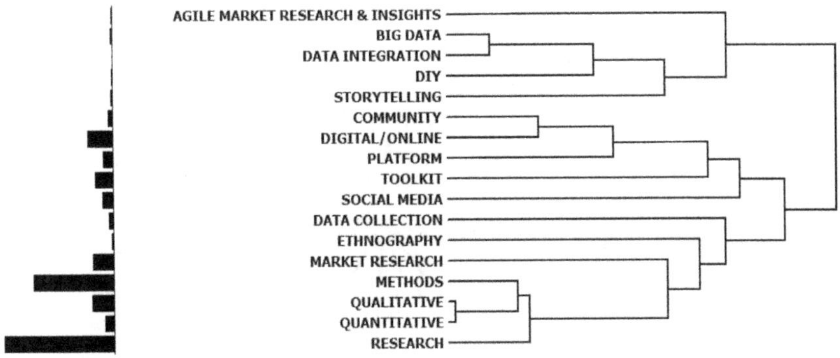

Figure 16.8 Dendrogram (agglomeration order) excerpts from data mining of 2020-2021 sensory and consumer science data.

When looking at the dendrogram from 2020 to 2021, we see that "research" is the phrase with the strongest connections (Figure 16.8). It is closely related to words seen in the topic extraction (qualitative, quantitative, data collection, platform, digital/online, emerging. . .), and this time we picked up on the word "ethnography." Ethnography has been mentioned by several researchers as their best tool of choice for deeper understanding of consumers. Since the early days of this millennium, the value of moving from a lab setting to more experiential settings has gathered strength and the fundamentals embraced by the field of ethnography has been adapted. Whether the ethnography is a hybrid ("contextual research") or practiced by those specifically trained in ethnographic practice, the idea of understanding culture, the language of people, their habits and practices, and design of behavior models has been found to be highly useful in providing understanding around the nuances surrounding humans. For those looking for a good summary, Chapter 5 in Methods in Consumer Research by Ares et al. may prove instructive. We also see more mentions of "qualitative" than "quantitative," which we hypothesize is due to the increased prevalence of tools that enable researchers to rapidly convert qual to quant data. For some, qualitative and ethnography can be considered on a continuum.

In a link analysis dive into the connections and conversations around "Sensory & Consumer Science," a proximity plot and link analysis of several words of interest was performed (Figure 16.9). The top related word to "Sensory & Consumer Science" (aside from "Sensory" and "Consumer Science") was "Consumer Insights." Second was the discussion around "Sensory vs. Instrumental Measures."

16.4 Emerging topic: democratization of data

When exploring the connections around "Consumer Insights" an important new topic we picked up on was about accessibility and democratization [of data] (Figure 16.10). This is especially a topic of interest in the more computer- and data-based consumer insights communities as consumer data and data mining technologies continue to grow. There is a push to make data more accessible so

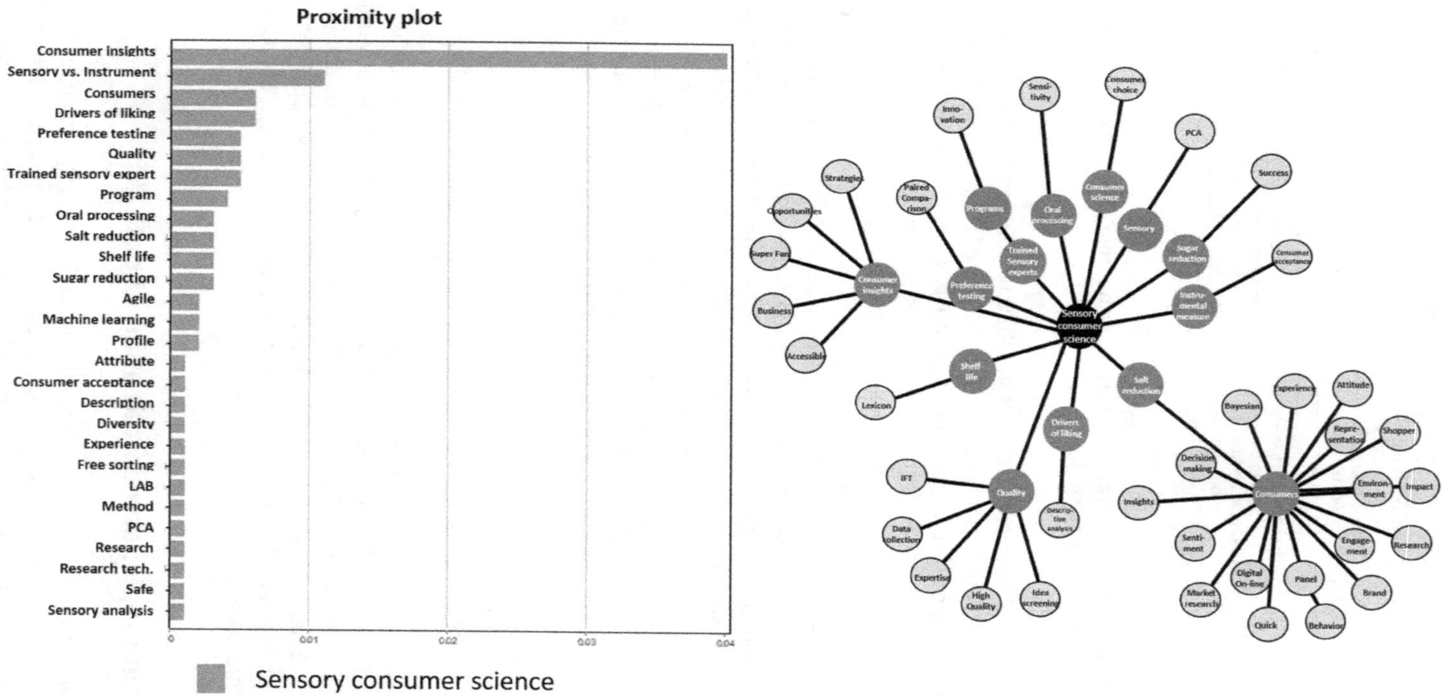

Figure 16.9 Proximity plot (left) and link analysis (right) link analysis of dictionary words related to "Sensory & Consumer Science" from data mining of 2020-Q1 2021 sensory and consumer science data.

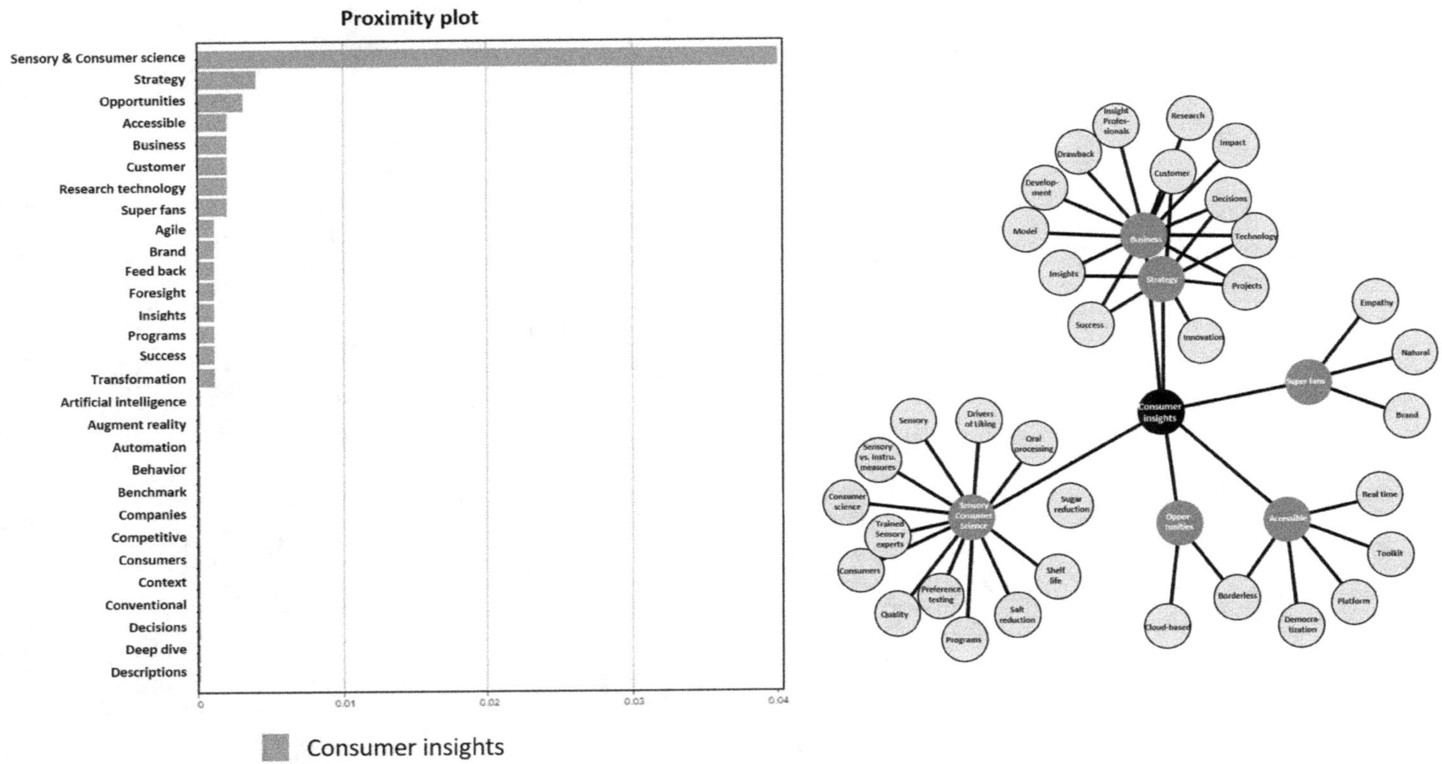

Figure 16.10 Proximity plot (left) and link analysis (right) link analysis of dictionary words related to "Consumer Insights" from data mining of 2020-Q1 2021 sensory and consumer science data.

that researchers can learn new ways to use and interpret the data from each other and internal stakeholders for rapid activation. There has been a lot of advancement in machine learning and AI programs over the past several years, but as some believe, "there is more to be done [to advance data modeling and text analytics] ahead of us than there is behind us" (N. Peladeau, personal communication, December 19, 2019). As fast as these tools have developed, there is still an amazing capacity for growth. One belief is that the democratization of data (getting it into more hands to "play with") is essential in speeding up the advancement of data mining technologies for the betterment of the whole consumer insights community.

16.5 Comparison to 2010–2019 consumer and sensory dive analysis

When comparing the findings from 2020-Q1 2021 data to findings from the 2019 data dive on sensory journals and Pangborn International Conference data between 2010 and 2019 (Speight et al., 2019), several longitudinal trends that back up the findings from the 2020-Q1 2021 data were discovered. Correspondence analysis and cross table displays from both sensory journals and Pangborn conference datasets (hereto referred as journal and conference data, respectively) were analyzed separately and showed similar trends in the area of sensory and consumer research including a decline in discussion around older methods, such as "Procrustes analysis" and "free choice profiling," and increase in conversation around newer methods such as "immersive experience," "eye tracking," "temporal methods," "virtual reality," and "emotions." In addition, topics related to desired consumer benefits and trends such as sugar reduction and salt reduction gained more prominence.

The results of the correspondence analysis were presented using a two-dimensional map (Figures 16.11 and 16.12). Similar to the interpretation of a

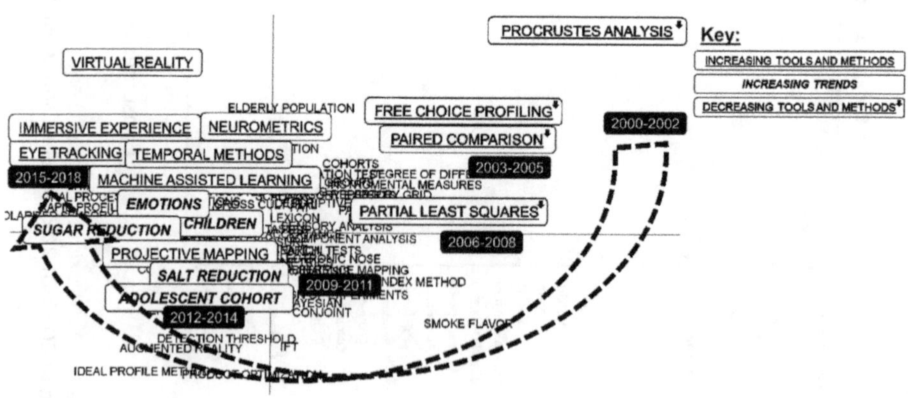

Figure 16.11 2-D Correspondence analysis plot of journal data by time periods (between 2000 and 2018) displaying correlation of sensory research terms in relationship to years.

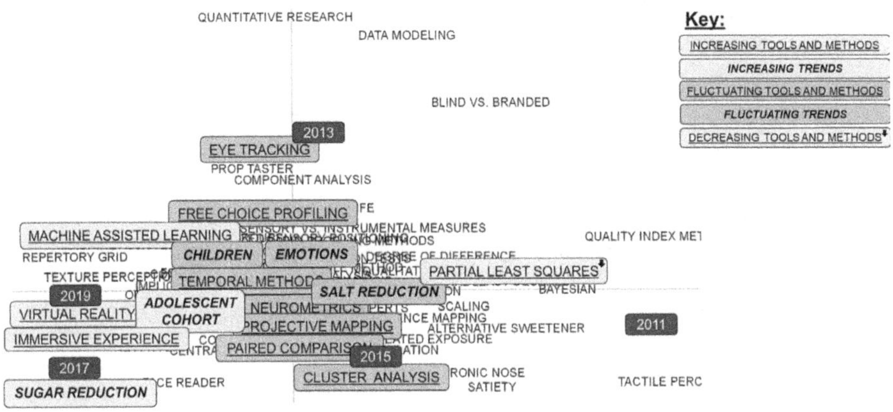

Figure 16.12 2-D Correspondence analysis plot of conference data by time periods (between 2011 and 2019) displaying correlation of sensory research terms in relationship to years.

principal component analysis, the farther a word from the center of a correspondence plot, the more discriminating it is, while the closer a word appears, the less likely it is to be distinct. The factor with the greatest impact on distribution is represented by the *x*-axis, and, in the case of the journal data, it confirms the discriminating role of the time period suggesting a change in topics across time. The year differentiation was less obvious for conference data however, due to the smaller dataset and fewer years. The results were still significant and showed the difference in term "popularity" across years.

Trends from journal and conference data were highlighted in the correspondence plots. Terms that increased over time are highlighted in green and included words revolving around cohort studies such as "children" and "adolescent cohort" as well as newer research method terms such as "emotions," "eye tracking," "immersive experience," "machine assisted learning," "neurometrics," "projective mapping," "temporal methods," and "virtual reality." Lastly, journal terms related to consumer health trends that showed an increase over time included "salt reduction" and "sugar reduction." The journal data also showed a decrease of occurrence over time for method related terms including "free choice profiling," "paired comparison," "partial least squares," and "Procrustes analysis." The conference data showed similar trends to the journal data but revealed more fluctuations in occurrences for "children," "emotions," "eye tracking," "free choice profiling," "neurometrics," "projective mapping," "salt reduction," and "temporal methods." This is likely due to the more recent and smaller time period for conference data. Furthermore, "Procrustes Analysis" did not appear in any of the conference titles analyzed, which corresponds to the trend found in the journal data.

Terms highlighted in the correspondence analysis were selected and graphically displayed via a bubble crosstab of word occurrence by *category percent* (or percentage of poster titles or articles that contain the word) across years (Figure 16.13). This confirmed the upwards, fluctuating, and downward trends of the selected terms.

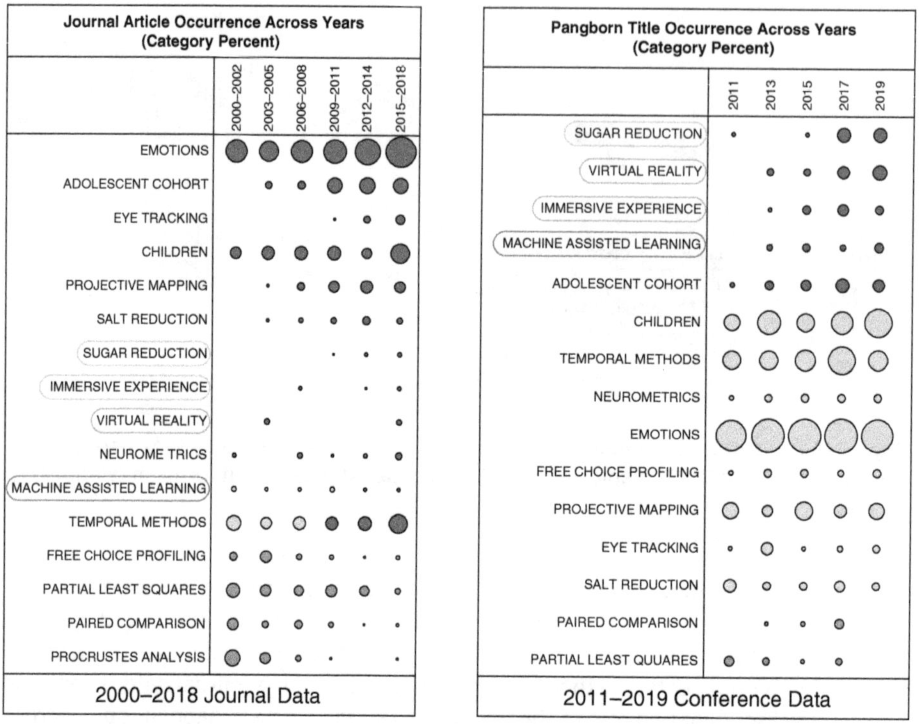

Figure 16.13 Crosstab bubble plot based on case occurrence and category percent of conference poster data and journal data displaying correlation of sensory research terms in relationship to years (2000-2019).

16.6 Conclusion

We were able to see several topics around experience, machine technology, and humans in the realm of sensory and consumer research from comparative analysis from two curated datasets pre-2019 and 2020-2021. AI will always be there, but it is how it is used that changes. Over the past year, there has been an increase in the number of programs that state the use AI/ML to turn large amounts of qualitative data into quantifiable resources. While these rapid tools will continue to thrive in the future, understanding consumers in deeper ways (ethnographies, in depth interviews) will remain. Protection of consumer and data (security and morals) as well as democratization of data will continue to resonate in the future. The rapid growth of available data for mining, new data collection, and rise of bots/fake people will continue to challenge CEs to sift real human interactions from made up connections. Lastly, it is important to remember that no matter how useful and advanced the machines, proper insights generation and appropriate results still depend on human content and software expertise.

Discussion questions

- What do you need to consider when using new technologies in consumer research?
- What research approaches transcend time and technologies?
- Why is it believed that AI and ML will increase in use in the years to come?

Note

1. TF-IDF is a statistical measure that evaluates how relevant a word is in a collection of documents and is scored by multiplying two metrics: how many times a word appears in a document, and the inverse document frequency of the word across a set of documents. (source: https://monkeylearn.com).

References

Barrett, L.F., Adolphs, R., Marsella, S., Martinez, A.M. and Pollak, S.D. (2019) "Emotional expressions reconsidered: challenges to inferring emotion from human facial movements". *Psychological Science in the Public Interest*, 20(3), 1-68. https://doi.org/10.1177/1529100619832930

Birchfield, D., Thornburg, H., Megowan-Romanowicz, M., Hatton, S., Mechtley, B., Dolgov, I. and Burleson, W. (2008) "Embodiment, multimodality, and composition: convergent themes across HCI and education for mixed reality learning environments". *Advances in Human Computer Interaction*, 2008, Article ID 874563, 19 pages. https://doi.org/10.1155/2008/874563.

Deng, Q., Hine, M., Ji, S. and Sur, S. (2017) Building an Environmental Sustainability Dictionary for the IT Industry. *50th Hawaii International Conference on System Sciences*. Hawaii, USA, 4-7 January 2017. https://doi.org/10.24251/HICSS.2017.112

Hanover, E. and McCune, S. (2020) Level Up Your Remote Research. *Talk presented at Insights Association NEXT* [Virtual conference], June 1-3 2020. Online.

Hearst, M. (1999). Untangling Text Data mining. Association for Computational Linguistics, *Proceedings of the 37th Annual Meeting of the Association for Computational Linguistics*. Association for Computational Linguistics, College Park, Maryland, USA, edition.

Information Tools Ltd. (2020) *Finding your 'Techquilibrium' in the Market Research Industry*. Infotools.com.

Kotu, V. and Deshpande, B. (2018) *Data Science: Concepts and Practice* (2nd edition). Cambridge, MA, USA: Morgan Kaufmann. Retrieved from https://learning.oreilly.com/library/view/Data+Science,+2nd+Edition/9780128147627/xhtml/chp009.xhtml#chp009titl

Paredes, D., Lopetcharat, K. and Peladeau, N. (2018) Qual-Quant Approach to Text Data for Deeper Consumer Insights and Knowledge. *Panel Discussion and Workshop at SenseAsia, The 3rd Asian Sensory and Consumer Research Symposium*, May 14, 2018. Kuala Lumpur, Malaysia.

Speight, K., Teratanavat, R., Paredes, D. and Peladeau, N. (2019) Using Advanced Text Analytics Tool for Deeper Insights in Big Data. *Talk Presented at 13th Pangborn Sensory Science Symposium*, July 31 2019. Edinburgh, Scotland.

Xu, Z. and Bengston, D. (1997) "Trends in national forest values among forestry professionals, environmentalists, and the news media, 1982-1993". *Society & Natural Resources*, 10(1), 43-49. Retrieved from https://www.fs.fed.us/nrs/pubs/jrnl/1997/nc_1997_xu_001.pdf

Chapter 17

Future Trends and Direction

Kannapon Lopetcharat, Dulce Paredes, and Jacqueline H. Beckley

17

During our inaugural book (2012), Product Innovation Toolbox: A Field Guide to Consumer Understanding and Research, we indicated in our last chapter that *"the future is happening now." What we see today in terms of consumer and product trends and research tools provide the framework for the future. We interviewed thought leaders and reviewed what our contributors have written and other discussions on predicting future applications and trends and what it means for Consumer Explorers (CEs) as insight leaders and strategic innovators'*. We saw six directions building for the future:

(1) Technology will continue to be a driver. IT CONTINUES TO DO SO.
(2) The people we seek to understand will continue to be more engaged in products that matter to them. IT HAS HAPPENED.
(3) Fun in the form of playfulness and gaming will increase in importance in enhancing engagement with people for deeper understanding. WE HAVE OBSERVED THIS.
(4) Nontraditional "data" (hybrids) will emerge due to the intersection of technology and people connections. FOR SURE, THIS HAS HAPPENED.
(5) Translational research - the more we have of "data" the more we will want to really understand and will learn that our too deep love of "data" without pairing it with appropriate technology and human centric tools for analysis - can leads us down paths that lack a future. YES, THIS HAS HAPPENED.
(6) When everyone feels they understand data, who really will understand data? GREAT OBSERVATION.

D.G. McNeil Jr. (2020), a *New York Times* reporter covering future predictions, might summarize direction #6 with the following quote: "unknown unknowns?... And there's no way I can prove that I have the facts right - because they do not exist yet."

Whether we look at the last 10 years or the last 20 years, there have been a lot of changes. While "technology" gets a lot of credits either in the form of devices and software or innovations like social media, the Anthropocene (age of humans) shows that people have the ability to stabilize or destabilize a lot of world

Product Innovation Toolbox: A Field Guide to Consumer Understanding and Research, Second Edition. Edited by Kannapon Lopetcharat, Dulce Paredes and Jacqueline H. Beckley.

events. Modern humans continue to evolve as millions move between nations leading to an unprecedented rate of genetic exchanges and mixtures (Hawks, 2021). Besides physical evolution, humans continue to adapt to a rapidly changing world brought by (geo) political, social, environmental, and technological forces. The 2020 UN Global Migration Report indicated that analysts and commentators referred to the current unprecedented pace of change as "the age of accelerations," "the fourth industrial revolution," and the "age of change" which brings both unexpected and unwanted impacts (International Organization for Migration, 2020).

For this book, we often used the word "disruptive" in a positive light especially for innovation and technologies. The emotionally negative disruptions from the COVID-19 pandemic are still raging as we write this chapter and its repercussions will resonate for years. History tells us this disruption has been the path with all global plagues of the past. With the devastating effects on global health and economies, we have seen opportunities even during rapid and abrupt changes. With this whole world event as a backdrop, we provide new suggestions about the future. Here are our next five directions that will shape the future:

(1) Pandemic influenced acceleration of technology
(2) Evolution of the hybrid model
(3) The rise of the individual, the human
(4) Nature influenced adoption
(5) Social forces for change

17.1 Pandemic influenced acceleration of technology

The pandemic accelerated technology adoption and digital transformation that were ready to go, yet companies and/or customers were not ready for them. The impact has been substantial and will continue. Examples are the following:

- The rise of e-commerce across all demographics and businesses
- Remote work and the viability of nonconcrete workplaces, school, and services
- Robotics to minimize human presence in production plants
- Commercial space vagaries; and
- Accelerated development and execution of new therapies and vaccines which are leading to dramatic moves in multiple channels of medical change and discoveries.

Human behaviors and attitudes shifted as people cope with the changing living environment. All these forced variations are projected to shape our world in years to come as aptly described by Professor Klaus Schwab, founder and executive chairman, World Economic Forum, when he said, "The pandemic represents a rare but narrow opportunity to reflect, reimagine and reset our world" (Schwab, n.d.).

There is an awakening that the civil discord within countries that has been occurring for the past 20 years prior to the SARS-CoV-2 pandemic and then the subsequent upheavals of all forms of social interactions resulting from the virus, are moving societies toward highly emotionally charged situations. Consumer Explorers (CEs) within any country may wish to isolate themselves from these events, but doing so may impact the relevance of their work.

The interface between life and technology is unavoidable. Some Asian countries may choose to limit influences of technology on their young population; yet, just as we could not limit how a virus particle 1000 to 2000 times smaller than a dust mite, could reshape the health and well-being of our world, technology will continue to spill into most aspects of our lives.

This world-wide reset applies to both the norms and benchmarks that we are accustomed to and the enormous knowledge base and data sets that we have collected and relied on for predictive modeling and future projections. So where do we go from here? How can we participate in a reimagined future?

17.2 The hybrid model of consumer understanding evolves

17.2.1 Back to the view

We chose the imagery of the field with an aquifer to stage the chapters in the book from early exploration, deep probing, and validation. Nothing stays the same in a farming field.

While we humans have practiced farming for millennia, shaping the world we live in today, farmers had to be some of the first agile people in the food sector, because they are directly accountable to nature and its vagaries. Seeds that create the plants we and our animals use and eat are absolutely essential to our future. Weeds can proliferate if not tended, and you may not know about the aquifer below that provides the water unless you probe.

17.2.2 How do we see *the view* changing?

Let us build on what we predicted in 2011: Blurring lines between qualitative-quantitative research tools, increasing use of digital technology and the rise of hybrid data.

17.2.2.1 What remains the same? Understanding consumer experience

As Consumer Explorers, companies ask us to help deliver innovation that matters to people (i.e. new products, technologies, ingredients, services). What remains the same is the role of the CE as it has always been about capturing behaviors and experiences of people and their responses to the stimuli we give them within context. More than ever, CEs need to pay attention to cultural diversity, beyond demographics, and multiple selves. Culture and cultural diversity expand beyond ethnicity. It covers cultural values, community affiliation, or tribes, and the broad range of actions that create, sustain, and dissolve culture. Beyond demographics, learning can start from the individual (idiographs which include bottom-up clustering) or with groups that they are in (top-down clustering). Multiple selves, a term we coined in the first edition, refers to the same individual being typed as having different mindsets depending on the situation and/or category they are in. Idiograph is the pathing of an individuals' internal or external "conversation" related to some experience, whether simple or complex.

The current research field is not static. It remains fluid and we assume nothing stays the same. Changing facts and spread of misinformation become more pronounced as we generate more data than before. The Consumer Explorer must be grounded on the facts and situations otherwise they can delve into authoritative misinformation or spread of mistruths by loud and influential people. The task to pay attention to details and not be distracted will require the CE to rely on credible and, better, certifiable sources, and robust processes. Having the ability to think about history and culture along with science and data will be needed.

The rise of digital technologies has made us "ultimate observers." We can participate in the consumers' experiences through intimate observations and conversations. It is important, then, to look for quality of information rather than simply volume. It may be more meaningful to gather information from a small group of people who represent a myriad of experiences rather than large surveys with minimum impact. It is the way to get to the why of an action or interaction. But here is the problem - our humanness loves fast and easy. We are wired for it. How does one balance the desire for quick and simple (some would call it shallow) yet needing to drink deeply in this opportunity for more consumer/people intimacy? It starts by being aware of the mixed desires and then being very mindful of the actions taken. Precision listening doesn't just happen - it means learning, practicing, and seeking to refine the skills. This is not a "one and done" approach. Wanting the experience of true connection is desirable. As a result, not everyone will be able to enjoy this future.

We can learn from Bill Moyers, a seasoned journalist and chronicler of people's lives. In his interview with Jon Stewart (Moyers, 2011), Moyers said he wants to interview people who "want to" reveal their thinking. He tries to get the meaning behind the words and that's when an interview becomes a conversation where both the interviewer and the interviewee become comfortable in exploring the topic.

That dialogue becomes effective when we experience a person's life and become emphatic to what that person is feeling, seeing, and doing in real time - in experience time. The CE must be skilled in keen observation and listening and understanding what must be done with the information being gathered. Successful CEs are able to persuade consumers to "want to" participate in their research activities.

Our content analysis of sensory and consumer research conference topics from January 2020 through March 2021 (Chapter 16) showed more mentions of "qualitative" than "quantitative" which we hypothesize is due to the increased prevalence of (1) online vs. in person qualitative, and (2) content analysis tools to rapidly convert qualitative to quantitative data, thereby being a proxy for larger-scale quant. Our observation is not alone. To quote a recent interview with Jim Longo from Discuss.io (2021):

> Despite the pandemic, the industry saw an increase in online qualitative research - in fact, the entire industry is at an inflection point now. While much of this represents the culmination of gradual trends, people are also adopting online methodologies, which we've been pushing for several years now, because traditional methods have suddenly been rendered inadequate. We saw the need to connect more than ever when COVID-19 hit. Though the shift to online qualitative was partially forced on us out of necessity, it opened up the methodology to those who have resisted it for years. They quickly saw that there was real value in online qual and, if anything, it opened up a new way to connect with consumers that took the hassle out of setting up qualitative sessions.

Agile tools have enabled real time conversations with groups of people simultaneously followed by instantaneous data analysis. The approach satisfies many quantitative researchers. These approaches do not have face to face, one-on-one interaction between CE and interviewee, but may be valuable for hypothesis expansion and validation and some concept screening.

17.2.2.2 What grows? Data and increasingly hybrid data, non-western perspectives

In 2017, *The Economist* called data "the fuel of the future." The constant flow of massive amounts of digital information which can be extracted, analyzed, valued, and sold in many ways has made data the new oil. This data economy is powered through computers that extract their value by identifying patterns, predictions, and suggested insights. There is no accurate count on the amount of digital data collected every day except that it is projected to grow exponentially every year. The World Economic Forum predicts that by 2025, the amount of data generated each day is expected to reach 463 exabytes globally (an exabyte is 1000 bytes to the sixth power) (Valeta, 2021).

And just like any currency, it is expected that data will be protected, governed, and taxed in the future. Technology, sadly, will not only accelerate data generation, analysis, and tracking. It will increasingly make it difficult to pay attention to "details" if we want to thwart authoritative misinformation.

It is not just the volume of research data that is increasing. The type of data being collected within the same study has also expanded. Hybrids have the allure of allowing us to introduce something familiar with something new. A common example is combining numerical data from surveys (quantitative), answers to open ends (qualitative, text), and observations of consumer behaviors (qualitative, tagged text). Hybrids can also include combining qual-quant data from primary research (company sponsored), secondary research (curated databases, news media reports), and unsolicited data (social media). The growth of social media sources, open source network analysis, and companies providing data mining, tracking, and analysis have made social media as another source of consumer understanding.

In 2012, we defined the various forms of digital data as hybrid data (text, voice, visual, and video). To these, we add virtual and augmented or mixed reality forms as part of immersive experiences (Chapter 9.3). Another form of hybrid data we discussed in our first edition involved the combination of data from conscious, rational decision-making, and nonconscious decision-making formed from emotions, impulses, reflex actions, habits, memories, and instincts that occur quickly and automatically with little awareness or feeling of effort (Sturgess, 2012). Chapter 9.2 describes the research tools associated with nonconscious decision-making. While there is broad discussion around explicit, implicit, and tacit forms of decision making by humans, this area will advance. It has the potential to be both positive to understanding and also misleading when used for less than "pure" intentions.

As we incorporate hybrid data we couldn't collect in the past, we also need the discipline to validate and move them forward in an orderly way. We need to be able to go back to the data streams and understand why and how we are collecting the information. The format for successful hybrid data will be a continuum of data that has a meaningful relationship to each other and is based on solid principles of science provided by a trusted source. An evolutionary disciplined hybridization will:

- Acknowledge the need to triangulate understanding so that you have context and are as sure of the findings as possible
- Use the most resource preserving approaches available
- Use tools that build upon each other, having enough overlap to form a continuum or iterate
- Integrate the advancing understanding of neuroscience and how it blends with acceptance of the influence of emotion
- Implement rapid storytelling to allow the findings and their output to be utilized quickly and then to provide redeploying of the resources that focused on the project
- Acknowledge the non-western perspective and expand its value in framing and understanding that point of view.

If we do all these points along with being trustworthy, hybrid data will gain more function and have the potential, building over time, to become more valuable.

17.3 The rise of the individual, the human. Moving from consumption as an end goal to understanding the whole person

17.3.1 AI + Human

The current versions of artificial intelligence (AI) increasingly influence and interact with people in their daily lives (mobile phone functions, news anchor in their favorite video clips, etc.) and business decision making (e.g. which product to offer, automated insights). Emerging generations, Z, alpha and beyond, will grow up with AIs and seamlessly assimilate AI in many aspects of their lives especially the formation of their attitudes, norms, beliefs, and decision making. Even at the time of writing this book, one can put a query into a search engine and the person's social network accounts (from separate companies) will keep feeding advertisements about the query for several days (or even weeks). This phenomenon, we named as "AI-looping influence," puts Consumer Explorers in a dilemma on how to discern true human-cognitive behavior from the AI-induced behavior in the big data world.

Current AI (specifically Machine Learning) thrives on spotting patterns in the data ocean and promotes it through the AI-looping influence that, consequently, creates a "new trend" in social network societies that, undeniably, have real impacts on larger economies and societies. Unfortunately, the AI-induced trends (such as Twitter's hashtag trends of the day) do not persist long enough for any company, especially physical product producers, to "react" properly. Recognizing this AI-looping influence, CEs, product developers, marketers, and entrepreneurs can focus their attention to the right place that is "product innovation."

Current AI can propagate the popularity and offer choices to consumers from what have been offered to the market like amplifying the amplitude and frequency of ripples in a pond after a stone has been cast. This is the inherent disadvantage of current AI technology that AI cannot discern cause and effect especially the infrequent events (Bergstein, 2020). In the new product development world, the infrequent events are known as disruptive type and low-cost disruption, and the human is still in control in creating these innovations. As long as the AI cannot formulate, produce or create physical products, the "human" still holds the power of creativity and authority to "pick and cast the stone" into the pond to create the first ripple.

17.3.2 Human + AI

In addition to data-collection and human-labor cost reduction, current research automation has evolved for streamlining sampling, questionnaire design, data collection, and insights reporting (Dean, 2020). Speed in delivering results has been in the forefront of research automation but, during the COVID-19 pandemic, "agility" emerged as the important factor for the survival of the research industry as online research became mainstream. Agility encompasses both speed (fast) and flexibility (ability to change direction) and that evolve to "data democratization" (Thomke, 2020) (see Chapters 12 and 16).

Beyond data collection, current research automation enables effective data democratization as it allows team members throughout the organization to access appealing and easy to understand results, and have the ability to organize results and extract insights in real time (automated insights). The team members can see the direct impact of their ideas and initiatives and make decisions quickly. However, there are certain risks associated with data democratization, namely, drawing wrong conclusions due to lack of understanding of underlying biases in the system (this happens frequently in automated qualitative data) and quick spread of misleading information due to some nuance or mistakes in the data (Dean, 2020).

With research automation technologies, Consumer Explorers or CEs can prevent these pitfalls by acting as human-machine connectors in the organization where the human-machine automation system allows the humans to review and adjust for errors (before they ever make it out to the broader organization). The human-machine teams will create positive synergy in decision-making and information dissemination (Johnson, 2020). Therefore, in the very near future and beyond, in addition to being the insights gatekeeper, the roles of CEs will include experiment designer, experimentation ambassador, human-machine connector, knowledge/insight disseminator and knowledge management curator.

The future is not just AI but all types of digital transformation aimed at narrowing the gaps between human reasoning and machine intelligence (Amit, 2021), augmenting human experiences from wearables to implants (Resnick, 2021), low code platforms for nondata scientists, and mainstream real-time data processing (Tucci, 2021). For future innovation, Consumer Explorers will interact with both users and avatars in the digital 3D realm or metaverse to bring real-world products into virtual spaces and to create "virtual first" products and services (Pringle, 2021). These are enough to keep Consumer Explorers on their toes!

17.4 Nature influenced adoption

What COVID-19 pandemic taught us is that "disruption can happen anytime and anywhere without any clear warning." While technology has been advancing many facets of our life and world, it did not reliably predict the 2020 global pandemic. We know that Mother Nature is still and will always be the most unpredictable and influential force upon humans, and this is the true meaning of nature influenced adoption.

What we are seeing and need to monitor and innovate toward is climate (how fast it changes) and the unrelenting effects of humans (Anthropocene) impacting choices, behaviors, and actions of people around the world. Whether how designers are choosing to upcycle, waste-free product packaging and move to a circular material systems, or evolving employment, food, or travel decisions, we need to understand how radical and impactful natural occurrence of hurricanes and flooding, sea and temperature rising, or droughts and fires, slowly and effectively alter how individuals and groups begin to live, work, and maintain communities.

The repertory of methods discussed in these past chapters can help us understand how those we seek to understand and/or market to are being changed by forces that ultimately may win against the humans.

The pandemic taught us the need to enhance our focus on humanity and empathy. Therefore, companies need to treat their consumers as "people" and switch from "looking at big data sets and then talking about consumers," to "empathizing with individuals at the human level." This is a deeper interpretation of "consumerism" than the traditional meaning as businesses are authentically trying to connect and understand humans in their complexity (McCoy and Joe, 2020). From now to the future, Consumer Explorers must always be vigilant to provide human-level insights through "perpetual human-centered research and development" platforms, so companies can "react" quickly to the next disruption. Nobody knows when the next major disruption will happen or if we will continue to be in a rate of influx from multiple sources.

17.5 Social forces for change

At the beginning of the new millennium, who would have thought that a simple communication device or some internet tool could be identified as being the reason the Arab Spring would occur (2010–2012), or influence a whole developing country to remove members of their country out of that country (Rohingya out of Myanmar)? Whether it is how we search for information today (Google, Bing), supply household goods (voice activated tools like Alexa from Amazon or Siri from Apple); deal with money-like tools (cryptocurrency) or new dancing routines (Tiktok); or monitor our health (blood pressure app like Kardia), the use of social platforms are impacting numerous social interactions across all generations and geographies.

David Brooks is a social and political commentator for a number of media outlets in the United States. In an opinion piece for the *New York Times*, he wrote about how technology and science broaden our knowledge about the interface between the senses and our thinking as humans. While not a consumer insight or market researcher, we felt that Brooks (2021) captures the moment we are in as consumer observers:

> We grew up believing that 'imagining' and 'seeing' describe different mental faculties. But as we learn more about what's going on in the mind, these concepts get really blurry really fast.
>
> This is happening all over the place. Over the centuries, humans have come up with all sorts of concepts to describe different thinking activities: memory, perception, emotion, attention, decision-making. But now, as scientists develop greater abilities to look at the brain doing its thing, they often find that the activity they observe does not fit the neat categories our culture has created and which we rely on to understand ourselves.

We could not agree more!

17.6 Conclusion

The pace of change continues to move fast. Consumer Explorers need an agile and resilient process such as the Iterative Experimentation Qualitative-Quantitative Research (IEQQR) model (see Chapter 4) that allows them to manage current project deliverables, keep an eye on changing environments and prepare for future needs. It may have been acceptable for CEs 20 or 30 years ago to be ignorant of world events and local situations, but not anymore. If one is going to be a successful data analyst today and, in the future, they need to "get" the world. The IEQQR model emphasizes learning from small pilots with built-in experimentation (scientific thinking as a way to sift facts from fiction) that allows CEs to be agile and resilient when they create a research program. The model encourages CEs to use experts who will bring mastery that allows the research program to move faster with precision and accuracy and emphasizes on emotional intelligence of CEs to disseminate human-centric approach and perpetual learning throughout the organization. This will help companies to transform invention to innovation (finding meaningful applications for the invention) fast and prevent themselves from misdirecting research programs due to buzzwords. Coupled with tapping experts in certain corresponding fields, the advancement of data science, machine learning and AI will make digital transformation, proliferation of DIY tools for data collection, and data analysis and understanding the underpinnings (whys) easier and more comprehensive. And these can be accomplished by human-machine synergy where the human is still the creator.

There is no insight if there is no starting hypothesis. Life is really about continuous experimentation. There is no right or wrong answer when you start with a hypothesis. We learn, we get better and maybe along the way we look for purpose. We are in this exploration together so that our innovations and insights can steer us to good places.

References

Amit, G. (2021, September 1) *Simply Science* Weekly Digest. The Economist. https://www.economist.com/simplyscience/ (accessed 1 September 2021).

Bergstein, B. (2020, February 19) *What AI Still Can't Do*. MIT Technology Review. https://www.technologyreview.com/2020/02/19/868178/what-ai-still-cant-do/ (accessed 8 March 2022).

Brooks, D. (2021, September 3) *You Are Not Who You Think You Are*. New York Times: New York Edition, A23.

Dean, E. (2020, December 16) *3 Things Every Market Researcher Should Know About Automation*. Research World. https://www.researchworld.com/3-things-every-market-researcher-should-know-about-automation/ (accessed 8 March 2022).

Discuss.io (2021, March 21) *Qualitative Research Trends of the Future: A Talk with Jim Longo*. https://www.discuss.io/qualitative-research-trends-of-the-future-a-talk-with-jim-longo/?utm_source=informa&utm_medium=email&utm_campaign=tmre-home (accessed 8 March 2022).

Hawks, J. (2021) "Still evolving (after all these years)". *Scientific American Special Edition*, 2021(Fall), 22-27.

International Organization for Migration (2020) *World Migration Report 2020*. United Nations. https://www.un.org/sites/un2.un.org/files/wmr_2020.pdf (accessed 8 March 2022).

Johnson, B. (2020, February 26) *The Professionals Who Predict the Future for a Living*. MIT Technology Review. https://www.technologyreview.com/2020/02/26/905703/professionals-who-predict-the-future-for-a-living-forecasting-futurists/ (accessed 8 March 2022).

McCoy, G. and Joe, K. (2020, December 28) *Gabriela McCoy of Bacardi on Priorities and Opportunities for 2021*. Research World. https://www.researchworld.com/gabriela-mccoy-of-bacardi-on-priorities-and-opportunities-for-2021/ (accessed 8 March 2022).

McNeil, D.G. (2020, August 27) *A Tall Task: Reporting the Future*. New York Times: New York Edition, A2.

Moyers, B. (2011) *Interview by Jon Steward, The Daily Show* (1st edition). http://www.thedailyshow.com/watch/wed-june-1-2011/bill-moyers-pt-2.

Pringle, N. (2021, September 6) *Why the 'Metaverse' Will Prove to be more than a Buzzword*. fastcompany.com (accessed 8 March 2022).

Resnick, M. (2021, August 30) *The 7 Key Trend Areas to Factor Into Your Strategic Planning* [video]. Gartner Webinar. https://www.gartner.com/en/webinars/4003979/the-7-key-trend-areas-to-factor-into-your-strategic-planning?srcId=1-4582955171&ref=btem&&utm_campaign=communication_viewer_followup&utm_medium=email&utm_source=brighttalk-transact (accessed 8 March 2022).

Schwab, K. (n.d.) *The Great Reset*. World Economic Forum. https://www.weforum.org/focus/the-great-reset (accessed 8 March 2022).

Sturgess, D. (2012) "Chapter 6.2: Insight through immersion". In J. Beckley, D. Paredes, K. Lopetcharat (eds). *Product Innovation Toolbox: A field Guide to Consumer Understanding and Research* (1st edition). Hoboken, New Jersey: John Wiley & Sons. pp. 91–99.

Tucci, L. (2021, March 10) *Ultimate Guide to Digital Transformation for Enterprise Leaders*. https://searchcio.techtarget.com/feature/Ultimate-guide-to-digital-transformation-for-nterprise-leaders?? (accessed 8 March 2022).

The Economist (2017, May 6) *Data is Giving Rise to a New Economy*. https://www.economist.com/briefing/2017/05/06/data-is-giving-rise-to-a-new-economy (accessed 8 March 2022).

Thomke, S.H. (2020) *Experimentation Works: The Surprising Power of Business Experiments*. Boston, MA: Harvard Business Review Press.

Valeta, B. (2021, January 28) *How Much Data Is Created Every Day? [27 Staggering Stats]*. https://seedscientific.com/how-much-data-is-created-every-day (accessed 8 March 2022).

Index

Product Innovation Toolbox: A Field Guide to Consumer Understanding and Research, Second Edition. Edited by Kannapon Lopetcharat, Dulce Paredes and Jacqueline H. Beckley.

www.ingramcontent.com/pod-product-compliance
Lightning Source LLC
LaVergne TN
LVHW080903111025
822255LV00012B/297
* 9 7 8 1 1 1 9 7 1 2 8 4 8 *